London Mathematical Society Lecture Note Series. 153

L-functions and Arithmetic

Edited by
J. Coates
University of Cambridge
and
M.J. Taylor
UMIST

The right of the University of Cambridge to print and sell all manner of books was granted by Henry VIII in 1534. The University has printed and published continuously since 1584.

CAMBRIDGE UNIVERSITY PRESS
Cambridge
New York Port Chester Melbourne Sydney

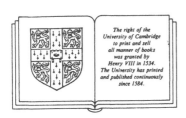

Published by the Press Syndicate of the University of Cambridge
The Pitt Building, Trumpington Street, Cambridge CB2 1RP
40 West 20th Street, New York, NY 10011, USA
10, Stamford Road, Oakleigh, Melbourne 3166, Australia

First published 1991

Library of Congress cataloguing in publication data available

British Library cataloguing in publication data available

ISBN 0 521 38619 5

Transferred to digital printing 2004

CONTENTS

PREFACE

This book is the fruit of the London Mathematical Society Symposium on 'L-functions and Arithmetic', which was held at the University of Durham from June 30 until July 11, 1989. The Symposium attempted to bring together the many and diverse aspects of number theory and arithmetical algebraic geometry which are currently being studied under the general theme of the connections between L-functions and arithmetic. In particular, there were series of lectures on each of the following topics:

(*i*) descent theory on elliptic curves;

(*ii*) automorphic L-functions;

(*iii*) Beilinson's conjectures;

(*iv*) p-adic cohomology and the Bloch-Kato conjectures on the Tamagawa numbers of motives;

(*v*) Iwasawa theory of motives;

(*vi*) l-adic representations attached to automorphic forms;

(*vii*) L-functions and Galois module structure of rings of integers and groups of units.

While the editors were not able to persuade all the lecturers to write up the material of their lectures,the present volume represents a major portion of the proceedings of the Symposium. The editors wish to express their warm appreciation to all the lecturers. The Symposium was sponsored by the London Mathematical Society, and benefited from the generous financial support of the Science and Engineering Research Council. We are grateful to Tony Scholl and the Department of Mathematics of the University of Durham for their help in organising the meeting; we particularly wish to thank Grey College for its kind hospitality.

J. Coates
M.J. Taylor

PARTICIPANTS

A. Agboola, Columbia University
J.V. Armitage, University of Durham
J. Arthur, University of Toronto
R. Blasubramanian, CIT, Madras
P.N. Balister, Trinity College, Cambridge
L. Barthel, Weizmann Institute
E. Bayer, Université de Besançon
B.J. Birch, University of Oxford
D. Blasius, ENS, Paris
J. Boxall, Université de Caen
J. Brinkhuis, Erasmus University, Rotterdam
D.A. Burgess, University of Nottingham
D.J. Burns, University of Cambridge
C.J. Bushnell, King's College, London
N. Byott, UMIST
H. Carayol, Université Louis Pasteur, Strasbourg
P. Cassou-Noguès, Université de Bordeaux 1
R. Chapman, Merton College, Oxford
T. Chinburg, University of Pennsylvania
L. Clozel, Université de Paris-Sud
J.H. Coates, University of Cambridge
J. Cougnard, Université de Besançon
J.E. Cremona, University of of Exeter
C. Denninger, Universität Munster
F. Diamond, Ohio State University
L. Dodd, University of Nottingham
H. Dormon, Harvard University
B. Erez, Université de Genève
G. Everest, University of East Anglia
D.E. Feather, University of Nottingham
M. Flach, St John's College, Cambridge
J-M. Fontaine, Université Paris-Sud
A. Fröhlich, University of Cambridge
Imperial College, London
S. Gelbart, Weizmann Institute
R. Gillard, Université de Grenoble 1
C. Goldstein, Université Paris-Sud
D. Grant, University of Cambridge
R. Greenberg, University of Washington
B.H. Gross, Harvard University
S. Haran, University of Jerusalem
G. Harder, Universität Bonn
M.C. Harrison, Jesus College, Cambridge
D.R. Hayes, Imperial College, London
G. Henniart, Université Paris-Sud
H. Hida, University of California,
Los Angeles
U. Jannsen, Max-Planck Institut, Bonn
B. Jordan, Baruch College, CUNY
M.A. Kenku, University of Lagos
M.A. Kervaire, Université de Genève
P. Kutzko, University of Iowa
R. Ledgard, University of Manchester
S. Lichtenbaum, Cornell University
Li Guo, University of Washington
S. Ling, University of California, Berkeley
W.G. McCallum, University of Arizona
M. MacQuillen, Harvard University
L.R. McCulloh, University of Illinois, Urbana
H. Maennel, Max-Planck Institut, Bonn
C. Matthews, University of Cambridge
B. Mazur, Harvard University
J. Neukirch, Universität Regensburg
J. Oesterlé, Université Paris VI
R. Odoni, University of Glasgow
D.B. Penman, University of Cambridge
R.G.E. Pinch, University of Cambridge
A.J. Plater, University of Cambridge
R.J. Plymen, University of Manchester
D. Ramakrishnan, California Institute
of Technology
R.A. Rankin, University of Glasgow
K.A. Ribet, University of California, Berkeley
G. Robert, Max-Planck Institut, Bonn
A. Roberston, University of Oxford
J.D. Rogawski, University of California,
Los Angeles
K. Rubin, Ohio State University
P. Satgé, Université de Caen
N. Schappacher, Max-Planck Institut, Bonn
C-G. Schmidt, RUG, Groningen
P. Schneider, Universität Köln
A.J. Scholl, University of Durham
J-P. Serre, Collège de France
Seymour Kim, Imperial College, London
E. de Shalit, University of Jerusalem
V. Snaith, McMaster University
D. Solomon, UMIST
M. Starkings, University of Nottingham
A. Srivastav, Université de Bordeaux 1
H.P.F. Swinnerton-Dyer, University Funding
Council
M.J. Taylor, UMIST
R. Taylor, University of Cambridge
J. Tilouine, UCLA
E. Todd, King's College, London
S. Ullom, University of Illinois, Urbana
N.J. Walker, University of Oxford
L.C. Washington, University of Maryland
U. Weselmann, Universität Bonn
E. Whitley, University of Exeter
J.W. Wildeshaus, King's College, Cambridge
A. Wiles, Princeton University
S.M.J. Wilson, University of Durham
K. Wingberg, Universität Erlangen-Nurnberg
C.F. Woodcock, University of Kent
R.I. Yager, Macquarie University
C.S. Yogananda, CIT, Madras

Lectures on automorphic L-functions

JAMES ARTHUR AND STEPHEN GELBART

PREFACE

This article follows the format of five lectures that we gave on automorphic L-functions. The lectures were intended to be a brief introduction for number theorists to some of the main ideas in the subject. Three of the lectures concerned the general properties of automorphic L-functions, with particular reference to questions of spectral decomposition. We have grouped these together as Part I. While many of the expected properties of automorphic L-functions remain conjectural, a significant number have now been established. The remaining two lectures were focused on the techniques which have been used to establish such properties. These lectures form Part II of the article.

The first lecture (§I.1) is on the standard L-functions for GL_n. Much of this material is familiar and can be used to motivate what follows. In §I.2 we discuss general automorphic L-functions, and various questions that center around the fundamental principle of functoriality. The third lecture (§I.3) is devoted to the spectral decomposition of $L^2(G(F) \setminus G(\mathbb{A}))$. Here we describe a conjectural classification of the spectrum in terms of tempered representations. This amounts to a quantitative explanation for the failure of the general analogue of Ramanujan's conjecture.

There are three principal techniques that we discuss in Part II. The lecture §II.1 is concerned with the trace formula approach and the method of zeta-integrals; it gives only a skeletal treatment of the subject. The lecture §II.2, on the other hand, gives a much more detailed account of the theory of theta-series liftings, including a discussion of counterexamples to the general analogue of Ramanujan's conjecture. We have not tried to relate the counterexamples given by theta-series liftings with the conjectural classification of §I.3. It would be interesting to do so.

These lectures are really too brief to be considered a survey of the subject. There are other introductory articles (references [A.1], [G], [B.1] for Part I) in which the reader can find further information. More detailed discussion is given in various parts of the Corvallis Proceedings and in many of the other references we have cited.

PART I

1 STANDARD L-FUNCTIONS FOR GL_n

Let F be a fixed number field. As usual, F_v denotes the completion of F with respect to a (normalized) valuation v. If v is discrete, $\mathfrak{o}_v$ stands for the ring of integers in F_v, and q_v is the order of the corresponding residue class field. We shall write $\mathbb{A} = \mathbb{A}_F$ for the adèle ring of F.

In this lecture, G will stand for the general linear group GL_n. Then $G(\mathbb{A})$ is the restricted direct product, over all v, of the groups $G(F_v) = GL_n(F_v)$. Thus, $G(\mathbb{A})$ is the topological direct limit of the groups

$$G_S = \prod_{v \in S} G(F_v) \cdot \prod_{v \notin S} G(\mathfrak{o}_v),$$

in which S ranges over all finite sets of valuations of F containing the set S_∞ of Archimedean valuations.

One is interested in the set $\Pi(G(\mathbb{A}))$ of equivalence classes of irreducible, admissible representations of $G(\mathbb{A})$. (Recall that a representation of $G(\mathbb{A})$ is admissible if its restriction to the maximal compact subgroup

$$K = \prod_{v \text{ complex}} U(n, \mathbb{C}) \times \prod_{v \text{ real}} O(n, \mathbb{R}) \times \prod_{v \text{ discrete}} GL_n(\mathfrak{o}_v)$$

contains each irreducible representation of K with only finite multiplicity.) Similarly, one has the set $\Pi(G(F_v))$ of equivalence classes of irreducible admissible representations of $G(F_v)$. It is known [F] that any $\pi \in \Pi(G(\mathbb{A}))$ can be decomposed into a restricted tensor product

$$\bigotimes_v \pi_v, \qquad \pi_v \in \Pi(G(F_v)),$$

of irreducible, admissible representations of the local groups.

The *unramified principal series* is a particularly simple subset of $\Pi(G(F_v))$ to describe. Suppose that the valuation v is discrete. One has the Borel subgroup

$$B(F_v) = \{b = \begin{pmatrix} b_1 & \cdots & * \\ & \ddots & \vdots \\ O & & b_n \end{pmatrix}\} \subseteq G(F_v)$$

of $G(F_v)$, and for any n-tuple $z = (z_1, \ldots, z_n) \in \mathbb{C}^n$,

$$b \longrightarrow \chi_z(b) = |b_1|_v^{z_1} \cdots |b_n|_v^{z_n}$$

gives a quasi-character on $B(F_v)$. Let $\tilde{\pi}_{v,z}$ be the representation of $G(F_v)$ obtained by inducing χ_z from $B(F_v)$ to $G(F_v)$. (Recall that $\tilde{\pi}_{v,z}$ acts on the space of locally constant functions ϕ on $G(F_v)$ such that

$$\phi(bx) = \chi_z(b)(\prod_{i=0}^{n-1} |b_i|_v^{\frac{n-1}{2}-i})\phi(x), \qquad b \in B(F_v),\ x \in G(F_v),$$

and that

$$(\tilde{\pi}_{v,z}(y)\phi)(x) = \phi(xy)$$

for any such ϕ.) We shall assume that

$$Re(z_1) \geq Re(z_2) \geq \cdots \geq Re(z_n).$$

It is then a very special case of the Langlands classification [B-W, XI, §2] that $\tilde{\pi}_{v,z}$ has a unique irreducible quotient $\pi_{v,z}$. The representations $\{\pi_{v,z}\}$ obtained in this way are the unramified principal series. They are precisely the representations in $\Pi(G(F_v))$ whose restrictions to $G(\mathfrak{o}_v)$ contain the trivial representation. If π_v is any representation in $\Pi(G(F_v))$ which is equivalent to some $\pi_{v,z}$, it makes sense to define a semisimple conjugacy class

$$\sigma(\pi_v) = \begin{pmatrix} q_v^{-z_1} & & 0 \\ & \ddots & \\ 0 & & q_v^{-z_n} \end{pmatrix}$$

in $GL_n(\mathbb{C})$. For $\sigma(\pi_v)$ does depend only on the equivalence class of π_v; conversely, if two such representations are inequivalent, the corresponding conjugacy classes are easily seen to be distinct.

Suppose that $\pi = \bigotimes_v \pi_v$ is a representation in $\Pi(G(\mathbb{A}))$. Since π is admissible, almost all the local constituents π_v belong to the unramified principal series. Thus, π gives rise to a family

$$\sigma(\pi) = \{\sigma_v(\pi) = \sigma(\pi_v) :\ v \notin S\}$$

of semisimple conjugacy classes in $GL_n(\mathbb{C})$, which are parametrized by the valuations outside of some finite set $S \supseteq S_\infty$. Bearing in mind that a semisimple conjugacy class in $GL_n(\mathbb{C})$ is determined by its characteristic polynomial, one defines the local L-functions

$$L_v(s,\pi) = L(s,\pi_v) = \det(1 - \sigma_v(\pi)q_v^{-s})^{-1}, \qquad s \in \mathbb{C},\ v \notin S.$$

The global L-function is then given as a formal product

$$L_S(s,\pi) = \prod_{v \notin S} L_v(s,\pi).$$

If the global L-function is to have interesting arithmetic properties, one needs to assume that π is automorphic. We shall first review the notion of an automorphic representation, and then describe the properties of the corresponding automorphic L-functions.

The group $G(F)$ embeds diagonally as a discrete subgroup of

$$G(\mathbb{A})^1 = \{g \in G(\mathbb{A}) : \ |\det \ g| = 1\}.$$

The space of *cusp forms* on $G(\mathbb{A})^1$ consists of the functions $\phi \in L^2(G(F) \setminus G(\mathbb{A})^1)$ such that

$$\int_{N_P(F)\setminus N_P(\mathbb{A})} \phi(nx)dn \ = \ 0$$

for almost all $x \in G(\mathbb{A})^1$, and for the unipotent radical N_P of any proper, standard parabolic subgroup P. (Recall that standard parabolic subgroups are subgroups of the form

$$P(\mathbb{A}) = \left\{p = \begin{pmatrix} p_1 & \cdots & * \\ & \ddots & \vdots \\ O & & p_r \end{pmatrix} : p_k \in GL_{n_k}\right\},$$

where $(n_1,\ldots,n_r)$ is a partition of n.) The space of cusp forms is a closed, right $G(\mathbb{A})^1$-invariant subspace of $L^2(G(F) \setminus G(\mathbb{A})^1)$, which is known to decompose into a discrete direct sum of irreducible representations of $G(\mathbb{A})^1$. A representation $\pi \in \Pi(G(\mathbb{A}))$ is said to be *cuspidal* if its restriction to $G(\mathbb{A})^1$ is equivalent to an irreducible constituent of the space of cusp forms. We note that such a representation need not be unitary; indeed, if π is cuspidal, so are all the representations $\{\pi \otimes |\det|^z : \ z \in \mathbb{C}\}$. Now, suppose that $p \in P(\mathbb{A})$ is as above, with P a given standard parabolic subgroup, and that for each i, $1 \leq i \leq r$, π_i is a cuspidal automorphic representation of $GL_{n_i}(\mathbb{A})$. Then

$$p \longrightarrow \pi_1(p_1) \otimes \cdots \otimes \pi_r(p_r)$$

is a representation of $P(\mathbb{A})$, which we can induce to $G(\mathbb{A})$. The automorphic representations of $G(\mathbb{A})$ are the irreducible constituents of induced representations of this form [L.4]. We shall denote the subset of automorphic representations in $\Pi(G(\mathbb{A}))$ by $\Pi_{\mathrm{aut}}(G)$.

Suppose that $\pi \in \Pi_{\mathrm{aut}}(G)$. It is easily seen that the infinite product for $L_S(s,\pi)$ converges in some right half plane. Moreover, Godement and Jacquet

[G-J] have shown that $L_S(s,\pi)$ has analytic continuation to a meromorphic function of $s \in \mathbb{C}$ which satisfies a functional equation. Their method is a generalization from GL_1 of the method of Tate's thesis [T.1], and will be described in §II.1.

It is useful to consider certain subsets of $\Pi_{\text{aut}}(G)$. Let $\Pi_{\text{disc}}(G)$ denote the set of irreducible, *unitary* representations of $G(\mathbb{A})$ whose restriction to $G(\mathbb{A})^1$ occurs discretely in $L^2(G(F)\setminus G(\mathbb{A})^1)$. This contains the set $\Pi_{\text{cusp}}(G)$ of irreducible, *unitary* cuspidal representations of $G(\mathbb{A})$. We shall then write $\Pi(G)$ simply for the set of irreducible representations of $G(\mathbb{A})$ obtained by inducing representations

$$p \longrightarrow \pi_1(p_1)\otimes\cdots\otimes\pi_r(p_r), \qquad\qquad \pi_i \in \Pi_{\text{disc}}(GL_{n_i}),$$

from standard parabolic subgroups. (It is a peculiarity of GL_n that such unitary induced representations are already irreducible. In general, one must define $\Pi(G)$ to be the set of irreducible constituents of these induced representations.) The representations $\Pi(G)$ are precisely the ones which occur in the spectral decomposition of $L^2(G(F)\setminus G(\mathbb{A}))$. This deep fact is a consequence of the theory of Eisenstein series, initiated by Selberg [S], and established for general groups by Langlands [L.3]. A second major consequence of the theory of Eisenstein series is that the representations in $\Pi_{\text{disc}}(G)$ and $\Pi(G)$ are automorphic. Taking this fact for granted, we obtain an embedded sequence

$$\Pi_{\text{cusp}}(G)\subset\Pi_{\text{disc}}(G)\subset\Pi(G)\subset\Pi_{\text{aut}}(G)\subset\Pi(G(\mathbb{A}))$$

of families of irreducible representations of $G(\mathbb{A})$.

The representations in $\Pi(G)$ have a striking rigidity property.

Theorem The map

$$\pi \longrightarrow \sigma(\pi)=\{\sigma_v(\pi):\ v\notin S\}, \qquad\qquad \pi\in\Pi(G),$$

from $\Pi(G)$ to families of semisimple conjugacy classes in $GL_n(\mathbb{C})$, is injective. In other words, a representation in $\Pi(G)$ is completely determined by the associated family of conjugacy classes.

For cuspidal representations, this theorem is closely related to the original multiplicity one theorem [Sh]. The extension to $\Pi(G)$ follows from analytic properties of the corresponding L-functions [J-S, Theorem 4.4] and the recent classification [M-W] by Moeglin and Waldspurger of the discrete spectrum of GL_n.

The theorem is reminiscent of a similar rigidity property of representations of Galois groups. Suppose that

$$r: Gal(\overline{F}/F) \longrightarrow GL_n(\mathbf{C})$$

is a continuous representation of the Galois group of the algebraic closure $\overline{F}$ of F. For every valuation v outside a finite set $S \supseteq S_\infty$, there is an associated Frobenius conjugacy class in the image of the Galois group. This gives a semisimple conjugacy class $\sigma_v(r)$ in $GL_n(\mathbf{C})$. It is an immediate consequence of the Tchebotarev density theorem that r is completely determined by the family

$$\sigma(r) = \{\sigma_v(r) : v \notin S\}.$$

We should also recall the local and global Artin L-functions

$$L_v(s,r) = \det(1 - \sigma_v(r)q_v^{-s})^{-1}, \qquad s \in \mathbf{C},\ v \notin S,$$

and

$$L_S(s,r) = \prod_{v \notin S} L_v(s,r),$$

attached to r.

Some years ago, Langlands conjectured [L.1] that the similarity between the two types of L-function was more than just formal.

Conjecture (Langlands) For any continuous representation

$$r: Gal(\overline{F}/F) \longrightarrow GL_n(\mathbf{C})$$

of the Galois group there is an automorphic representation $\pi \in \Pi(G)$, necessarily unique, such that $\sigma_v(\pi) = \sigma_v(r)$ for all v outside some finite set $S \supseteq S_\infty$. In particular

$$L_S(s,\pi) = L_S(s,r).$$

The conjecture represents a fundamental problem in number theory. The case $n = 1$ is just the Artin reciprocity law, which is of course known, but highly nontrivial. There has also been significant progress in the case $n = 2$. If the image of r in $GL_2(\mathbf{C})$ is a dihedral group, the conjecture follows from the converse theorem of Hecke theory [J-L] or from the properties of the Weil representation [S-T]. (See §II.1.) If r is an irreducible 2-dimensional representation which is not dihedral, its image in $PGL_2(\mathbf{C}) \cong SO(3,\mathbf{C})$ will be either tetrahedral, octahedral or icosohedral. These cases are much deeper, but the first two have been solved [L.6], [Tu]. The essential new ingredient was

Langlands' solution of the (cyclic) base change problem for GL_2. For general n, the base change problem was solved recently by Arthur and Clozel [A-C]. This leads to an affirmative answer to the conjecture for any representation r whose image is nilpotent.

We shall describe the base change theorem in more detail. Suppose that E/F is a Galois extension, with cyclic Galois group

$$Gal(E/F) = \{1, \gamma, \gamma^2, \ldots, \gamma^{\ell-1}\}$$

of prime order ℓ. Then there is a short exact sequence

$$1 \longrightarrow Gal(\overline{E}/E) \longrightarrow Gal(\overline{F}/F) \longrightarrow Gal(E/F) \longrightarrow 1$$

of Galois groups. If r is a representation of $Gal(\overline{F}/F)$, let r_E be the restriction of r to the subgroup $Gal(\overline{E}/E)$. Suppose that $v \notin S$ is a valuation which is unramified for both r and E/F, and that $v_E = v \circ \mathrm{Norm}_{E/F}$ is the associated function on E. It is easy to check that the conjugacy classes in $GL_n(\mathbb{C})$ are related by

$$\sigma_{V_i}(r_E) = \sigma_v(r)$$

if $v_E = V_1 \cdots V_\ell$ splits completely in E, and

$$\sigma_V(r_E) = \sigma_v(r)^\ell$$

if $v_E = V$ remains prime in E. Another way to say this is that

$$L_{S_E}(s, r_E) = \prod_{j=1}^{\ell} L_S(s, r \otimes \varepsilon^j),$$

where S_E is the set of valuations of E over S, and ε is the one dimensional representation

$$\gamma^k \longrightarrow e^{\frac{2\pi i k}{\ell}}, \qquad k = 1, \ldots, \ell,$$

of $Gal(E/F)$. Thus, the map $r \longrightarrow r_E$ from n-dimensional representations of $Gal(\overline{F}/F)$ to n-dimensional representations of $Gal(\overline{E}/E)$ is determined in a simple way by its behaviour on Frobenius conjugacy classes. Moreover, it is easy to check that an arbitrary n-dimensional representation R of $Gal(\overline{E}/E)$ is of the form r_E if and only if the conjugate R^γ of R by γ is equivalent to R.

Langlands' conjecture suggests that there should be a parallel operation on automorphic representations. Let $G_E = GL_{n,E}$ denote the general linear group, regarded as an algebraic group over E.

Theorem [A-C] There is a canonical map $\pi \to \pi_E$ from $\Pi(G)$ to $\Pi(G_E)$ such that

$$\sigma_{V_i}(\pi_E) = \sigma_v(\pi)$$

if $v_E = V_1 \cdots V_\ell$ splits completely in E and

$$\sigma_V(\pi_E) = \sigma_v(\pi)^\ell$$

if $v_E = V$ remains prime in E. In particular

$$L_{S_E}(s, \pi_E) = \prod_{j=1}^{\ell} L_S(s, \pi \otimes (\eta \circ \det)^j),$$

where η is the Grössencharacter of F associated to ε by class field theory. Moreover, an automorphic representation $\Pi \in \Pi(G_E)$ is of the form π_E if and only if $\Pi^\gamma \cong \Pi$.

We have already mentioned that the theorem was proved for $n = 2$ by Langlands [L.6], who built on earlier work of Saito and Shintani. The proof for general n in [A-C] actually applies only to a subset of $\Pi(G)$, namely representations 'induced from cuspidal'. However, the general case follows easily from this and the classification [M-W] of the discrete spectrum of GL_n. The proof of base change relies in an essential way on the trace formula for GL_n.

We have so far treated the simplest case of unramified primes. We should say a few words about the ramified places before we go on to more general L-functions. In [G-J], Godement and Jacquet define a local L-function $L(s, \pi_v)$ and ε-factor $\varepsilon(s, \pi_v, \psi_v)$ for any admissible representation $\pi_v \in \Pi(G(F_v))$ and any nontrivial additive character ψ_v of F_v. They then define the global L-function

$$L(s, \pi) = \prod_v L(s, \pi_v)$$

and ε-factor

$$\varepsilon(s, \pi) = \prod_v L(s, \pi_v)$$

as products over all places v. Here $\pi \in \Pi_{\text{aut}}(G)$ is any automorphic representation, and $\psi = \otimes_v \psi_v$ is a nontrivial additive character on $\mathbb{A}/F$. Since the local root numbers are trivial at unramified places, the global root number is defined as a finite product. It is independent of ψ. The main result of [G-J] is

Theorem [G-J] Suppose that $\pi \in \Pi_{\mathrm{cusp}}(G)$ is a cuspidal representation with contragradient $\tilde{\pi}$. Then $L(s,\pi)$ can be analytically continued as a meromorphic function of $s \in \mathbb{C}$ which satisfies the functional equation

$$L(s,\pi) = \varepsilon(s,\pi)L(1-s,\tilde{\pi}).$$

The function $L(s,\pi)$ is entire unless $n = 1$ and π is an unramified character. □

2 GENERAL AUTOMORPHIC L-FUNCTIONS.

From now on, G will be an arbitrary reductive algebraic matrix group defined over F. The objects $G(\mathbb{A})$, $G(\mathbb{A})^1$, $\Pi(G(F_v))$, $\Pi(G(\mathbb{A}))$, etc., are defined essentially as above. Using the standard parabolic subgroups of G as we did for GL_n, we can also define the families

$$\Pi_{\mathrm{cusp}}(G) \subset \Pi_{\mathrm{disc}}(G) \subset \Pi(G) \subset \Pi_{\mathrm{aut}}(G) \subset \Pi(G(\mathbb{A}))$$

of irreducible representations of $G(\mathbb{A})$. By the general theory of Eisenstein series [L.3], $\Pi(G)$ is precisely the set of irreducible representations which occur in the spectral decomposition of $L^2(G(F)\setminus G(\mathbb{A}))$.

If $\pi = \bigotimes_v \pi_v$ is any representation in $\Pi(G(\mathbb{A}))$, it can be shown that π_v is unramified for all v outside a finite set $S \supset S_\infty$. (This means that G is quasi-split over F_v and split over an unramified extension, and that π_v has a fixed vector under a hyperspecial maximal compact subgroup of $G(F_v)$. As with GL_n, any unramified representation is a constituent of the representation induced from an unramified quasi-character, essentially uniquely determined, on a Borel subgroup defined over F_v. See [B, §10.4].) What plays the role for general G of the conjugacy classes $\sigma(\pi_v)$ in $GL_n(\mathbb{C})$?

To take the place of $GL_n(\mathbb{C})$, Langlands [L.1] introduced a certain complex, nonconnected group. In its simplest form, this *L-group* is a semi-direct product

$${}^LG = \hat{G} \rtimes Gal(E/F),$$

where $\hat{G}$ is a complex reductive group which is 'dual' to G, and E/F is any finite Galois extension over which G splits. The action of $Gal(E/F)$ on $\hat{G}$ is determined in a canonical way, up to inner automorphism, from the action of the Galois group on the Dynkin diagram of G. Rather than define the L-group precisely, we shall simply note that it comes with some extra structure, which in essence determines it uniquely. Suppose that $T \subset B \subset G$ and $\hat{T} \subset \hat{B} \subset \hat{G}$ are maximal tori, embedded in Borel subgroups of G and $\hat{G}$. The L-group is then equipped with an isomorphism from $\hat{T}$ onto the complex dual torus $X^*(T) \otimes \mathbb{C}^*$ of T, which maps the simple roots $\hat{\Delta}$ of $(\hat{B}, \hat{T})$ onto the simple co-roots $\Delta^\vee$ of (G, T), and which is compatible with the canonical actions of $Gal(E/F)$. (See [K.1, §1].) The simplest examples of pairs $(G, \hat{G})$ are $(GL_n, GL_n(\mathbb{C}))$, $(SL_n, PGL_n(\mathbb{C}))$, $(PGL_n, SL_n(\mathbb{C}))$, $(SO_{2n+1}, Sp_{2n}(\mathbb{C}))$, $(Sp_{2n}, SO_{2n+1}(\mathbb{C}))$, $(SO_{2n}, SO_{2n}(\mathbb{C}))$. In each of these cases, G is already split, and the field E may be taken to be F.

The unramified representations have the following striking characterization in terms of the L-group [B, §10.4]. For almost all places v, G is quasi-split over F_v and split over an unramified extension, and $G(\mathfrak{o}_v)$ is a hyperspecial maximal compact subgroup. For any such v, the representations $\pi_v \in \Pi(G(F_v))$ which have a $G(\mathfrak{o}_v)$-fixed vector are in one-to-one correspondence with the semisimple conjugacy classes $\sigma(\pi_v)$ in ${}^LG$ whose projection onto the factor $Gal(E/F)$ equals the Frobenius class at v. Thus, a representation $\pi = \bigotimes_v \pi_v$ in $\Pi(G(\mathbb{A}))$ gives rise to a family

$$\sigma(\pi) = \{\sigma_v(\pi) = \sigma(\pi_v) : v \notin S\}$$

of semisimple conjugacy classes in ${}^LG$.

In order to define an automorphic L-function, one needs to take a finite dimensional representation

$$r : {}^LG \longrightarrow GL_n(\mathbb{C})$$

of the L-group as well as an automorphic representation $\pi \in \Pi_{\text{aut}}(G)$. This gives rise to a family

$$\{r(\sigma_v(\pi)) : \ v \notin S\}$$

of semisimple conjugacy classes in $GL_n(\mathbb{C})$. The general automorphic L-function is then defined as the product

$$L_S(s, \pi, r) = \prod_{v \notin S} \det(1 - r(\sigma_v(\pi) q_v^{-s})^{-1}, \quad s \in \mathbb{C}.$$

It is not hard to verify that the product converges in some right half plane. Again, one expects the L-functions to have analytic continuation and functional equation, although this is still far from known in general. Bear in mind that we are free to let the extension E/F be arbitrarily large. Therefore, the general automorphic L-functions include both the Artin L-functions and the standard L-functions for GL_n discussed in §1.

Examples

1. Suppose that E/F is a cyclic extension of prime order ℓ. Take G to be $Res_{E/F}(GL_{n,E})$, the group obtained from the general linear group over E by restriction of scalars. This is perhaps the simplest example after GL_n itself. The L-group is given by

$$^LG = \underbrace{(GL_n(\mathbb{C}) \times \cdots \times GL_n(\mathbb{C}))}_{\ell} \rtimes Gal(E/F),$$

where the cyclic Galois group acts by permuting the factors. There is a canonical representation

$$r : {}^LG \longrightarrow GL_{n\ell}(\mathbb{C}),$$

in which $\hat{G}$ is embedded diagonally, and $Gal(E/F)$ is mapped into the obvious group of permutation matrices. Since $G(\mathbb{A}) \cong GL_n(\mathbb{A}_E)$, an automorphic representation $\pi \in \Pi_{\mathrm{aut}}(G)$ can be identified with an automorphic representation $\Pi \in \Pi_{\mathrm{aut}}(GL_{n,E})$ of the general linear group over E. One can check that

$$L_S(s, \pi, r) = L_{S_E}(s, \Pi).$$

2. Suppose that $G = GL_n \times GL_m$. There is a canonical representation

$$r : GL_n(\mathbb{C}) \times GL_m(\mathbb{C}) \longrightarrow GL_{nm}(\mathbb{C}).$$

An automorphic representation $\pi \in \Pi_{\mathrm{aut}}(G)$ is a tensor product of automorphic representations $\pi_1 \in \Pi_{\mathrm{aut}}(GL_n)$ and $\pi_2 \in \Pi_{\mathrm{aut}}(GL_m)$. The corresponding L-function $L_S(s, \pi, r)$ equals the general Rankin–Selberg product $L_S(s, \pi_1 \times \tilde{\pi}_2)$. Its analytic continuation and functional equation have been established by Jacquet, Piatetskii-Shapiro, and Shalika [J-P-S], and will be discussed in §II.1.

3. Suppose that G is one of the classical groups SO_{2n+1}, Sp_{2n} or SO_{2n}. Then $\hat{G}$ equals $Sp_{2n}(\mathbb{C}), SO_{2n+1}(\mathbb{C})$, or $SO_{2n}(\mathbb{C})$ respectively. In each case, there is a standard embedding r of $\hat{G}$ into a complex general linear group. The corresponding L-functions $L_S(s, \pi, r)$ have been studied by Piatetskii-Shapiro and Rallis [P-R], and will also be discussed in §II.1.

Underlying everything is the fundamental problem of establishing Langlands' functoriality principle [L.1], [B]. This pertains to maps $\rho : {}^LG \longrightarrow {}^LG'$ between two L-groups. We shall say that such a map is an *L-homomorphism* if E' is a subfield of E, and if the composition of ρ with the projection of ${}^LG'$ onto $Gal(E'/F)$ equals the canonical map of $Gal(E/F)$ onto $Gal(E'/F)$.

Conjecture (Langlands) Suppose that G and G' are reductive groups over F, that G' is quasi-split, and that $\rho : {}^LG \longrightarrow {}^LG'$ is an L-homomorphism between their L-groups. Then for any automorphic representation $\pi \in \Pi_{\text{aut}}(G)$, there is an automorphic representation $\pi' \in \Pi_{\text{aut}}(G')$ such that $\rho(\sigma_v(\pi)) = \sigma_v(\pi')$ for all v outside a finite set $S \supset S_\infty$. In particular,

$$L_S(s, \pi, r \circ \rho) = L_S(s, \pi', r)$$

for any finite dimensional representation r of ${}^LG'$.

Remarks

1. Suppose that π belongs to the subset $\Pi(G)$ of $\Pi_{\text{aut}}(G')$. Then one should be able to choose π' in $\Pi(G')$. This question is related to the discussion in §3.
2. Suppose that $G = \{1\}$ and $G' = GL_n$. Then an L-homomorphism between their L-groups is an n-dimensional representation of $Gal(E/F)$. The functoriality principle becomes the conjecture stated in §1, relating Artin L-functions to the automorphic L-functions of GL_n. We have already discussed the limited number of cases in which it has been solved. This apparently simple case illustrates the depth of the general functoriality principle.

Examples

1. (a) (Base change). Let E/F be a cyclic extension of prime degree ℓ. Set $G = GL_n$ and $G' = Res_{E/F}(GL_{n,E})$. Define an L-homomorphism $\rho : {}^LG \longrightarrow {}^LG'$ by taking the diagonal embedding

$$\hat{G} = GL_n(\mathbb{C}) \hookrightarrow \underbrace{(GL_n(\mathbb{C}) \times \cdots \times GL_n(\mathbb{C}))}_{\ell} = \hat{G}'.$$

Functoriality in this case asks for a correspondence from $\Pi_{\text{aut}}(G)$ to

$$\Pi_{\text{aut}}(G') = \Pi_{\text{aut}}(GL_{n,E}).$$

This follows easily from the base change theorem stated in §1.

(b) (Automorphic induction). Let G' be as above and take G'' to be $GL_{n\ell}$. Let

$$\rho' : {}^LG' \longrightarrow GL_{n\ell}(\mathbb{C}) = {}^LG''$$

be the representation defined in the previous set of examples. This case of functoriality is also known. It was proved (in a slightly different form) in [A-C], as a consequence of base change.

2. Set $G = GL_n \times GL_m$, $G' = GL_{nm}$ and

$$\rho : {}^LG \cong GL_n(\mathbf{C}) \times GL_m(\mathbf{C}) \longrightarrow GL_{nm}(\mathbf{C}) \cong {}^LG'.$$

It would be extremely valuable to have functoriality for this example. However, it is very deep, and is far from being proved.

3. Suppose that ${}^LG$ is one of the complex classical groups $Sp_{2n}(\mathbf{C})$, $SO_{2n+1}(\mathbf{C})$ or $SO_{2n}(\mathbf{C})$, and that ρ is the standard representation. Functoriality is not known in this case, but one hopes that it will eventually follow from the twisted trace formula for the general linear group, and the developing theory of endoscopy. There is also another approach, using L-functions, which will be discussed briefly in §II.1.

We have continued to emphasize mainly the unramified places and the associated conjugacy classes. This is the simplest way to describe things, at least initially. However, it eventually becomes necessary to deal with the ramified places as well.

To this end, we first recall that the local Weil group W_{F_v} is defined for any valuation v. It fits into the commutative diagram

$$\begin{array}{ccc} W_{F_v} & \longrightarrow & Gal(\overline{F}_v/F_v) \\ \downarrow & & \downarrow \\ W_F & \longrightarrow & Gal(\overline{F}/F) \end{array}$$

of locally compact groups, in which W_F denotes the global Weil group. The vertical embeddings are determined only up to conjugacy. One can also define the (local) Weil–Deligne group [K.1, §12] in the form

$$L_{F_v} = \begin{cases} W_{F_v}, & \text{if } v \text{ is Archimedean,} \\ W_{F_v} \times SU(2,\mathbf{R}), & \text{if } v \text{ is discrete.} \end{cases}$$

Langlands and Deligne have defined local L-functions $L(s, r_v)$ and root numbers $\varepsilon(s, r_v, \psi_v)$ for any finite dimensional representation r_v of either W_{F_v} or L_{F_v}, and for any nontrivial additive character ψ_v on F_v. (See [T.2].)

It is actually best to define the L-group of G as

$${}^LG = \hat{G} \rtimes W_F,$$

where W_F acts on $\hat{G}$ through its projection onto $Gal(E/F)$. One can also define local L-groups

$$^LG_v = \hat{G} \rtimes W_{F_v}.$$

These come equipped with embeddings

$$^LG_v \hookrightarrow {}^LG,$$

which are determined up to conjugacy in ^LG. The notion of L-homomorphism can be extended in the obvious way to maps between local or global L-groups, or indeed to any maps between locally compact groups which both fibre over W_{F_v} or W_F.

The local Langlands conjecture can be stated informally as follows.

Conjecture (Langlands [L.1], [B]) There is a partition of the admissible representations $\Pi(G(F_v))$ into finite disjoint subsets Π_{ϕ_v} which are parametrized by the $\hat{G}$-orbits of admissible L-homomorphisms

$$\phi_v : L_{F_v} \longrightarrow {}^LG_v.$$

The definition of an *admissible* L-homomorphism ϕ_v, which we have omitted, is straightforward. One simply imposes several natural conditions, the most significant one being that if the image of ϕ_v is contained in a parabolic subgroup of ^LG, then the corresponding parabolic subgroup of G must be defined over F [B, §8.2]. The conjectured partition should have a number of natural properties [B, §10]. For example, if ϕ_v is unramified in the obvious sense, and corresponds to a semisimple conjugacy class Φ in ^LG_v, then Π_{ϕ_v} should consist of the set of unramified representations $\pi_v \in \Pi(G(F_v))$ such that $\sigma(\pi_v) = \Phi$. However, the expected properties as they are presently conceived are not strong enough to determine the partition uniquely. The local Langlands conjecture has been established in the following cases.

(i) G is a torus [L.2].
(ii) F is Archimedean [L.7].
(iii) $G = GL_n$, and n is prime to q_v [My].
(iv) $G = GL_p$, with p prime [K-M].

Assume for the moment that the local conjecture has been established. Then any representation $\pi_v \in \Pi(G(F_v))$ belongs to a unique packet Π_{ϕ_v}. In this context, the functoriality principle can be described as follows. Suppose that G' is quasi-split, and that

$$\rho : {}^LG \longrightarrow {}^LG'$$

is an L-homomorphism. For each v this determines a commutative diagram

$$\begin{array}{ccc} {}^LG_v & \xrightarrow{\rho_v} & {}^LG'_v \\ \downarrow & & \downarrow \\ {}^LG & \xrightarrow{\rho} & {}^LG'. \end{array}$$

Suppose that $\pi = \bigotimes_v \pi_v$ is a representation in $\Pi_{\mathrm{aut}}(G)$. We are assuming that each π_v belongs to a uniquely determined packet Π_{ϕ_v}. For each v, $\rho_v \circ \phi_v$ is then an admissible homomorphism from L_{F_v} to ${}^LG'_v$. The functoriality principle is that there is an automorphic representation $\pi' = \bigotimes_v \pi'_v$ in $\Pi_{\mathrm{aut}}(G')$ such that for each v, π'_v belongs to the packet $\pi_{\rho_v \circ \phi_v}$.

Suppose that

$$r : {}^LG \longrightarrow GL_n(\mathbb{C})$$

is a finite dimensional representation of the global L-group, and that

$$\pi = \bigotimes_v \pi_v, \qquad \pi_v \in \Pi_{\phi_v},$$

belongs to $\Pi_{\mathrm{aut}}(G)$. If r_v is the composition of the embedding ${}^LG_v \hookrightarrow {}^LG$ with r, $r_v \circ \phi_v$ is an n-dimensional representation of L_{F_v}. One can thus define the local L-functions

$$L(s, \pi_v, r_v) = L(s, r_v \circ \phi_v)$$

and ε-factors

$$\varepsilon(s, \pi_v, r_v, \psi_v) = \varepsilon(s, r_v \circ \phi_v, \psi_v)$$

for all places v. The global L-functions

$$L(s, \pi, r) = \prod_v L(s, \pi_v, \rho_v)$$

and ε-factors

$$\varepsilon(s, \pi, r) = \prod_v \varepsilon(s, \pi_v, r_v, \psi_v)$$

can then be defined as products over all v. As before, $\psi = \bigotimes_v \psi_v$ is a nontrivial additive character on $\mathbb{A}/F$. The global automorphic L-functions are expected to have analytic continuation, and to satisfy the functional equation

$$L(s, \pi, r) = \varepsilon(s, \pi, r) L(1 - s, \pi, \tilde{r}).$$

When he first defined these objects [L.1], Langlands pointed out that the analytic continuation and functional equation would follow from the general functoriality principle and the theorem for GL_n that was later proved by Godement and Jacquet. One would just apply functoriality with $G' = GL_n$, and with

$$\rho = r \times 1 \; : \; {}^LG = \hat{G} \rtimes W_F \longrightarrow GL_n(\mathbb{C}) \times W_F.$$

3 UNIPOTENT AUTOMORPHIC REPRESENTATIONS

We shall conclude with a conjectural explanation for the failure of the analogue of Ramanujan's conjecture for general G. This amounts to a classification of the representations in $\Pi(G)$ in terms of tempered representations.

The problem can be motivated from a different point of view. For general G, the strong multiplicity one theorem fails. In other words, the map

$$\sigma : \pi \longrightarrow \sigma(\pi) = \{\sigma_v(\pi) : v \notin S\},$$

from $\Pi(G)$ to families of semisimple conjugacy classes in ${}^L G$, is not injective. (Let us agree to identify two elements $\sigma(\pi)$ and $\sigma(\pi')$ in the image if $\sigma_v(\pi) = \sigma_v(\pi')$ for almost all v.) One could look for some equivalence relation on $\Pi(G)$, defined without reference to σ, whose classes are contained in the fibres of σ. The original idea for such an equivalence relation is due to Langlands, and is now part of the theory of endoscopy.

Suppose that

$$\phi : W_F \longrightarrow {}^L G$$

is an L-homomorphism from the global Weil group into ${}^L G$ which is admissible; that is, each of the corresponding local maps

$$\phi_v : L_{F_v} \longrightarrow\!\!\!\!\rightarrow W_{F_v} \longrightarrow {}^L G_v$$

is admissible. Then according to the local Langlands conjecture, there are finite packets Π_{ϕ_v} in $\Pi(G(F_v))$, and from these one can form a global packet

$$\Pi_\phi = \{\pi = \bigotimes_v \pi_v \in \Pi(G(\mathbb{A})) : \pi_v \in \Pi_{\phi_v}\}.$$

It would be a consequence of the general functoriality principle that Π_ϕ actually contains a representation in $\Pi_{\mathrm{aut}}(G)$. However, one would ultimately like to have more precise information. Suppose that the image of ϕ in $\hat{G}$ is *bounded*. (If $G = GL_n$, this means that ϕ corresponds to a *unitary* representation of W_F.) Then there is a conjectural formula, which is implicit in the paper [L-L] of Labesse and Langlands, for the multiplicity with which any representation $\pi \in \Pi_\phi$ occurs in $L^2(G(F) \setminus G(\mathbb{A}))$. Observe that the admissibility of the representations $\pi \in \Pi_\phi$ is built into the definition. This implies that for almost all v, π_v is the unique representation in Π_{ϕ_v} which has a $G(\mathfrak{o}_v)$-fixed vector. It follows that the map σ is constant on Π_ϕ.

We shall not recall the definition of a tempered representation, beyond noting that a global packet Π_ϕ should consist of tempered representations precisely when the image of ϕ in $\hat{G}$ is bounded. This is one of the required properties of the conjectural local correspondence. The classical Ramanujan conjecture can be regarded as an assertion that certain representations of GL_2 are tempered. Now the multiplicity formulas in [L-L] were intended only for tempered representations. But it is known that for general G there are many representations in $\Pi_{\mathrm{cusp}}(G)$ which are not tempered. Such examples were first discovered for Sp_4, by Kurokawa [Ku] and Howe and Piatetskii-Shapiro [H-P]. How can one account for these objects?

We shall begin by describing the theorem of Moeglin and Waldspurger, which gives a classification of the discrete spectrum of GL_n in terms of cuspidal representations.

Theorem [M-W] There is a bijection between the set of representations $\pi \in \Pi_{\mathrm{disc}}(GL_n)$ and the set of pairs (d, τ), in which d is a divisor of n and τ is a representation in $\Pi_{\mathrm{cusp}}(GL_d)$.

For a given pair (d, τ), the corresponding $\pi \in \Pi_{\mathrm{disc}}(GL_n)$ is constructed as follows. Set

$$P(\mathbb{A}) = \{p = \begin{pmatrix} p_1 & \cdots & * \\ & \ddots & \vdots \\ 0 & & p_m \end{pmatrix} : \quad p_i \in GL_d(\mathbb{A})\}$$

where $n = dm$, and let $\tilde{\pi}$ be the representation obtained by inducing the representation

$$p \longrightarrow \tau(p_1)|\det p_1|^{\frac{m-1}{2}} \otimes \tau(p_2)|\det p_2|^{\frac{m-3}{2}} \otimes \cdots \otimes \tau(p_m)|\det p_m|^{-\frac{(m-1)}{2}}$$

from $P(\mathbb{A})$ to $G(\mathbb{A})$. Then π is the unique irreducible quotient of $\tilde{\pi}$. In particular,

$$L(s, \pi) = L(s + \tfrac{m-1}{2}, \tau) L(s + \tfrac{m-3}{2}, \tau) \cdots L(s - \tfrac{m-1}{2}, \tau).$$

Notice that this formula provides an extension of the analyticity assertion of Godement–Jacquet to representations in the discrete spectrum of GL_n; $L(s, \pi)$ will be entire unless $d = 1$ and τ is an unramified character. Now, it is not hard to see that π is nontempered if $m > 1$. Conversely if $m = 1$, so that π belongs to $\Pi_{\mathrm{cusp}}(GL_n)$, then π is expected to be tempered. This is the generalized Ramanujan conjecture, which is believed to hold for GL_n. The theorem can therefore be interpreted as a description of $\Pi_{\mathrm{disc}}(GL_n)$ in terms of tempered representations. It is this interpretation which should carry over to other groups.

The classification of $\Pi_{\text{disc}}(GL_n)$ has a description in terms of the global parameters $\phi : W_F \to {}^LG$. In the case of GL_n, a packet Π_ϕ should contain only one representation, and this in turn should belong to $\Pi_{\text{cusp}}(GL_n)$ precisely when the parameter ϕ, regarded as an n-dimensional representation of W_F, is irreducible and unitary. However, the map $\phi \to \pi$ is not surjective. There are many representations $\pi \in \Pi_{\text{cusp}}(GL_n)$ which do not correspond to any n-dimensional representation of the global Weil group. With this difficulty in mind, Langlands [L.5, §2] suggested that the tempered representations in the sets

$$\Pi(GL_n), \qquad n = 1, 2, \cdots,$$

might possibly form a tannakian category. This (together with the generalized Ramanujan conjecture) would imply the existence of a locally compact group L_F, whose irreducible, unitary n-dimensional representations parametrize all of $\Pi_{\text{cusp}}(GL_n)$. We shall assume that L_F exists in what follows. We shall also assume that there are injections $L_{F_v} \hookrightarrow L_F$, determined up to conjugacy, as well as a canonical surjective map $L_F \twoheadrightarrow W_F$ with compact connected kernel. (See [K.1, §12].) For us, the introduction of L_F is primarily for book-keeping. The reader can pretend that L_F is the Weil group, or even the Galois group of a finite Galois extension E.

In [L.5, §2], Langlands also introduced the collection of *isobaric* representations, a subset of $\Pi_{\text{aut}}(GL_n)$ which in turn contains $\Pi(GL_n)$. The significance of the isobaric representations is that they are in bijective correspondence with the (equivalence classes of) all semisimple representations of L_F of dimension n. In other words, they are parametrized by maps

$$w \longrightarrow \phi_1(w)\,|w|^{r_1} \oplus \phi_2(w)|w|^{r_2} \oplus \cdots \oplus \phi_m(w)|w|^{r_m}, \qquad w \in L_F,$$

where each $r_i \in \mathbf{R}$, ϕ_i is an irreducible unitary representation of L_F of dimension d_i, and $d_1 + \cdots + d_m = n$. What are the maps that correspond to the subset $\Pi_{\text{disc}}(GL_n)$ of all the isobaric representations? They are just the maps with $\phi_1 = \phi_2 = \cdots = \phi_m = \phi$, and $r_1 = \frac{m-1}{2}$, $r_2 = \frac{m-3}{2}, \ldots, r_m = -\frac{m-1}{2}$. Now there is a nice way to characterize these particular maps within the set of all n-dimensional representations of L_F. Given an irreducible unitary representation ϕ of L_F of dimension d, with $n = md$, let ρ_m be the unique irreducible representation of $SL(2, \mathbf{C})$ of dimension m. Then

$$\psi(w, u) = \phi(w) \otimes \rho_m(u), \qquad w \in L_F,\ u \in SL(2, \mathbf{C}),$$

is an irreducible n-dimensional representation of $L_F \times SL(2, \mathbf{C})$. Having constructed ψ, we can define

$$\phi_\psi(w) = \psi\left(w, \begin{pmatrix} |w|^{\frac{1}{2}} & 0 \\ 0 & |w|^{-\frac{1}{2}} \end{pmatrix}\right), \qquad w \in L_F.$$

In other words,

$$\phi_\psi(w) \;=\; \phi(w)|w|^{\frac{m-1}{2}} \oplus \cdots \oplus \phi(w)|w|^{-\frac{m-1}{2}}.$$

The maps corresponding to $\Pi_{\text{disc}}(GL_n)$ are thus obtained from irreducible representations of $L_F \times SL(2,\mathbb{C})$ whose restriction to L_F is unitary. In a similar vein, one sees that the maps corresponding to $\Pi(GL_n)$ are obtained by allowing the representations ψ to be reducible.

We return now to the case that G is arbitrary. Let $\Psi(G)$ denote the set of $\hat{G}$-orbits of admissible L-homomorphisms

$$\psi : L_F \times SL(2,\mathbb{C}) \longrightarrow {}^LG,$$

such that the projection of $\psi(L_F)$ onto $\hat{G}$ is bounded. To any such ψ one can associate a finite group $\mathcal{S}_\psi$ [A.2, §8]. We will not reproduce the definition in general. However, if G is split over F, $\mathcal{S}_\psi$ equals $\pi_0(S_\psi/Z(\hat{G}))$, where $\pi_0(\;)$ denotes the group of connected components, S_ψ is the centralizer of $\psi(L_F \times SL(2,\mathbb{C}))$ in $\hat{G}$, and $Z(\hat{G})$ is the center of $\hat{G}$. One can also attach to ψ a certain sign character

$$\varepsilon_\psi : \mathcal{S}_\psi \longrightarrow \{\pm 1\}.$$

Conjecture [A.2, §6, §8], [A.3, §4] For every $\psi \in \Psi(G)$ there exist

(i) finite local packets $\Pi_{\psi_v} \subset \Pi(G(F_v))$, which for almost all v contain precisely one representation with a $G(\mathfrak{o}_v)$-fixed vector; and

(ii) finite dimensional characters

$$s \longrightarrow < s, \pi > = \prod_v < s, \pi_v >, \qquad\qquad s \in \mathcal{S}_\psi,$$

defined for each representation in the global packet

$$\Pi_\psi \;=\; \{\pi \;=\; \bigotimes_v \pi_v \in \Pi(G(\mathbb{A})) : \pi_v \in \Pi_{\psi_v}\},$$

such that the multiplicity in $L^2(G(F) \setminus G(\mathbb{A}))$ of any representation $\pi \in \Pi(G(\mathbb{A}))$ equals

$$|\mathcal{S}_\psi|^{-1} \sum_{\{\psi \in \Psi(G) : \pi \in \Pi_\psi\}} \; \sum_{s \in \mathcal{S}_\psi} \varepsilon_\psi(s) < s, \pi > .$$

Remarks

1. The multiplicity formula implies that any representation $\pi \in \Pi(G)$ belongs to one of the packets Π_ψ . The nature of the parameters ψ then suggests that there is a Jordan decomposition for the elements in $\Pi(G)$ which is parallel to the Jordan decomposition for conjugacy classes in $G(F)$. For example, a parameter ψ can be called *unipotent* if the projection of $\psi(L_F)$ onto $\hat{G}$ equals $\{1\}$. The *unipotent automorphic representations* are then the constituents of sets $\Pi(G) \cap \Pi_\psi$, in which ψ is a unipotent parameter. The trivial one dimensional representation of $G(\mathbb{A})$ is the simplest example of such a representation. It corresponds to the principal unipotent class in $\hat{G}$. More interesting examples of unipotent automorphic representations have been constructed for (split) classical groups by Moeglin [Mg].

2. There are two conditions, which often hold for a given ψ, under which the multiplicity formula simplifies. Suppose that Π_ψ is disjoint from all the other packets $\Pi_{\psi'}$, $\psi' \neq \psi$, and that each of the functions
$$s \longrightarrow < s, \pi_v >, \qquad \pi_v \in \Pi_{\psi_v},$$
is a one dimensional abelian character. Then the multiplicity for any $\pi = \otimes_v \pi_v \in \Pi_\psi$ is 1 if the product $\prod_v < s, \pi_v >$ equals the sign character ε_ψ, and is 0 otherwise.

3. Suppose that G is quasi-split. Then
$$\phi_\psi(w) = \psi(w, \begin{pmatrix} |w|^{\frac{1}{2}} & 0 \\ 0 & |w|^{-\frac{1}{2}} \end{pmatrix}), \qquad w \in L_F,$$
is an admissible L-homomorphism of L_F into ${}^LG$. The global packet Π_ψ should then contain the global Langlands packet Π_{ϕ_ψ}.

4. For maps ψ which are trivial on $SL(2, \mathbb{C})$, the conjecture contains nothing which is not already implicit in [L-L]. In this case the sign character ε_ψ is trivial. In general, however, ε_ψ is defined in terms of symplectic root numbers. Suppose that G is split. A parameter ψ determines a finite dimensional representation
$$R(s, w, u) = Ad(s\psi(w, u)), \qquad s \in S_\psi/Z(\hat{G}),\ w \in L_F,\ u \in SL(2, \mathbb{C}),$$
of $(S_\psi/Z(\hat{G})) \times L_F \times SL(2, \mathbb{C})$ on the Lie algebra of $\hat{G}$. Let
$$R = \bigoplus_{i \in I} (\lambda_i \otimes \mu_i \otimes \nu_i)$$

be the decomposition of R into irreducible representations. Observe that the determinant of an irreducible representation λ_i of $S_\psi/Z(\hat{G})$ can be regarded as a function on $\mathcal{S}_\psi = \pi_0(S_\psi/Z(\hat{G}))$. Observe also that for an irreducible representation μ_i of L_F, one can define the L-function $L(s, \mu_i)$ and root number $\varepsilon(s, \mu_i)$ from the embeddings $L_{F_v} \hookrightarrow L_F$. Let J be the subset of indices $i \in I$ such that the representation μ_i is symplectic, and such that $\varepsilon(\frac{1}{2}, \mu_i) = -1$. Then

$$\varepsilon_\psi(s) = \prod_{i \in J} \det(\lambda_i(s)), \qquad s \in \mathcal{S}_\psi.$$

This formula for the sign character is strongly suggested by the spectral side of the trace formula. (See [A.3, especially §6].)

5. The parameters $\psi \in \Psi(G)$ have a direct bearing on our principal theme, automorphic L-functions, through the theory of Shimura varieties. They come up in the important problem of expressing the zeta function of a Shimura variety in terms of automorphic L-functions. Roughly speaking, the $SL(2, \mathbb{C})$ factor in a parameter ψ is the same object as the group $SL(2, \mathbb{C})$ that comes from the Lefschetz hyperplane section in cohomology. This allows one to determine the various degrees of cohomology to which a given ψ will contribute. (See [A.2, §9], [K.2, §8-10].) At the end of §10 of [K.2], Kottwitz states a conjectural decomposition of the λ-adic cohomology which implies a formula for the zeta function of a Shimura variety in terms of automorphic L-functions.

PART II

1 THE TRACE FORMULA, AND THE METHOD OF ZETA-INTEGRALS

We begin by taking a closer look at the two fundamental problems of Langlands' theory of automorphic L-functions.

Conjecture A Every general automorphic L-function

$$L(s, \pi, r) = \prod_{v \notin S} \det(1 - r(\sigma_v(\pi) q_v^{-s})^{-1},$$

initially defined as an Euler product convergent in some half-plane, continues to a meromorphic function in all of $\mathbb{C}$, with only finitely many poles, and a functional equation relating its values at s and $1 - s$. (As remarked in Part I, a *precise* functional equation can be expected only after local factors $L(s, \pi_v, r)$ have been defined at the 'bad primes' v in S as well.)

Conjecture B Functoriality with respect to the L-group Suppose we are given reductive algebraic groups G and G' as in Section I.2, and a homomorphism

$$\rho : {}^L G \longrightarrow {}^L G' .$$

Then for each π in $\Pi_{aut}(G)$ there is a π' in $\Pi_{aut}(G)$ such that

$$\rho\{\sigma_v(\pi)\} = \{\sigma_v(\pi')\}$$

for all v outside some finite set $S \supset S_\infty$. In particular, for any $r : {}^L G' \to GL_d(\mathbb{C})$, this transfer of '$L$-packets' is such that

$$(*) \qquad L_S(s, \pi', r) = L_S(s, \pi, r \circ \rho).$$

Remarks

1. Suppose we take G arbitrary, $G' = GL_d$, ρ any L-group representation $\rho : {}^L G \to GL_d(\mathbb{C})$, and $r : {}^L G' \to GL_d(\mathbb{C})$ the standard representation (taking g to itself). Then $(*)$ above reads

$$L_S(s, \pi', St) = L_S(s, \pi, \rho),$$

with the left-hand side L-function a standard Godement–Jacquet L-function on GL_d. Thus Conjecture B indeed implies Conjecture A, and the immediate impression is that work on Conjecture A might be superfluous. This impression is misleading, however, since in practice it is often easier to establish Conjecture A directly, rather than appeal to the relevant form of Conjecture B.

2. Conjecture A has been successfully attacked in general using two different methods – the explicit construction of zeta-integrals, and the Langlands–Shahidi method using Eisenstein series (and their Fourier coefficients). A detailed survey of these methods – and their range of applicability – is the subject matter of [GeSh]. Our present state of knowledge concerning Conjecture A is roughly the following.
   (a) The existence of a meromorphic continuation is known for the L-function of almost any reductive quasi-split group and the 'standard' representation of its L-group;
   (b) except for $GL(n)$ (and its standard L-function), there are almost always non-trivial problems encountered in establishing an exact functional equation for $L(s, \pi, r)$, or the finiteness of the number of its poles; and
   (c) the best results have recently been obtained via the method of explicit zeta-integral representations – in fact, this method should ultimately prove most useful in number theory, since it makes possible

an analysis of the location (and possible arithmetic significance) of the poles of $L(s,\pi,r)$.

Examples illustrating (a)–(c) will be discussed at end of this lecture (Concluding Remarks and Theorems).

3. As we shall soon see, attempts to prove Conjecture A or B have drawn freely from *three* principal tools of the theory of automorphic forms: the trace formula, explicit zeta-integral representations of L-functions, and the theory of Θ-series liftings. We now proceed to discuss these topics in earnest.

Trace formula methods The prototype example here is the theory of base change already discussed in Part I. This example is also the most general to date, in that functorial lifting is proved for two quasi-split groups of rank n. Other examples involve either a 'compact' form of the trace formula (see [BDKV] and [Ro.1], which concern liftings between division algebras and GL_n), or else lower rank groups. See [La.3] for further discussion and references.

Unitary group examples ([Ro.2]) These examples concern functorial lifting from $U(2)$ to $U(3)$, and 'base change for $U(3)$'. Unlike the example of base change for $GL(n)$, one obtains here genuinely new information on the automorphic L-functions in question, as we now explain.

Let E denote a quadratic extension of the global field F, and V a three-dimensional Hermitian space which is defined over E and possesses an isotropic vector. Then set G equal to the unitary group $U(V)$ ('the' quasi-split unitary group in three variables), H the 'endoscopic' subgroup

$$U(2)\times U(1) = \left\{\begin{bmatrix} * & 0 & * \\ 0 & * & 0 \\ * & 0 & * \end{bmatrix} \text{ in } U(V)\right\},$$

and $\tilde{G}$ the group $Res_F^E G$. The corresponding L-groups are ${}^LG = GL_3(\mathbf{C}) \rtimes W_F$, where w acts via $g \to \begin{pmatrix} & & 1 \\ & -1 & \\ 1 & & \end{pmatrix} {}^tg^{-1} \begin{pmatrix} & & 1 \\ & -1 & \\ 1 & & \end{pmatrix}$ if its image in $Gal(E/F)$ is non-trivial, ${}^LH = (GL_2(\mathbf{C}) \times GL_1(\mathbf{C})) \rtimes W_F$, with a similar action, and ${}^L\tilde{G} = (GL_3(\mathbf{C}) \times GL_3(\mathbf{C})) \rtimes W_F$, where w permutes the coordinates of ${}^L\tilde{G}^o$ (if the image of w in $Gal(E/F)$ is nontrivial). Finally, we consider the following L-group homomorphisms: let $\psi_G : {}^LG \to {}^L\tilde{G}$ denote the standard base-change imbedding $\psi_G(g,w) = (g,g,w)$, and let $\psi_H = {}^LH \to {}^LG$ extend the natural embedding of ${}^LH^\circ$ into ${}^LG^\circ$ via the formula

$$\psi_H(w) = \begin{pmatrix} 1 & & \\ & 1 & \\ & & -1 \end{pmatrix} \rtimes w$$

if the image of w in $Gal(E/F)$ is nontrivial, and otherwise,

$$\psi_H(w) = \begin{pmatrix} \mu(w) & & \\ & 1 & \\ & & \mu(w) \end{pmatrix} \rtimes w,$$

with μ a character of C_E whose restriction to C_F is $w_{E/F}$ (the quadratic character determined by E via class field theory).

Theorem ([Ro.2])

(a) There exist functorial transfers of L-packets corresponding to ψ_H and ψ_G.

(b) If π is cuspidal automorphic on $U(3)$, then its lift (through ψ_G) is again cuspidal (on $GL(3)$ over E) if and only if π is not the lift (through ψ_H) of any τ in $\Pi_{aut}(H)$.

Corollary For π any automorphic cuspidal representation of $U(V)$, and ξ any unitary idele-class character of E, set

$$L(s, \pi \otimes \xi) \equiv L(s, \psi_G(\pi) \otimes \xi),$$

the L-function on the right being the standard Godement–Jacquet L-function for $GL(3)$ over E of the lift of π. Then $L(s, \pi \otimes \xi)$ is always entire (for any ξ) if and only if π is not a ψ_H-transfer of some τ on H. Moreover, if π *is* of the form $\psi_H(\tau)$, then for some 'twist' ξ, $L(s, \pi \otimes \xi)$ has a pole on the line $Re(s) = 1$ if τ is cuspidal, and on $Re(s) = 3/2$ if τ is one-dimensional.

Remarks

1. The L-function $L(s, \pi \otimes \xi)$ may be interpreted as the 'standard' degree six L-function on $U(3)$. More precisely, let $G' = U(3) \times Res_F^E GL(1)$, so that

$${}^LG' = (GL_3(\mathbb{C}) \times \mathbb{C}^* \times \mathbb{C}^*) \rtimes W_F$$

with the obvious action of W_F on the connected component of ${}^LG'$. Let ρ denote the 6 dimensional representation of ${}^LG'$ induced from the 'standard' representation $St_3 \otimes St_1 \otimes 1$ of the (index two) normal subgroup

$GL_3(\mathbb{C}) \times \mathbb{C}^x \times \mathbb{C}^x$. Given π and ξ as above, we obtain the following conjugacy classes in ${}^LG'$ for each unramified v:

$$\sigma_v(\pi) \times \xi_w(\tilde{\omega}_w) \times \xi_{w'}(\tilde{\omega}_{w'}) \rtimes \sigma, \quad \text{if} \quad w, w'|v; \qquad \text{and}$$

$$\sigma_v(\pi) \times \xi_w(\tilde{\omega}_w) \times 1 \rtimes \sigma, \quad \text{if} \quad v \text{ is inert.}$$

Thus the L-function $L(s, \pi \otimes \xi)$ introduced above is precisely the 'standard' L-function $L(s, \pi \otimes \xi, \rho)$ on G', all of whose analytic properties, as predicted by Conjecture A (including finiteness of poles and an exact functional equation), now follow directly from the Godement–Jacquet theory for $GL(n)$.

2. Our experience with $GL(n)$ does not prepare us for the existence of poles of automorphic L-functions to the right of the line $Re(s) = 1$ (the right-hand boundary of the critical strip). Indeed, it is a (non-trivial) fact that for π (resp. π') any automorphic cuspidal unitary representation of $GL_n(\mathbb{A})$ (resp. $GL_m(\mathbb{A})$), even the Euler product

$$\prod L(s, \pi_v \times \pi'_v)$$

defining $L_S(s, \pi \times \pi')$ *converges absolutely* for $Re(s) > 1$ (see [J-S], I). In particular $L_S(s, \pi \times \pi')$ has neither zeros *nor* poles in this half-plane, generalizing a classical result for Dirichlet's L-functions.

On the other hand, for groups other than $GL(n)$ – typically $Sp(n)$, we encounter a quite new phenomenon, namely the existence of cuspidal π whose standard L-functions can have poles at 3/2, 2, etc. The explanation for this fact is intimately tied up with the theory of theta-series liftings. This phenomenon is also related to the existence of counterexamples to the generalized Ramanujan conjecture, and the fact that many cuspidal π (outside $GL(n)$) fail to possess Whittaker models. In particular, for $U(3)$, those cuspidal π such that $L(s, \pi \otimes \xi)$ can have a pole to the right of $Re(s) = 1$ are non-tempered almost everywhere, possess no Whittaker model, and have the same L-function as an Eisenstein series (i.e., are CAP representations in the sense of [P-S.3]). All these matters will be discussed in detail at the end of the next section.

3. The discussion above suggests a question about $L(s, \pi \otimes \xi)$ on $U(3)$ left unanswered by the trace formula analysis of [Ro.2]. If π is such that its L-function can have a pole, can we characterize π by the peculiarity of its Fourier expansion? This question we shall discuss in earnest in §II.2 after first reviewing the use of explicit zeta-integrals in the theory of L-functions.

Zeta-integral methods Once again, the basic game plan is simple: it follows the lines introduced in Tate's analysis of the Dirichlet L-functions $L(s,\chi)$, and generalized by Godement–Jacquet for the standard $GL(n)$ L-functions $L(s,\pi)$.

For simplicity, we review the program for $GL(1)$ only. Given a grössen-character χ (an automorphic representation of $GL(1)$), we consider the global zeta-integral

$$\mathcal{Z}(s,f,\chi) = \int_{\mathbb{A}_F^\times} f(x)\chi(x)|x|_{\mathbb{A}}^s d^\times x,$$

where f is a Schwartz–Bruhat function on $\mathbb{A}$. The program we have in mind consists of the following five steps:

(1) Express the global zeta-integral as an Euler product of *local zeta-integrals* (in this case $\mathcal{Z}(s,f_v,\chi_v)$);
(2) Analyse the meromorphic behavior and functional equation of the *global* zeta-integrals;
(3) Do Step (2) for the *local* zeta-integrals as well;
(4) Interpret the *unramified* local zeta-integrals (in this case, meaning f_v is the characteristic function of the ring of v-adic integers, and χ_v is unramified) as an appropriate Langlands factor $[\det(I - r(\sigma_v(\pi))q_v^{-s})]^{-1}$ (in this case $1-\chi(v)q_v^{-s}$); and
(5) Establish additional basic properties of the local zeta-integrals, if possible introducing $L(s,\chi_v)$ *at the ramified primes* as a g.c.d. of the local zeta-integrals.

Because of Step (4), we say that $\mathcal{Z}(s,f,\chi)$ 'interpolates' $L(s,\chi)$, and we use all five steps of this 'L-function machine' to deduce the expected analytic properties of $L(s,\chi)$ from those of $\mathcal{Z}(s,f,\chi)$. For general groups, an account of how this method is carried out for $L(s,\pi,\rho)$ is found in [GeSh]; we content ourselves here only with some important examples which either motivate or are required in the discussion of Θ-series liftings in §II.2.

Hecke theory for $GL(n)$ Fix $\pi = \otimes\pi_v$ an irreducible unitary (not necessarily automorphic) representation of $GL(n,\mathbb{A}_F)$. For $GL(2)$, the fact that the L-function

$$L(s,\pi) = \prod_v L(s,\pi_v)$$

is 'nice' characterizes π as an automorphic cuspidal representation; more precisely the 'converse theorem' states that if every $L(s,\pi\otimes\chi)$ continues to an *entire* function in $\mathbb{C}$ which is bounded in vertical strips and satisfies the expected functional equation, then π is automorphic cuspidal. This result is proved using not a Godement–Jacquet zeta-integral, but rather a Hecke-type zeta-integral

$$\mathcal{Z}(s,\varphi,\chi) = \int_{\mathbb{Q}^x\backslash\mathbb{A}^x} \varphi_\pi\begin{pmatrix} a & 0 \\ 0 & 1 \end{pmatrix}\chi(a)|a|^{s-\frac{1}{2}}d^x a$$

where φ_π belongs to the space of π. As is well-known, this result is useful in proving examples of Langlands functoriality. In particular, it implies that the Langlands reciprocity conjecture mentioned in Lecture §I.1 is true for two-dimensional Galois representations provided Artin's conjecture is true.

Throughout the 1970s, much effort was expended in obtaining converse theorems for $GL(n),\ n > 2$. For $GL(3)$, it was found that a $GL(2)$-type result remains valid, and that it once again implies many interesting instances of Langlands functoriality. One such application is base change 'induction' for *non*-normal cubic extensions, proved in [J-P-S.2] and used by Tunnell in his proof of Artin's conjecture for arbitrary two-dimensional octahedral Galois representations. Another application of this theorem is (the proof of) Conjecture B for the symmetric square map $Ad:\ GL(2) \to GL(3)$. Complete trace formula proofs of these particular liftings have yet to be published.

For $n > 3$, converse theorems for $GL(n)$ apparently require twistings by automorphic representations of $GL(m),\ m > 1$, i.e., the theory of generalized Rankin–Selberg convolutions discussed below.

The method of Rankin–Selberg, generalized The starting point is Jacquet's 1972 analysis of the zeta-integral interpolating the standard L-function $L(s,\pi_1 \times \pi_2,\rho)$ on $GL(2) \times GL(2)$ (i.e., $\rho:\ GL_2(\mathbb{C}) \times GL_2(\mathbb{C}) \to GL_4(\mathbb{C})$ is just the outer tensor product). His Rankin–Selberg type integral looks like

$$\int_{H(F)\backslash H(\mathbb{A})} \varphi_1(g)\varphi_2(g)E(g,s)dg,$$

where $H = GL_2$ diagonally embedded in $G = GL_2 \times GL_2$, φ_i is in the space of π_i, and $E(g,s)$ is an Eisenstein series on GL_2. Subsequently, similar zeta-integrals were introduced by Jacquet, Piatetski-Shapiro and Shalika to interpolate the automorphic L-functions $L(s,\pi_1\times\pi_2)$ on $GL(n)\times GL(m)$. As mentioned already in the first lecture, the results thus obtained for $L(s,\pi_1 \times \pi_2)$ yield a strong multiplicity one result for automorphic representations $\Pi(G)$ of $GL(n)$, and are used in the theory of base change for $GL(n)$. To be more precise, the results required are those describing if and when $L(s,\pi_1\times\pi_2)$ has a pole on the line $Re(s)=1$. Whereas explicit zeta-integrals are used to prove these results in [J-S](I), an alternate, shorter proof has recently been given in [W.1], using the Langlands–Shahidi theory of Eisenstein series.

Another possible application of these Rankin–Selberg convolutions is to the theory of converse theorems for $GL(n)$. The simplest type of theorem to state is the following: suppose Π is an irreducible unitary representation of $GL_n(\mathbb{A}_F)$ with the property that for *any* automorphic generic irreducible unitary representation τ of $GL_{n-1}(\mathbb{A}_F)$, the L-function $L(s,\Pi\times\tau)$ is 'nice', i.e., entire, bounded in vertical strips, and satisfies a functional equation of the form

$$L(s,\pi\times\tau)=\varepsilon(s,\pi\times\tau)L(1-s,\hat{\pi}\times\hat{\tau}).$$

Then Π itself is automorphic cuspidal. In practice, one usually needs a converse theorem with weaker hypotheses, for example, $L(s,\pi\times\tau)$ need be 'nice' only for a restricted class of τ's (but then, as is well-known for the case $n=2$, the conclusion is also weaker). These (and related) results are presently being developed by Piatetski-Shapiro and co-workers; see [P-S.1] and the discussion of functorial lifting below.

The general L-functions of Piatetski-Shapiro, Rallis, et al. In 1983–4, Piatetski-Shapiro and Rallis discovered a general type of zeta-integral which in one fell swoop generalized the Godement–Jacquet zeta-integral from $GL(n)$ to an arbitrary simple classical group, and paved the way for a vast generalization of the Rankin–Selberg method as well.

The idea is to consider zeta-integrals of the form

$$\text{(I)}\qquad \int_{H(F)\backslash H(\mathbb{A})} \varphi_G(h)E_H(h,s)dh$$

or

$$\text{(II)}\qquad \int_{G(F)\backslash G(\mathbb{A})} \varphi_G(g)E_H(g,s)dg,$$

where φ is a cusp form on some group G (belonging to an automorphic cuspidal representation π whose L-function we are trying to interpolate), and E is an Eisenstein series on some group H closely related to G. In both cases, the analytic properties of the relevant Eisenstein series must be established in order to obtain the meromorphic continuation of the corresponding automorphic L-functions. The key difference between Case I and II is that in the first case, $G \supset H$ (so that $\varphi(h)$ represents the restriction of φ to H), whereas in Case II, $G \subset H$; as we shall see in the few examples below, this distinction leads to different obstacles in trying to apply the L-function machine to these zeta-integrals.

Case I examples $H =$ *split* SO_{2n} (with an n-dimensional isotropic subspace), $G = SO_{2n+1} \supset H$, π is a *generic* cuspidal representation of $G(\mathbb{A})$ (i.e., each φ in π has nontrivial Fourier coefficients with respect to the standard maximal unipotent subgroup of G), and τ is a (generic) cuspidal representation of GL_n (regarded as a representation of the Levi component of the maximal parabolic subgroup P of H). Gelbart and Piatetski-Shapiro have applied several steps of the L-function machine to the zeta-integrals

$$\int_{H(F)\backslash H(\mathbb{A})} \varphi(h) E_\tau(h,s)dh \sim L(s, \pi \times \tau, r),$$

where $E_\tau(h,s)$ is an Eisenstein series on $H(\mathbb{A})$ attached to the parabolically induced representation $ind_{P(\mathbb{A})}^{H(\mathbb{A})} \tau|\det|^s$, and $L(s,\pi\times\tau,r)$ is the automorphic L-function on $G \times GL_n$ attached to the standard (tensor product) representation r of ${}^LG \times GL_n(\mathbb{C}) = Sp_n(\mathbb{C}) \times GL_n(\mathbb{C})$; see III.1.5 of [GeSh]

In a very interesting recent development, D. Ginsburg, Piatetski-Shapiro, D. Soudry, *et al.* have modified this method in order to cover the case of $G \times GL_k$, with k arbitrary relative to n; see [Gi] for the case $1 \leq k \leq n$. The significance of this general case is that it should *eventually* yield a functorial lifting from $G = SO_{2n+1}$ to GL_{2n} corresponding to the L-group homomorphism $\rho : {}^LG \to Sp_n(\mathbb{C}) \subset GL_{2n}(\mathbb{C})$; indeed, the converse theorem for GL_{2n} requires twistings $L(s, \pi \times \tau)$ with τ on GL_{2n-1}. Such an application of the theory of zeta-integrals to prove Conjecture B would be very exciting and should generalize to other classical groups. However, at present this work represents mostly a program for future research; much remains to be done, especially concerning the archimedean analysis in Steps 3 and 5 of the L-function machine, and the precise analytic properties of the global Eisenstein series. Moreover, even when complete, this work will give functoriality only for generic automorphic representations.

Case II examples: Rankin triple L-functions ([Ga] and [P-R.2]) Here $G = GL_2 \times GL_2 \times GL_2 \subset H = Sp_6$, so we are in Case II. Given a triplet of cuspidal representations of $GL(2)$, which we regard as a single representation of $GL(2) \times GL(2) \times GL(2)$, we derive the analytic properties of $L(s, \pi_1 \times \pi_2 \times \pi_3, \rho)$, where $\rho : {}^LG \to GL_8(\mathbb{C})$ is given by an outer tensor product, through the zeta-integrals

$$\int_{G(F)\backslash G(\mathbb{A})} \varphi_1(g_1)\varphi_2(g_2)\varphi_3(g_3)E_H(g_1, g_2, g_3, s)dg_1dg_2dg_3 \,.$$

Here $E_H(h, s)$ is an Eisenstein series on H induced from the character $|\det|^s$ of the maximal parabolic subgroup P of H (whose Levi component is isomorphic to GL_3, and whose unipotent radical is abelian). The crucial point is that $P\backslash H$ has only one open orbit under the (right) action of $G \subset H$; this is the 'orbit yoga' which makes possible the appropriate factorization of these global zeta-integrals into local ones.

Other examples of zeta-integrals of type II (where $G \subset H$) also require an analysis of the orbit structure of $P\backslash H/G$ in order to obtain the necessary Euler product factorization (Step 1 of the L-function-machine). Two of the most striking applications of this method are the following:

Theorem [P-R.1] Given a simple classical group (such as SO_n, Sp_n), its standard L-function $L_S(s, \pi, St)$ has a meromorphic continuation and functional equation.

Theorem [PS-R-S] Suppose $G = G_2 \times GL(2)$, π (resp. τ) is an automorphic cuspidal *generic* representation of G_2 (resp. GL_2), and $\rho : G_2(\mathbb{C}) \times GL_2(\mathbb{C}) \to GL_{14}(\mathbb{C})$ is the standard representation of ${}^LG$ obtained from taking the tensor product of the standard embedding of $G_2(\mathbb{C}) \subset SO_7(\mathbb{C})$ in $GL_7(\mathbb{C})$ with the standard representation of $GL_2(\mathbb{C})$. Then $L_S(s, \pi \times \tau, \rho)$ has a meromorphic continuation and functional equation.

Concluding remarks

(a) The last example above is of special interest for the following reason. Groups like G_2 lie outside the range of applicability of the method of Langlands–Shahidi, since this method works only for groups which can be embedded as the Levi component of a parabolic subgroup of some larger reductive group. Although the method of explicit zeta-integrals has no such *a priori* limitations, it was nevertheless an open problem for years whether L-functions attached to G_2 could be analyzed via zeta-integrals.

(b) One of the main problems in the general theory of zeta-integrals comes from the difficulty in executing Step 5 of the L-function machine – i.e., in controlling the zeta-integrals at the 'bad' places. For example, in proving Theorem [P-R.1], one encounters an identity of the form

$$\int\limits_{G\times G(F)\backslash G\times G(\mathbb{A})} \varphi_1(g_1)\widetilde{\varphi_2}(g_2)E_H(g_1,g_2,s)dg_1dg_2$$

$$= \left(\prod_{v\in S} Z_v(s,\varphi_1,\varphi_2,E)\right) L_S(s,\pi,St).$$

Now one does know exactly where the poles of the Eisenstein series E on the left side are located (see [P-R.1] and [P-R.2]). In general, however, one does not yet have complete control of the non-vanishing of the local zeta-integrals on the right-hand side of this identity; thus one can not conclude that the (finitely many?) poles of $L(s,\pi,\rho)$ are among those of the Eisenstein series. (See Lecture §II.2 for an alternate approach to this finiteness of poles result, at least for $G = Sp_{2n}$.) In the case of Rankin triple products, one *does* have sufficient control of the non-vanishing of the archimedean integrals, and hence precise information on the (finitely many) poles of $L(s,\pi_1\times\pi_2\times\pi_3)$. However, the rest of Step 5 – expressing the local integrals (at infinite primes) in terms of gamma factors – remains an open problem (except for the special case of holomorphic cusp forms; see [Ga]).

(c) Suppose we can prescribe exactly the location of the possible poles of a particular L-function. The method of zeta-integrals as described thus far still gives no information about characterizing those π for which these poles occur, nor does it give information about special values or non-vanishing properties of these L-functions. Such results seem to be accessible only via the theory of Θ-series liftings to be described in Section II.2 below. Typical of the results we wish to discuss are the following:

Theorem ([P-R.3]) Suppose $G = Sp(4)$, π is an automorphic cuspidal representation of G, and $L(s,\pi,\rho)$ is the degree 5 L-function associated to the standard embedding $\rho: SO_5(\mathbb{C}) \to GL_5(\mathbb{C})$. Then $L(s,\pi,\rho)$ is holomorphic for $Re(s) > 2$, has a meromorphic continuation to $\mathbb{C}$, and a simple pole at $s = 2$ if and only if π is a certain Θ-series 'in two variables'.

Theorem ([W.2-3]) Suppose $G = PGL_2$, π is a cuspidal representation of $G(\mathbb{A})$ coming from a holomorphic cusp form f of even integral weight k, and $L(s,\pi)$ is the standard Hecke–Jacquet–Langlands L-function attached to π (so that $L(s,\pi)$ is entire in $\mathbb{C}$). Then $L(\frac{1}{2},\pi\otimes\chi_b) \neq 0$ for some quadratic character χ_b, if and only if f is the Shimura correspondence image of some cusp form $\tilde{f}$ of weight $\frac{k}{2}+\frac{1}{2}$ (in which case the non-vanishing of the χ_b-twisted L-function is related to the non-vanishing of the appropriate bth Fourier coefficient of $\tilde{f}$).

2 HOWE'S CORRESPONDENCE AND THE THEORY OF THETA-SERIES LIFTINGS

In his 1964 *Acta* paper, Weil gives a representation-theoretic formulation of the Siegel theory of theta-functions. In this theory, theta-functions comprise a space of functions on the so-called metaplectic group, functions which under right translation realize the metaplectic (oscillatory, or Weil) representation.

In more detail, first fix a local field F (not of characteristic 2), and a non-trivial additive character ψ of F. Let W denote a $2n$-dimensional symplectic vector space over F (equipped with antisymmetric form $< >$), and let $Sp(W)$ denote the symplectic group of W. Then Weil's metaplectic group $Mp(W)$ may be introduced as a (certain) group of unitary operators on some space S, fitting into the exact sequence

$$1 \longrightarrow T \longrightarrow Mp(W) \longrightarrow Sp(W) \longrightarrow 1,$$

with $T = \{z \in \mathbb{C} : |z| = 1\}$. Because the action of these operators on S depends on ψ, we should denote the metaplectic group by $Mp_\psi(W)$. We recall that:

(i) if $W = X \oplus X^\vee$ with X an n-dimensional isotropic ('Lagrangian') subspace of W, then S may be described as the (unitary completion of the) Schwartz–Bruhat space $S(X)$; moreover, the operators of Mp_ψ preserve $S(X)$;

(ii) $Mp_\psi(W)$ determines a *non-trivial* central extension of $Sp(W)$ by T, and hence does *not* produce an *ordinary* representation of $Sp(W)$; on the other hand (see [Rao]), there is a canonical splitting of $Mp_\psi(W)$ over the (unique) two-fold cover $\overline{Sp}$ of $Sp(W)$.

Similarly, given an $\mathbb{A}$-field F, a non-trivial additive character $\psi = \Pi\psi_v$ of $\mathbb{A}/F$, and a global symplectic space W, there is a metaplectic extension

$$1 \longrightarrow T \longrightarrow Mp_\psi(W_\mathbb{A}) \longrightarrow Sp(W_\mathbb{A}) \longrightarrow 1,$$

where $Mp_\psi(W_\mathbb{A})$ is now a group of unitary operators which preserves $S(X_\mathbb{A}) = \otimes S(X_v)$ and is compatible with the local metaplectic groups. The connection between this metaplectic group and the theory of automorphic forms derives from the fact that $Mp(W_{\mathbb{A}})$ splits (and then again canonically) over the subgroup of rational points $Sp(W)_F$ in $Sp(W_{\mathbb{A}})$. This splitting $r_F: Sp(W)_F \to Mp(W_{\mathbb{A}})$ is determined by the condition that for each γ in $Sp(W)_F$,

$$(*) \qquad \sum_{\xi \in X_F} (r_F(\gamma)\Phi)(\xi) = \sum_{\xi \in X_F} \Phi(\xi), \quad \forall\, \Phi \in \mathcal{S}(X_{\mathbb{A}}).$$

In particular, we can (and shall) regard $Sp(W)_F$ as a subgroup of $Mp_\psi(W_{\mathbb{A}})$, i.e., as a group of operators on $\mathcal{S}(X_{\mathbb{A}})$. If we let $\Theta: \mathcal{S}(X_{\mathbb{A}}) \to \mathbf{C}$ denote the functional $\Phi \to \Theta(\Phi) = \sum_{\xi \in X_F} \Phi(\xi)$, then $(*)$ simply says that this 'theta-functional' is $Sp(W)_F$-invariant.

Henceforth, we shall understand by 'Weil's metaplectic group' *either* $\overline{Sp}$ or the group of operators Mp_ψ. This latter group of operators determines an ordinary representation of $\overline{Sp}$ (or Mp_ψ itself) in the space S. We shall denote this representation by ω_ψ and refer to it as 'Weil's representation of the metaplectic group'.

By an automorphic form on the metaplectic group we understand a 'smooth' function on $Sp(W)_F \backslash Mp_\psi(\mathbb{A})$ or $Sp(W)_F \backslash \overline{Sp}(W_{\mathbb{A}}))$ satisfying the usual conditions of moderate growth, K-finiteness, etc. We refer the reader to [B-J] for the general definition of automorphic forms (which makes sense even for covering groups of algebraic groups). The theta-functional above gives an intertwining operator from the space of the Weil representation to the space of automorphic forms on $Mp_\psi(W_{\mathbb{A}})$ (or $\overline{Sp}$), namely

$$\begin{aligned} \Phi \to \Theta_\Phi(m_g) &= \Theta(m_g\, \Phi) \\ &= \sum_{\xi \in X_F} \omega_\psi(m_g)\Phi(\xi). \end{aligned}$$

The resulting automorphic forms are called theta-functions because, when $F = \mathbf{Q}$, $Sp(W) = SL_2$, and Φ is properly chosen, $\Theta_\Phi(g)$ essentially reduces to the classical theta-series

$$\Theta(z) = \sum_{n=-\infty}^{\infty} e^{2\pi i n^2 z}\,.$$

In general, $\Theta_\Phi(g)$ is a *genuine* function of $Mp_\psi(W)$, in the sense that

$$\Theta(\lambda g) = \lambda\Theta(g)$$

for all λ in T. Moreover, $\Theta_\Phi(g)$ still retains a basic distinguishing characteristic of classical theta-series, namely that most of the Fourier coefficients of Θ_Φ are zero. This fact plays a crucial role in the theory of L-functions (by way of facilitating the Shimura type zeta-integral constructions described below). However, to understand the full impact of the metaplectic representation in the theory of automorphic forms, one needs first to review Howe's correspondence and the theory of theta-series liftings.

Dual reductive pairs and Howe's correspondence ([Ho]) Howe's theory is a refinement of Weil's theory which converts the construction of automorphic forms via theta-functions into a machinery for lifting automorphic forms on one group to automorphic forms on another.

Definition A *dual reductive pair* in $Sp(W)$ is a pair of reductive subgroups G, H in $Sp(W)$ which comprise each other's centralizers in $Sp(W)$.

Examples

(I)
$$G = Sp(W_1), \quad H = O(V_1) \subset Sp(W),$$

where $W = W_1 \otimes V_1$, W_1 is a symplectic space, V_1 is a quadratic space (with orthogonal group $O(V_1)$), and W is the symplectic space whose antisymmetric form is the tensor product of the forms on W_1 and V_1. Analogously, there are the Hermitian dual pairs $U(V_1), U(V_2) \subset Sp(V_1 \otimes V_2)$, where each V_i is a Hermitian space over some quadratic extension E of F, and the symplectic form on the Hermitian space $V_1 \otimes V_2$ is obtained by viewing $V_1 \otimes V_2$ as an F-vector space with $<\ ,\ >$ the 'imaginary' part of the form $(\ ,\ )_1(\ ,\ )_2$.

(II)
$$G = GL(X_1), \quad H = GL(X_2) \subset Sp(W),$$

where $W = (X_1 \otimes X_2) \oplus (X_1 \otimes X_2)^*$ and

$$< (x, x^*), (y, y^*) >= y^*(x) - x^*(y).$$

Facts

(1) These examples exhaust the set of all *irreducible* dual reductive pairs; for a precise statement of the classification, see Chapter 1 of [MVW].

(2) Given a dual pair G, H in $Sp(W)$, the metaplectic extension $Mp_\psi(W)$ splits over G and H in all cases *except* when $G = Sp(W_1)$ is paired with an *odd*-dimensional orthogonal group $H = O(V_1)$, in which case $Mp_\psi(W)$ splits over H but *not* over G. This is a non-trivial fact whose proof is discussed in Chapter 3 of [MVW].

Philosophy for the duality correspondence Suppose G, H is a dual reductive pair in $Sp(W)$, and consider the *restriction* of the Weil representation ω_ψ of $Mp_\psi(W)$ to $G \times H$ (or rather, to a subgroup of $Mp_\psi(W)$ which is isomorphic to $G \times H$; we ignore the fact that this might not always be possible if $G = Sp(W_1)$). Because G and H are each others' mutual centralizers in $Sp(W)$, $\omega_\psi \big|_{G\times H}$ should decompose into irreducible representations of the form $\pi_1 \otimes \pi_2$, with π_2 an irreducible representation of H determined by the irreducible representation π_1 of G. In other words, but still roughly speaking, each π_1-isotypic component of $\omega_\psi \big|_{G\times H}$ should provide an irreducible $G \times H$-module of the form $\pi_1 \otimes \pi_2$, with $\pi_2 = \Theta_\psi(\pi_1)$ the 'Howe correspondence' image of π_1 on H; symbolically,

$$\omega_\psi \Big|_{G\times H} = \bigoplus_{\pi_1 \text{ in } G^\wedge} \pi_1 \otimes \Theta_\psi(\pi_1),$$

where the sum is over those π_1 in $G^\wedge$ which 'occur' in ω_ψ.

More precisely, we say that π_1 occurs in ω_ψ if $Hom_G(\omega_\psi, \pi_1) \neq \{0\}$, in which case we set

$$S(\pi_1) = \bigcap \ ker\ f\ , \quad \text{where } f \text{ runs through } \ Hom_G(\omega_\psi, \pi_1)$$

and

$$S[\pi_1] = S/S(\pi_1) \quad (\text{where } S \text{ is the space of } \omega_\psi).$$

The space $S(\pi_1)$ is G-stable (since each *ker* f is), and H-stable (since H and G commute). By passage to the quotient, one obtains a representation of $G \times H$ in $S[\pi_1]$ which must be of the form $\pi_1 \otimes \pi'_2$ for some smooth (not necessarily irreducible) representation π'_2 of H. *Howe's conjectured duality correspondence* amounts to the assertion that there exists a unique irreducible quotient of π'_2, i.e., a unique invariant subspace of π'_2 whose quotient produces an *irreducible* representation π_2 of H. Assuming this quotient exists, we call π_2 the Howe image of π_1, and denote it by $\Theta_\psi(\pi_1)$.

Remark The Howe correspondence just described is symmetric in G and H, i.e., it doesn't matter which group we take as the 'domain' group. Thus the correspondence $\pi \to \Theta_\psi(\pi)$ goes in both directions!

From the work of Howe (see [W.4] and [MVW]), we have the following (local) result:

Theorem

(a) The Howe correspondence exists (at least for $p \neq 2$).

(b) If π_1 is unramified, so is $\Theta_\psi(\pi_1)$; in fact, whenever possible, this correspondence should be functorial with respect to the L-group, in a sense to be explained below, and in further examples.

Examples

1. (O_n, Sp_n) ([Li]). Consider the dual pair (Sp_n, O_n) where O_n is the orthogonal group of a non-degenerate quadratic form of dimension n, and *n is even.* Then Howe's duality correspondence gives rise to an injection from the unitary dual of O_n to the unitary dual of Sp_n. Assume that the quadratic form defining O_n is split and the character ψ defining Weil's representation is unramified. Then Howe's correspondence takes any unramified π in $O_n^\wedge$ to an unramified representation $\Theta_\psi(\pi)$ of $Sp_n(F)$; moreover, these representations are functorially related as follows: there is a natural map $\rho : {}^L O_n \to {}^L Sp_n \approx SO_{2n+1}(\mathbf{C})$ such that the conjugacy class in ${}^L Sp_n$ parametrizing $\Theta_\psi(\pi)$ is just $\rho(\sigma(\pi))$; in terms of local L-functions, this relation reads

$$L(s, \Theta_\psi(\pi)) = \zeta(s) \prod_{i=1}^{n/2} \zeta(s+i)\zeta(s-i)L(\pi, s),$$

a very special case of Theorem 6.1 of [Ra.1]. The injection $\Theta_\psi : O_n^\wedge \to Sp_n^\wedge$ is also a special case of 'explicit Howe duality in the stable range'; see [Ho.2], [Li.2] and [So.2] for more general results.

2. *Shalika–Tanaka Theory* ([S-T]). Let F be a local field, $H = SL_2(F) = Sp(F^2)$, where $F^2 = F \oplus F$ is equipped with the form $<(x, x'), (y, y')> = xy' - x'y$, and $G = SO(E)$, the (special) orthogonal group of the quadratic space E over F (so that $G = E^1 =$ the norm 1 group of E). Modulo the fact that G is not the full orthogonal group, (G, H) is a dual reductive pair in $Sp(W)$, where $W = F^2 \otimes E = X \oplus X^v$, and $X \approx F \otimes E \approx E$. In this case, ω_ψ acts in $L^2(E)$, $\omega_\psi \Big|_{G \times H}$ is an ordinary representation, and $\omega_\psi|_G$ acts through the 'regular representation' of E^1 in $L^2(E)$. Thus every character χ in $G^\wedge$ 'occurs in ω_ψ' (in fact discretely). Moreover, if F is nonarchimedean, it turns out that $\pi_2 = \pi_2(\chi) = \Theta_\psi(\chi)$ is an irreducible supercuspidal representation of $H = SL_2(F)$, *unless* $\chi^2 = 1$.

The caveat $\chi^2 \neq 1$ is necessary here because we are dealing with $SO(2)$ in place of $O(2)$. Indeed, when χ is the unique character of order 2, $\Theta_\psi(\chi)$ is the sum of *two* irreducible supercuspidal representations π^+

and π^- and when $\chi = 1$, $\Theta_\psi(\chi)$ is an irreducible principal series representation (which is class 1 whenever E and ψ are unramified). Moreover, $\chi^2 \neq 1$ implies $\Theta_\psi(\chi) = \Theta_\psi(\chi^{-1})$; thus the correspondence $\chi \to \Theta_\psi(\chi)$ is two-to-one for such χ. On the other hand, *if we take $O(2)$ in place of $SO(2)$*, we get a one-to-one correspondence $\rho_\chi \leftrightarrow \Theta_\psi(\rho_\chi)$ between the irreducible (two-dimensional) representations of $O(2)$ (excluding the unique non-trivial character of $O(2)/SO(2)$) and certain *irreducible* representations $\Theta_\psi(\rho_\chi)$ of SL_2; namely, $\rho_\chi \to \Theta_\psi(\chi)$ when $\chi^2 \neq 1$ and $\rho_\chi = ind_{SO(2)}^{O(2)}\chi$, $\rho_\chi \to \pi^+$ or π^- when ρ_χ is a character of $O(2)$ non-trivial on $SO(2)$, and $\rho_\chi \to \theta_\psi(1)$ if ρ_χ is the trivial character.

Locally, the example $(SL_2, SO(2))$ is of interest because it provides an explicit construction of supercuspidal representations of $SL_2(F)$. Globally, it gives a generalization of the classical construction of cusp forms (both holomorphic and real-analytic) due to Hecke and Maass. More precisely, given a character $\chi = \Pi\chi_v$ of $E^1_{\mathbb{A}}$ trivial on E^x, we may consider the irreducible representation $\pi = \pi(\chi) = \otimes\pi_{\psi_v}(\chi_v)$ of $SL_2(\mathbb{A})$. (When E_v remains a field, $\pi(\chi_v)$ is as described above; otherwise, $\pi(\chi_v)$ is a principal series representation of $SL_2(F_v)$, unramified if χ_v is; in any case, $\pi(\chi_v)$ is class 1 for almost every v, and therefore $\pi(\chi)$ is a well-defined element of $\Pi(SL_2(\mathbb{A}))$. In fact, $\pi(\chi)$ is an automorphic representation of $SL_2(\mathbb{A})$, a result which Shalika and Tanaka establish by realizing $\pi(\chi)$ directly in the space of χ-isotypic theta-functions

$$f_\chi^\Phi(g) = \int_{E^x\backslash E^1_{\mathbb{A}}} \sum_{\xi\in E} \omega_\chi(tg)\Phi(\xi)\chi(t)d^xt.$$

This method generalizes, as we shall now see.

The global Howe correspondence Suppose $\pi_1 = \otimes\pi_v$ is an irreducible unitary representation of $G(\mathbb{A})$ which occurs in ω_ψ, i.e., for each fixed v, $Hom(\omega_\psi, \pi_v) \neq 0$. Then we can form the irreducible representation

$$\pi_2 = \Theta_\psi(\pi_1) = \otimes\Theta_\psi(\pi_v)$$

of $H(\mathbb{A})$, where for each v, $\Theta_\psi(\pi_v)$ is the local Howe image of π_v, and for almost every v, π_v and $\Theta_\psi(\pi_v)$ are class 1. We call π_2 the *Howe lift* of π_1.

Conjecture The irreducible representation $\pi_2 = \Theta_\psi(\pi_1)$ of $H(\mathbb{A})$ is (usually) automorphic if π_1 is. Moreover, the lift $\pi \to \Theta_\psi(\pi)$ should be functorial whenever possible.

Remarks

(1) This conjecture has deliberately been stated in a vague way, since not enough is known yet to justify making a more precise statement, and there are already some delicate counter examples, some of which will be described below.

(2) Although one might one day be able to establish the automorphy of $\Theta_\psi(\pi)$ in general using the trace formula, at present the best way to attack Howe's conjecture is by way of the theory of Θ-series liftings, generalizing what we described already in Example 2 above.

The theory of Θ-series liftings This theory makes it possible to prove the automorphy of $\Theta_\psi(\pi)$ by directly constructing a realization of $\Theta_\psi(\pi)$ inside the space of automorphic forms on $H(\mathbb{A})$. To simplify the exposition, let us suppose that π is an automorphic *cuspidal* representation of $G(\mathbb{A})$ and H_π is an irreducible subspace of $L_0^2(G(F)\backslash G(\mathbb{A}))$ realizing π. Then we can consider a space of functions on $H(\mathbb{A})$ given by the integrals

$$f_\varphi(h) = \int_{G(F)\backslash G(\mathbb{A})} \Theta_\Phi^\psi(g,h)\varphi(g)dg \tag{2.1}$$

where $\varphi \in H_\pi$ and $\Theta_\Phi^\psi(g,h)$ is the restriction to $G(\mathbb{A}) \times H(\mathbb{A})$ of any theta-function Θ_Φ on $Mp(W)_\mathbb{A}$. Because Θ_Φ^ψ is known from Weil's theory to be a slowly-increasing continuous function on $H \times G(F)\backslash H \times G(\mathbb{A})$, and because φ is cuspidal (and hence rapidly decreasing) on $G(F)\backslash G(\mathbb{A})$, it follows that each of these integrals converges absolutely, and defines an automorphic form on $H(F)\backslash H(\mathbb{A})$. Let us denote by $\Theta(\pi,\psi)$ the space of functions f_φ^Φ so generated on $H(F)\backslash H(\mathbb{A})$.

What is the relation between this new $H(\mathbb{A})$-module $\Theta(\pi,\psi)$ (*the theta-series lifting of* π), and the $H(\mathbb{A})$-module $\Theta_\psi(\pi)$ (the Howe lifting of π)? The answer should be that $\Theta(\pi,\psi)$ realizes $\Theta_\psi(\pi)$ *provided* $\Theta(\pi,\psi)$ is not identically zero! The problem is that the non-vanishing of $\Theta(\pi,\psi)$ is subtle to detect.

Prototype example ([W.2]) Fix $G = SL(2)$, and $H = SO(V)$, where V is the 3-dimensional space of trace zero 2×2 matrices equipped with the quadratic form $q(X) = -\det(X)$. In this case, Howe's correspondence relates representations of $SO(V)$ (which is isomorphic to $PGL(2)$) with representations of $SL(2)$ (or rather the two-fold cover of $SL(2)$, since $M_{p_\psi}(F^2 \otimes V)$ does not split over $SL(2)$). In particular, the local correspondence establishes a bijection between the set of irreducible admissible *genuine* representations of $\overline{SL(2)}$ which possess a ψ-Whittaker model and the set of all irreducible admissible representations of $PGL(2)$. Globally, we are dealing with an integral of the form (2.1), where G is $SL(2)$; although both $\Theta(g,h)$ and $\varphi(g)$ are now *genuine* functions on the metaplectic cover of $SL(2)$, their *product* is naturally defined on $SL(2)$.

Now consider the question whether π automorphic on $H(\mathbb{A})$, say, implies $\Theta_\psi(\pi)$ automorphic on $G(\mathbb{A})$, and if so, is $\Theta_\psi(\pi)$ realizable in $\Theta(\pi,\psi)$? The answer (in the direction $PGL_2 \to \overline{SL_2}$) is that $\Theta_\psi(\pi)$ is automorphic *if and only if* $\varepsilon(\pi,\frac{1}{2}) = 1$; here $\varepsilon(\pi,s)$ is the ε-factor in the functional equation

$$L(s,\pi) = \varepsilon(\pi,s)L(1-s,\pi)$$

satisfied by the Hecke–Jacquet–Langlands L-function of π. Moreover, even when the condition $\varepsilon(\pi,\frac{1}{2}) = 1$ is satisfied, $\Theta_\psi(\pi)$ will be realizable in $\Theta_\psi(\pi,\psi)$ only if this latter space is non-zero; this happens if and only if the stronger condition

$$L(\frac{1}{2},\pi) \neq 0$$

is satisfied.

In the reverse direction, $\overline{SL_2} \to PGL_2$, Howe's correspondence is equally subtle. Here it turns out that $\Theta_\psi(\sigma)$ is always automorphic on PGL_2 if σ is automorphic, but $\Theta(\sigma,\psi)$ is non-zero (and then realizes $\Theta_\psi(\sigma)$) if and only if the ψth Fourier coefficients

$$f_\sigma^\psi(g) = \int f_\sigma\left(\begin{pmatrix} 1 & x \\ 0 & 1 \end{pmatrix} g\right)\overline{\psi(x)}dx$$

do not vanish identically for f_σ in the space of σ. There is also an intriguing characterization of the non-vanishing of these Fourier coefficients in terms of special values of L-functions; this we shall describe later after introducing some general zeta-integrals involving theta-series (zeta-integrals of Shimura type).

Examples involving functoriality The construction of Shalika–Tanaka relating χ and $\pi(\chi)$ was independently obtained in [J-L] with E and $GL(2)$ in place of E^1 and $SL(2)$. At this level, it is easy to check that the lifting $\chi \to \pi(\chi)$ is 'functorial' with respect to the natural L-group morphism $\Psi :\ (\mathbb{C}^x \times \mathbb{C}^x) \rtimes W_F \to GL_2(\mathbb{C}) \times W_F$. Another example involving functoriality is the following.

Let $G = GSp_2$, the group of symplectic similitudes of a 4-dimensional symplectic space W_1, and let $GSO(V_{3,3})$ denote the group of orthogonal similitudes of a 6-dimensional orthogonal space $V_{3,3} = \Lambda^2(F^4)$ equipped with the inner product

$$(W, W') = W \wedge W' \in \Lambda^4(V) \approx F.$$

Although we are dealing here with groups of similitudes inside $GSp(W_1 \otimes V_{3,3})$, a suitable modification of the theory of dual reductive pairs leads us to liftings $\pi \leftrightarrow \Theta_\psi(\pi)$ and $\pi \leftrightarrow \Theta(\pi, \psi)$ between GSp_2 and $GSO(V)$.

Remarks

(1) When dealing with groups of *similitudes*, it turns out that Howe's lifting (or the theta-series lifting) is *independent* of ψ. (The subscript ψ may therefore be suppressed in these cases, though we shall usually refrain from doing so.)

(2) There is a natural injection of $GL_4/\mathbb{Z}_2$ into $GSO(V)$. Therefore, the Θ-series correspondence $\pi \to \Theta(\pi, \psi)$ may be viewed as a correspondence between representations of GSp_2 and GL_4 (by restricting functions in $\Theta(\pi, \psi)$ to $GL_4/\mathbb{Z}_2$); similarly, Howe's correspondence for the pair $(GSp_2, GSO(6))$ naturally defines a correspondence between representations of GSp_2 and $H = GL_4$ which we again denote by $\Theta_\psi(\pi)$. With these remarks in mind, we state the following:

Theorem ([J-P-S.3])

(a) Suppose π is an automorphic cuspidal representation of $G(\mathbb{A})$; then $\Theta(\pi, \psi) \not\equiv \{0\}$ *if and only if* π *is globally generic*, i.e., possesses a standard Whittaker model (on the space of its ψ-Fourier coefficients). In this case, $\Theta(\pi, \psi)$ realizes the Howe lift $\Theta_\psi(\pi)$, and the lifting is compatible with Langlands' functoriality in the following sense: if $\rho : {}^LG \to {}^LH = GL_4(\mathbb{C})$ denotes the standard embedding of $GSp_2(\mathbb{C})$ in $GL_4(\mathbb{C})$, and $\{\sigma_v(\pi)\}$ is the collection of conjugacy classes in ${}^LH$ determined by $\pi = \otimes\pi_v$, then $\rho\{\sigma_v(\pi)\}$ coincides with the collection of conjugacy classes in $GL_4(\mathbb{C})$ determined by $\Theta_\psi(\pi)$.

(b) Suppose Π is an automorphic cuspidal representation of $H(\mathbb{A})$. Then Π is the Θ-series lift of some (globally generic) π on $G(\mathbb{A})$ as in (a) if and only if the degree 6 L-function

$$L(s, \Pi \otimes \chi, \Lambda^2)$$

has a pole at $s = 1$ for some grössen-character χ.

Concluding remarks

1. The example just given shows how Conjecture B can be proved using the theory of Θ-series liftings, at least for generic π. One should also be able to establish this lifting (again for generic π) as a special case of the 'converse theorem' program outlined in Section I.1; for the case at hand, this requires an analysis of the L-functions $L(s, \Pi \times \tau)$ on $GSp_2 \times GL_2$ already studied in [PS-So]. In general, for arbitrary π, one must also eventually be able to establish this lifting using the trace formula. In any case, the functorial identity

$$L(s, \Pi, \Lambda^2) = L(s, \pi, \rho o \Lambda^2),$$

where $\rho : {}^L G \rightarrow {}^L H$, already implies that (some twisting of) $L(s, \Pi, \Lambda^2)$ must have a pole at $s = 1$ if $\Pi = \Theta_\psi(\pi) = \rho(\pi)$, since $\rho o \Lambda^2$ contains a one-dimensional subrepresentation (and therefore $L(s, \pi \otimes \chi, \rho o \Lambda^2)$ will – for some χ – contain the Riemann zeta-function as a factor).

2. As the preceding examples confirm, instances of functorial lifting can be established using the trace formula, L-functions, the theory of theta-series liftings, or any combination thereof. In some cases, such as the Shalika–Tanaka example $\chi \rightarrow \pi(\chi)$, each one of these methods provides an (alternate) proof; the L-function method was applied in §12 of [J-L] (albeit at the level of $GL(2)$), whereas the trace formula approach was developed in [L-L], and led to new and provocative results.

Zeta-integrals of Shimura type Although of interest in its own right, our lengthy detour through the theory of Θ-series was motivated entirely by our interest in the analytic properties of automorphic L-functions. The connection between these subjects is provided by zeta-integrals of Shimura type, themselves modifications of Rankin–Selberg integrals involving Θ-series.

To explain the general construction, we fix $G = Sp_n$, and we consider zeta-integrals of the type

$$\zeta(s, \varphi, \Phi, F) = \int_{G(F)\backslash G(\mathbb{A})} \varphi(g) \Theta_T^\Phi(g) E(g, F, s) dg. \tag{2.2}$$

Here $\varphi(g)$ is a cusp form in the subspace of $L_0^2(G(F)/G(\mathbb{A}))$ realizing the irreducible cuspidal representation π of $G(\mathbb{A})$; T is an $n \times n$ symmetric non-degenerate matrix, which we confuse with the n-dimensional orthogonal space V_T it determines; $\Theta_T^\Phi(g)$ is (the value at $(g, h) = (g, 1)$ of) the theta-kernel

$$\Theta_T^\Phi(g, h) = \sum_{\xi \in M_{n,n}(F)} (\omega_\psi(g, h)\Phi)(\xi)$$

corresponding to the dual pair $(Sp_n, O(V_T))$ in Sp_n and the choice of Schwartz–Bruhat function Φ on the Lagrangian subspace $F^n \otimes V_T$ (viewed as $n \times n$ matrix space); finally, $E(g, F, s)$ is the Eisenstein series on $G(\mathbb{A})$ of the form

$$\sum_{\gamma \in P(F) \backslash G(F)} F_s(\gamma g)$$

where $P = \{\begin{pmatrix} * & * \\ 0 & * \end{pmatrix}\}$ in G is the parabolic subgroup whose Levi component M is isomorphic to GL_n, and $F_s(g)$ belongs to $Ind_{P(\mathbb{A})}^{G(\mathbb{A})} |\det_M m(p)|^{s+\frac{n+1}{2}}$. For convenience, *we assume n is even* (and therefore the metaplectic group never need occur here).

The zeta-integral $\zeta(s, \varphi, \Phi, F)$ interpolates the standard L-function of degree $2n+1$ for $G(\mathbb{A})$, and the desired properties of this L-function are thus obtained by an application of the L-function machine to the above zeta-integrals. For example, we have the following result, first discussed at the end of the last Lecture.

Theorem [P-R.3] Suppose π is an automorphic cuspidal (not necessarily generic) representation of $Sp_2(\mathbb{A})$, and suppose T is a symmetric invertible 2×2 matrix in $M_{2,2}(F)$ such that for some φ in the space of π, the 'T-Fourier coefficient'

$$\varphi_T(g) = \int_{S(F) \backslash S(\mathbb{A})} \varphi\left(\begin{pmatrix} 1 & X \\ 0 & 1 \end{pmatrix} g\right) \overline{\psi(tr\, TX)} dX$$

does not vanish (there $S = \{X\}$ denotes the F-vector space consisting of symmetric 2×2 matrices). Let ρ denote the standard 5-dimensional representation of the L-group of $SO_5(\mathbb{C})$ of Sp_2 and let $\chi_T(a)$ denote the quadratic character $(a, -\det T)$. Then the L-function

$$L_S(s, \pi \otimes \chi_T, \rho), \qquad Re(s) \gg 0,$$

has meromorphic continuation to $\mathbb{C}$, with only finitely many poles, and is analytic for $Re(s) > 2$.

Sketch of proof Everything follows from the basic identity

$$\zeta(s, \varphi, \Phi, F) = G_\infty(s) L_S(s + \frac{1}{2}, \pi \otimes \chi_T, \rho), \tag{2.3}$$

where $F_s(g)$ is suitably normalized, i.e., multiplied by a suitable product of zeta-functions, such that (1) the identity (2.3) holds *and* (2) the resulting (normalized) Eisenstein series has only finitely many poles (in this case, at certain half-integral or integral points of the interval $[-\frac{3}{2}, \frac{3}{2}]$). The function $G_\infty(s)$, which depends on the local data Φ_v and $F_{v,s}$, is a meromorphic function in $\mathbb{C}$ which can be chosen to be non-zero at any one of the possible poles of $E(s,g,F)$. From this it is clear that $L(s+\frac{1}{2}, \pi \otimes \chi_T, \rho)$ can have a pole at $s = s_0$ only if $E(s,g,F)$ does. Moreover, a pole of $E(s,g,F)$ will produce a pole of $L(s', \pi \otimes \chi_T, \rho)$ only if this pole survives integration against $\varphi(g)\Theta(g)$. This circumstance brings us full circle back to the theory of Θ-series liftings. In particular, we have the following important Corollaries to the proof just sketched:

Theorem [Li] The L-function $L_S(s, \pi \otimes \chi_T, \rho)$ has a pole at $s = 2$ if and only if the Fourier expansion of any φ in π has only one 'orbit' of non-zero Fourier coefficients φ_T, i.e., π is T-distinguished in the sense of [Li].

Remark There is always *some* nondegenerate T such that $\varphi_T \neq 0$; this was proved earlier by Li in his thesis [Li.3]. But then for $T' = {}^tA^{-1}TA^{-1}$, with A in GL_n, we have

$$\varphi_{T'}(g) = \varphi_T\left(\begin{bmatrix} A^{-1} & 0 \\ 0 & {}^tA \end{bmatrix} g\right). \tag{2.4}$$

Thus $\varphi_{T'} \neq 0$ for all T' in the 'orbit' of T. The meaning of π being distinguished is that there is a T such that $\varphi_{T'} \neq 0$ if and only if T' is in the orbit of T. Li's characterization of distinguished π follows from (the analog for Sp_2 of) the first half of the following result.

Theorem [P-R.3] and [So] Consider the similitude group $G = GSp_2$ in place of Sp_2 (so now ${}^LG = GSp_2(\mathbb{C})$, with five dimensional representation ${}^LG \to PGSp_2(\mathbb{C}) \approx SO_5(\mathbb{C}) \subset GL_5(\mathbb{C})$). Then $L_S(s, \pi \otimes \chi_T, \rho)$ has a simple pole at $s = 2$ if and only if either one of the following conditions are satisfied: (1) the theta-series lifting of π to $GO(T)$ is non-zero, i.e., $\Theta(\pi, \psi) \neq 0$; or (2) π is a CAP representation with respect to either the Borel subgroup of G, or the parabolic subgroup Q which preserves a line.

Remarks

1. The first condition follows immediately from the fact that $Res_{s=3/2}E(g,s,F)$ is independent of g, and therefore $L(s', \pi \otimes \chi_T, \rho)$ has a pole at $s' = 2$ iff

$$Res_{s=3/2}\zeta(s, \varphi, \Phi, F) = c \int_{G(F)\backslash G(\mathbb{A})} \varphi(g)\Theta(g,1)dg \neq 0$$

i.e., iff $\Theta(\pi, \psi)$ is not identically zero on $GO(T)$. (Again, some modification of the theory of dual pairs is required for the similitude groups $GO(T)$ and GSp_2; in this case, as already mentioned in a recent remark, $\Theta(\pi, \psi)$ no longer depends on ψ.)

2. We shall come back to CAP representations at the end of Lecture 4. Suffice it now to give the definition: a cuspidal π is C(uspidal), A(ssociated to a) P(arabolic) if there exists a proper parabolic subgroup $P = MU$, and a cuspidal automorphic representation τ of $M(\mathbb{A})$, such that for almost all v, π_v is a constituent of $ind_{P_v}\tau_v$. Put more colorfully, π is in the 'shadow' of an Eisenstein series!

3. A similar characterization of those π such that $L(s, \pi \otimes \chi_T, \rho)$ might have a pole at a different singularity of $E(s, g, F)$ depends (at least) on the possibility of identifying the remaining residues of $E(s, g, F)$ via some kind of Siegel–Weil formula. For example, for the point $s = \frac{1}{2}$, such a Siegel–Weil formula is developed in [Ku-Ra-So], and applied to give a characterization of the pole of $L(s, \pi \otimes \chi_T, \rho)$ at $s = 1$ (analogous to the theorem above).

Comments on the Shimura zeta-integral

(1) Theorem [P-R.3] above (and also Li's theorem) generalizes to Sp_n with n *even*. It gives the finiteness of poles result missing from the authors' earlier work on $L(s, \pi, \rho)$ via the Godement–Jacquet type integral of [PR.1]. (Recall that for the latter zeta-integral, the required non-vanishing of the bad local integrals has yet to be established in full generality.) This result also improves on Shahidi's theory, which gives the finiteness of poles result only for *generic* π.

(2) The analysis of the poles of the Eisenstein series $E(g, s, F)$ is a delicate business involving intertwining operators for the induced representation $Ind|\det|^{s'}$; this is the subject matter of §4 of [P-R.2] as well as §3 of [P-R.1]. A still more difficult problem is the analogous analysis of intertwining operators and Eisenstein series for the induced (from cuspidal!) representations $Ind_p^G \tau|\det|^{s'}$ on, say, $G = SO_{2n+1}$; until this problem is resolved, the finiteness of poles result for the Rankin–Selberg L-functions on $G \times GL(n)$ will remain incomplete. For a discussion of what must be proved, see Chapters II and III of [Ge-Sh].

(3) For Sp_n, with *n odd*, these methods must be modified to involve Eisenstein series on the metaplectic group. Indeed $\Theta_T(g)$ is now a genuine function on the metaplectic cover of Sp_n, and hence must be multiplied by an Eisenstein series of the same 'genuine' type (so that the resulting integrand in the Shimura integral (2.2) will be naturally defined on Sp_n). For arbitrary n, the theory has not yet been worked out. However, for $n = 1$ we encounter the zeta-integral

$$\int_{SL_2(F)\backslash SL_2(\mathbb{A})} \varphi(g)\Theta^{\Phi}(g)E(g, F, s)dg$$

interpolating the degree 3 L-function for $SL(2)$ corresponding to the L-group homomorphism

$$\rho = Ad: \ PGL_2(\mathbb{C}) \longrightarrow SO_{2,1}(\mathbb{C}) \subset GL_3(\mathbb{C}).$$

In this case, Gelbart and Jacquet have shown that $L(s, \pi \otimes \chi, Ad)$ is entire for all twists χ, unless the theta-series lifting of π to an isotropic $G' = SO(2)$ is non-trivial. If this is so, there can be a pole at $s = 1$, again for reasons of Langlands functoriality. Indeed, such a π is then the Shalika–Tanaka lift of some π' on $SO(2)$ coming from an L-group homomorphism $\Psi : {}^LG' \to PGL_2(\mathbb{C})$ with the property that $\Psi o \rho$ contains the identity representation, i.e., $L(s, \pi, \rho) = L(s, \pi', \rho o \Psi)$ has a degree 1 factor producing a pole at $s = 1$. In any event, an application of the converse theorem for $GL(3)$ yields the Gelbart–Jacquet lifting from $GL(2)$ to $GL(3)$.

(4) A further modification of the zeta-integral (2.2) comes from replacing φ by an automorphic cusp form of half-integral weight, i.e., an automorphic cusp form on the metaplectic group. In case $n = 1$, this reduces to precisely the work [Shim] which got this entire business started, and which ties in with the work of Waldspurger discussed below. For general n, the theory is not yet developed.

(5) The method of the Shimura zeta-integral works for L-functions on groups other than Sp_n, for example $Sp_n \times GL_n$ and $U(3)$ ([GeRo]), where it gives results completely analogous to Theorem [P-R.3]. The latter results for $U(3)$ complement those which can be obtained via the trace formula, since they give intrinsic characterizations in terms of Θ-series liftings and Fourier expansions – of those π whose (twisted) degree six L-functions are not always entire.

Waldspurger's work (on the non-vanishing of Fourier coefficients and special values of L-functions) and counterexamples to Ramanujan's Conjecture Consider again the dual pair $(\overline{SL}_2, PGL_2)$. Recall that in this case, the theory of theta-series liftings gives a bijection between cuspidal σ on $\overline{SL}_2$ with non-vanishing ψth Fourier coefficients and cuspidal π on PGL_2 with $L(\pi, \frac{1}{2}) \neq 0$. But the question remains: how can we characterize the non-vanishing of this Fourier coefficient intrinsically in terms of σ? Waldspurger's answer comes from combining a Siegel–Weil formula with the theory of the Shimura integral.

Given any automorphic cuspidal representation σ on $\overline{SL}_2$, define its *Shimura image* on PGL_2 by

$$\pi = Sh_\psi(\sigma) \equiv \Theta(\sigma, \psi_a) \otimes \chi_a \, .$$

Here $\psi^a(x) = \psi(ax)$ is any character such that $\Theta(\sigma, \psi^a) \not\equiv \{0\}$, and χ_a is the corresponding quadratic character $x \to (a, x)$. In [W.2] it is shown that $Sh_\psi(\sigma)$ depends only on ψ, and not on the choice of a in F^x. From the bijective properties of the correspondence $\Theta(\cdot, \psi)$ recalled above, it then also follows that, for any automorphic cuspidal representation π of PGL_2,

$$\begin{gathered} \pi = Sh_\psi(\sigma) \text{ for some cuspidal } \sigma \text{ on } \overline{SL}_2 \\ \text{if and only if} \\ L(\tfrac{1}{2}, \pi \otimes \chi_a) \neq 0 \text{ for some } a \text{ in } F^x. \end{gathered}$$

It is interesting to note that this same characterization of the image of Sh_ψ has been sketched recently by Jacquet using the 'relative' trace formula (see [Ja]). However, the following remarkable result seems to lie deeper, and is (thus far) proved only in [W.3].

Theorem Suppose $\sigma = \otimes \sigma_v$ is an automorphic cuspidal representation of $\overline{SL}_2$, and f is any cusp form in the space of σ. Then f admits a non-zero ψth Fourier coefficient if and only if:
(i) for all v, σ_v has a ψ_v-Whittaker model; and
(ii) $L_S(\frac{1}{2}, Sh_\psi(\sigma)) \neq 0$.

Sketch of proof The only if direction is easy, since the non-vanishing of the ψ-Fourier coefficient immediately implies both the existence of a global (and hence local) Whittaker model, and the non-vanishing of the theta-series lift $\Theta(\sigma, \psi) = \pi = Sh_\psi(\sigma)$ (hence $L(\frac{1}{2}, \pi) = L(\frac{1}{2}, Sh_\psi(\sigma)) \neq 0$).

So suppose now that σ satisfies (i) and (ii). By the properties of the bijection $\sigma \leftrightarrow \Theta(\sigma, \psi)$, it will suffice to show that $\Theta(\sigma, \psi) \neq \{0\}$ (since this is equivalent to the non-vanishing of the ψth Fourier coefficients of f in σ); i.e., we must prove that for some choice of φ, Φ, and g,

$$\zeta(g) \equiv \int\limits_{SL_2(F)\backslash SL_2(\mathbb{A})} \varphi(h)\Theta_\psi^\Phi(g,h)dh \neq 0 \qquad \text{(see (2.1))}.$$

Note that for any $\varphi \neq 0$, there is some a in F^x so that $\varphi_{\psi^a} \neq 0$. If $a \in (F^x)^2$, then the fact that φ_ψ and $\varphi_{\psi\lambda a}$ are related in an elementary fashion (see (2.4)) implies that we are done, i.e., $\varphi_\psi \neq 0$. Thus we may assume $a \notin (F^x)^2$. In this case, recall that X is the three-dimensional quadratic space on which $PGL_2 \approx SO(2,1)$ acts, fix $x_a = \begin{bmatrix} 0 & -a \\ 1 & 0 \end{bmatrix}$ in X, and decompose X into the line generated by x_a and the orthocomplement X'_a. We may take Φ of the form $\phi(\lambda x_a + x') = \phi_1(\lambda x_a)\phi_2(x')(x' \in X'_a)$, so that for t in the stabilizer T of x_a in PGL_2, we have

$$\zeta(t) = \int \varphi(h)\Theta_\psi^{\Phi_1}(h)\Theta^{\Phi_2}(t,h)dh$$

i.e., for $(t,h) \in T \times SL_2$, the Weil representation (or theta-kernel) decomposes in this simple way.

Let us now assume $\zeta(t) \equiv 0$ and show this leads to a contradiction of our hypotheses. Let $K = F(\sqrt{a})$, and view T as the anisotropic form of norm 1 elements of K^x. Since $T_F\backslash T_\mathbb{A}$ is compact, we can integrate the integral expression for $\zeta(t)$ with respect to $T(F)\backslash T(\mathbb{A})$, and interchange the order of integration to obtain

$$0 \equiv \int\limits_{SL_2(F)\backslash SL_2(\mathbb{A})} \varphi(h)\Theta_\psi^{\Phi_1}(h) \left(\int\limits_{T(F)\backslash T(\mathbb{A})} \Theta^{\Phi_2}(t,h)dt \right) dh.$$

The Siegel–Weil type formula proved by Waldspurger asserts that the integral in parentheses above equals the value at $s = \frac{1}{2}$ of the Eisenstein series

$$E^{\phi_2}(h,s) = \sum_{\gamma \in B(F)\backslash SL_2(F)} f_s(\gamma h).$$

(Here $f_s(h) = L(s+\frac{1}{2},\chi_a)|\alpha_h|^{s-\frac{1}{2}}(\omega_\psi^2(h)\Phi_2)(0)$, where ω_ψ^2 denotes the Weil representation associated to the dual pair $SO(K) \times SL_2 \subset Sp_4$, and h in SL_2 has the Iwasawa decomposition $h = \begin{pmatrix} \alpha_h & * \\ 0 & \alpha_h^{-1} \end{pmatrix} n$, with $u \in K_\mathbb{A}$.) Since f_s belongs to $Ind_{B(\mathbb{A})}^{SL_2(\mathbb{A})}|\alpha_h|^s\chi_a$, and its value at $h = e$ is essentially $L(s+\frac{1}{2},\chi_a)$, $E^{\phi_2}(h,s)$ is a familiar *normalized Eisenstein series* on SL_2. For s

sufficiently large, this Eisenstein series converges absolutely, and hence we can write

$$0 = \operatorname{value}_{s=\frac{1}{2}} \zeta^*(s), \qquad \text{where}$$

$$\zeta^*(s) = \int_{SL_2(F)\backslash SL_2(\mathbb{A})} \varphi(h)\Theta_\psi^{\Phi_1}(h)E^{\Phi_2}(h,s)dh$$

To complete the proof, we note that $\zeta^*(s)$ is precisely the kind of Shimura-type zeta integral introduced in [GePS] to interpolate the L-function $L(s, Sh_\psi(\sigma))$! Thus it is not surprising that we end up with the basic identity

$$\zeta^*(s) = \left(\prod_{v\in S} \zeta_v^*(s)\right) L_S(s, Sh_\psi(\sigma)).$$

Here $\zeta_v^*(s)$ is a local zeta-integral which does not vanish identically at $s = \frac{1}{2}$ *if and only if* σ_v has a ψ_v-Whittaker model. Thus the theorem follows.

Remarks concerning the statement and proof of Waldspurger's Theorem

(a) The idea of the proof of Waldspurger's theorem can be summarized in one sentence: in manipulating the expression for a theta-series lifting, one runs smack up against a (special value of a) particular zeta-integral; thus the non-vanishing of the lifting must indeed be related to the non-vanishing of a special value of the automorphic L-function interpolated by this zeta-integral. An attempt to generalize this phenomenon was made in 1982 by Rallis, who considered higher dimensional orthogonal groups in place of PGL_2, and computed the L^2-norm of the resulting lifting in terms of a (then) new kind of zeta-integral of Rankin–Selberg type; see [Ra.2]. It was precisely these calculations which eventually gave rise to the general Rankin–Selberg–Godement–Jacquet zeta-integrals used by Piatetski-Shapiro and Rallis in their treatment of the standard automorphic L-functions attached to the simple classical groups (Theorem [P-R.1] of §II.1). It is therefore clear that the theory of theta-series liftings is inextricably linked with the theory of automorphic L-functions.

(b) Let us call a cuspidal representation ψ – *globally generic* if (ψ is a non-trivial character of the maximal unipotent homogeneous space $U(F)\backslash U(\mathbb{A})$ and) the space of ψth Fourier coefficients of cusp forms in the space of σ does not vanish identically. According to the statement of Waldspurger's theorem, σ need not be globally ψ-generic even though its factors σ_v are locally ψ_v-generic. In fact, explicit examples have been given (by Gelbart and Soudry) of cuspidal representations of $\overline{SL_2}(\mathbb{A})$ which are everywhere locally ψ_v-generic and hence possess 'abstract' ψ-Whittaker models globally, but are not globally ψ-generic. Clearly it will be interesting to resolve the following:

Problem. If π is a cuspidal representation of an *algebraic* reductive group G, and each π_v is ψ_v-generic, prove that π is globally ψ-generic.

A special instance of this problem for GSp_2 arises in the work of [Bl-Ra]. Some progress towards an affirmative solution has recently been made in the work of [Ku-Ra-So]; here $G = Sp_2$, and it is shown that local ψ-generic implies global ψ-generic *provided* the degree five L-function of π is non-vanishing at $Re(s) = 1$.

(c) The statement (and proof) of Waldspurger's theorem is not exactly true as stated since a few special cusp forms on the metaplectic group fail to lift to *cusp* forms on PGL_2. Examples of such cusp forms include the 'elementary theta-functions' on $\overline{SL}_2$ arising from the dual pair $(SL_2, O(1))$. The explanation for this phenomenon is a part of Rallis' theory of 'towers of Θ-series liftings', which we now briefly describe in this special context.

Consider the following sequence of orthogonal groups paired dually with SL_2:

$$\begin{array}{ccc} & & SO(3,2) \\ & \nearrow & \\ SL_2 & \rightarrow & SO(2,1) \\ & \searrow & \\ & & O(1) \end{array}$$

For each $j = 0,1,2$, let I_j denote the subspace of (genuine) cusp forms on $\overline{SL}_2(\mathbb{A})$ whose Θ-series lifts to $O(n+1,n)$ are zero for $n < j$, but not for $n = j$. According to [Ra.3]:

(i) $I_0 \oplus I_1 \oplus I_2$ *exhausts* the space of genuine cusp forms on $\overline{SL}_2$, and
(ii) if $\sigma \subset I_j$, then the theta-series lift of σ from $\overline{SL}_2$ to its dual pair partner $0(j+1,j)$ is automatically *cuspidal* (and non-zero); however, the theta-lift of this same σ to any 'larger' $O(n+1,n)$ is *non*-cuspidal (and non-zero).

In particular, for Waldspurger's dual pair $(\overline{SL}_2, PGL_2)$, the ψ theta-lift of σ will be non-cuspidal precisely when the theta-lift of this same σ to $O(1)$ is non-zero. Such cuspidal σ have non-vanishing ψ-Fourier coefficients and generalize the classical theta-series

$$\Theta_\chi(z) = \Sigma\chi(n)n^\nu e^{2\pi i n^2 z}$$

where $\nu = 0$ or 1, $\chi(-1) = (-1)^\nu$, and Θ_χ is a classical cusp form of weight $\frac{1}{2} + \nu$. It is precisely these cuspidal σ which spoil the Waldspurger bijection between certain cusp forms on $\overline{SL}_2$ and PGL_2, and hence must be removed. To sum up: the correct space of cusp forms for Waldspurger's bijective correspondence is precisely Rallis' space I_1.

What about the space I_2? It turns out that I_2 is just the space of genuine cusp forms on $\overline{SL}_2$ with vanishing ψ-Fourier coefficients. For example, if $\psi'(x) = \psi_t(x) = \psi(tx)$, with $t \notin (F^x)^2$, then the cusp forms on $\overline{SL}_2$ which are ψ'-theta lifts from $O(1)$ will have *vanishing* ψ-Fourier coefficients. (Classically, these lifts correspond to theta-series of the form $\Sigma\chi(n)n^\nu e^{2\pi i n^2 tz}$.) The space of all cusp forms in I_2 is the space which Piatetski-Shapiro isolated for study in [P-S.2] and showed to be of such great interest in connection with Ramanujan's conjecture. Indeed, let π on $SO(3,2) \approx PGSp_4$ be the ψ-theta-series lift of a cuspidal space σ in I_2. From Rallis' theory of towers, it follows that π is automatically cuspidal. What Piatetski-Shapiro proves in [P-S.2] is that each such π also 'satisfies' the following unusual properties:

(1) π provides a counterexample to the generalized Ramanujan conjecture; in fact, such π's contain the counterexamples of [Ku] if σ does not come from *any* theta-series attached to a quadratic form in 1-variable, and the counterexamples of [H-P] otherwise;

(2) π is a CAP (*cuspidal associated to a parabolic*) representation;

(3) the standard (degree 4) L-function of π is *not* entire, and in fact has a pole to the right of the line $Re(s) = 1$;

(4) π is not globally ψ-generic for any ψ; and

(5) π has a 'unipotent' component in the sense of §I.3.

It is this last property which seems to be at the root of the problem of extending Ramanujan's conjecture to groups beyond the context of $GL(n)$ (where properties (2)–(5) are never satisfied by cuspidal representations, and hence one still believes in the truth of the Conjecture). Note that the theory of towers works in the opposite direction as well. For example, corresponding to the diagram

$$\begin{array}{ccc} O(2) & \rightarrow & Sp_2 \\ & \searrow & \\ & & Sp_1 = SL_2 \end{array}$$

one concludes that 'cusp forms' on $O(2)$ which lift to zero on Sp_1, i.e., do not play a role in the Shalika–Tanaka construction of cusp forms on SL_2 from $O(2)$, are precisely the forms which lift to (non-zero) *cusp* forms on Sp_2. (It is crucial now that we deal with $O(2)$ in place of $SO(2)$.) Locally, at a place where the quadratic form is anisotropic, we saw earlier that there is just one representation of O_2 missing from the pairing with SL_2, namely the non-trivial character which is *trivial* on $SO(2)$. The resulting lifted cusp forms on Sp_2 are precisely the [H-P] counterexamples to Ramanujan's conjecture mentioned above, and in §3 of Part I.

REFERENCES FOR PART I

[A.1] J. Arthur, *Automorphic representations and number theory*, Canad. Math. Soc. Conf. Proc. **1** (1981), pp. 3–51.

[A.2] ———, *Unipotent automorphic representations: Conjectures*, to appear in Astérisque.

[A.3] ———, *Unipotent automorphic representations: Global motivation.* to appear in Perspectives in Mathematics, Academic Press.

[A-C] J. Arthur and L. Clozel, *Simple Algebras, Base Change, and the Advanced Theory of the Trace Formula*, Annals of Math. Studies **120**, Princeton Univ. Press, 1989.

[B] A. Borel, *Automorphic L-functions*, in *Automorphic Forms, Representations and L-Functions,* Proc. Sympos. Pure Math. **33**, Part II, A.M.S., Providence, 1979, pp. 27–61.

[B-W] A. Borel and N. Wallach, *Continuous Cohomology, Discrete Subgroups and Representations of Reductive Groups,* Annals of Math. Studies **94**, Princeton Univ. Press, 1980.

[F] D. Flath, *Decomposition of representations into tensor products,* in *Automorphic Forms, Representations and L-functions,* Proc. Sympos. Pure Math. **33**, Part I, A.M.S., Providence, 1979, pp. 179–84.

[G] S. Gelbart, *An elementary introduction to the Langlands program,* Bull. Amer. Math. Soc. **10** (1984), pp. 177–219.

[G-J] R. Godement and H. Jacquet, *Zeta Functions of Simple Algebras,* Lecture Notes in Math. **260**, Springer-Verlag, 1972.

[H-P] R. Howe and I. Piatetskii-Shapiro, *A counterexample to the "generalized Ramanujan conjecture" for (quasi-) split groups*, in *Automorphic Forms, Representations and L-functions*, Proc. Sympos. Pure Math. **33**, Part I, A.M.S., Providence, 1979, pp. 315–22.

[J-L] H. Jacquet and R. Langlands, *Automorphic Forms on* $GL(2)$, Lecture Notes in Math. **114**, Springer-Verlag, 1970.

[J-P-S] H Jacquet, I. Piatetskii-Shapiro, and J. Shalika, *Rankin-Selberg convolutions,* Amer. J. Math., **105** (1983), pp. 367–464.

[J-S] H. Jacquet and J. Shalika, *On Euler products and the classification of automorphic representations II*, Amer. J. Math., **103** (1981), pp. 777–815.

[K.1] R. Kottwitz, *Stable trace formula: cuspidal tempered terms*, Duke Math. J. **51** (1984), pp. 611–50.

[K.2] ——, *On the* λ*-adic representations associated to some simple Shimura varieties*, to appear in Perspectives in Mathematics, Academic Press.

[K-M] P. Kutzko and A. Moy, *On the local Langlands conjecture in prime dimension*, Ann. of Math. **121** (1985), pp. 495–516.

[Ku] N. Kurokawa, *Examples of eigenvalues of Hecke operators on Siegel cusp forms of degree two,* Invent. Math. **49** (1978), pp. 149–65.

[L-L] J.-P. Labesse and R. Langlands, *L-indistinguishability for* $SL(2)$, Canad. J. Math. **31** (1979), pp. 726–85.

[L.1] R. Langlands, *Problems in the theory of automorphic forms,* in *Lectures in Modern Analysis and Applications,* Lecture Notes in Math. **170**, Springer-Verlag 1970, pp. 18–86.

[L.2] ——, *Representations of abelian algebraic groups*, Yale Univ. 1968 (preprint).

[L.3] ——, *On the Functional Equations Satisfied by Eisenstein Series,* Lecture Notes in Math. **544**, Springer-Verlag, 1976.

[L.4] ———, *On the notion of an automorphic representation*, in *Automorphic Forms, Representations and L-functions*, Proc. Sympos. Pure Math. **33**, Part I, A.M.S., Providence, 1979, pp. 203–7.

[L.5] ———,*Automorphic representations, Shimura varieties and motives. Ein Märchen*, in *Automorphic Forms, Representations and L- functions*, Proc. Sympos. Pure Math. **33** Part II, A.M.S., Providence, 1979, pp. 205–46.

[L.6] ———, *Base Change for $GL(2)$*, Annals of Math. Studies **96**, Princeton Univ. Press, 1980.

[L.7] ———, *On the classification of irreducible representations of real reductive groups*, in *Representation Theory and Harmonic Analysis on Semisimple Lie Groups*, Math. Surveys and Monographs, **31**, A.M.S., Providence, 1989, pp. 101–70.

[M-W] C. Moeglin and J.-L. Waldspurger, *Le spectre residuel de $GL(n)$*, pre- print.

[Mg] C. Moeglin, *Orbites unipotentes et spectre discrèt non ramifíe*, pre-print

[My] A. Moy, *Local constants and the tame Langlands correspondence*, Amer. J. Math. **108** (1986), 863–929.

[P-R] I. Piatetskii-Shapiro and S. Rallis, *L-functions for the classical groups*, notes prepared by J. Cogdell, in *Explicit Constructions of Automorphic L-functions*, Lecture Notes in Math., **1254**, Springer-Verlag, 1987.

[S] A. Selberg, *Harmonic analysis and discontinuous groups in weakly symmetric Riemannian spaces with applications to Dirichlet series*, J. Indian Math. Soc. **20** (1956), 47–87.

[Sh] J. Shalika, *The multiplicity one theorem for $GL(n)$*, Annals of Math. **100** (1974), pp. 171–93.

[S-T] J. Shalika and S. Tanaka, *On an explicit construction of a certain class of automorphic forms,* Amer. J. Math. **91** (1969), pp. 1049–76.

[T.1] J. Tate, *Fourier analysis in number fields and Hecke's zeta function,* in *Algebraic Number Theory,* Academic Press, New York, 1968, pp. 305–47.

[T.2] ———, *Number theoretic background,* in *Automorphic Forms, Representations and L-functions,* Proc. Sympos. Pure Math. **33**, Part II, A.M.S., Providence, 1979, pp. 3–26.

[Tu] J. Tunnell, *Artin's conjecture for representations of octahedral type,* Bull. Amer. Math. Soc. **5** (1981), pp. 173–5.

REFERENCES FOR PART II

[A] J. Arthur, 'Automorphic Representations and Number Theory'. In *Canadian Math. Society Conference Proceedings,* **1**, Providence, R.I., 1981, pp. 3–51.

[A-C] J. Arthur and L. Clozel, 'Simple Algebras, Base Change and the Advanced Theory of the Trace Formula'. *Annals of Math. Studies* **120**, Princeton Univ. Press, 1989.

[BDKV] J. Bernstein, P. Deligne, D. Kazhdan, M.-F. Vignéras, *Représentations des groupes réductifs sur un corps local.* Hermann, Paris, 1984.

[Bl-Ra] D. Blasius and D. Ramakrishnan, 'Maass forms and Galois representations'. Preprint 1988.

[Bö] S. Böcherer, 'Siegel modular forms and theta series'. In *Proceedings of Symposia in Pure Math.*, **49**, A.M.S., Providence, R.I. 1989.

[B] A. Borel, 'Automorphic *L*-functions'. In *Proc. Sympos. Pure Math.* **33**, Part II, A.M.S., Providence, 1979, pp. 27–61.

[B-J] A. Borel and H. Jacquet, 'Automorphic forms and automorphic representations'. In *Proceedings of Symposia in Pure Math.*, **33**, Part 1, A.M.S., Providence, R.I. 1979, pp. 189–202.

[Bu] D. Bump, 'The Rankin–Selberg Method: A Survey'. In *Number Theory, Trace Formulas, and Discrete Groups*, Symposium in honor of Atle Selberg, Oslo 1987, Academic Press, 1989.

[CL] L. Clozel, 'Motifs et formes automorphes: applications du principe de functorialité'. To appear in *Proceedings of the 1988 Summer Conference on Automorphic Forms and Shimura Varieties*, Michigan, Academic Press.

[Ga] P. Garrett, 'Decomposition of Eisenstein series: Rankin triple products'. *Annals of Math.*, **125** (1987), pp. 209–35.

[GePS] S. Gelbart and I. Piatetski-Shapiro, 'On Shimura's correspondence for modular forms of half-integral weight'. In *Automorphic Forms, Representation Theory and Arithmetic* (Bombay, 1979), Tata Institute of Fundamental Research Studies in Math., **10**, Bombay, 1981, pp. 1–39; see also 'Some remarks on metaplectic cusp forms and the correspondences of Shimura and Waldspurger', *Israel J. Math.*, **44**, No. 2, 1983, pp. 97–126.

[GeRo] S. Gelbart and J. Rogawski, 'On the Fourier–Jacobi coefficients of automorphic forms on $U(3)$'. In preparation.

[GeSh] S. Gelbart and F. Shahidi, 'The Analytic Properties of Automorphic L-functions'. *Perspectives in Math.* **6**, Academic Press, 1988.

[Gi] D. Ginsburg, 'L-functions for $SO_n \times GL_k$'. *Journal für die reine und angewandte Mathematik.* To appear.

[G-J] R. Godement and H. Jacquet, 'Zeta Functions of Simple Algebras'. *Lecture Notes in Math.* **260**, Springer-Verlag, 1972.

[Ho] R. Howe, 'Θ-series and invariant theory'. In *Proc. Symp. Pure Math.*, **33**, Part 1, A.M.S., 1979, pp. 275–86.

[Ho.2] —, 'L 2-duality in the stable range', preprint, Yale University.

[H-P] R. Howe and I. Piatetski-Shapiro, 'A counterexample to the 'generalized Ramanujan conjecture' for (quasi-) split groups'. In *Automorphic Forms, Representations and L-functions*, Proc. Sympos. Pure Math. **33**, Part I, A.M.S., Providence, 1979, pp. 315–22.

[Ja] —, 'On the vanishing of some L-functions'. *Proceedings Indian Acad. Sciences (Ramanujan Centenary Volume)*, **97** (Special), Dec. 1987, No.1–3, pp. 117–55.

[J-L] H. Jacquet and R. Langlands, 'Automorphic forms on $GL(2)$'. *Lecture Notes in Math.* **114**, Springer-Verlag, 1970.

[J-P-S.1] H. Jacquet, I. Piatetski-Shapiro and J. Shalika, 'Rankin–Selberg convolutions'. *Amer. J. Math.* **109** (1983), pp. 367–464.

[J-P-S.2] —, 'Relévement cubique nonnormal'. *C.R. Acad. Sci. Paris*, Sér. I Math. **292** (1981), pp. 13–18.

[J-P-S.3] —, 'The Θ-correspondence from $GSp(4)$ to $GL(4)$'. In preparation.

[J-S] H. Jacquet and J. Shalika, 'On Euler products and the classification of automorphic representations, I and II'. *Amer. J. Math.* **103** (1981), pp. 499–558 and 777–815.

[KuRaSo] S. Kudla, S. Rallis and D. Soudry, 'On the second pole of the L-function for Sp_2'. Preprint.

[Ku] N. Kurokawa, 'Examples of eigenvalues of Hecke operators on Siegel cusp forms of degree two'. *Invent. Math.* **49** (1978), pp. 149–65.

[L-L] J.-P. Labesse and R. Langlands, 'L-indistinguishability for $SL(2)$'. *Canad. J. Math.* **31** (1979), pp. 726–85.

[L.1] R. Langlands, 'Problems in the theory of automorphic forms'. In *Lectures in Modern Analysis and Applications*, Lecture Notes in Math. **170**, Springer-Verlag, 1970, pp. 18–86.

[L.2] —, 'On the functional equations satisfied by Eisenstein series'. *Lecture Notes in Math.* **544**, Springer-Verlag, 1976.

[L.3] —, 'Eisenstein series, the trace formula and the modern theory of automorphic forms'. In *Number Theory, Trace Formulas and Discrete Groups*, Symposium in honor of A. Selberg, Oslo (1987), Academic Press, 1989.

[Li] J.-S. Li, 'Distinguished cusp forms are theta-series'. Preprint, 1987.

[Li.2] —, 'Singular unitary representations of classical groups'. *Inventiones Math.*, 97 (1989), pp. 237–55.

[Li.3] —, 'Theta-series and distinguished representations of symplectic groups'. Thesis, Yale University, 1987.

[MVW] C. Moeglin, M.-F. Vignéras and J.L. Waldspurger, 'Correspondences de Howe sur un corps p-adique'. *Lecture Notes in Mathematics* **1291**, Springer-Verlag, N.Y.1987.

[M-W] C. Moeglin and J.-L. Waldspurger, 'Le spectre residuel de $GL(n)$'. Preprint.

[P-S.1] I. Piatetski-Shapiro, 'Zeta-functions of $GL(n)$'. Preprint, Univ. of Maryland, Technical Report TR 76-46, November, 1976.

[P-S.2] —, 'On the Saito–Kurokawa lifting'. *Inventiones Math.* **71** (1983), pp. 309–38; see also 'Cuspidal automorphic representations associated to parabolic subgroups and Ramanujan's conjecture', in *Number Theory Related to Fermat's Last Theorem*, Neal Koblitz, Editor, Progress in Mathematics, **26**, Birkhäuser, Boston, 1982.

[P-R.1] I. Piatetski-Shapiro and S. Rallis, 'L-functions for the classical groups'. Notes prepared by J. Cogdell, in *Explicit Constructions of Automorphic L- functions*, Lecture Notes in Math. **1254**, Springer-Verlag, 1987.

[P-R.2] —, 'Rankin triple L-functions'. *Compositio Math.* **64** (1987), pp. 31–115.

[P-R.3] —, 'A new way to get Euler products'. *Journal für die reine und angewandte Mathematik* **392** (1988), pp. 110–24.

[PS-R-S] I. Piatetski-Shapiro, S. Rallis and G. Schiffmann, 'L-functions for G_2'. To appear.

[PS-So] I. Piatetski-Shapiro and D. Soudry, 'L and ε functions for $GSp(4) \times GL(2)$'. *Proc. Nat. Acad. Sci.*, U.S.A., **82**, June 1984, pp. 3924–27.

[Ra.1] S. Rallis, 'Langlands functoriality and the Weil representation'. *Amer. J. Math.* **104** (1982), pp. 469–515.

[Ra.2] —, 'L-functions and the oscillator representation'. *Lecture Notes in Mathematics* **1245**, Springer-Verlag, 1988.

[Ra.3] —, 'On the Howe duality conjecture'. *Comp. Math.* **51** (1984), pp. 333–99.

[Rao] R. Ranga Rao, 'On some explicit formulas in the theory of the Weil representation'. Preprint, 1977–78.

[Ro.1] J. Rogawski, 'Representations of $GL(n)$ and division algebras over a p-adic field'. *Duke Math. J.* **50** (1983), pp. 161–96.

[Ro.2] —, 'Automorphic representations of unitary groups in three variables'. To appear *Annals of Math. Studies*, Princeton U. Press.

[S] A. Selberg, 'Harmonic analysis and discontinuous groups in weakly symmetric Riemannian spaces with application to Dirichlet series'. *J. Indian Math. Soc.* **20** (1956), 47–87.

[Shah] F. Shahidi, 'On the Ramanujan conjecture and finiteness of poles for certain L-functions'. *Annals of Math.* **127** (1988), pp. 547–84.

[S-T] J. Shalika and S. Tanaka, 'On an explicit construction of a certain class of automorphic forms'. *Amer. J. Math.* **XCI**, No. 4, 1969, pp. 1049–76.

[Shim] G. Shimura, 'On modular forms of half-integral weight'. *Annals of Math.* **97** (1973), pp. 440–81.

[So] D. Soudry, 'The CAP representations of $GSp(4, \mathbb{A})$'. *J. reine angew. Math.* **383** (1988), pp. 87–108.

[So.2] —, 'Explicit Howe duality in the stable range'. *J. reine angew. Math.* 396 (1989), pp. 70–86.

[T.1] J. Tate, 'Fourier analysis in number fields and Hecke zeta functions'. In *Algebraic Number Theory*, Academic Press, New York, 1968, pp. 305–47.

[Tu] J. Tunnell, 'Artin's conjecture for representations of octahedral type'. *Bull. Amer. Math. Soc.* **5** (1981), pp. 173–5.

[W.1] J.-L. Waldspurger, 'Põles des fonctions L de paires pour $GL(N)$'. Preprint, 1989.

[W.2] —, 'Correspondance de Shimura'. *J. Math. Pures Appl.* **59** (1980), pp. 1–133.

[W.3] —, 'Correspondance de Shimura et quaternions'. Preprint, 1982.

[W.4] —, 'Demonstration d'une conjecture de dualite de Howe dans le cas p-adique, $p \neq 2$. Preprint, 1988.

Gauss sums and local constants for GL(N)

COLIN J. BUSHNELL

Let F be a non-Archimedean local field, $G = GL_N(F)$ for some integer $N \geq 1$, and π a smooth irreducible representation of G on some complex vector space $\mathcal{V}$. (Thus, if $N = 1$, we have $\mathcal{V} \cong \mathbf{C}$ and we may view π as a quasicharacter of $F^\times = GL_1(F)$.) In [6], Godement and Jacquet define zeta-integrals, an L-function $L(\pi, s)$ and a local constant $\varepsilon(\pi, s, \psi)$ attached to π, completely generalising Tate's treatment [11] of the case $N = 1$ except in one respect. Tate gives an 'explicit formula' for $\varepsilon(\pi, s, \psi)$ in terms of a Gauss sum $\tau(\pi, \psi)$ essentially identical to a classical Gauss sum. The object of this article is to give an analogous formula in the general case.

Apart from a few remarks, there is nothing really new here: all material beyond [6] has already appeared in [2], [1] or [8]. However, this unified account may slightly ease the lot of any potential user.

Notation We use the following notation throughout.

F = a non-Archimedean local field;
$\mathfrak{o}_F$ = the discrete valuation ring in F;
$\mathfrak{p}_F$ = the maximal ideal of $\mathfrak{o}_F$;
$q_F = (\mathfrak{o}_F : \mathfrak{p}_F)$ (group index).
$\| \cdot \|_F$ = the absolute value on F such that

$$\| x \|_F = (\mathfrak{o}_F : x\mathfrak{o}_F)^{-1}, \quad x \in F^\times.$$

(We frequently omit the subscript F here, as F is fixed for long periods.)
ψ^F = a nontrivial continuous character of the additive group of F.
$N \geq 1$ is an integer;
$A = M_N(F), \quad G = GL_N(F)$;
$\psi = \psi^A = \psi^F \circ \mathrm{tr}_{A/F}$, where tr denotes the matrix trace.

1 THE GODEMENT–JACQUET FUNCTIONAL EQUATION

We start with a very brief summary of the main definitions and results of [6] pertaining to the non-Archimedean local case. The survey article [7] is also a convenient reference for this material.

We write $\mathcal{S}(A)$ for the space of locally constant compactly-supported functions $\Phi : A \to \mathbf{C}$. If $(\pi, \mathcal{V})$ is a smooth irreducible (hence admissible) representation of G, we write $(\tilde{\pi}, \tilde{\mathcal{V}})$ for the contragredient or smooth dual of $(\pi, \mathcal{V})$, and $\langle\ ,\ \rangle$ for the evaluation pairing $\mathcal{V} \times \tilde{\mathcal{V}} \to \mathbf{C}$. We denote by $\mathcal{M}(\pi)$ the space of functions on G spanned by the 'matrix coefficients'

$$g \mapsto \langle \pi(g)v, \tilde{v} \rangle, \quad g \in G,$$

for $v \in \mathcal{V}, \tilde{v} \in \tilde{\mathcal{V}}$. Choose a Haar measure $d^\times g$ on G, and let s be a complex variable. Define

$$Z(\Phi, f, s) = \int_G \Phi(g) f(g) |g|^s d^\times g, \quad \Phi \in \mathcal{S}(A), \quad f \in \mathcal{M}(\pi),$$

where $|g| = \|\det g\|$. There exists $\sigma_0 = \sigma_0(\pi) \in \mathbf{R}$ such that this integral converges for $\mathrm{Re}(s) > \sigma_0$. From [7] (1.2) we have:

(1.1) There exists a unique function $L(\pi, s)$ satisfying
 (i) $L(\pi, s) = P(q^{-s})^{-1}$, for some $P(X) \in \mathbf{C}[X]$ with $P(0) = 1$;
 (ii) $L(\pi, s)^{-1} Z(\Phi, f, s) \in \mathbf{C}[q^{-s}, q^s]$ for all $\Phi \in \mathcal{S}(A)$, and all $f \in \mathcal{M}(\pi)$;
 (iii) there exist $\Phi \in \mathcal{S}(A)$, $f \in \mathcal{M}(\pi)$ such that $L(\pi, s) = Z(\Phi, f, s)$.
Moreover, if π is supercuspidal and $N \geq 2$, then $L(\pi, s) = 1$.

We now identify A with its Pontrjagin dual $\hat{A}$ via the pairing $(x, y) \mapsto \psi(xy)$, $x, y \in A$, and let dx be the self-dual Haar measure on A for this identification. Thus, if $\Phi \in \mathcal{S}(A)$ and we put

$$\hat{\Phi}(y) = \int_A \Phi(x) \psi(xy) dx, \quad y \in A,$$

then $\hat{\Phi} \in \mathcal{S}(A)$ and $\hat{\hat{\Phi}}(x) = \Phi(-x)$ for $x \in A$. Also, for $f \in \mathcal{M}(\pi)$, define $\tilde{f} \in \mathcal{M}(\tilde{\pi})$ by

$$\tilde{f}(g) = f(g^{-1}), \quad g \in G.$$

(1.2) (See [7], (1.3.7)) There exists $\varepsilon(\pi, s, \psi)$ such that

$$\frac{Z(\hat{\Phi}, \tilde{f}, 1 - s + (N-1)/2)}{L(\tilde{\pi}, 1 - s)} = \varepsilon(\pi, s, \psi) \frac{Z(\Phi, f, s)}{L(\pi, s)}$$

for all $\Phi \in \mathcal{S}(A)$, $f \in \mathcal{M}(\pi)$. Moreover, we have

$$\varepsilon(\pi, s, \psi).\varepsilon(\tilde{\pi}, 1-s, \psi) = \omega_\pi(-1),$$

where ω_π is the central quasicharacter of π. In particular,

$$\varepsilon(\pi, s, \psi) = C(\pi, \psi).q^{-ms},$$

for some $C(\pi, \psi) \in \mathbf{C}$ of absolute value 1 and some $m = m(\pi) \in \mathbf{Z}$.

Now let ψ_1^F be some other nontrivial continuous additive character of F, and put $\psi_1 = \psi_1^F \circ \mathrm{tr}$. There exists $c \in F^\times$ such that $\psi_1(x) = \psi(cx)$, $x \in A$, and we have

$$(1.3) \qquad \varepsilon(\pi, s, \psi_1) = \omega_\pi(c)|c|^{(s-\frac{1}{2})N}.\varepsilon(\pi, s, \psi).$$

This completes the list of basic definitions we require. Notice that $L(\pi, s)$ is known when π is supercuspidal, the case $N = 1$ being given by [11], and the others by (1.1). Jacquet shows in his survey article [7] how $L(\pi, s)$ can be computed in the general case, using the supercuspidal one and the Zelevinsky classification of the representations of G. Using the same procedure, we need only compute $\varepsilon(\pi, s, \psi)$ when π is supercuspidal. Indeed, we need only treat the case $N \geq 2$, appealing to [11] for the abelian one $N = 1$. In fact, our method is more general than this.

2 HEREDITARY ORDERS AND STRATA

It is for the moment more convenient to identify our algebra A with $\mathrm{End}_F(V)$, where V is some F-vector space of finite dimension $N \geq 1$. An *$\mathfrak{o}_F$-lattice chain in V* is a set $\mathcal{L} = \{L_i : i \in \mathbf{Z}\}$, where each L_i is an $\mathfrak{o}_F$-lattice in V (i.e., a finitely generated $\mathfrak{o}_F$-submodule of V which spans V over F) such that

(i) $L_i \supset L_{i+1}$, $\quad L_i \neq L_{i+1}$, $\quad i \in \mathbf{Z}$, and
(ii) there exists $e \in \mathbf{Z}$ such that $L_{i+e} = \mathfrak{P}_F L_i$, $i \in \mathbf{Z}$.

The integer $e = e(\mathcal{L})$ is uniquely determined. We attach a ring $\mathfrak{U} = \mathfrak{U}(\mathcal{L})$ to the lattice chain $\mathcal{L}$ by

$$\mathfrak{U} = \bigcap_{i \in \mathbf{Z}} \mathrm{End}_{\mathfrak{o}_F}(L_i) = \bigcap_{i=0}^{e-1} \mathrm{End}_{\mathfrak{o}_F}(L_i).$$

This is an $\mathfrak{o}_F$-lattice, hence an $\mathfrak{o}_F$-order, in A. We can recover the lattice chain $\mathcal{L}$ from $\mathfrak{U}(\mathcal{L})$, up to a shift of index, since $\mathcal{L}$ is precisely the set of all left $\mathfrak{U}$-lattices in V. Thus we may view the 'period' e of the lattice chain as a function of $\mathfrak{U}$, $e = e(\mathfrak{U})$. Orders $\mathfrak{U}$ in A obtained in this manner we refer to as *hereditary orders.* For a full account of these, see [10]. There is a convenient summary of the facts needed for the present situation in [2]. Now, for $n \in \mathbf{Z}$, we define

$$\mathfrak{P}^n = \mathfrak{P}^n_{\mathfrak{U}} = \{x \in A : xL_i \subset L_{i+n}, \quad i \in \mathbf{Z}\}.$$

Then $\mathfrak{P} = \mathfrak{P}^1$ is just the Jacobson radical of $\mathfrak{U}$, and $\mathfrak{P}^n$ is the nth power of $\mathfrak{P}$. The meaning of this is obvious if $n \geq 0$; otherwise, we recall that $\mathfrak{P}$ is an invertible $(\mathfrak{U}, \mathfrak{U})$-bimodule. In particular, we have the relation

$$\mathfrak{p}_F \mathfrak{U} = \mathfrak{P}^e, \quad \text{where } e = e(\mathfrak{U}).$$

The powers $\mathfrak{P}^n$ are $\mathfrak{o}_F$-lattices in A, so we may define compact open subgroups $U^i(\mathfrak{U})$ of $G = \mathrm{Aut}_F(V)$ by

$$U^0(\mathfrak{U}) = \mathfrak{U}^\times, \quad \text{and} \quad U^i(\mathfrak{U}) = 1 + \mathfrak{P}^i, \quad \text{for } i \geq 1.$$

Also, we put

$$\mathfrak{K}(\mathfrak{U}) = \mathrm{Aut}(\mathcal{L}) = \{x \in G : x\mathcal{L} = \mathcal{L}\}.$$

This is an open subgroup of G, and it is compact mod centre, i.e., $\mathfrak{K}(\mathfrak{U})/F^\times$ is a compact subgroup of $G/F^\times$, where we identify $F^\times$ with the centre of G.

The groups $U^i(\mathfrak{U})$ are all normal subgroups of $\mathfrak{K}(\mathfrak{U})$. Moreover, we have

$$U^0(\mathfrak{U})/U^1(\mathfrak{U}) \cong \prod_{i=0}^{e-1} \mathrm{Aut}_{\mathfrak{o}_F/\mathfrak{p}_F}(L_i/L_{i+1}),$$

while

$$U^i(\mathfrak{U})/U^{i+1}(\mathfrak{U}) \cong \mathfrak{P}^i/\mathfrak{P}^{i+1}$$

for $i \geq 1$, via the map $x \mapsto x - 1$.

We now recall our additive character $\psi = \psi^F \circ \mathrm{tr}_{A/F}$ of A. It is easy to work out the effect of changing the basic character ψ^F, as we shall see below, and it is convenient for our present purposes to assume that *the conductor of* ψ^F *is* $\mathfrak{p}_F$. Thus ψ^F is trivial on $\mathfrak{p}_F$ but not on $\mathfrak{o}_F$. If M denotes some $\mathfrak{o}_F$-lattice in A, define

$$M^* = \{x \in A : \psi(xm) = 1, \quad m \in M\}.$$

With our choice of ψ, we get

$$(\mathfrak{P}^n)^* = \mathfrak{P}^{1-n}, \quad n \in \mathbf{Z}.$$

This gives us a very useful description of the Pontrjagin duals of the finite groups U^i/U^{i+1}.

(2.1) *Proposition* The map $b \mapsto \psi_b$, where ψ_b is the character of U^i/U^{i+1} given by

$$\psi_b(1+x) = \psi(bx), \quad x \in \mathfrak{P}^i,$$

establishes an isomorphism

$$\mathfrak{P}^{-1}/\mathfrak{P}^{1-i} \cong (U^i/U^{i+1})^\wedge$$

for any $i \geq 1$.

We write $\theta \mapsto \delta_\theta$ for the inverse of the isomorphism of (2.1).

A *stratum in* A is a triple $(\mathfrak{U}, n, \theta)$, where $\mathfrak{U}$ is a hereditary order in A, n is a non-negative integer, and θ is an irreducible representation of $U^n(\mathfrak{U})/U^{n+1}(\mathfrak{U})$. Thus, in the case $n = 0$, θ is of the form

$$\theta = \theta_0 \otimes \ldots \otimes \theta_{e-1},$$

where $e = e(\mathfrak{U})$ and θ_i is an irreducible representation of the finite general linear group $\mathrm{Aut}(L_i/L_{i+1})$ over $\mathfrak{o}_F/\mathfrak{p}_F$. We say that the stratum $(\mathfrak{U}, n, \theta)$ is *fundamental* if

(2.2) (i) $n = 0$, and each θ_i is a cuspidal representation, or
(ii) $n \geq 1$ and δ_θ does not contain a nilpotent element of A.

Now let π be some smooth irreducible representation of $G = \mathrm{Aut}_F(V)$. We say that π *contains* the stratum $(\mathfrak{U}, n, \theta)$ if the restriction of π to the subgroup $U^n(\mathfrak{U})$ contains the representation θ (inflated to a representation of U^n in the obvious way). The definition of smoothness implies that π contains some stratum.

(2.3) *Theorem* Let π be a smooth irreducible representation of G, and let $(\mathfrak{U}, n, \theta)$ be a stratum contained in π. The following conditions are equivalent:

(i) for any other stratum $(\mathfrak{U}', n', \theta')$ contained in π, we have $n/e(\mathfrak{U}) \leq n'/e(\mathfrak{U}')$ and, if $n = n' = 0$, then $e(\mathfrak{U}) \geq e(\mathfrak{U}')$;
(ii) $(\mathfrak{U}, n, \theta)$ is fundamental.

This result was originally conjectured by A. Moy. It is proved in [1]. It follows immediately from (2.3) that a smooth irreducible representation π of G contains some fundamental stratum. This fundamental stratum is not unique in any absolute sense, but has strong and useful uniqueness properties. For the moment, let H_i be some subgroup of G, and ρ_i an irreducible representation of H_i, $i = 1, 2$. Recall that an element $x \in G$ *intertwines* θ_1 with θ_2 if there is a nontrivial $x^{-1}H_1x \cap H_2$-homomorphism from θ_1^x to θ_2. If we do not wish to specify the intertwining element x , we simply say that θ_1 and θ_2 *intertwine in* G. Any two strata which are contained in the same irreducible π must intertwine, and we have a very strong necessary condition for two fundamental strata to intertwine. To state this, we first need another definition.

Let $(\mathfrak{U}, n, \theta)$ be some fundamental stratum with $n \geq 1$, and let $b \in \delta_\theta$, so that $\delta_\theta = b + \mathfrak{P}^{1-n}$. Fix some prime element ω_F of F, set $g = \gcd(n, e)$, $e = e(\mathfrak{U})$, define $y \in \mathfrak{U}/\mathfrak{P}$ by

$$y = \omega_F^{n/g}.b^{e/g},$$

and let $\phi_\theta(t) \in (\mathfrak{o}_F/\mathfrak{p}_F)[t]$ be the characteristic polynomial of y viewed as an endomorphism of, say, $L_\circ/\omega_F L_\circ$. This definition is independent of the choice of $b \in \delta_\theta$ and depends only trivially on the choice of ω_F. Further, choosing a different basic character ψ^F would replace δ_θ by $c\delta_\theta$, for some $c \in F^\times$. We would have to define y as $\omega_F^m b^{e/g}$, with the integer m chosen to ensure $y \in \mathfrak{U}, \notin \mathfrak{P}$. All this has no material effect on the relevant properties of $\phi_\theta(t)$. Note that $\phi_\theta(t)$ is not a power of t , since the stratum is fundamental.

(2.4) *Proposition* Let $(\mathfrak{U}, n, \theta)$, $(\mathfrak{U}', n', \theta')$ be fundamental strata, and assume that θ, θ' intertwine in G. Then

$$\frac{n}{e(\mathfrak{U})} = \frac{n'}{e(\mathfrak{U}')}.$$

If $n = n' = 0$, then $e(\mathfrak{U}) = e(\mathfrak{U}')$ and the cuspidal tensor factors of θ' are the same as those of θ up to isomorphism, possibly after a renumbering. Otherwise, we have

$$\phi_\theta(t) = \phi_{\theta'}(t).$$

When $n \geq 1$, we now have two cases to consider. A fundamental stratum $(\mathfrak{U}, n, \theta)$ is called *split* if the polynomial $\phi_\theta(t)$ has at least two distinct irreducible factors (over the field $\mathfrak{o}_F/\mathfrak{p}_F$).

(2.5) *Theorem* Let π be an irreducible smooth representation of G, and suppose that π contains a split fundamental stratum. Then π is not supercuspidal.

This version is proved in [8]. However, in recent joint work [3] of the author and P. C. Kutzko, it is shown that a representation π containing a split fundamental stratum is in fact *irreducibly* induced from a maximal proper parabolic subgroup of G. It is therefore the other case, in which $\phi_\theta(t)$ is a power of some irreducible polynomial, which will concern us. We neglect the case in which $n = 0$. This is actually easier, but requires a different panoply of definitions.

Now let $\mathfrak{A} = \mathfrak{A}(\mathcal{L})$ be some hereditary order in A, with $\mathcal{L} = \{L_i\}$ as before. The lattice chain $\mathcal{L}$ is called *uniform* if the index $(L_i : L_{i+1})$ is constant, independent of i. A hereditary order $\mathfrak{A} = \mathfrak{A}(\mathcal{L})$ is called *principal* if the lattice chain $\mathcal{L}$ is uniform. There are various equivalent formulations of this. For example, $\mathfrak{A}$ is principal if and only if $\mathfrak{P} = \mathfrak{A}x$, for some $x \in G$. Moreover, $\mathfrak{A}$ is principal if and only if $\mathfrak{K}(\mathfrak{A})$ is a maximal compact-mod-centre subgroup of G . We refer to [2] for the details of this.

Now let $(\mathfrak{A}, n, \theta)$ be some fundamental stratum with $n \geq 1$. We say that it is *simple* if there exists an element $\alpha \in \delta_\theta$ satisfying

(i) $E = F[\alpha]$ is a field;
(ii) $E^\times \subset \mathfrak{K}(\mathfrak{A})$ (i.e., $E^\times$ conjugates $\mathfrak{A}$ into itself);
(iii) α is minimal over F.

In (iii), the term *minimal* means the following:

(a) $\gcd(n, e(E \mid F)) = 1$, and
(b) for a prime element ϖ_F of F, the residue class of $\alpha^{e(E|F)}.\varpi_F^{-\nu(\alpha)}$ (modulo $\mathfrak{p}_E$) generates the residue class field of E over that of F.

We note that a simple stratum $(\mathfrak{A}, n, \theta)$ is automatically not split. Moreover, we have $\delta_\theta \subset \mathfrak{K}(\mathfrak{A})$. The next result is also taken from [8], but note that simple strata are there called *alfalfa*.

(2.6) *Theorem* Let π be a smooth irreducible representation of G, and assume that π contains a non-split fundamental stratum $(\mathfrak{A}, n, \theta)$ with $n \geq 1$. Then π contains a simple stratum.

In particular, we note that if π is supercuspidal, then it must contain a simple stratum (or a stratum $(\mathfrak{A}, 0, \theta)$).

3 NON-ABELIAN CONGRUENCE GAUSS SUMS

Now let $\mathfrak{A}$ be a *principal order* in $A = \mathrm{End}_F(V)$. Thus $\mathfrak{A} = \mathfrak{A}(\mathcal{L})$ for some uniform lattice chain $\mathcal{L} = \{L_i\}$ in the vector space V. So, there exists $x \in G$ such that $xL_i = L_{i+1}$ for all $i \in \mathbb{Z}$. For any such x, we have $\mathfrak{P} = \mathfrak{A}x = x\mathfrak{A}$, where, as before, $\mathfrak{P}$ denotes the Jacobson radical of $\mathfrak{A}$. Indeed, we get $\mathfrak{P}^n =$

$\mathfrak{U}x^n = x^n\mathfrak{U}$ for all $n \in \mathbf{Z}$. Further, the compact-mod-centre subgroup $\mathfrak{K}(\mathfrak{U})$ of G is the semidirect product of $\mathfrak{U}^\times$ with the infinite cyclic group generated by any such x. See [2] for details.

Let ρ be a smooth irreducible representation of $\mathfrak{K}(\mathfrak{U})$ on some complex vector space $\mathcal{W}$. Then $\mathcal{W}$ is finite-dimensional, and the restriction of ρ to the centre $F^\times$ of $\mathfrak{K}(\mathfrak{U})$ is a multiple of some quasicharacter ω_ρ of $F^\times$. Further, there is an integer $f \geq 0$ such that ρ is null on the open normal subgroup $U^f(\mathfrak{U})$ of $\mathfrak{K}(\mathfrak{U})$. Take f minimal for this property, and put

$$\mathfrak{f}(\rho) = \mathfrak{P}^f. \tag{3.1}$$

We also use the (slightly disreputable) convention from the abelian theory of Gauss sums that

$$1 + \mathfrak{f}(\rho) = U^f(\mathfrak{U}). \tag{3.2}$$

Now let $\psi = \psi^F \circ \mathrm{tr}_{A/F}$ be our additive character of A, as before. We no longer make any particular assumption about the conductor of ψ^F. Define

$$T(\rho, \psi) = \sum_x \rho(c^{-1}x)\psi(c^{-1}x) \in \mathrm{End}_{\mathbf{C}}(\mathcal{W}), \tag{3.3}$$

where, in the summation, x ranges over a set of representatives of $\mathfrak{U}^\times$ modulo $1+\mathfrak{f}(\rho)$, and c is any element of $\mathfrak{K}(\mathfrak{U})$ such that $c^{-1}\mathfrak{U} = (\mathfrak{f}(\rho))^*$. Schur's Lemma shows that the definition of $T(\rho, \psi)$ is independent of all these choices, and also that there exists $\tau(\rho, \psi) \in \mathbf{C}$ such that

$$T(\rho, \psi) = \tau(\rho, \psi).1_{\mathcal{W}}. \tag{3.4}$$

This $\tau(\rho, \psi)$ is the *non-abelian congruence Gauss sum* of the representation ρ. As we shall see below, this Gauss sum has many properties in common with the abelian congruence Gauss sum defined analogously in the case $N = \dim_F(V) = 1$. There is one serious difference, however, in that, for $N \geq 1$, $\tau(\rho, \psi)$ can vanish. We now describe this phenomenon.

Suppose for the moment that $\mathfrak{f}(\rho) = \mathfrak{P}^f$ with $f \geq 2$, and consider the restriction of ρ to $U^{f-1}(\mathfrak{U})$. This restriction decomposes as a direct sum of abelian characters θ of U^{f-1}/U^f, and we may consider the sets δ_θ defined in §2. We say that ρ is *nondegenerate* if

$$\delta_\theta \subset \mathfrak{K}(\mathfrak{U})$$

for some (equivalently any) θ occurring in $\rho \mid U^{f-1}(\mathfrak{U})$. Moreover, the set δ_θ is contained in $\mathfrak{K}(\mathfrak{U})$ if and only if it meets it nontrivially. Immediately from the definitions we have

(3.5) *Proposition* Let ρ be a smooth irreducible representation of $\mathfrak{K}(\mathfrak{U})$ with conductor $\mathfrak{P}^f$, $f \geq 2$. Let θ be a character occurring in $\rho \mid U^{f-1}(\mathfrak{U})$, and suppose that the stratum $(\mathfrak{U}, f-1, \theta)$ is simple. Then ρ is nondegenerate.

We also say that ρ *is nondegenerate* if its conductor is $\mathfrak{f}(\rho) = \mathfrak{U}$ or $\mathfrak{P}$. Now we have

(3.6) *Proposition* The representation ρ of $\mathfrak{K}(\mathfrak{U})$ is nondegenerate if and only if $\tau(\rho, \psi)$ is nonzero.

(3.7) *Theorem* Let π be a smooth irreducible representation of G and $\mathfrak{U}$ a principal order in A. Suppose that π satisfies the following conditions:

(i) $L(\pi, s) = 1$, and
(ii) the restriction $\pi \mid \mathfrak{K}(\mathfrak{U})$ contains a nondegenerate representation ρ.

Then

$$\varepsilon(\pi, s, \psi) = ((\mathfrak{f}(\rho)^* : \mathfrak{U})^{(1/2-s)/N}.\tau(\tilde{\rho}, \psi).(\mathfrak{U} : \mathfrak{f}(\rho))^{-1/2},$$

where $\tilde{\rho}$ denotes the contragredient of ρ.

The hypotheses (i) and (ii) hold when π is supercuspidal.

Remarks If π is supercuspidal (and $N \geq 2$), we know from (2.6) that it contains a simple stratum $(\mathfrak{U}, n, \theta)$ or a stratum $(\mathfrak{U}, 0, \theta)$. (It is easy to see, in the latter case, that we may choose $\mathfrak{U}$ to be principal: indeed, we can take $\mathfrak{U} \cong M_N(\mathfrak{o}_F)$ here.) There is certainly an irreducible representation ρ of $\mathfrak{K}(\mathfrak{U})$ occurring in π and containing this θ. Any such ρ is nondegenerate in the above sense. This proves the last assertion of (3.7), and gives the desired formula for the local constant in the supercuspidal case. Note that it is an exact generalisation of Tate's formula in the case $N = 1$.

A different proof of the last assertion is given in [1], involving a more detailed analysis of the functional equation. The one we have given here, via Kutzko's theorem (2.6), seems more direct and its ideas have proved more useful.

The hypotheses of (3.7) hold for more general representations π. It follows from the forthcoming [3] that a representation π for which they fail has either to be irreducibly induced from a maximal parabolic subgroup of G, or to have a fixed vector for an Iwahori subgroup of G.

The conditional version of (3.7) holds also when $G = GL_N(F)$ is replaced by $GL_M(D)$, where D is a finite-dimensional central F-division algebra of dimension d^2, $dM = N$, except that a factor $(-1)^{M(d-1)}$ has to be inserted in the formula for ε. However, in this degree of generality, no results of the type (2.6), or even (2.5), are known.

Finally, it has to be remarked that the proof of (3.7) in [2] apparently demands that the field F be of characteristic zero. However, once one defines, as here, a Gauss sum attached to an arbitrary additive character ψ^F, the proof goes through in all characteristics.

A different description of the exponential factor of $\varepsilon(\pi, s, \psi)$ has been given by Jacquet/Piatetski–Shapiro/Shalika in Math. Ann. **256** (1981) 199–214.

4 ARITHMETICAL PROPERTIES OF GAUSS SUMS

Again let ρ be an irreducible smooth representation of the normaliser $\mathfrak{K}(\mathfrak{U})$ of some principal order $\mathfrak{U}$ in the algebra $A \cong M_N(F)$, and $\tau(\rho, \psi)$ its Gauss sum, as above. We note first that this behaves very simply under change of the additive character ψ. If ψ' is any other additive character of A of the requisite type, then there exists some $b \in F^\times$ such that $\psi'(x) = \psi(bx)$, $x \in A$. We find

(4.1) $$\tau(\rho, \psi') = \tau(\rho, \psi).\omega_\rho(b)^{-1},$$

where ω_ρ is the central quasicharacter of ρ.

Now let χ be an unramified quasicharacter of G. Thus $\chi = \chi_\circ \circ \det$ for some unramified quasicharacter $\chi_\circ$ of $F^\times$, and we may view χ as a quasicharacter of $\mathfrak{K}(\mathfrak{U})$ by restriction. It is then immediate that

$$\tau(\rho \otimes \chi, \psi) = \chi(c^{-1}).\tau(\rho, \psi),$$

for any $c \in \mathfrak{K}(\mathfrak{U})$ such that $c\mathfrak{U} = (\mathfrak{f}(\rho))^*$ (where 'star' is defined relative to ψ as in §2). Thus, in many questions concerning Gauss sums, it is enough to treat the case in which the central quasicharacter ω_ρ *is of finite order.*

(4.2) *Proposition* Let ρ be an irreducible representation of $\mathfrak{K}(\mathfrak{U})$, as above, and let $\tilde{\rho}$ be the contragredient of ρ. Let p be the residual characteristic of F, and let $\mathfrak{f}(\rho) = \mathfrak{P}^f$. Then

$$\tau(\rho, \psi).\tau(\tilde{\rho}, \psi) = \omega_\rho(-1).\nu(\rho),$$

where $\nu(\rho)$ is given by

(i) 0 if ρ is degenerate;
(ii) $(\mathfrak{U} : \mathfrak{f}(\rho))$ if ρ is nondegenerate and $f \geq 2$;
(iii) p^k for some nonnegative integer k such that p^k divides $(\mathfrak{U} : \mathfrak{f}(\rho))$ otherwise.

(4.3) *Corollary* Suppose that the central quasicharacter ω_ρ of ρ has finite order. Then the absolute value of $\tau(\rho, \psi) \in \mathbf{C}$ is $\nu(\rho)^{1/2}$.

For a detailed interpretation of case (4.2)(iii), see [2] and [5]. If we stick to the case of representations ρ attached to supercuspidal representations of G, we can neglect some of these possibilities. If $\mathfrak{f}(\rho) = \mathfrak{P}$ or $\mathfrak{U}$, we say ρ is *nondeficient* if $\nu(\rho) = (\mathfrak{U} : \mathfrak{f}(\rho))$. As a supplement to (3.7) we have

(4.4) *Lemma* Let ρ be an irreducible representation of the normaliser $\mathfrak{K}(\mathfrak{U})$ of a principal order $\mathfrak{U}$ such that $\mathfrak{f}(\rho) = \mathfrak{U}$ or $\mathfrak{P}$. Suppose also that ρ occurs in some irreducible representation π of G with $L(\pi, s) = 1$. Then $\mathfrak{f}(\rho) = \mathfrak{P}$ and ρ is nondeficient.

This is proved in [2], and provides some justification for regarding nondeficient nondegenerate representations as somewhat more interesting than the others. In particular, the property that the absolute value of the Gauss sum of such a representation is the square root of the norm of the conductor is exactly what one would expect from the classical case.

Now suppose that F *has characteristic zero.* Thus F is a finite extension of $\mathbf{Q}_p$. Let $\Omega_\mathbf{Q}$ denote the Galois group $\mathrm{Gal}(\overline{\mathbf{Q}}/\mathbf{Q})$ of an algebraic closure $\overline{\mathbf{Q}}/\mathbf{Q}$. We write χ_p for the p-adic character $\Omega_\mathbf{Q} \to \mathbf{Z}_p^\times$ given by

$$\zeta^{\sigma \cdot \chi_p(\sigma)} = \zeta,$$

for $\sigma \in \Omega_\mathbf{Q}$ and any p-power root of unity ζ in $\overline{\mathbf{Q}}$.

If our representation ρ has ω_ρ of finite order, then it is easy to see that the image $\rho(\mathfrak{K}(\mathfrak{U}))$ is finite. It follows that ρ may be realised over the ring of integers in some algebraic number field. We draw two conclusions. First, we can 'twist' ρ by Galois automorphisms $\sigma \in \Omega_\mathbf{Q}$ and, second, $\tau(\rho, \psi)$ is an algebraic integer. We then get a formula which, although very easy to prove, is nontheless remarkable.

(4.5) *Proposition* Let ρ be an irreducible representation of $\mathfrak{K}(\mathfrak{U})$ as above, and suppose that ω_ρ has finite order. Then

$$\tau(\rho^{\sigma^{-1}}, \psi)^\sigma = \tau(\rho, \psi).\omega_\rho(\chi_p(\sigma)), \quad \text{for all } \sigma \in \Omega_{\mathbf{Q}}.$$

There is a strikingly similar formula for the so-called *Galois Gauss sums.* Let Ω_F be the Galois group of some algebraic closure of F (which we continue to assume to be a finite extension of $\mathbf{Q}_p$). It is also convenient to take our basic character ψ^F of F to be the Iwasawa–Tate character. This is the composite of the field trace $F \to \mathbf{Q}_p$ with the obvious embedding of $\mathbf{Q}_p/\mathbf{Z}_p$ in the unit circle. In the formula for $\varepsilon(\pi, s, \psi)$ given by (3.7), the factor $((\mathfrak{f}(\rho)^* : \mathfrak{U})$ is then just

$$((\mathfrak{f}(\rho)^* : \mathfrak{U}) = (\mathfrak{U} : \mathfrak{D}_F\mathfrak{f}(\rho)),$$

where $\mathfrak{D}_F$ is the absolute different of F.

Now let μ be a continuous finite-dimensional representation of Ω_F, and $\tau(\mu, \psi^F)$ its Galois Gauss sum (with ψ^F as above). (See, for example, [9] for a treatment of Galois Gauss sums.) This Gauss sum is related to the Langlands–Deligne local constant (see [4] or [12]) $\varepsilon(\mu, s, \psi^F)$ by the formula

$$\varepsilon(\mu, s, \psi^F) = (\mathfrak{o}_F : \mathfrak{f}(\mu)\mathfrak{D}_F^{\dim(\mu)})^{(1/2-s)}.\tau(\tilde{\mu}, \psi^F).(\mathfrak{o}_F : \mathfrak{f}(\mu))^{-1/2}$$

where $\mathfrak{f}(\mu)$ is the Artin conductor of μ. Apart from the similarity between this formula and (3.7), we get a Galois action formula for the Galois Gauss sum:

$$\tau(\mu^{\sigma^{-1}}, \psi)^\sigma = \tau(\mu, \psi).\det_\mu(\chi_p(\sigma))$$

for all $\sigma \in \Omega_{\mathbf{Q}}$. Here $\det_\mu$ denotes the character of $F^\times$ associated to the one-dimensional representation $\det(\mu)$ of Ω_F via class field theory. Note here that $F^\times$ contains $\mathbf{Z}_p^\times$ and hence the values of χ_p. Comparison of this formula with (4.5) is particularly interesting in the context of the local Langlands conjectures.

REFERENCES

1 *C. J. Bushnell*, Hereditary orders, Gauss sums and supercuspidal representations of GL_N, J. reine angew. Math. **375/376** (1987) 184–210.

2 *C. J. Bushnell and A. Fröhlich*, Non-abelian congruence Gauss sums and p-adic simple algebras, Proc. London Math. Soc.(3) **50** (1985) 207–64.

3 *C. J. Bushnell and P. C. Kutzko*, The admissible dual of GL_N via compact open subgroups, in preparation.

4 *P. Deligne*, Les constantes des équations fonctionelles des fonctions L, Lectures Notes in Math. **349**, Springer 1973, 501–95.

5 *A. Fröhlich*, Tame representations of local Weil groups and of chain groups of local principal orders, Heidelberger Akad. Wiss., Springer 1986, 1–100.

6 *R. Godement and H. Jacquet*, Zeta functions of simple algebras, Lecture Notes in Math. 260, Springer 1972.

7 *H. Jacquet*, Principal L-functions of the linear group, Proc. Symposia Pure Math. **33**, Amer. Math. Soc. 1977, 63–86.

8 *P. C. Kutzko*, Towards a classification of the supercuspidal representations of GL_N, J. London Math. Soc.(2) **37** (1988) 265–74.

9 *J. Martinet*, Character theory and Artin L-functions, Algebraic Number Fields (*A. Fröhlich, ed.*), Academic Press 1977, 1–87.

10 *I. Reiner*, Maximal Orders, Academic Press 1972.

11 *J. Tate*, Fourier analysis on number fields and Hecke's zeta functions, thesis, Princeton University 1950. *Also Algebraic number theory* (*J. W. S. Cassels & A. Fröhlich ed.*) Academic Press 1967.

12 *J. Tate*, Local constants, Algebraic Number Fields (*A. Fröhlich ed.*), Academic Press 1977, 89–132.

L-functions and Galois modules

PH. CASSOU-NOGUÈS, T. CHINBURG, A. FRÖHLICH, AND M. J. TAYLOR

Notes by

D. BURNS AND N. P. BYOTT

0 INTRODUCTION

Let N/K be a finite Galois extension of number fields and let $\Gamma = Gal(N/K)$ denote its Galois group. In these lectures we study the structures of certain 'Galois modules' arising in this context such as the ring of algebraic integers $\mathcal{O}_N$ of N, the multiplicative group $\mathcal{U}_N$ of units of $\mathcal{O}_N$ modulo torsion elements, and the ideal class group Cl_N of N. A striking feature of the theory discussed is the close interplay between the module structure and the arithmetic of the number fields. In particular (Artin) L-functions play a fundamental rôle in various places. What results however is much more than theorems and conjectures relating two very abstract and sophisticated points of view – the module theoretic aspect is frequently very down to earth, and indeed some indication of the power of the theory lies in its capacity to provide concrete and explicit information.

There are several approaches to these Galois module theoretic problems and we shall discuss different approaches in separate paragraphs.

The case which is best understood is that of $\mathcal{O}_N$ when N/K is at most tamely ramified. Here, $\mathcal{O}_N$ is a (locally-free) $\mathbf{Z}[\Gamma]$-module and in §1 we discuss the fundamental result which shows that its structure as such is determined, up to stable isomorphism, by the root numbers occuring in the functional equation of the Artin L-functions of N/K attached to the irreducible symplectic characters of Γ. The material of this section is also useful in providing motivation for many of the approaches adopted in more general situations.

In §2 we again restrict to Galois extensions which are at most tamely ramified but here consider the converse problem of recovering the symplectic root number values from the Galois module structures attached to N. The $\mathbf{Z}[\Gamma]$-structure of $\mathcal{O}_N$ alone is not sufficient for this. However the root numbers are completely determined by the isometry class of the Hermitian $\mathbf{Z}[\Gamma]$-module consisting of $\mathcal{O}_N$ together with the trace form, and this is the subject of §2.

For wildly ramified Galois extensions there is even today no general theory analogous to that discussed in §1 and §2 for extensions which are at most tamely ramified. In §3 we recall the notion of factor-equivalence between modules which provides additional insight into the module structure of $\mathcal{O}_N$ for wildly ramified extensions. Furthermore factor-equivalence allows one to exhibit some remarkable, and at the present time unexplained, similarities between the 'additive' structure theory of $\mathcal{O}_N$ and the 'multiplicative' structure theory of $\mathcal{U}_N$ and this is an underlying theme throughout §3. In particular the results discussed here raise integral variants of theorems, problems and conjectures relating to Tate's formulation of the Stark conjectures.

A basic technique in studying arithmetic Galois modules is to compare them with given 'standard' modules which have better understood structures. In §1 and §3 this is done by constructing suitable module invariants. In §4 we continue with this general approach but now use exact sequences and comparison of the resulting cohomology theories. In particular we discuss the construction of certain cohomological invariants of a Galois extension which simultaneously reflect both the 'additive' and 'multiplicative' Galois module structures attached to N. The invariants of §4 are constructed in the context of general global fields N/K and without any hypotheses on ramification (of N/K). They are elements of the locally-free class group $Cl(\mathbf{Z}[\Gamma])$ of $\mathbf{Z}[\Gamma]$. Conjecturally at least these invariants are closely related to other aspects of the arithmetic of N/K and, in particular, to the root number classes introduced in §1. The Stark conjectures (again in Tate's formulation) also have interesting implications for the material of this section.

Finally, in §5, we consider the problem of finding explicit Galois generators for $\mathcal{O}_N$ (and other related modules) for certain special classes of extensions. The classical example of this type of result is Leopoldt's Theorem which gives such a generator for any abelian extension of the field of rationals $\mathbf{Q}$. We discuss how, under suitable hypotheses, an elliptic curve with complex multiplication can be used to provide analogous generators for the rings of integers of extensions arising from its torsion points. In particular we give a result which may be considered as an integral contribution to the famous 'Jugendtraum' of Kronecker.

We mention here two of the conventions in use throughout this article. Let K be a number field and let $\mathcal{O}_K$ denote its ring of integers. Let G denote an arbitrary finite group, with $\mathcal{A}$ any $\mathcal{O}_K$-order in the K-algebra $K[G]$. With these notations all $\mathcal{A}$-modules to be considered in this article are *finitely generated right* $\mathcal{A}$-modules. By a 'prime divisor', or equivalently a 'place', of K, we shall mean an equivalence class of non-trivial valuations of K. Such a prime divisor (place) is 'finite' if it comes from some non-zero prime ideal of $\mathcal{O}_K$. Otherwise it is said to be 'infinite', or equivalently, 'archimedean'.

1 THE TAME ADDITIVE THEORY

In this paragraph our concern is with the Galois structure of $\mathcal{O}_N$, primarily in the context of tamely ramified Galois extensions. Of course $\mathcal{O}_N$ naturally has the structure of $\mathcal{O}_K[\Gamma]$-module but for our purposes it is more fruitful to restrict scalars and consider $\mathcal{O}_N$ *qua* $\mathbf{Z}[\Gamma]$-module.

At the field level then, the Normal Basis Theorem implies an isomorphism of $\mathbf{Q}[\Gamma]$-modules

$$N \cong K[\Gamma] \tag{1.1}$$

and hence, on the integral level, it is natural to compare the $\mathbf{Z}[\Gamma]$-module structures of $\mathcal{O}_N$ and $\mathcal{O}_K[\Gamma]$. More specifically, writing $\sim_{\mathbf{Z}[\Gamma]}$ for the relation of $\mathbf{Z}[\Gamma]$-genus equivalence (that is, isomorphic Γ-module structure after tensoring with $\mathbf{Z}_p$ for each rational prime p) one might first ask for the conditions under which $\mathcal{O}_N \sim_{\mathbf{Z}[\Gamma]} \mathcal{O}_K[\Gamma]$. In the sequel we shall say that an extension is 'tame' if it is at most tamely ramified.

Theorem 1 (Noether, 1932) Let N/K be a finite Galois extension of number fields of Galois group Γ. Then $\mathcal{O}_N \sim_{\mathbf{Z}[\Gamma]} \mathcal{O}_K[\Gamma]$ if, and only if, N/K is tame.

But $\mathcal{O}_K[\Gamma]$ is a free $\mathbf{Z}[\Gamma]$-module and hence, for a tame extension N/K it is a natural problem to determine the global structure of the locally-free $\mathbf{Z}[\Gamma]$-module $\mathcal{O}_N$. In the 1970s a theory was developed in response to this problem, the 'tame additive theory', and is now more or less complete. In this paragraph we shall nevertheless give a brief survey of certain of its aspects which will provide us with a guide for the underlying strategy and for possible results to aim for in the general case.

1.1 The locally-free class group

We must first introduce a suitable classifying group $Cl(\mathbf{Z}[\Gamma])$, the 'locally-free class group' of $\mathbf{Z}[\Gamma]$. If X and Y are any two locally-free $\mathbf{Z}[\Gamma]$-modules such that $X \otimes_{\mathbf{Z}} \mathbf{Q} \cong Y \otimes_{\mathbf{Z}} \mathbf{Q}$ then $Cl(\mathbf{Z}[\Gamma])$ will classify them to within stable $\mathbf{Z}[\Gamma]$-isomorphism. That is, X and Y will be represented by the same element of $Cl(\mathbf{Z}[\Gamma])$ if, and only if, there is an isomorphism of $\mathbf{Z}[\Gamma]$-modules

$$X \oplus \mathbf{Z}[\Gamma] \cong Y \oplus \mathbf{Z}[\Gamma].$$

Remark 1.2 Under certain conditions on (the absolutely irreducible complex valued characters of) Γ – satisfied for example if Γ is either abelian, dihedral, or of odd order – stable $\mathbf{Z}[\Gamma]$-isomorphism implies $\mathbf{Z}[\Gamma]$-isomorphism but this is not true in general.

Let $K_0(\mathbf{Z}[\Gamma])$ denote the Grothendieck group of the category of locally-free $\mathbf{Z}[\Gamma]$-modules modulo relations given by direct sums. Each locally-free $\mathbf{Z}[\Gamma]$-module X has a rank $n(X) \in \mathbf{N}$ defined by

$$X \otimes_{\mathbf{Z}} \mathbf{Q} \cong (\mathbf{Q}[\Gamma])^{n(X)}$$

and extending the map $X \longmapsto n(X)$ to a homomorphism

$$rk : K_0(\mathbf{Z}[\Gamma]) \longrightarrow \mathbf{Z}$$

one defines $Cl(\mathbf{Z}[\Gamma])$ by the exactness of the sequence

$$0 \longrightarrow Cl(\mathbf{Z}[\Gamma]) \longrightarrow K_0(\mathbf{Z}[\Gamma]) \xrightarrow{rk} \mathbf{Z} \longrightarrow 0. \tag{1.3}$$

Equivalently, defining an injective group homomorphism

$$0 \longrightarrow \mathbf{Z} \xrightarrow{fr} K_0(\mathbf{Z}[\Gamma])$$

by

$$n \overset{f_\Gamma}{\longmapsto} n(\mathbf{Z}[\Gamma]), \quad n > 0$$

where here (X) denotes the class of a locally-free $\mathbf{Z}[\Gamma]$-module X in $K_0(\mathbf{Z}[\Gamma])$, one has an exact sequence

$$0 \to \mathbf{Z} \xrightarrow{f_\Gamma} K_0(\mathbf{Z}[\Gamma]) \to Cl(\mathbf{Z}[\Gamma]) \to 0. \tag{1.4}$$

Both descriptions (1.3) and (1.4) are however inappropriate for arithmetical computations and so we must work in terms of a description, the 'Hom-description' of $Cl(\mathbf{Z}[\Gamma])$, in terms of functions on characters of Γ. For this we let E denote any Galois extension of $\mathbf{Q}$ (contained in the algebraic closure $\mathbf{Q}^c$ of $\mathbf{Q}$ in $\mathbf{C}$) over which all absolutely irreducible complex representations of Γ can be realised. For convenience we also assume that $N \subseteqq E$. We write $\mathcal{J}(E)$ for the multiplicative group of ideles of E, and R_Γ for the additive group of (complex) virtual characters of Γ. For any number field K we shall write Ω_K for the absolute Galois group $Gal(\mathbf{Q}^c/K)$. For $K = \mathbf{Q}$ we shall frequently abbreviate this by $\Omega = \Omega_{\mathbf{Q}}$. Then there is an exact sequence

$$Hom_\Omega(R_\Gamma, \mathcal{J}(E)) \xrightarrow{\pi_\Gamma} Cl(\mathbf{Z}[\Gamma]) \longrightarrow 0 \tag{1.5}$$

where for an explicit description of the projection map $\pi_\Gamma = \pi_{\Gamma,E}$ the reader is referred, for example, to Fröhlich (1983). For the reader's convenience however we shall now briefly recall, for any given locally-free rank one $\mathbf{Z}[\Gamma]$-module X, the recipe for constructing a homomorphism $f \in Hom_\Omega(R_\Gamma, \mathcal{J}(E))$ such that $\pi_\Gamma(f)$ is equal to the element $(X)_{\mathbf{Z}[\Gamma]}$ of $Cl(\mathbf{Z}[\Gamma])$ corresponding to X.

Consider a representation

$$T : \Gamma \longrightarrow GL_n(E).$$

Let B denote either $\mathbf{Q}$ or $\mathbf{Q}_p$ for some rational prime p. The representation T extends by B-linearity to give a B-algebra homomorphism

$$T : B[\Gamma] \longrightarrow M_n(E \otimes_{\mathbf{Q}} B)$$

which in turn induces a homomorphism

$$Det_T : (B[\Gamma])^* \xrightarrow{T} GL_n(E \otimes_{\mathbf{Q}} B) \xrightarrow{det} (E \otimes_{\mathbf{Q}} B)^* .$$

In fact $Det_T = Det_\chi$ depends only upon the character χ of the representation T. Choose now a free generator x of $X \otimes_{\mathbf{Z}} \mathbf{Q}$ over $\mathbf{Q}[\Gamma]$ and, for each rational prime divisor p choose a free generator x_p of $X \otimes_{\mathbf{Z}} \mathbf{Z}_p$ over $\mathbf{Z}_p[\Gamma]$. Both x and x_p are free generators of $X \otimes_{\mathbf{Z}} \mathbf{Q}_p$ over $\mathbf{Q}_p[\Gamma]$ and hence

$$x_p = x\lambda_p \ , \ \ \lambda_p \in (\mathbf{Q}_p[\Gamma])^*.$$

The map $f \in Hom_\Omega(R_\Gamma, \mathcal{J}(E))$ defined componentwise at each character χ of Γ by

$$(f(\chi))_p = Det_\chi(\lambda_p)$$

for each rational prime divisor p, then satisfies

$$\pi_\Gamma(f) = (X)_{\mathbf{Z}[\Gamma]}.$$

The advantage of this 'Hom-description' is that it allows us, as we shall presently see, to relate the algebraic and arithmetical aspects of the theory and so to state and prove the basic results. Apart from this it is also useful for computation and is a powerful tool for the derivation of functorial properties under change of group and so on.

Before proceeding we note that the double rôle of number fields apparent here – both as the objects of study and as the ranges in which the classifying functions take their values – is a typical feature of the theory. The Hom-description also gives us a first indication of the absolutely fundamental rôle played throughout the theory by functions on Galois characters.

1.2 The arithmetic theory

Now given a tame Galois extension N/K a central problem is therefore to find a homomorphism $f = f_{N/K} \in Hom_\Omega(R_\Gamma, \mathcal{J}(E))$ with

$$\pi_\Gamma(f) = (\mathcal{O}_N)_{\mathbf{Z}[\Gamma]}.$$

Quite remarkably, in order to describe such a homomorphism we must turn to the Artin L-function attached to each complex character χ of Γ. For the moment we may also relax the restriction that N/K is tame.

For any character χ of Γ we write $\tilde{L}(s, \chi)$ for the extended Artin L-function attached to N/K and to χ. For each such χ we denote by $\overline{\chi}$ its contragredient. Then $\tilde{L}(s, \chi)$ has a meromorphic continuation to the entire complex plane and satisfies a functional equation

$$\tilde{L}(s, \chi) = w(\chi)A(\chi)^{\frac{1}{2}-s}\tilde{L}(1-s, \overline{\chi}) \tag{1.6}$$

where $A(\chi) = A(N/K, \chi) > 0$ is a positive real constant and $w(\chi) = w(N/K, \chi)$ is a complex constant of absolute value one. In the literature $w(\chi)$ is known as the '(global) Artin root number' of χ. In particular then if χ is real valued then $\chi = \overline{\chi}$ and $w(\chi) = \pm 1$. To decide this parity question it is convenient to consider two sub-classes of real-valued characters. We call a character χ 'orthogonal' (respectively 'symplectic') if it is attached to a representation that factors through the group of orthogonal (respectively symplectic) matrices. Orthogonal and symplectic characters are real-valued and, conversely, any real-valued character can be written as a $\mathbb{Z}$-linear combination of orthogonal and symplectic characters. If the character χ is orthogonal then in fact $w(\chi) = +1$. On the other hand if χ is symplectic then $w(\chi)$ may be either $+1$ or -1 and, if N/K is tame, the values of $w(\chi)$ for all symplectic characters χ of Γ actually determine the element $(\mathcal{O}_N)_{\mathbb{Z}[\Gamma]}$!

To make this striking remark more precise we define an element

$$W_{N/K} \in Hom_\Omega(R_\Gamma, \mathcal{J}(E))$$

in terms of the values $w(\chi)$. (This definition was originally due to Ph.Cassou-Noguès). For this we need an 'adjusted' root number defined at each absolutely irreducible character χ of Γ by

$$w'(\chi) = \begin{cases} w(\chi), & \text{if } \chi \text{ is symplectic;} \\ 1, & \text{otherwise.} \end{cases}$$

and extended to R_Γ by linearity. In fact if N/K is tame then for each $\eta \in \Omega$ and character χ of Γ one has

$$w'(\chi^\eta) = w'(\chi) \tag{1.7}$$

so that the homomorphism

$$\chi \longmapsto w'(\chi)$$

is Ω-equivariant. Note however that equation (1.7) is not in general true for wild (that is, non-tame) extensions. Next, for each virtual character χ of Γ we define an idele

$$W_{N/K}(\chi) \in \mathcal{J}(\mathbb{Q}) \subset \mathcal{J}(E)$$

via its local components:

$$\begin{aligned} (W_{N/K}(\chi))_\infty &= 1\ ; \\ (W_{N/K}(\chi))_p &= w'(\chi), \text{ all finite } p. \end{aligned}$$

We write $W_{N/K}$ for the corresponding element of $Hom_\Omega(R_\Gamma, \mathcal{J}(E))$ (for the Ω-equivariance of $W_{N/K}$ recall (1.7)). Then the main theorem in this section gives, for any tame Galois extension N/K, a complete classification of the global structure of the locally-free $\mathbb{Z}[\Gamma]$-module $\mathcal{O}_N$ in terms of the homomorphism $W_{N/K}$. The result of this theorem was originally conjectured by Fröhlich.

Theorem 2 (Taylor (1981)) If N/K is a tame Galois extension, then

$$(\mathcal{O}_N)_{\mathbf{Z}[\Gamma]} = \pi_\Gamma(W_{N/K}).$$

Of course if Γ has no irreducible symplectic characters then $W_{N/K}$ is trivial. Furthermore under these conditions stable $\mathbf{Z}[\Gamma]$-isomorphism is equivalent to $\mathbf{Z}[\Gamma]$-isomorphism (see remark 1.2) and therefore

Corollary 3 If N/K is a tame Galois extension which has Galois group possessing no irreducible symplectic character, then

$$\mathcal{O}_N \cong (\mathbf{Z}[\Gamma])^{|K:\mathbf{Q}|}.$$

In particular then if Γ is abelian, or dihedral, or of odd order then $\mathcal{O}_N$ is a free $\mathbf{Z}[\Gamma]$-module. This justifies to a considerable degree our claim that the theory produces explicit concrete information.

Thus for tame extensions the Artin root numbers determine the $\mathbf{Z}[\Gamma]$-module structure of $\mathcal{O}_N$. However as the following example shows the converse is not true.

Example 1.8 For any integer m we write H_{4m} for the generalised quaternion group of order $4m$. Then $Cl(\mathbf{Z}[H_8])$ has order two, and (as first demonstrated by Martinet) there exist tamely ramified extensions $N/\mathbf{Q}$ with $Gal(N/\mathbf{Q}) \cong H_8$ and $(\mathcal{O}_N)_{\mathbf{Z}[\Gamma]} = +1$, and also tamely ramified extensions $N/\mathbf{Q}$ with $Gal(N/\mathbf{Q}) \cong H_8$ and $(\mathcal{O}_N)_{\mathbf{Z}[\Gamma]} = -1$. On the other hand however, $Cl(\mathbf{Z}[H_{16}])$ also has order two, and whilst there exist tamely ramified extensions $N/\mathbf{Q}$ such that $Gal(N/\mathbf{Q}) \cong H_{16}$ and the symplectic root numbers are of each possible value, one knows that for any such extension $(\mathcal{O}_N)_{\mathbf{Z}[\Gamma]} = +1$.

Hence, in order to determine each of the root numbers $w(\chi)$ by means of the Galois module structures attached to N, one must consider more than just the Galois module structure of $\mathcal{O}_N$. In fact the root numbers are completely determined by the $\mathbf{Z}[\Gamma]$-structure of the $\mathcal{O}_K[\Gamma]$-Hermitian module $(\mathcal{O}_N, T_{N/K})$ where

$$T_{N/K} : N \times N \longrightarrow K[\Gamma]$$

is derived from the trace pairing

$$Tr_{N/K} : N \times N \longrightarrow K,$$

which is given by

$$Tr_{N/K}(x, y) = trace_{N/K}(xy), \quad \text{all } x, y \in N.$$

We shall discuss this more fully in §2.

1.3 The basic techniques

For the remainder of §1 we shall discuss some of the ideas behind the proof of Theorem 2. This will in fact provide useful motivation for many of the approaches to be adopted in subsequent sections.

At the heart of the proof of Theorem 2 are invariants constructed to interpret the connections between the module theory and the arithmetic of tame Galois extensions N/K. In fact these invariants, or'resolvents', generalise the classical notion of Lagrange resolvents to the case of non-abelian characters of Γ. For a complex character χ of Γ, corresponding to the representation T, and any element a of N such that $a.K[\Gamma] = N$ one defines the resolvent of a and χ to be

$$(a \mid \chi) = det\left(\sum_{\gamma\in\Gamma} a^{\gamma} T(\gamma^{-1})\right) \in E^*$$

where here E is as before any 'sufficiently large' Galois extension of $\mathbf{Q}$. (Of course one can similarly define a resolvent attached to any complex valued character and normal basis generator of a finite Galois extension of local fields). This definition of resolvent is extended to virtual characters of Γ by linearity. As we are investigating the $\mathbf{Z}[\Gamma]$-structure of $\mathcal{O}_N$ (i.e. not the $\mathcal{O}_K[\Gamma]$-structure) we are forced to work with the 'Norm-resolvent' defined for each a and χ by

$$\mathcal{N}_{K/\mathbf{Q}}(a \mid \chi) = \prod_{\sigma : K \hookrightarrow \mathbf{Q}^c} (a \mid \chi^{\sigma^{-1}})^{\sigma}$$

where the product is taken over a set of coset representatives $\{\sigma\}$ for Ω_K in Ω. This definition does depend upon the choice of representatives $\{\sigma\}$ but only to within a root of unity.

Working now at the integral level there exists an idele

$$\alpha = (\alpha_{\mathcal{P}})_{\mathcal{P}} \in \mathcal{J}(N)$$

such that for each prime divisor $\mathcal{P}$ of N

$$(\mathcal{O}_N)_{\mathcal{P}\cap K} = \alpha_{\mathcal{P}}.(\mathcal{O}_K)_{\mathcal{P}\cap K}[\Gamma] \; . \tag{1.9}$$

Thus for each virtual character χ of Γ one can define an idelic resolvent $(\alpha \mid \chi) \in \mathcal{J}(E)$, by

$$((\alpha \mid \chi))_{\mathcal{P}'} = (\alpha_{\mathcal{P}'\cap N} \mid \chi)$$

for each prime divisor $\mathcal{P}'$ of E. Similarly one defines an idelic norm resolvent

$$\mathcal{N}_{K/\mathbf{Q}}(\alpha \mid \chi) = \prod_{\sigma:K\hookrightarrow\mathbf{Q}^c} (\alpha \mid \chi^{\sigma^{-1}})^{\sigma}$$

which again only depends upon the choice of coset representatives $\{\sigma\}$ (of Ω_K in Ω) to within a root of unity.

Theorem 4 (Fröhlich (1976)) Let N/K be a tame Galois extension and let the elements $a \in N$ and $\alpha \in \mathcal{J}(N)$ be chosen as above. Then the map

$$\chi \longmapsto \mathcal{N}_{K/\mathbf{Q}}(\alpha \mid \chi).\mathcal{N}_{K/\mathbf{Q}}(a \mid \chi)^{-1}$$

is an element of $Hom_{\Omega}(R_\Gamma, \mathcal{J}(E))$ which has image under the projection map π_Γ of (1.5) equal to $(\mathcal{O}_N)_{\mathbf{Z}[\Gamma]}$.

This theorem is the main reason for introducing resolvents into the subject and also justifies the use of the 'Hom-language' for classgroups.

A second basic ingredient of the theory is the notion of the Galois–Gauss sum. To introduce this we may again relax the restriction that N/K be tame and work in complete generality. We shall deal initially with the local case and define a complex valued function τ_p on the set $\mathcal{E}_p$ of pairs $(E/F, \chi)$ where F runs through the set of finite extensions of $\mathbf{Q}_p$, E through the finite Galois extensions of F, and χ through the group $R_{Gal(E/F)}$ of virtual characters of $Gal(E/F)$. Suppose for the moment that χ is an abelian character of $Gal(E/F)$. Then, by local class field theory, there is a multiplicative character θ_χ of F^* corresponding to χ and we write $\mathcal{F}(\chi)$ for the conductor of θ_χ. Let

$$\psi_F : F(additive) \longrightarrow \mathbf{C}^*$$

denote the canonical (Tate–Iwasawa) additive character of F. The Galois–Gauss sum of the pair $(E/F, \chi)$ is then

$$\tau_p(E/F,\chi) = \begin{cases} \sum_u \theta_\chi(uc^{-1})\psi_F(uc^{-1}), & \text{if } \mathcal{F}(\chi) \neq \mathcal{O}_F, \\ \theta_\chi(\mathcal{D}_F^{-1}), & \text{if } \mathcal{F}(\chi) = \mathcal{O}_F, \end{cases} \tag{1.10}$$

where $\mathcal{D}_F$ is the different of the extension $F/\mathbf{Q}_p$ and, if $\mathcal{F}(\chi) \neq \mathcal{O}_F$, then $c \in F^*$ is chosen such that

$$(c) = \mathcal{F}(\chi)\mathcal{D}_F \ ,$$

and u runs through a complete set of representatives of $\mathcal{O}_F^*$ modulo $1 + \mathcal{F}(\chi)$. (This definition is independent of the choices of u and c within the stated conditions). By a reformulation of the theorem of Langlands proving the existence of local constants, there exists a function τ_p defined on $\mathcal{E}_p$ which agrees with (1.10) for the abelian characters and satisfies in addition that

$$\tau_p(E/F, \chi_1 + \chi_2) = \tau_p(E/F, \chi_1).\tau_p(E/F, \chi_2), \tag{1.11}$$

and that, if $L \supseteqq M$ are intermediate fields of E/F with $\Lambda = Gal(E/L)$ and $\Sigma = Gal(E/M)$, and ϕ is any character of Λ of degree 0, then

$$\tau_p(E/L, \phi) = \tau_p(E/M, ind_\Lambda^\Sigma(\phi)) \tag{1.12}$$

where here ind_Λ^Σ is the induction map. Moreover, as a consequence of Brauer's induction theorem, formulae (1.10), (1.11) and (1.12) will together determine the function τ_p uniquely and this then is the local Galois–Gauss sum. One now defines a global Galois–Gauss sum by means of local components. Let then N/K denote a Galois extension of number fields (not necessarily tame) with $\Gamma = Gal(N/K)$. Let $\wp$ be a prime ideal of K (of residue characteristic p say) with $\mathcal{P}$ a prime ideal of N lying above $\wp$. Taking completions one has a Galois extension of fields $N_\mathcal{P}/K_\wp$ whose group $\Gamma_\mathcal{P}$ embeds into Γ as the stabiliser of $\mathcal{P}$, with this embedding uniquely determined by $\wp$ to within conjugacy. The local component $\chi_\wp$ of a character χ of Γ is then simply the restriction of χ to $\Gamma_\mathcal{P}$. For each character χ of Γ one defines the global Galois–Gauss sum to be

$$\tau(\chi) = \tau(N/K, \chi) = \prod_\wp \tau_p(N_\mathcal{P}/K_\wp, \chi_\wp) \tag{1.13}$$

where the product (taken over all prime ideals $\wp$ of K) is convergent since $\tau(N_\mathcal{P}/K_\wp, \chi_\wp) = 1$ for almost all $\wp$.

For any complex character χ of Γ we write $\mathcal{F}(\chi)$ for its Artin conductor and $N\mathcal{F}(\chi)$ for the absolute norm of $\mathcal{F}(\chi)$. From the aforementioned existence theorem of Langlands one has a canonical decomposition of the root number $w(\chi)$ as a product over all places $\wp$ of K of the so-called local root numbers (or 'local constants') $w_\wp(\chi)$, each a complex number of absolute value 1. Combining this decomposition with the theorem of Tate (1950) on abelian characters one has the formula

$$w(\chi) = N\mathcal{F}(\chi)^{-\frac{1}{2}} w_\infty(\chi)\tau(\overline{\chi}) \tag{1.14}$$

where here $w_\infty(\chi)$ is the product of $w_\wp(\chi)$ over all archimedean places $\wp$ of K (in fact $w_\wp(\chi)$ is a 4th root of unity) and $N\mathcal{F}(\chi)^{\frac{1}{2}} \in (\mathbf{Q}^c)^*$ is the positive square root of $N\mathcal{F}(\chi)$. From equation (1.14) it is apparent that the Galois–Gauss sums arise essentially from the functional equation of the extended

Artin L-function. We stress that (1.14) remains valid even for wildly ramified extensions N/K. The fundamental connection between the resolvents and the Galois–Gauss sums is provided by

Theorem 5 (Fröhlich (1976)) Assume the conditions and notation of Theorem 4. Then

(1) The map $\chi \longmapsto \mathcal{N}_{K/\mathbf{Q}}(\alpha \mid \chi)/\tau(\chi)$ belongs to $Hom_{\Omega}(R_\Gamma, \mathcal{J}(E))$.
(2) If χ is a symplectic character of Γ and $\mathcal{P}$ is an archimedean place of E then

$$(\mathcal{N}_{K/\mathbf{Q}}(\alpha \mid \chi)/\tau(\chi))_{\mathcal{P}}$$

is real and has the same sign as $(W_{N/K}(\chi))_{\mathcal{P}}$.
(3) $\mathcal{N}_{K/\mathbf{Q}}(\alpha \mid \chi)/\tau(\chi)$ is a unit idele for each character χ of Γ.

This theorem lies at the very core of the tame theory, implying an entirely unsuspected interpretation of the Galois–Gauss sum, and hence of the functional equation of the extended Artin L-function, in terms of the module structure of the rings of integers. Indeed if N/K is a tame extension then by Theorem 4 the class $(\mathcal{O}_N)_{\mathbf{Z}[\Gamma]}$ is equal to

$$\pi_\Gamma\left(\chi \longmapsto \frac{\mathcal{N}_{K/\mathbf{Q}}(\alpha \mid \chi)}{\tau(\chi)}\right).\pi_\Gamma\left(\chi \longmapsto \frac{\tau(\chi)}{\mathcal{N}_{K/\mathbf{Q}}(a \mid \chi)}\right). \tag{1.15}$$

But the second map in (1.15) is actually one from R_Γ to the multiplicative group E^* and any such map can be shown to lie in $ker(\pi_\Gamma)$. Hence

Corollary 6 Assume the conditions and notations of Theorem 4. Then

$$(\mathcal{O}_N)_{\mathbf{Z}[\Gamma]} = \pi_\Gamma\left(\chi \longmapsto \frac{\mathcal{N}_{K/\mathbf{Q}}(\alpha \mid \chi)}{\tau(\chi)}\right).$$

The proof of Theorem 5 depends critically upon the explicit decomposition (1.13) of the Galois–Gauss sum $\tau(\chi)$ into local factors. This analysis is indeed a major part of the tame theory but we do not discuss it here since at present no trace of a possible generalisation has become apparent. We shall also not discuss the proof of Theorem 2 in any further detail save to say that, given the above ingredients, it is proved by combining (refined) congruences for the Galois–Gauss sums with Taylor's logarithm map for group rings. For more details the reader is referred to either Fröhlich (1983) or to the original paper of Taylor (1981).

To end this section we note a further consequence of the above techniques. Let $\mathcal{M}$ denote any maximal $\mathbf{Z}$-order in $\mathbf{Q}[\Gamma]$ with $\mathcal{M} \supseteq \mathbf{Z}[\Gamma]$. In the case that N is a Galois extension of $\mathbf{Q}$ (with $\Gamma = Gal(N/\mathbf{Q})$) Martinet had conjectured that the $\mathcal{M}$-module $\mathcal{O}_N\mathcal{M}$ generated by $\mathcal{O}_N$ was always a stably-free $\mathcal{M}$-module. In fact for N/K tame, and even with $K \neq \mathbf{Q}$, this is an easy consequence of the Hom-description of $Cl(\mathcal{M})$ (corresponding to (1.5)) together with Corollary 6 and Theorem 5 (parts (2) and (3)).

Corollary 7 Let N/K be a tame Galois extension of group Γ. If $\mathcal{M}$ is any maximal $\mathbf{Z}$-order in $\mathbf{Q}[\Gamma]$ with $\mathcal{M} \supseteqq \mathbf{Z}[\Gamma]$, then $\mathcal{O}_N\mathcal{M}$ is a stably-free $\mathcal{M}$-module.

Even in case $K = \mathbf{Q}$, the result of Corollary 7 is in general not true without the tameness hypothesis. However, we shall later derive an analogue of Corollary 7 for certain wildly ramified Galois extensions (§3 (3.23)).

There is an alternative way of stating the result of Corollary 7 which is useful for later reference. For any maximal $\mathbf{Z}$-order $\mathcal{M}$ as above, tensor product with $\mathcal{M}$ over $\mathbf{Z}[\Gamma]$ induces a homomorphism

$$\pi_{\mathcal{M}} : Cl(\mathbf{Z}[\Gamma]) \longrightarrow Cl(\mathcal{M})$$

and we let $D(\mathbf{Z}[\Gamma])$ denote the kernel of the map $\pi_{\mathcal{M}}$. In fact, $D(\mathbf{Z}[\Gamma])$ is independent of the choice of maximal order $\mathcal{M}$ as above. Now if $X \sim_{\mathbf{Z}[\Gamma]} \mathbf{Z}[\Gamma]$ then $X\mathcal{M}$ naturally identifies with $X \otimes_{\mathbf{Z}[\Gamma]} \mathcal{M}$, and hence an equivalent formulation of Corollary 7 is

Corollary 7′ Let N/K be a tame Galois extension of group Γ. Then

$$(\mathcal{O}_N)_{\mathbf{Z}[\Gamma]} \in D(\mathbf{Z}[\Gamma]).$$

2 ARTIN ROOT NUMBERS AND HERMITIAN GALOIS MODULES

Let N/K be a finite tame Galois extension of number fields of Galois group Γ. In this paragraph we examine more closely the connection between the structure of $\mathcal{O}_N$ as a module over the integral group ring $\mathbf{Z}[\Gamma]$ and the values of the Artin root numbers $w(\chi)$ attached to the symplectic characters χ of Γ.

From Theorem 2 of §1 one knows that the $\mathbf{Z}[\Gamma]$-structure of $\mathcal{O}_N$ is determined (up to stable isomorphism) by the root numbers $w(\chi)$ of the irreducible symplectic characters of Γ. On the other hand the Example 1.8 indicates that the $\mathbf{Z}[\Gamma]$-structure of $\mathcal{O}_N$ alone is not sufficient to determine the symplectic root numbers and hence we shall consider the finer structure of the Hermitian module $(\mathcal{O}_N, T_{N/K})$ as mentioned in §1.

The idea of relating the Hermitian structure to the symplectic root numbers originated with Fröhlich, who made the following conjecture:

Conjecture 2.1 For a tame Galois extension N/K of number fields (respectively local fields) the global (respectively local) Hermitian module structure of $(\mathcal{O}_N, T_{N/K})$ determines the global (respectively local) symplectic Artin root numbers.

In this paragraph we shall describe the precise form in which Conjecture (2.1) was proved. However for this we must first be more precise about the notion of $\mathbf{Z}[\Gamma]$-Hermitian modules and also introduce some general notation and algebraic preliminaries.

2.1 Hermitian modules

From now on K is a finite extension of $\mathbf{Q}$ or of the p-adic field $\mathbf{Q}_p$ for some prime number p. We let G denote an arbitrary finite group. If M is any right $\mathcal{O}_K[G]$-module then we shall write M_K for the $K[G]$-module $M \otimes_{\mathcal{O}_K} K$, and for an element $a = \sum_{g \in G} a_g g \in K[G]$ we shall write $\overline{a} = \sum_{g\in G} a_g g^{-1}$.

Definition 2.2 A Hermitian $\mathcal{O}_K[G]$-module (M, h) consists of a locally-free right $\mathcal{O}_K[G]$-module M and a map $h : M_K \times M_K \longrightarrow K[G]$ such that

(i) h is K-linear in each variable;
(ii) h is non-degenerate;
(iii) $h(x, yg) = h(x, y)g$ for $x, y \in M_K$ and $g \in G$;
(iv) $h(y, x) = \overline{h(x, y)}$ for $x, y \in M_K$.

Remark 2.3(i) Giving such a map is equivalent to giving a non-degenerate G-invariant symmetric K-bilinear form $h' : M_K \times M_K \longrightarrow K$, the relation between h and h' being that

$$h(x, y) = \sum_{g \in G} h'(x, yg^{-1})g \quad \text{for each } x, y \in M_K .$$

Remark 2.3(ii) The map h is defined at field level and need not map $M \times M$ into $\mathcal{O}_K[G]$. The non-degeneracy condition means that the discriminant of h is a unit of $K[G]$ – it need not be a unit in, or even an element of, $\mathcal{O}_K[G]$.

Example 2.4(i) The standard Hermitian $\mathcal{O}_K[G]$-module is $(\mathcal{O}_K[G], \mu_K)$ where $\mu_K : K[G] \times K[G] \longrightarrow K[G]$ is given by $\mu_K(x, y) = \overline{x}y$.

Example 2.4(ii) If N/K is a tame Galois extension of number fields with Galois group Γ then $(\mathcal{O}_N, T_{N/K})$ is a Hermitian $\mathcal{O}_K[\Gamma]$-module, where $T_{N/K} : N \times N \longrightarrow K[\Gamma]$ is derived from the trace pairing $Tr_{N/K} : N \times N \longrightarrow K$; explicitly,

$$T_{N/K}(x,y) = \sum_{\gamma \in \Gamma} (Tr_{N/K}(x, y\gamma^{-1}))\gamma \quad \text{for each } x,y \in N.$$

2.2 The Hermitian class group

We shall now outline the theory of the Hermitian class group as developed in Fröhlich (1984). To classify Hermitian $\mathcal{O}_K[G]$-modules one can certainly form the Grothendieck group $K_0H(\mathcal{O}_K[G])$ of isometry classes of such modules with relations given by orthogonal sums, but this group is too big for our purposes. To obtain an invariant for the Hermitian Galois modules which is more useful than the class in $K_0H(\mathcal{O}_K[G])$ we will define a suitable notion of the discriminant of such a module. We introduce the Hermitian class group $HCL(\mathcal{O}_K[G])$ as the group in which these discriminants lie. This group has a 'Hom-description' analogous to that described in §1.1 for the locally-free class group of $\mathbf{Z}[G]$-modules. For this we must recall some standard notation.

We let R_G^s denote the subgroup of the group R_G of virtual characters of G that is generated by the symplectic characters. For any virtual character $\chi \in R_G$ we set $Tr(\chi) = \chi + \overline{\chi}$: then $Tr(\chi) \in R_G^s$ and $w(Tr(\chi)) = +1$. As in §1 we shall also fix a 'sufficiently large' number field E and let $\mathcal{J}(E)$ denote its idele group.

For the purposes of this exposition, we now restrict attention to the global case and, for simplicity, assume that $K = \mathbf{Q}$. The reader is referred to Fröhlich (1984), in particular to chapter 2, for a more general and a more detailed treatment of the construction of the Hermitian class group. Again we abbreviate the absolute Galois group $Gal(\mathbf{Q}^c/\mathbf{Q})$ by Ω. We will define $HCL(\mathbf{Z}[G])$ in terms of certain subgroups of $Hom_\Omega(R_G, \mathcal{J}(E))$.

We set

$$U(\mathbf{Z}[G]) = \prod_p (\mathbf{Z}_p[G])^*,$$

the 'unit-ideles' of $\mathbf{Z}[G]$, where here the product is taken over all places of $\mathbf{Q}$ (with $\mathbf{Z}_\infty = \mathbf{R}$). For $u = (u_p)_p \in U(\mathbf{Z}[G])$ we may define as in §1.1 an element $Det(u) \in Hom_\Omega(R_G, \mathcal{J}(E))$. We thus obtain a subgroup $Det(U(\mathbf{Z}[G]))$ of $Hom_\Omega(R_G, \mathcal{J}(E))$. Now let

$$\Delta : Hom_\Omega(R_G, E^*) \longrightarrow \frac{Hom_\Omega(R_G, \mathcal{J}(E))}{Det(U(\mathbf{Z}[G]))} \times Hom_\Omega(R_G^s, E^*)$$

be the homomorphism defined by

$$\Delta(f) = (f^{-1} \text{ modulo } Det(U(\mathbf{Z}[G])), f^{s})$$

where f^{s} denotes the restriction of f to R^{s}_{G}. We then make the

Definition 2.5 The Hermitian class group $HCL(\mathbf{Z}[G])$ of $\mathbf{Z}[G]$ is the quotient group $coker(\Delta)$.

The group $HCL(\mathcal{O}_K[G])$ is defined similarly and, in general, the Hermitian class group has good functorial behaviour with respect to changes of group or of base field.

Our next task is to relate $HCL(\mathbf{Z}[G])$ to $K_0H(\mathbf{Z}[G])$ by means of a discriminant map

$$d : K_0H(\mathbf{Z}[G]) \longrightarrow HCL(\mathbf{Z}[G]) .$$

Given any class in $K_0H(\mathbf{Z}[G])$ we choose a representative (M, h) of that class. Assuming for simplicity that M has $\mathbf{Z}[G]$-rank 1, we take v a basis of $M_{\mathbf{Q}}$ over $\mathbf{Q}[G]$, and set $c = h(v, v)$. It follows easily from the definition of a Hermitian module that $c = \bar{c} \in \mathbf{Q}[G]^*$. Now let χ be any symplectic character of G and let $T : G \longrightarrow GL(V)$ be a representation of G, affording χ on an E-vector space V. As T is symplectic there is a non-degenerate G-invariant skew-symmetric E-bilinear form β on V. Taking adjoints with respect to β gives an involution $\jmath$ on $GL(V)$: $\jmath$ is determined by the rule

$$\beta(xA, y) = \beta(x, yA^{\jmath}) \text{ for each } x, y \in V,\ A \in GL(V).$$

Moreover, as β is G-invariant, we have that $T(z)^{\jmath} = T(\bar{z})$ for all $z \in \mathbf{Q}[G]^*$, so that in particular $T(c)^{\jmath} = T(c) \in GL(V)$. For any $A \in GL(V)$ with $A^{\jmath} = A$ we can define a new non-degenerate skew-symmetric E-bilinear form $\alpha = \alpha(A)$ on V by setting

$$\alpha(x, y) = \beta(xA, y) \text{ for } x, y \in V.$$

Since all such forms are equivalent there is an automorphism $P = P(A)$ of V, unique up to composition with a symplectic automorphism, transforming β into α. Thus

$$\alpha(x, y) = \beta(xP, yP) \text{ for all } x, y \in V,$$

$$\text{i.e., } \beta(xA, y) = \beta(xPP^{\jmath}, y) \text{ for all } x, y \in V,$$

and we deduce that $A = PP^{\jmath}$.

Applying this to $A = T(c)$ we have $T(c) = PP^{\jmath}$ for some $P \in GL(V)$, unique up to a symplectic automorphism. We define the Pfaffian $Pf_\chi(c)$ of χ at c to be $det(P)$. Since any symplectic automorphism has determinant $+1$ this is indeed independent of the choice of P and so gives a canonical square root of $det(T(c))$. We define $Pf(c) \in Hom_\Omega(R_G^s, E^*)$ as the map $\chi \longmapsto Pf_\chi(c)$.

Finally, we choose an idele $\alpha = (\alpha_p)_p \in \mathcal{J}(\mathbf{Q}[G])$ such that, for each place p of $\mathbf{Q}$,

$$M_p = \alpha_p V.\mathbf{Z}_p[G] .$$

We now define the discriminant map d by setting

$$d([(M,h)]) = \text{ the class of } (Det(\alpha), Pf(c)) \in HCL(\mathbf{Z}[G]).$$

The next stage is to use d to construct a subgroup $K_0'H(\mathbf{Z}[G])$ of $K_0H(\mathbf{Z}[G])$. For this we first define

$$H(G) = Hom_\Omega(R_G^s/Tr(R_G), \pm 1) \times Hom_\Omega(R_G^s, \mathbf{Q}^*).$$

Given $(f,g) \in H(G)$ let $\tilde{f}$ be the element of $Hom_\Omega(R_G, \mathcal{J}(E))$ defined by

$$\tilde{f}(\chi)_p = \begin{cases} f(\chi), & \text{if } \chi \text{ is symplectic and } p \text{ is finite,} \\ 1, & \text{otherwise,} \end{cases}$$

and let $\tilde{g}$ be the element of $Hom_\Omega(R_G^s, E^*)$ obtained by composing g with the natural embedding $\mathbf{Q}^* \longrightarrow E^*$. Now define

$$h_G : H(G) \longrightarrow HCL(\mathbf{Z}[G])$$

by

$$(f,g) \longmapsto \text{ the class of } (\tilde{f}, \tilde{g}) .$$

The map h_G can be shown to be injective and we use it to regard $H(G)$ as a subgroup of $HCL(\mathbf{Z}[G])$. The required subgroup of $K_0H(\mathbf{Z}[G])$ is then obtained by setting

$$K_0'H(\mathbf{Z}[G]) = d^{-1}(H(G)) .$$

Finally, composing d with the projections of $H(G)$ onto its two factors, we obtain homomorphisms

$$\theta : K_0'H(\mathbf{Z}[G]) \longrightarrow Hom_\Omega(R_G^s/Tr(R_G), \pm 1)$$

and

$$\eta : K_0'H(\mathbf{Z}[G]) \longrightarrow Hom_\Omega(R_G^s, \mathbf{Q}^*) .$$

2.3 The arithmetic case

We now return to the Hermitian $\mathcal{O}_K[\Gamma]$-module $(\mathcal{O}_N, T_{N/K})$, where N/K is a finite tame Galois extension of number fields of Galois group Γ. We want to compare this module with the standard Hermitian module $(\mathcal{O}_K[\Gamma], \mu_K)$, regarding both as Hermitian modules over $\mathbf{Z}[\Gamma]$. Thus we define

$$X_{N/K} = Res([(\mathcal{O}_N, T_{N/K})] - [(\mathcal{O}_K[\Gamma], \mu_K)]) \in K_0H(\mathbf{Z}[\Gamma])$$

where here $Res : K_0H(\mathcal{O}_K[\Gamma]) \longrightarrow K_0H(\mathbf{Z}[\Gamma])$ is the homomorphism given by restriction of scalars.

The map

$$\chi \longmapsto w(\chi), \qquad \chi \in R_\Gamma^s$$

induces an element of $Hom_\Omega(R_\Gamma^s/Tr(R_\Gamma), \pm 1)$. For any symplectic character χ of Γ we let $w_\infty(\chi)$ and $N\mathcal{F}(\chi)^{\frac{1}{2}} \in \mathbf{Q}^*$ be as in equation (1.14) of §1. We then have

Theorem 1 (Cassou-Noguès and Taylor (1983a))

(i) $X_{N/K} \in K_0'H(\mathbf{Z}[\Gamma])$;
(ii) $\theta(X_{N/K})$ is the map induced by $\chi \longmapsto w(\chi)$;
(iii) $\eta(X_{N/K})$ is the map given by $\chi \longmapsto N\mathcal{F}(\chi)^{\frac{1}{2}} w_\infty(\chi)$.

Remark 2.6 Recall that, for each symplectic character χ of Γ, the complex algebraic numbers $w_\infty(\chi)$ and $N\mathcal{F}(\chi)^{\frac{1}{2}}$ were defined immediately prior to equation (1.14) of §1.

In particular then the class $X_{N/K}$ determines the symplectic root numbers so that we have a precise statement (and a proof) of the global version of Conjecture (2.1). A similar result also holds for the local version of the conjecture (Cassou-Noguès and Taylor (1983b)).

The two main ingredients of the proof of Theorem 1 are the algebraic fact that the map

$$h_\Gamma : H(\Gamma) \longrightarrow HCL(\mathbf{Z}[\Gamma])$$

is injective, and the comparison of the arithmetic behaviour of Norm-resolvents and Galois–Gauss sums. As we have seen the second of these is also at the heart of the proof of Theorem 2 of §1, and indeed Theorem 2 of §1 can be recovered from the above Theorem 1 by applying the forgetful functor $K_0H(\mathbf{Z}[\Gamma]) \longrightarrow K_0(\mathbf{Z}[\Gamma])$.

2.4 The isometry class of $(\mathcal{O}_N, T_{N/K})$

If, for example, Γ has no irreducible symplectic characters then, by Theorem 2 of §1, $\mathcal{O}_N$ is known to be isomorphic as a $\mathbf{Z}[\Gamma]$-module to $\mathcal{O}_K[\Gamma]$. On the other hand for example the factor $N\mathcal{F}(\chi)^{\frac{1}{2}}$ in Theorem 1(iii) shows that we do not, however, have an isometry of Hermitian modules between $(\mathcal{O}_N, T_{N/K})$ and $(\mathcal{O}_K[\Gamma], \mu_K)$. We now consider the problem of finding a Hermitian module, related to the group ring, which is isometric to $(\mathcal{O}_N, T_{N/K})$.

We begin by comparing $(\mathcal{O}_N, T_{N/K})$ and $(\mathcal{O}_K[\Gamma], \mu_K)$ at the field level. Thus we ask whether there is an isometry of Hermitian $K[\Gamma]$-modules between $(N, T_{N/K})$ and $(K[\Gamma], \mu_K)$. This is equivalent to asking whether there exists an element c of N such that $\{c\gamma : \gamma \in \Gamma\}$ is a basis for N/K satisfying the condition

$$trace_{N/K}(c.c\gamma) = \begin{cases} 1, & \text{if } \gamma = 1; \\ 0, & \text{otherwise.} \end{cases}$$

Such an element, if it exists, is called a self-dual normal basis of N/K. If Γ has odd order then the answer to our question is given by

Theorem 2 (Bayer–Fluckiger and Lenstra (1989)) If Γ has odd order then N/K has a self-dual normal basis.

Remark 2.7 If Γ has a quotient of order any power of 2 then there is no self-dual normal basis. In the cases $\Gamma = A_4$ and $\Gamma = A_5$ Serre has shown that the existence of a self-dual normal basis is connected with the Witt invariant of the trace form.

From now on we assume that Γ has odd order. Thus the Hermitian $\mathcal{O}_K[\Gamma]$-modules $(\mathcal{O}_N, T_{N/K})$ and $(\mathcal{O}_K[\Gamma], \mu_K)$ become isometric on extension of scalars to $K[\Gamma]$. They are not however in general isometric at integral level, even locally, since $(\mathcal{O}_K[\Gamma], \mu_K)$ is unimodular whereas $(\mathcal{O}_N, T_{N/K})$ is not (its discriminant being the relative discriminant of the extension N/K). To take account of the ramification in N/K we therefore replace $\mathcal{O}_K[\Gamma]$ by a so-called Swan module.

For simplicity, we assume that the prime divisors of K lying above 2 do not ramify in the extension N/K. Let F be a quadratic extension of K such that every prime ideal of $\mathcal{O}_K$ which ramifies in N/K also ramifies in F/K. For each prime ideal $\mathcal{P}$ of $\mathcal{O}_F$ we let $\wp$ be the prime ideal of $\mathcal{O}_K$ below $\mathcal{P}$ and let $\mathcal{P}'$ be some prime of $\mathcal{O}_N$ lying over $\wp$. We write $I_{\mathcal{P}'}$ for the inertia group of $\mathcal{P}'$ in Γ.

We now define the Swan module (S_F, μ_F) to be the Hermitian $\mathcal{O}_F[\Gamma]$-submodule of $(\mathcal{O}_F[\Gamma], \mu_F)$ which has local completions $S_{F,\wp}$ (at each prime ideal of $\mathcal{O}_K$) given by

$$S_{F,\wp} = \begin{cases} \mathcal{O}_{F,\wp}[\Gamma], & \text{if } \wp \text{ is unramified in } N/K; \\ \{\mathcal{P},\ \sum_{\delta \in I_{\mathcal{P}'}} \delta\}\mathcal{O}_{F,\wp}[\Gamma], & \text{otherwise.} \end{cases}$$

The module S_F therefore depends on the choices of the prime ideals $\mathcal{P}'$ of $\mathcal{O}_N$ lying above the ramified prime ideals $\wp$ of $\mathcal{O}_K$, but the isometry class of (S_F, μ_F) is independent of these choices.

By extending scalars from $\mathcal{O}_K$ to $\mathcal{O}_F$ we obtain from $(\mathcal{O}_N, T_{N/K})$ a Hermitian $\mathcal{O}_F[\Gamma]$-module $(\mathcal{O}_N, T_{N/K})_F$ and so we can compare the classes $[(\mathcal{O}_N, T_{N/K})_F]$ and $[(S_F, \mu_F)]$ in $K_0H(\mathcal{O}_F[\Gamma])$. Writing

$$Res : K_0H(\mathcal{O}_F[\Gamma]) \longrightarrow K_0H(\mathbf{Z}[\Gamma])$$

for the homomorphism given by restriction of scalars we set

$$Y_{N/K} = Res([(\mathcal{O}_N, T_{N/K})_F] - [(S_F, \mu_F)]) \in K_0H(\mathbf{Z}[\Gamma]).$$

We then have

Theorem 3 (Taylor (1989)) $Y_{N/K} = 0$.

Remark 2.8(i) If Γ has even order then one can still obtain a result about the isometry class of $(\mathcal{O}_N, T_{N/K})$ (Cassou-Noguès and Taylor (1989)).

Remark 2.8(ii) Instead of allowing for the ramification of N/K by replacing $\mathcal{O}_K[\Gamma]$ with a Swan module, one could replace $\mathcal{O}_N$ by a suitable fractional $\mathcal{O}_N$-ideal. In this context a special case of a theorem of Erez and Morales (1989) gives the following result.

Let $N/\mathbf{Q}$ be a tame abelian extension of odd degree, and let $\mathcal{D}_{N/\mathbf{Q}}^{-1}$ denote its inverse different. Then there is a fractional ideal $\mathcal{A}$ of $\mathcal{O}_N$ such that $\mathcal{A}^2 = \mathcal{D}_{N/K}^{-1}$, and the Hermitian $\mathbf{Z}[\Gamma]$-module $(\mathcal{A}, T_{N/\mathbf{Q}})$ is isometric to $(\mathbf{Z}[\Gamma], \mu_{\mathbf{Q}})$.

3 ON SOME PARALLEL RESULTS IN ADDITIVE AND MULTIPLICATIVE GALOIS MODULE THEORY

In this paragraph we shall continue the study of the Galois structure of rings of algebraic integers but in the context of wildly ramified extensions. We shall also discuss some remarkable, and at the present time unexplained, similarities between this 'additive' theory and the 'multiplicative' theory of the Galois structure of groups of units. Indeed the close analogies between these additive and multiplicative theories is an underlying motivation and theme for much of the material in this paragraph. We shall work in the setting of arbitrary finite groups. In the particular case that Γ is abelian proofs for most of the results discussed here can be found in Fröhlich (1989).

We first introduce the elementary but nevertheless powerful notion of factorisability. There are a number of variants of this notion, some of which are needed elsewhere. We shall use only that which is best for the present lecture – this is, for example, stronger than that discussed in Fröhlich (1988).

3.1 Factorisability and factor-equivalence

Let G denote an arbitrary finite group (which need not arise as a Galois group). Each subgroup H of G gives rise to a G-set $H \setminus G$ which in turn defines a permutation representation of G of character ρ_H. We set

$$\mathcal{S}(G) = \{G\text{-isomorphism classes of } H \setminus G : H \leq G\}.$$

We write $G^\dagger$ for the set of absolutely irreducible complex valued characters of G, so that $G^\dagger$ gives a $\mathbf{Z}$-basis of R_G. For a number field F let $\mathcal{I}(F)$ denote the group of fractional $\mathcal{O}_F$-ideals. For number fields F' and F with $F' \subset F$ we shall identify $\mathcal{I}(F')$ with a subgroup of $\mathcal{I}(F)$ in the natural way. We consider functions on $\mathcal{S}(G)$ taking values in $\mathcal{I}(\mathbf{Q})$. Such a function f will be said to be 'factorisable' if for some 'sufficiently large' number field E which is Galois over $\mathbf{Q}$ (and hence for every sufficiently large such E) there exists a homomorphism $g : R_G \longrightarrow \mathcal{I}(E)$ i.e.,

$$g(\chi_1 + \chi_2) = g(\chi_1)g(\chi_2) \qquad (3.1)(a)$$

such that, for each subgroup $H \leq G$,

$$f(H \setminus G) = g(\rho_H) \qquad (3.1)(b)$$

and also, for each $\chi \in G^\dagger$ and $\eta \in Gal(E/\mathbf{Q})$,

$$g(\chi^\eta) = g(\chi)^\eta \ . \qquad (3.1)(c)$$

Note that, writing $< \, , \, >$ for the $\mathbb{Z}$-bilinear pairing on R_G defined on elements of $G^\dagger$ by

$$< \chi, \phi >= \begin{cases} 1, & \text{if } \chi = \phi, \\ 0, & \text{otherwise,} \end{cases}$$

then condition (3.1)(b) is equivalent to the condition that, for each subgroup $H \leq G$,

$$f(H \setminus G) = \prod_{\chi \in G^\dagger} g(\chi)^{<\rho_H, \chi>} . \qquad (3.1)(b')$$

To introduce a general procedure for producing suitable maps f on $\mathcal{S}(G)$ we let L and M denote finitely generated $\mathbb{Z}[G]$-lattices (that is, $\mathbb{Z}[G]$-modules that are $\mathbb{Z}$-torsion free) such that there exists an isomorphism of $\mathbb{Q}[G]$-modules

$$\imath : L \otimes_{\mathbb{Z}} \mathbb{Q} \cong M \otimes_{\mathbb{Z}} \mathbb{Q}. \qquad (3.2)$$

For each subgroup $H \leq G$ let L^H (respectively M^H) denote the $\mathbb{Z}[G]$-sublattice of L (respectively M) consisting of those elements invariant under the action of each element of H. Define a map $f = f_{L,M,\imath}$ on $\mathcal{S}(G)$ by

$$f(H \setminus G) = [(\imath L)^H : M^H]_{\mathbb{Z}} \, , \quad \text{all } H \leq G$$

where here $[:]_{\mathbb{Z}}$ denotes the $\mathbb{Z}$-module index as defined for $\mathbb{Z}$-lattices that span the same $\mathbb{Q}$-space.

Definition 3.3 Two $\mathbb{Z}[G]$-lattices L and M will be said to be G-factor-equivalent, written $L \wedge_G M$, if for some choice of isomorphism $\imath$ as in (3.2) the function $f_{L,M,\imath}$ is factorisable.

In fact if $\jmath$ is any other isomorphism as in (3.2) then $f_{L,M,\imath}$ is factorisable if, and only if, $f_{L,M,\jmath}$ is factorisable. One can also show easily that $\wedge_G$ is an equivalence relation on the set of $\mathbb{Z}[G]$-lattices (that span the same $\mathbb{Q}[G]$-space). Furthermore $\wedge_G$ is a weakening of the equivalence relation of $\mathbb{Z}[G]$-genus equivalence – i.e., if $L \sim_{\mathbb{Z}[G]} M$ then $L \wedge_G M$.

Example 3.4 Let K be a number field. If G is abelian and $\mathcal{M}(\mathcal{O}_K, G)$ denotes the maximal $\mathcal{O}_K$-order in $K[G]$ then $\mathcal{M}(\mathcal{O}_K, G) \wedge_G \mathcal{O}_K[G]$ if, and only if, G is cyclic.

Thus the notion of factor equivalence is far from trivial. Despite this, and without any hypothesis on ramification, one has the following theorem. To state this and subsequent theorems we shall say that a group is 'admissible' if it satisfies a certain condition on its complex representations about which we shall say more presently. At the moment it is not known whether all finite groups are admissible but, for example, this is certainly the case for all abelian, dihedral, and (generalised) quaternion groups and also for all p-groups.

Theorem 1 (Fröhlich) Let N/K be a finite Galois extension of number fields of Galois group Γ. If Γ is admissible then $\mathcal{O}_N \wedge_\Gamma \mathcal{O}_K[\Gamma]$.

Corollary 2 Assume the notation and conditions of Theorem 1, and in addition that Γ is abelian but not cyclic. If $\mathcal{M}(\mathcal{O}_K, \Gamma)$ denotes the maximal $\mathcal{O}_K$-order in $K[\Gamma]$, then $(\mathcal{O}_N)\mathcal{M}(\mathcal{O}_K, \Gamma) \nsubseteq \mathcal{O}_N$.

Proof of Corollary 2 From Theorem 1 and Example 3.4 one deduces immediately that $\mathcal{M}(\mathcal{O}_K, \Gamma) \wedge_\Gamma \mathcal{O}_N$. Hence, *a fortiori*, $\mathcal{M}(\mathcal{O}_K, \Gamma) \not\cong_{\mathcal{O}_K[\Gamma]} \mathcal{O}_N$, and the claim follows by a standard property of $\mathcal{M}(\mathcal{O}_K, \Gamma)$-lattices. □

If N/K is wildly ramified then it is natural to compare $\mathcal{O}_N$ not with $\mathcal{O}_K[\Gamma]$ but with the full set $\mathcal{A}(N/K)$ of elements of $K[\Gamma]$ that induce endomorphisms of $\mathcal{O}_N$ (see Theorem 1 of §1). This set $\mathcal{A}(N/K)$ is an $\mathcal{O}_K$-order in $K[\Gamma]$, the 'associated order' of $\mathcal{O}_N$ in $K[\Gamma]$, and is strictly bigger than $\mathcal{O}_K[\Gamma]$. Theorem 1 also gives some information on the old vexed question of when $\mathcal{O}_N$ is locally-free as an $\mathcal{A}(N/K)$-module. Indeed if $\mathcal{O}_N$ is a locally-free $\mathcal{A}(N/K)$-module then necessarily $\mathcal{A}(N/K) \wedge_\Gamma \mathcal{O}_K[\Gamma]$. But given an explicit description of $\mathcal{A}(N/K)$ this is a purely computational question. Even better, if Γ is abelian and no prime ideal (of $\mathcal{O}_K$) which ramifies wildly in the extension N/K divides the different of the extension $K/\mathbb{Q}$ then under certain conditions the relation $\mathcal{A}(N/K) \wedge_\Gamma \mathcal{O}_K[\Gamma]$ is actually *sufficient* to ensure that $\mathcal{O}_N$ is locally-free as an $\mathcal{A}(N/K)$-module. For more details of this last result see Burns (1989).

Before discussing the proof of Theorem 1 we shall briefly consider the multiplicative theory. Here our aim is to understand the Galois structure of the multiplicative group $\mathcal{U}_N$ of units of $\mathcal{O}_N$ modulo torsion elements, and again it is fruitful to compare this arithmetical lattice with a lattice which has a better understood structure.

For any number field F we let $S_\infty(F)$ denote the set of archimedean places of F. We let $Y_F = Div_{\mathbb{Z}}(S_\infty(F))$ denote the additive group of $\mathbb{Z}$-divisors supported on $S_\infty(F)$ and write X_F for the subgroup of Y_F consisting of divisors of degree 0. That is, X_F is defined by means of the exactness of the sequence

$$0 \longrightarrow X_F \longrightarrow Y_F \xrightarrow{\epsilon} \mathbb{Z} \longrightarrow 0,$$

in which $\epsilon : Y_F \longrightarrow \mathbf{Z}$ is the homomorphism defined at each place $v \in S_\infty(F)$ by

$$v \overset{\epsilon}{\longmapsto} 1.$$

By a theorem of Herbrand one knows that $X_N \otimes_{\mathbf{Z}} \mathbf{Q}$ and $\mathcal{U}_N \otimes_{\mathbf{Z}} \mathbf{Q}$ are isomorphic as $\mathbf{Q}[\Gamma]$-modules. Thus choosing a $\mathbf{Q}[\Gamma]$-isomorphism

$$\kappa : X_N \otimes_{\mathbf{Z}} \mathbf{Q} \cong \mathcal{U}_N \otimes_{\mathbf{Z}} \mathbf{Q}, \tag{3.5}$$

one can define a function $f_\kappa = f_{\mathcal{U}_N, X_N, \kappa}$ on $\mathcal{S}(\Gamma)$ by

$$f_\kappa(\Delta \setminus \Gamma) = [\kappa(X_N)^\Delta : (\mathcal{U}_N)^\Delta]_{\mathbf{Z}} . h_{N^\Delta} \ , \qquad \Delta \leq \Gamma$$

where, for any number field F, h_F denotes the class number $ord(Cl_F)$ of F. In analogy to Theorem 1 one has

Theorem 3 (Fröhlich) Assume the notation of Theorem 1. If N is totally real and Γ is an admissible group of odd order then, for any choice of isomorphism κ as in (3.5), the function f_κ is factorisable.

Remark 3.6 In the statement of Theorem 3 the hypotheses that N is totally real and that Γ is of odd order are made purely for the sake of simplicity – there is a more general theorem of which Theorem 3 is a specialisation.

Apart from the assertion of factorisability in Theorems 1 and 3 we shall also presently see that a great deal of interest lies in the precise arithmetic nature and in the interpretation of the respective factors $g(\chi)$ (see (3.1)) giving the factorisations of the functions $f_{\mathcal{U}_N, X_N, \kappa}$ and $f_{\mathcal{O}_N, \mathcal{O}_K[\Gamma], \imath}$ for any suitable isomorphisms κ and $\imath$. But to discuss this further we must introduce invariants that connect the arithmetic and the module theory. Recall that in the tame additive theory, as discussed in §1, the necessary invariants were the resolvents. The invariants to be introduced here both for the wild additive theory (again labelled 'resolvents') and also for the multiplicative theory (labelled 'regulators') can be considered as generalisations of the fractional ideals generated by the 'tame' resolvents. Indeed the resolvents and regulators to be introduced here are each defined by the same formal procedure. We shall deal firstly with the additive theory.

3.2 The generalised resolvents

Writing S_N for the set of field embeddings $N \hookrightarrow \mathbf{C}$, one has a non-degenerate $\mathbf{Q}$-bilinear Γ-pairing

$$b : N \times Div_{\mathbf{Q}}(S_N) \longrightarrow \mathbf{C}$$

defined by

$$(x,s) \longmapsto x^s \quad x \in N, s \in S_N.$$

Let E again denote a number field, Galois over $\mathbf{Q}$, over which all elements of $\Gamma^\dagger$ can be realised and in addition such that $N \subset E$. Thus to each actual character ϕ of Γ there corresponds an $E[\Gamma]$-module V_ϕ. Associated to b and to V_ϕ there is a non-degenerate E-bilinear Γ-pairing

$$b_{V_\phi} : Hom_{E[\Gamma]}(V_\phi, E \otimes_{\mathbf{Q}} N) \times V_\phi \otimes_{E[\Gamma]} (E \otimes_{\mathbf{Q}} Div_{\mathbf{Q}}(S_N)) \longrightarrow \mathbf{C}$$

defined by

$$(h, v \otimes_{\mathbf{Q}} s) \longmapsto h(v)^s$$

for each $h \in Hom_{E[\Gamma]}(V_\phi, E \otimes_{\mathbf{Q}} N)$, $v \in V_\phi$, and $s \in S_N$. For any $\mathcal{O}_E[\Gamma]$-lattice H spanning V_ϕ one defines the resolvent $R(H, \mathcal{O}_N)$ as the discriminant with respect to the pairing b_{V_ϕ} of the sub-lattices

$$Hom_{\mathcal{O}_E[\Gamma]}(H, \mathcal{O}_E \otimes_{\mathbf{Z}} \mathcal{O}_N) \subset Hom_{E[\Gamma]}(V_\phi, E \otimes_{\mathbf{Q}} N)$$

and

$$H \otimes_{\mathcal{O}_E[\Gamma]} (\mathcal{O}_E \otimes_{\mathbf{Z}} Div_{\mathbf{Z}}(S_N)) \subset V_\phi \otimes_{E[\Gamma]} (E \otimes_{\mathbf{Q}} Div_{\mathbf{Q}}(S_N)).$$

If these lattices are free over $\mathcal{O}_E$ then this discriminant is as usual the basis discriminant modulo $\mathcal{O}_E^*$. In general the discriminant is defined via localisations.

Let $\chi \in \Gamma^\dagger$. We assume that there exists an $\mathcal{O}_E[\Gamma]$-lattice T_χ spanning V_χ and with the following property: for every $\omega \in Gal(E/\mathbf{Q}(\chi))$ the ω-semilinear translate $(T_\chi)^\omega$ of T_χ is isomorphic to T_χ as an $\mathcal{O}_E[\Gamma]$-module. This is certainly the case if χ can be realised over $\mathbf{Q}(\chi)$. It is also true for all representations of both (generalised) quaternion groups and p-groups. In general we say that a group Γ is 'admissible' if this condition can be satisfied for each element of $\Gamma^\dagger$ (recall the remarks prior to Theorem 1). Assuming that Γ is admissible choose for every $\chi \in \Gamma^\dagger$ an $\mathcal{O}_E[\Gamma]$-lattice T_χ as above and furthermore such that for each $\omega \in Gal(\mathbf{Q}(\chi)/\mathbf{Q})$ and each $\chi \in \Gamma^\dagger$

$$T_{\chi^\omega} \cong (T_\chi)^\omega \tag{3.7}$$

where now the lattice on the right is unique to within isomorphism. For each actual character ϕ of Γ we now set

$$T_\phi = \prod_{\chi \in \Gamma^\dagger} T_\chi^{<\chi,\phi>}.$$

Definition 3.8 Let Γ be admissible. Then for any (complex) character ϕ of Γ the generalised resolvent of $\mathcal{O}_N$ with respect to ϕ is

$$R(\phi, \mathcal{O}_N) = R(T_\phi, \mathcal{O}_N)\,.$$

Each resolvent $R(\phi, \mathcal{O}_N)$ is therefore a fractional $\mathcal{O}_E$-ideal. From the explicit definition of $R(\phi, \mathcal{O}_N)$ in terms of the pairing b_{V_ϕ} it is not difficult to show that

$$R(\phi, \mathcal{O}_N) \;=\; \prod_{\chi \in \Gamma^\dagger} R(\chi, \mathcal{O}_N)^{<\chi,\phi>} \tag{3.9}$$

for each character ϕ of Γ. Via the product expression (3.9), the definition of a generalised resolvent can easily be extended to all virtual characters $\phi \in R_\Gamma$.

If Γ is abelian then the set $\{T_\chi\}_{\chi\in\Gamma^\dagger}$ can be chosen canonically. In general however this definition of each $R(\phi, \mathcal{O}_N)$ is dependent upon the choice of lattices $\{T_\chi\}_{\chi\in\Gamma^\dagger}$ but the theorems to be stated remain valid with any set of lattices chosen as above. Furthermore if N/K is tamely ramified and $\alpha \in \mathcal{J}(N)$ is chosen to satisfy equation (1.9) then, for each virtual character $\phi \in R_\Gamma$ one has

$$R(\phi, \mathcal{O}_N) = (\mathcal{N}_{K/\mathbf{Q}}(\alpha \mid \phi)).(d_K)^{\frac{1}{2}degree(\phi)} \tag{3.10}$$

where d_K is the absolute discriminant of the field K and equation (3.10) is to be interpreted as an equation between fractional $\mathcal{O}_E$-ideals.

3.3 The generalised regulators

Turning now to the multiplicative theory we shall introduce pairings and discriminants in an exactly analogous fashion.

Defining a $\mathbf{Z}$-lattice $\mathcal{G}_N$ by the exactness of the sequence

$$0 \longrightarrow \left(\sum_{s\in S_\infty(N)} s\right) \longrightarrow Y_N \longrightarrow \mathcal{G}_N \longrightarrow 0$$

one has a natural identification of $\mathcal{G}_N$ with the $\mathbf{Z}$-linear dual $Hom_{\mathbf{Z}}(X_N, \mathbf{Z})$. Thus $\mathcal{G}_N \otimes_{\mathbf{Z}} \mathbf{Q}$ and $Hom_{\mathbf{Q}}(\mathcal{U}_N \otimes_{\mathbf{Z}} \mathbf{Q}, \mathbf{Q})$ are isomorphic $\mathbf{Q}[\Gamma]$-modules (see (3.5)) and there exists a non-degenerate $\mathbf{Q}$-bilinear Γ-pairing

$$b' : \mathcal{U}_N \otimes_{\mathbf{Z}} \mathbf{Q} \times \mathcal{G}_N \otimes_{\mathbf{Z}} \mathbf{Q} \longrightarrow \mathbf{C}$$

given by

$$(u, v) \longmapsto log \parallel u \parallel_v \qquad u \in \mathcal{U}_N, \quad v \in S_\infty(N)$$

where here $|| \quad ||_v$ denotes the canonically normalised absolute value at the place v. For any (complex) character ϕ of Γ, corresponding to the $E[\Gamma]$-module V_ϕ, and any $\mathcal{O}_E[\Gamma]$-lattice H spanning V_ϕ one has a regulator $R(H,\mathcal{U}_N)$ defined just as in the additive case but here with respect to the extended pairing b'_{V_ϕ} and suitable sublattices of

$$Hom_{E[\Gamma]}(V_\phi, E \otimes_{\mathbf{Z}} \mathcal{U}_N)$$

and

$$V_\phi \otimes_{E[\Gamma]} (E \otimes_{\mathbf{Z}} \mathcal{G}_N).$$

(Note that in this case, if $K = \mathbf{Q}$ and $\phi = \epsilon$ is the identity character of Γ then

$$Hom_{E[\Gamma]}(V_\epsilon, E \otimes_{\mathbf{Z}} \mathcal{U}_N) \;=\; V_\epsilon \otimes_{E[\Gamma]} (E \otimes_{\mathbf{Z}} \mathcal{G}_N) \;=\; (0)$$

and we set $R(H,\mathcal{U}_N) = \mathcal{O}_E$).

Similarly, if Γ is admissible then, after making a choice of $\mathcal{O}_E[\Gamma]$-lattices $\{T'_\chi\}_{\chi\in\Gamma^\dagger}$ as above, one defines lattices T'_ϕ for each (complex) actual character ϕ of Γ, and then defines a regulator.

Definition 3.11 Let Γ be admissible. Then for any (complex) character ϕ of Γ the generalised regulator of $\mathcal{U}_N$ with respect to ϕ is

$$R(\phi,\mathcal{U}_N) = R(T'_\phi,\mathcal{U}_N).$$

Just as with the additive theory one has the product expression

$$R(\phi,\mathcal{U}_N) \;=\; \prod_{\chi\in\Gamma^\dagger} R(\chi,\mathcal{U}_N)^{<\chi,\phi>}$$

valid for each character ϕ of Γ, and this allows a natural extension of the notion of generalised regulator to all virtual characters $\phi \in R_\Gamma$.

Remark 3.12 Each regulator $R(\phi,\mathcal{U}_N)$ is a rank one $\mathcal{O}_E$-lattice in $\mathbf{C}$. Multiplication here therefore is that induced on the set of such lattices by the usual product in $\mathbf{C}$.

Whilst the definition of regulator given here is in general dependent upon the choice of lattices $\{T'_\chi\}_{\chi\in\Gamma^\dagger}$ the theorems to be stated remain valid for any choice as above. Of course if Γ is abelian then the definition can again be made canonical.

We add a number of remarks concerning the comparison of the above regulator with that introduced by Tate in his treatment of the Stark conjectures (see Tate (1984)). Our remarks are not very precise as this would require further lengthy definitions. Firstly our regulator is not as in Tate's approach a complex number modulo E^* but is a (transcendental) rank one $\mathcal{O}_E$-lattice. Such rank one lattices inside a given one dimensional E-subspace of $\mathbf{C}$ have divisibility and integrality properties and, as we shall presently see, this enables one to formulate integral variants of theorems, problems and conjectures relating to the Stark conjectures. Aside from this any rank one $\mathcal{O}_E$-lattice defines a class in Cl_E, its Steinitz class, and we shall later discuss an interpretation of the class of a regulator in terms of Galois modules. Moreover there is also an action of $Gal(E/\mathbf{Q})$; for $\eta \in Gal(E/\mathbf{Q})$ there is an η-semilinear isomorphism

$$R(H,\mathcal{U}_N) \xrightarrow{\eta} R(H^\eta,\mathcal{U}_N) \qquad (3.13)(a)$$

or

$$R(\phi,\mathcal{U}_N) \xrightarrow{\eta} R(\phi^\eta,\mathcal{U}_N). \qquad (3.13)(b)$$

For each (complex) character ϕ of Γ we now let $L(s,\phi)$ denote the Artin L-function obtained by omitting from $\tilde{L}(s,\phi)$ the Euler factors corresponding to all archimedean places of N.

Then the E-vector space in which $R(\phi,\mathcal{U}_N)$ lies is generated by the regulator of Tate at $\overline{\phi}$, and conjecturally (Stark's Conjecture) by the leading non-zero coefficient at $s = 0$ of $L(s,\overline{\phi})$. Moreover the $\mathcal{O}_E$-lattice spanned by (any choice of complex representative of) Tate's regulator can be recovered as a regulator $R(\phi,\kappa(X_N))$ for a suitable isomorphism κ as in (3.5).

3.4 On some parallel results in the additive and multiplicative theories

In the remainder of this paragraph we shall use the resolvents and regulators introduced here to make more precise some remarkable parallels that exist between the additive and multiplicative theories. With this aim in mind we shall first give a module theoretic interpretation of the generalised resolvents by using them to prove Theorem 1. But for this we shall need a reformulation of the notion of factor-equivalence.

Let G denote an arbitrary finite group with L and M finitely generated $\mathbf{Z}[G]$-lattices that span the same $\mathbf{Q}[G]$-space. As always E denotes a number field, Galois over $\mathbf{Q}$, over which each element of $G^\dagger$ can be realised. For each subgroup $H \leq G$ we write e_H for the idempotent of $\mathbf{Q}[G]$ defined by

$$e_H = \frac{1}{ord(H)} \sum_{h \in H} h.$$

Recall that, for each subgroup $H \leq G$, ρ_H denotes the character of the permutation representation of G obtained from the G-set $H \setminus G$. For each subgroup $H \leq G$, we now set

$$J_H(L) = [Hom_{\mathcal{O}_E[G]}(e_H\mathcal{O}_E[G], L \otimes_{\mathbf{Z}} \mathcal{O}_E) : Hom_{\mathcal{O}_E[G]}(T_{\rho_H}, L \otimes_{\mathbf{Z}} \mathcal{O}_E)]_{\mathcal{O}_E}$$

where here T_{ρ_H} is the $\mathcal{O}_E[G]$-lattice (spanning $e_H E[G]$) as defined immediately prior to Definition 3.8. For each subgroup $H \leq G$ we shall also set

$$J_H(L, M) = J_H(L).(J_H(M))^{-1} = J_{n,H}(L, M).(J_{d,H}(L, M))^{-1}$$

where, for each $H \leq G$,

$$J_{d,H}(L, M) = \\ [Hom_{\mathcal{O}_E[G]}(T_{(\rho_H)}, L \otimes_{\mathbf{Z}} \mathcal{O}_E) : Hom_{\mathcal{O}_E[G]}(T_{(\rho_H)}, M \otimes_{\mathbf{Z}} \mathcal{O}_E)]_{\mathcal{O}_E}$$

and

$$J_{n,H}(L, M) = \\ [Hom_{\mathcal{O}_E[G]}(e_H\mathcal{O}_E[G], L \otimes_{\mathbf{Z}} \mathcal{O}_E) : Hom_{\mathcal{O}_E[G]},(e_H\mathcal{O}_E[G], M \otimes_{\mathbf{Z}} \mathcal{O}_E)]_{\mathcal{O}_E} .$$

Lemma 3.14 Using the above notation, the function on $\mathcal{S}(G)$ defined for each subgroup H by

$$H \setminus G \longmapsto J_{d,H}(L, M)$$

is factorisable.

Proof For each subgroup $H \leq G$ one has

$$J_{d,H}(L, M) = \prod_{\chi \in G^\dagger} g(\chi)^{<\chi,\rho_H>}$$

where g is the homomorphism from R_G to $\mathcal{I}(E)$ defined at each $\chi \in G^\dagger$ by

$$g(\chi) = [Hom_{\mathcal{O}_E[G]}(T_\chi, L \otimes_{\mathbf{Z}} \mathcal{O}_E) : Hom_{\mathcal{O}_E[G]}(T_\chi, M \otimes_{\mathbf{Z}} \mathcal{O}_E)]_{\mathcal{O}_E} .$$

□

Furthermore, for each subgroup $H \leq G$,

$$J_{n,H}(L, M) = \left([L^H : M^H]_{\mathbf{Z}}\right) \mathcal{O}_E$$

and hence, defining a function $J(L, M)$ on $\mathcal{S}(G)$ by

$$H \setminus G \overset{J(L,M)}{\longmapsto} J_H(L, M)$$

one has the equivalence

$$L \wedge_G M \iff J(L, M) \text{ is factorisable.} \tag{3.15}$$

If now L and M span isomorphic (but unequal) $\mathbb{Q}[G]$-spaces then, for any $\mathbb{Z}[G]$-embeddings $\imath$ and $\jmath$ of L into $M \otimes_{\mathbb{Z}} \mathbb{Q}$ one has

$$J(\imath(L), M) = J(\jmath(L), M)$$

and the symbol $J(L, M)$ is to be interpreted as this function. Equivalence (3.15) is then still valid.

We now discuss the proof of Theorem 1, the notation of which we shall continue to use. Using equivalence (3.15) it is clearly sufficient to give a factorisation of the function $J(\mathcal{O}_K[\Gamma], \mathcal{O}_N)$. For this we shall introduce the 'adjusted' global Galois–Gauss sum, defined for any virtual character $\phi \in R_\Gamma$ by

$$\tilde{\tau}(K, \phi) = \tau(N/K, \phi)(d_K^{\frac{1}{2}})^{degree(\phi)}$$

where here $\tau(N/K, \phi)$ is the Galois–Gauss sum as defined in §1 (1.13) and $d_K^{\frac{1}{2}}$ is an appropriately normalised square root of the absolute discriminant d_K of the field K. In other words $\tilde{\tau}(K, \phi)$ is the Galois–Gauss sum of the character of $\Omega_{\mathbb{Q}}$ induced from the character ϕ of Ω_K. (Recall also equality (3.10)). We define a homomorphism $g^+_{N/K}$ on R_Γ taking values in $\mathcal{I}(E)$ by specifying its image at each $\chi \in \Gamma^\dagger$:

$$\chi \overset{g^+_{N/K}}{\longmapsto} \frac{\tilde{\tau}(K, \chi)}{R(\chi, \mathcal{O}_N)}.$$

We claim that $g^+_{N/K}$ satisfies the conditions (3.1) with respect to the function $J(\mathcal{O}_K[\Gamma], \mathcal{O}_N)$. Explicitly therefore, at each subgroup $H \leq G$, we claim that

$$J_H(\mathcal{O}_K[\Gamma], \mathcal{O}_N) = \prod_{\chi \in \Gamma^\dagger} g^+_{N/K}(\chi)^{<\rho_H, \chi>} \quad , \tag{3.16)(a}$$

while for each $\chi \in \Gamma^\dagger$ and $\eta \in Gal(E/\mathbb{Q})$, one has

$$g^+_{N/K}(\chi)^\eta = g^+_{N/K}(\chi^\eta) \tag{3.16)(b}$$

where here the Galois action on each $R(\chi, \mathcal{O}_N)$ is as in (3.13). Of these conditions we shall discuss only (3.16)(a) – the other being a fairly straightforward consequence of known results. Before this however we note that by means of (3.16)(a) together with the known localisation behaviour of the function $J(\mathcal{O}_K[\Gamma], \mathcal{O}_N)$ one can in fact say much about the $\mathcal{O}_E$-ideals $g^+_{N/K}(\chi)$:

Theorem 4 (Fröhlich) Let N/K denote a finite Galois extension of number fields of group Γ. Assume that Γ is admissible and let χ be a complex character of Γ. Then any prime ideal of $\mathcal{O}_E$ which occurs in the decomposition of the (fractional) $\mathcal{O}_E$-ideal $g^+_{N/K}(\chi)$ has the same residue characteristic as some prime ideal of K which wildly ramifies in the extension N/K. In particular therefore the support of $g^+_{N/K}(\chi)$ only involves prime divisors of $ord(\Gamma)$.

We shall later see a multiplicative analogue of Theorem 4.

Now to prove formula (3.16)(a) one uses two different product expressions. The first is of course (3.9) for the character $\phi = \rho_H$. But on the other hand one has by definition

$$\prod_{\chi \in \Gamma^\dagger} \tilde{\tau}(K, \chi)^{<\chi, \rho_H>} = \tilde{\tau}(K, \rho_H) \tag{3.17}$$

and, by direct computation,

$$\tilde{\tau}(K, \rho_H) = \mathcal{R}(e_H \mathcal{O}_E[\Gamma], \mathcal{O}_N).$$

Thus (3.16)(a) follows from (3.9) (with $\phi = \rho_H$) and (3.17) together with the obvious (formal) behaviour of discriminants under change of lattices.

Remark 3.18 With a strengthening of the notion of factorisabilty one can prove that the Galois Gauss sum ideals are actually uniquely determined by the factorisation (in the stronger sense) of the function $J(\mathcal{O}_K[\Gamma], \mathcal{O}_N)$. This result therefore gives a generalisation of Theorem 5 of §1 to the case of wildly ramified extensions, but lies outside the scope of this article.

To develop the analogy between the additive and multiplicative theories we now discuss Theorem 3 and a multiplicative analogue of Theorem 4. Thus until explicitly stated otherwise we shall assume the notations and conditions of Theorem 3. In particular therefore N is a totally real number field and $\Gamma = Gal(N/K)$ is an admissible group of odd order. In this context one can prove the existence of a homomorphism g on R_Γ which takes values in $\mathcal{I}(E)$ and satisfies conditions (3.1) but here with respect to the function defined on $\mathcal{S}(\Gamma)$ by

$$\Delta \backslash \Gamma \longmapsto J_\Delta(X_{N^\Delta}).J_\Delta(\mathcal{U}_{N^\Delta})^{-1}.h_{N^\Delta}$$

But, for each subgroup $\Delta \leq \Gamma$, one has $(X_N)^\Delta \cong X_{N^\Delta}$ and $(\mathcal{U}_N)^\Delta = \mathcal{U}_{N^\Delta}$, so that

$$J_\Delta(X_{N^\Delta}).J_\Delta(\mathcal{U}_{N^\Delta})^{-1} = J_\Delta(X_N).J_\Delta(\mathcal{U}_N)^{-1} = J_\Delta(X_N, \mathcal{U}_N)$$

and hence (using Lemma 3.14) the existence of such a homomorphism g is indeed sufficient to prove Theorem 3. However for a more natural interpretation of such a factorisation we consider the structure of Cl_N as a $\mathbf{Z}[\Gamma]$-module.

We write Cl_N^* for the maximal subgroup of Cl_N of order coprime to Γ and set $h_N^* = ord(Cl_N^*)$. We define a homomorphism h^* on R_Γ taking values in $\mathcal{I}(\mathbf{Q}(\chi)) \subset \mathcal{I}(E)$ by

$$\chi \stackrel{h^*}{\longmapsto} ord_{\mathbf{Z}[\chi]}\left((Cl_N^* \otimes_{\mathbf{Z}} \mathbf{Z}[\chi])^{(\chi)}\right) \qquad \chi \in \Gamma^\dagger$$

where here $\mathbf{Z}[\chi]$ denotes the ring of integers of $\mathbf{Q}(\chi)$, $ord_{\mathbf{Z}[\chi]}$ denotes the order ideal of a torsion $\mathbf{Z}[\chi]$-module, and a superscript (χ) denotes the χ-isotypic component of a $\mathbf{Z}[\chi]$-module.

Remark 3.19 Since Cl_N^* has order coprime to $ord(\Gamma)$ it is not difficult to show that h^* satisfies conditions (3.1) with respect to the function defined on $\mathcal{S}(\Gamma)$ by

$$\Delta \backslash \Gamma \longmapsto h^*_{(N^\Delta)}.$$

Theorem 5 (Fröhlich) Let N/K denote a finite Galois extension of number fields of group Γ. If N is totally real and Γ is an admissible group of odd order then there exists a homomorphism g on R_Γ which satisfies conditions (3.1) with respect to the function f_κ and is such that, for each character χ of Γ, the support of the fractional ideal

$$\left(\frac{g(\chi)}{h^*(\chi)}\right)$$

only involves prime divisors of $ord(\Gamma)$.

Remark 3.20 The reader should recall the result of Theorem 4.

As a specific example we take N to be an absolutely abelian field of odd degree with $K = \mathbf{Q}$. For each (complex) character ϕ of $\Gamma = Gal(N/\mathbf{Q})$ we write $L^*(0,\phi)$ for the leading non-zero coefficient in the Taylor expansion at $s = 0$ of the Artin L-function $L(s,\phi)$ as defined as the end of §3.3. In this case the homomorphism $g^\times_{N/\mathbf{Q}}$ defined on R_Γ by

$$\chi \longmapsto \frac{L^*(0,\overline{\chi})}{R(\chi,\mathcal{U}_N)}, \quad \text{all } \chi \in \Gamma^\dagger \tag{3.21}$$

satisfies conditions (3.1) with respect to the function f_κ (notation as in Theorem 3). However at the present time one knows only that any prime ideal occuring in the decomposition of the fractional ideals

$$\frac{g^{\times}_{N/\mathbf{Q}}(\chi)}{h^*(\chi)}$$

must divide $2.ord(\Gamma)$. (For more details of this special case see Fröhlich (1989)). This latter result is essentially a reinterpretation of the Gras conjecture, which is now known to be true (at least for any prime $p \neq 2$). But this new formulation has the advantage that it admits a natural conjectural generalisation in any case in which (some weak form of) Stark's conjectures are known to be true. That is, a natural conjectural generalisation exists for any class of extensions for which the function corresponding to (3.21) is known to satisfy condition (3.1)(c) – for example abelian extensions of quadratic imaginary fields or arbitrary Galois extensions of exponent 2. In this way one is therefore led to pose integral variants of conjectures and problems relating to Stark's conjectures.

We now turn to consider module theoretic interpretations for the factorising functions $g^+_{N/K}$ and $g^\times_{N/K}$. However for simplicity of exposition we shall restrict to the case that Γ has odd order. As usual we shall begin with the additive theory. Thus N/K is a finite Galois extension of number fields of group Γ which is now assumed to have odd order. As before we write $\mathcal{M}$ for an arbitrary maximal $\mathbf{Z}$-order in $\mathbf{Q}[\Gamma]$ satisfying $\mathcal{M} \supseteqq \mathbf{Z}[\Gamma]$. The locally-free class group $Cl(\mathcal{M})$ of $\mathcal{M}$ is a quotient of $Cl(\mathbf{Z}[\Gamma])$ and (following the ideal theoretic version of the Hom-description (1.5)) each class is represented by homomorphisms g on R_Γ satisfying for each character $\chi \in \Gamma^\dagger$

$$g(\chi) \in \mathcal{I}(\mathbf{Q}(\chi)), \qquad (3.22)(a)$$

whilst for each $\eta \in Gal(\mathbf{Q}(\chi)/\mathbf{Q})$

$$g(\chi)^\eta \;=\; g(\chi^\eta). \qquad (3.22)(b)$$

In particular, an element of $Cl(\mathcal{M})$ is trivial if, and only if, a representing homomorphism takes only principal ideal values (in the fields $\mathbf{Q}(\chi)$). For any finitely generated $\mathbf{Z}[\Gamma]$-lattice X we shall write $X^{\mathcal{M}}$ for the maximal $\mathbf{Z}[\Gamma]$-sublattice of X which is also an $\mathcal{M}$-module and let $(X^{\mathcal{M}})_{\mathcal{M}}$ denote the corresponding element of $Cl(\mathcal{M})$.

Theorem 6 (Fröhlich) Let N/K denote a finite Galois extension of number fields of group Γ. Assume that Γ is admissible and has odd order. Then the class $(\mathcal{O}_N{}^{\mathcal{M}})_{\mathcal{M}}$ is represented by the homomorphism $(g^{+}_{N/K})^{-1}$.

Thus one has an explicit module theoretic interpretation of a factorisation of the function $J(\mathcal{O}_K[\Gamma], \mathcal{O}_N)$. Moreover this has concrete arithmetical implications. For example, if Γ has prime power order (and is therefore admissible) then, as an immediate consequence of Theorem 4 together with the fact that all p-primary ideals in a cyclotomic field of p-power conductor are principal, one deduces that

$$(\mathcal{O}_N{}^{\mathcal{M}})_{\mathcal{M}} = 1, \tag{3.23}$$

i.e., that $\mathcal{O}_N{}^{\mathcal{M}}$ is a stably-free $\mathcal{M}$-module. Note that this result is related to that of Corollary 7 of §1. Indeed, if, for example, N/K is a tame abelian extension, then $\mathcal{O}_N{}^{\mathcal{M}}$ is isomorphic to $\mathcal{O}_N\mathcal{M}$ as an $\mathcal{M}$-module. In general however, and even for abelian extensions, neither $(\mathcal{O}_N{}^{\mathcal{M}})_{\mathcal{M}}$ or $(\mathcal{O}_N\mathcal{M})_{\mathcal{M}}$ is always trivial (c.f. Burns (1990)).

Yet again one can find an analogous multiplicative theorem. For this we impose the restrictions that N is a totally real absolutely abelian number field and that $\Gamma = Gal(N/\mathbf{Q})$ has odd order. We write T_Γ for the principal ideal of $\mathbf{Z}[\Gamma]$ generated by the trace element

$$\sum_{\gamma\in\Gamma} \gamma \in \mathbf{Z}[\Gamma].$$

We define the $\mathbf{Z}$-lattice $\mathcal{A}_\Gamma$ by the exactness of the sequence

$$0 \longrightarrow T_\Gamma \longrightarrow \mathbf{Z}[\Gamma] \longrightarrow \mathcal{A}_\Gamma \longrightarrow 0.$$

This lattice $\mathcal{A}_\Gamma$ is a $\mathbf{Z}$-order in the $\mathbf{Q}$-algebra $\mathbf{Q}[\Gamma]/(T_\Gamma)$, and $\mathcal{U}_N$ is a rank one $\mathcal{A}_\Gamma$-lattice. Let $\mathcal{M}'$ denote the maximal $\mathbf{Z}$-order in $\mathbf{Q}[\Gamma]/(T_\Gamma)$. By means of the Hom-language, elements of the locally-free classgroup $Cl(\mathcal{M}')$ of $\mathcal{M}'$ are again represented by homomorphisms on R_Γ satisfying the conditions (3.22).

Theorem 7 (Fröhlich) Let N denote an absolutely abelian number field of odd degree and set $\Gamma = Gal(N/\mathbf{Q})$. Then the class $(\mathcal{U}_N{}^{\mathcal{M}'})_{\mathcal{M}'}$ is represented by the homomorphism $(g^{\times}_{N/\mathbf{Q}})^{-1}$.

Thus, for example, if $Gal(N/\mathbf{Q})$ is an abelian group of odd prime power degree then the support of the class $(\mathcal{U}_N{}^{\mathcal{M}'})_{\mathcal{M}'}$ must lie above $2h^*_N$ – this can be a very strong restriction on possible module-structures. In general the question of the precise support of $(\mathcal{U}_N{}^{\mathcal{M}'})_{\mathcal{M}'}$ lying above 2 remains unanswered.

3.5 Open problems

We end this section by making explicit mention of three open problems.

Problem 1 Let N/K be a finite Galois extension of number fields with $\Gamma = Gal(N/K)$. For the additive theory, Theorem 2 of §1 implies that if Γ is abelian then $\mathcal{O}_N$ is a locally-free $\mathbf{Z}[\Gamma]$-module if, and only if, it is a free $\mathbf{Z}[\Gamma]$-module. We now raise a possible multiplicative analogue of this result. Assume then that N is a real absolutely abelian field and that $K = \mathbf{Q}$. The group Cl_N^* (as defined immediately prior to Remark 3.19) is an $\mathcal{A}_\Gamma$-module, the quotient of two locally-free $\mathcal{A}_\Gamma$-modules. Thus Cl_N^* defines in a natural way a class $(Cl_N^*)_{\mathcal{A}_\Gamma} \in Cl(\mathcal{A}_\Gamma)$. If $\mathcal{U}_N$ is locally-free as an $\mathcal{A}_\Gamma$-module, is the equality

$$(\mathcal{U}_N)_{\mathcal{A}_\Gamma}.(Cl_N^*)_{\mathcal{A}_\Gamma}^{-1} = 1$$

necessarily true? In all known examples when $\mathcal{U}_N$ is locally free over $\mathcal{A}_\Gamma$ (for example extensions $N/\mathbf{Q}$ of prime degree) this is certainly true (except possibly for the 2-primary parts), and the verification of this points to a systematic general method.

Problem 2 The additive theory discussed in this paragraph is more straightforward than the corresponding multiplicative theory because the localisation behaviour of the function which factorises $J(\mathcal{O}_K[\Gamma], \mathcal{O}_N)$ (i.e., $g^{+}_{N/K}$) is completely understood. Thus we ask, for the multiplicative theory (for example, N real absolutely abelian and $K = \mathbf{Q}$) can one obtain a factorisation of the function

$$\Delta \backslash \Gamma \longmapsto J_\Delta(X_N, \mathcal{U}_N) h_{N^\Delta} \quad , \qquad \Delta \leq \Gamma$$

each localisation of which has a natural module theoretic interpretation in terms perhaps of either local units or of local principal units? More specifically, can one decompose the function $g^{\times}_{N/K}$ into a product of local factors in any 'natural' way?

Problem 3 Again assume that N is a real absolutely abelian field with $K = \mathbf{Q}$ and suppose in addition that $N/\mathbf{Q}$ is of odd degree. In view of the discussion following Theorem 5 and, for example, the result of Theorem 7 it is of considerable interest to clarify the rôle of 2 in the decomposition of the fractional $\mathcal{O}_{\mathbf{Q}(\chi)}$-ideal $g^{\times}_{N/\mathbf{Q}}(\chi)/h^*(\chi)$ for each character $\chi \in \Gamma^{\dagger}$.

Remark 3.24 (added 10 December 1989) Chinburg has recently indicated how one can deduce the Gras conjecture at $p = 2$ from work of Greenberg, Gillard and Wiles (Chinburg (1989d)). By combining this argument of Chinburg together with the approach adopted by Fröhlich (see §6 and §7 of Fröhlich

(1989)), one can now satisfactorily answer all of the questions concerning behaviour at the prime $p = 2$ that were explicitly mentioned in this paragraph. For example, in the setting of Open Problem 3, one can now prove that no prime ideal which lies above 2 can occur in the decomposition of any fractional ideal $g^{\times}_{N/\mathbf{Q}}(\chi)/h^*(\chi)$. The point in all of this is that there are various differing definitions of cyclotomic units which, *a priori*, may affect the truth or otherwise of the 'Gras conjecture'-type result at $p = 2$ that is needed to pursue the Fröhlich approach. In fact one can check that the argument suggested by Chinburg does indeed work for the definition of units underlying Fröhlich's approach. For more details of Chinburg's argument see Remark 4.23.

Remark 3.25 (added 14 May 1990) J. Ritter and A. Weiss have now proved that all finite groups are admissible (in the sense of this paragraph).

4 ADDITIVE–MULTIPLICATIVE GALOIS STRUCTURES

Let N/K denote a finite Galois extension of number fields of Galois group Γ. (We shall later also consider global function fields). As in §3 we write $S_\infty(N)$ for the set of infinite places of N. Let S denote any finite Γ-stable set of places of N large enough to satisfy both of the following conditions:

(4.1) S contains $S_\infty(N)$ and the places of N which ramify over K, as well as at least one finite place of N, and
(4.2) The S-class number of each subfield of N containing K is equal to 1.

In this paragraph we study the Γ-module structure of objects which incorporate the Γ-module structures of both $\mathcal{O}_N$ and of the multiplicative group $U_S = U_{N,S}$ of S-units of N.

To be more precise we need more notation. We denote by N_v the completion of N at the place v (of N) and by N_v^* the multiplicative group of N_v. For any non-archimedean place v we let U_v denote the group of units of the valuation ring of N_v. The group $J_S = J_{N,S}$ of S-ideles of N is therefore

$$J_S = \prod_{v\in S} N_v^* \times \prod_{v\notin S} U_v .$$

The objects we consider in this section are exact sequences

$$0 \longrightarrow U_S \longrightarrow J_{S,f} \longrightarrow C_{S,f} \longrightarrow 0 \tag{4.3}$$

of certain finitely generated $\mathbf{Z}[\Gamma]$ modules constructed from $\mathcal{O}_N$, U_S and J_S, and such that the sequence (4.3) has the same cohomology as the basic exact sequence

$$0 \longrightarrow U_S \longrightarrow J_S \longrightarrow C \longrightarrow 0\,, \tag{4.4}$$

in which, as a consequence of condition (4.2), $C = C_N$ is the idele classgroup of N. In §4.1 we discuss how by means of an 'approximating sequence' (4.3) one can define elements $\Omega(N/K,1)$, $\Omega(N/K,2)$, and $\Omega(N/K,3)$ of the locally-free class group $Cl(\mathbf{Z}[\Gamma])$ which are associated to the Γ-module structures of $C_{S,f}$, $J_{S,f}$ and U_S respectively. Remarkably these elements do not depend on the precise choice of the sequence (4.3) or on any of the other arbitrary choices made in the course of their definitions and are therefore invariants of the extension N/K. Having defined the invariants, in §4.2 we discuss conjectures and theorems relating them to other aspects of the arithmetic of N/K. In particular we discuss the conjectured relationship between these invariants and the root number class defined in §1.

In fact much of the theory to be discussed here is also valid *mutatis mutandis* for the case in which N/K is a finite Galois extension of global function fields and in the text we shall make remarks to this effect whenever appropriate. Complete proofs of the new assertions made here about function fields will appear in Chinburg (1989c).

4.1 Definition of the $\Omega(N/K,i)$

Let N/K be a finite Galois extension of number fields of Galois group Γ. We shall write S_f for the set $S \setminus S_\infty(N)$ of finite places contained in S. By condition (4.1) therefore $S_f \neq \emptyset$. In order to construct the sequences (4.3) we need two technical lemmata. The first of these is not difficult to prove and the second is taken from Chinburg (1985).

Lemma 4.5 After enlarging S (if necessary), one can find an element $\alpha \in U_S \cap K$ and a free rank one $\mathcal{O}_K[\Gamma]$-submodule F of $\mathcal{O}_N$ with the following property:

($*$) α is a non-unit at each place in S_f and $\alpha^3\mathcal{O}_N \subset F \subset \alpha^2\mathcal{O}_N$.

Furthermore, for all F and α for which ($*$) is true the closure $\overline{1+F}$ of $1+F$ in $\prod_{v\in S_f} N_v^*$ has a filtration by the modules

$$T(m) = \overline{(1+\alpha^m F)}/\overline{(1+\alpha^{m+1}F)}$$

for $m \geq 0$, where each $T(m)$ is Γ-isomorphic to $F/\alpha F$. In particular, $\overline{1+F}$ is a cohomologically trivial Γ-module of finite index in $\prod_{v\in S_f} U_v$.

For each place v of N we write Γ_v for the decomposition group of v in Γ. In particular therefore if v is an infinite place then Γ_v has order 1 or 2.

Lemma 4.6 Assume the above notations. Then for each infinite place $v \in S_\infty(N)$ there exists a finitely generated $\mathbf{Z}[\Gamma_v]$-submodule W_v of N_v^* such that $U_S \subsetneqq W_v$, the quotient group W_v/U_S is $\mathbf{Z}$-torsion free, and the injection $W_v \hookrightarrow N_v^*$ induces an isomorphism in Γ_v-cohomology. Furthermore, for each element $\gamma \in \Gamma$ and place $v \in S_\infty(N)$ one may also require $\gamma(W_v) = W_{\gamma(v)}$.

Remark 4.7 Of course if Γ_v is trivial then one can take $W_v = U_S$ to satisfy the conditions of Lemma 4.6!

Using Lemmata 4.5 and 4.6 we can now define groups J'_S and $J_{S,f}$ that are 'approximations' to J_S by

$$\begin{array}{ccccccc} J_S & = & \prod_{v\notin S} U_v & \times & \prod_{v\in S_f} N_v^* & \times & \prod_{v\in S_\infty(N)} N_v^* \\ & & \downarrow & & \downarrow & & \| \\ J'_S & = & 1 & \times & \left(\dfrac{\prod_{v\in S_f} N_v^*}{1+F}\right) & \times & \prod_{v\in S_\infty(N)} N_v^* \\ & & \| & & \| & & \uparrow \\ J_{S,f} & = & 1 & \times & \left(\dfrac{\prod_{v\in S_f} N_v^*}{1+F}\right) & \times & \prod_{v\in S_\infty(N)} W_v \end{array} \qquad (4.8)$$

Note that with this definition $J_{S,f}$ is finitely generated and has the same cohomology as J_S. Indeed for the second and third factors in the product decomposition of J_S and $J_{S,f}$ in (4.8) the equality of cohomologies is clear from the assumptions on F and on each W_v. But if $v \notin S$ then v is unramified over K and hence U_v is a cohomologically trivial Γ_v-module so that the cohomologies do indeed agree.

By the assumptions on each W_v, the diagonal embedding $U_S \hookrightarrow J_S$ induces embeddings $U_S \hookrightarrow J_{S,f}$ and $U_S \hookrightarrow J'_S$. With respect to these embeddings, we define quotients C'_S and $C_{S,f}$ of J'_S and $J_{S,f}$ respectively by means of the exactness of all rows in the following commutative diagram:

$$\begin{array}{ccccccccc} 0 & \longrightarrow & U_S & \longrightarrow & J_S & \longrightarrow & C & \longrightarrow & 0 \\ & & \| & & \downarrow & & \downarrow & & \\ 0 & \longrightarrow & U_S & \longrightarrow & J'_S & \longrightarrow & C'_S & \longrightarrow & 0 \\ & & \| & & \uparrow & & \uparrow & & \\ 0 & \longrightarrow & U_S & \longrightarrow & J_{S,f} & \longrightarrow & C_{S,f} & \longrightarrow & 0 \end{array} \qquad \begin{array}{c} (4.9)(a) \\ \\ (4.9)(b) \\ \\ (4.9)(c) \end{array}$$

In diagram (4.9) all vertical homomorphisms, either up or down, are the natural projection maps and hence induce isomorphisms in cohomology with respect to all subgroups of Γ. Moreover each module in row (4.9)(c) is by definition finitely generated. In this way therefore one constructs 'approximating sequences' of the form (4.3).

In the above construction one could consider replacing $\overline{1+F}$ by an arbitrary cohomologically trivial Γ-submodule T of finite index inside $\prod_{v \in S_f} U_v$. The reason for considering T of the form $\overline{1+F}$ is that the invariants $\Omega(N/K,2)$ and $\Omega(N/K,1)$ which we define below using the sequence (4.9)(c) will then not depend on the choice of F. The rôle played by $\mathcal{O}_N$ in the above construction is thus to provide a family of T of the above kind which all lead to the same invariants $\Omega(N/K,2)$ and $\Omega(N/K,1)$, namely the family of all $\overline{1+F}$ with F as in Lemma 4.5.

We let Y_S denote the additive group $Div_{\mathbf{Z}}(S)$ of $\mathbf{Z}$-divisors supported on the set S with X_S the subgroup of Y_S consisting of divisors of degree 0. That is, X_S is defined by means of the exactness of the sequence

$$0 \longrightarrow X_S \longrightarrow Y_S \xrightarrow{\epsilon} \mathbf{Z} \longrightarrow 0 \tag{4.10}$$

in which $\epsilon : Y_S \longrightarrow \mathbf{Z}$ is the homomorphism defined at each place $v \in S$ by

$$v \overset{\epsilon}{\longmapsto} 1.$$

In particular therefore, recalling the notation of §3, $X_{S_\infty(N)} = X_N$ and $Y_{S_\infty(N)} = Y_N$.

In fact the canonical exact sequence (4.10) is closely related to the 'approximating sequence' (4.9)(c) as constructed above. Using the techniques of Tate (1966) this connection is expressed by means of the following exact diagram of $\mathbf{Z}[\Gamma]$-modules

$$\begin{array}{ccccccccccc}
 & & 0 & & 0 & & 0 & & 0 & & \\
 & & \downarrow & & \downarrow & & \downarrow & & \downarrow & & \\
0 & \longrightarrow & U_S & \longrightarrow & A_3 & \longrightarrow & B_3 & \longrightarrow & X_S & \longrightarrow & 0 \qquad (4.11)(a) \\
 & & \downarrow & & \downarrow & & \downarrow & & \downarrow & & \\
0 & \longrightarrow & J_{S,f} & \longrightarrow & A_2 & \longrightarrow & B_2 & \longrightarrow & Y_S & \longrightarrow & 0 \qquad (4.11)(b) \\
 & & \downarrow & & \downarrow & & \downarrow & & \downarrow{\scriptstyle \epsilon} & & \\
0 & \longrightarrow & C_{S,f} & \longrightarrow & A_1 & \longrightarrow & B_1 & \longrightarrow & \mathbf{Z} & \longrightarrow & 0 \qquad (4.11)(c) \\
 & & \downarrow & & \downarrow & & \downarrow & & \downarrow & & \\
 & & 0 & & 0 & & 0 & & 0 & &
\end{array}$$

in which the left hand column is the sequence (4.9)(c), the right hand column is the sequence (4.10), and in addition the following three conditions (4.12), (4.13) and (4.15) are satisfied.

Condition 4.12 The $\mathbf{Z}[\Gamma]$-modules A_i and B_i for $i = 1, 2$, and 3 are finitely generated and of finite projective dimension.

In the sequel we shall write $\hat{H}^i(\Gamma, M)$ for the ith Tate cohomology group of the Γ-module M. We write $H^i(\Gamma, M)$ for the corresponding ordinary cohomology group. Thus $\hat{H}^i(\Gamma, M) = H^i(\Gamma, M)$ for each integer $i \geq 1$.

Condition 4.13 The extension class

$$\alpha_{1,f} \in Ext^2_\Gamma(\mathbf{Z}, C_{S,f}) = \hat{H}^2(\Gamma, C_{S,f})$$

of the sequence (4.11)(c) is the pullback via diagram (4.9) of the canonical class in $\hat{H}^2(\Gamma, C)$ as defined in Tate (1966).

Before stating the third condition we must make some preliminary remarks. Since Y_S is $\mathbf{Z}$-torsion free one has $Ext^q_{\mathbf{Z}}(Y_S, J_S) = 0$ for $q > 0$. Hence the spectral sequence

$$H^p(\Gamma, Ext^q_{\mathbf{Z}}(Y_S, J_S)) \implies Ext^{p+q}_\Gamma(Y_S, J_S)$$

degenerates, and in particular

$$Ext^2_\Gamma(Y_S, J_S) = H^2(\Gamma, Hom_{\mathbf{Z}}(Y_S, J_S)).$$

But if S_0 denotes a set of representatives for the Γ-orbits of S then

$$Y_S \cong \bigoplus_{v \in S_0} Ind^\Gamma_{\Gamma_v}(\mathbf{Z}),$$

and hence by Shapiro's Lemma

$$Ext^2_\Gamma(Y_S, J_S) \cong \bigoplus_{v \in S_0} \hat{H}^2(\Gamma_v, J_S). \tag{4.14}$$

Condition 4.15 By means of the identification (4.14) the extension class

$$\alpha_{2,f} \in Ext^2_\Gamma(Y_S, J_{S,f})$$

of the sequence (4.11)(b) is the pullback via diagram (4.9) of the class

$$\alpha_2 = \bigoplus_{v \in S_0} i_v^*(\alpha_v) \in \bigoplus_{v \in S_0} \hat{H}^2(\Gamma_v, J_S)$$

where here, i_v^* is the map induced on cohomologies by the natural injection $N_v \hookrightarrow J_S$, and each α_v is the canonical class in $\hat{H}^2(\Gamma_v, N_v^*)$ as defined in Tate (1966).

One now uses diagram (4.11) to define the invariants $\Omega(N/K,i)$. For this we first note that any finitely generated $\mathbf{Z}[\Gamma]$-module X of finite projective dimension defines in a natural way a class $(X)_{\mathbf{Z}[\Gamma]} \in Cl(\mathbf{Z}[\Gamma])$. Indeed, by standard module theoretic results, for any such module X there exist finitely generated locally-free $\mathbf{Z}[\Gamma]$-modules X_1 and X_2 such that the sequence

$$0 \longrightarrow X_1 \longrightarrow X_2 \longrightarrow X \longrightarrow 0 \tag{4.16}$$

is exact and one defines

$$(X)_{\mathbf{Z}[\Gamma]} = (X_2)_{\mathbf{Z}[\Gamma]} - (X_1)_{\mathbf{Z}[\Gamma]} .$$

By Schanuel's Lemma this class $(X)_{\mathbf{Z}[\Gamma]}$ is indeed independent of the choices of locally-free $\mathbf{Z}[\Gamma]$-modules X_1 and X_2 satisfying (4.16).

Definition 4.17 For any Galois extension N/K of number fields one defines

$$\Omega(N/K,i) = (A_i)_{\mathbf{Z}[\Gamma]} - (B_i)_{\mathbf{Z}[\Gamma]}$$

for $i = 1, 2$ and 3.

The main theorem of this section is then

Theorem 1 (Chinburg (1985)) Assume the above notations. Then the classes $\Omega(N/K,i)$ for $i = 1, 2$ and 3 depend only upon the extension N/K. In particular they do not depend upon the choice of the diagram (4.11), the W_v for $v \in S_\infty(N)$, F, S_0 or S satisfying the conditions discussed above.

Clearly, as a direct consequence of diagram (4.11), for any Galois extension of number fields N/K one has the equality

Theorem 2 $\Omega(N/K,2) = \Omega(N/K,1) + \Omega(N/K,3)$.

Finally we briefly remark on the function field case. Thus N/K is now a finite Galois extension of global function fields with $\Gamma = Gal(N/K)$. Let p be the characteristic of K. Let $S_\infty = S_\infty(N)$ be any finite non-empty Γ-stable set of places of N which are split over K and which have sufficiently divisible residue field degrees over $\mathbf{Z}/p\mathbf{Z}$. With such a choice of places S_∞, we let $\mathcal{O}_N$

denote the ring of elements of N which are regular at all places outside S_∞, and we define $\mathcal{O}_K = K \cap \mathcal{O}_N$. Then the elements $\Omega(N/K, i)$ of $Cl(\mathbf{Z}[\Gamma])$ for $i = 1, 2$ and 3 can be defined exactly as in the number field case, and are again only dependent upon the extension N/K. (The assumption that the residue field degrees of the places in S_∞ are sufficiently divisible is needed to ensure that the invariants $\Omega(N/K, 2)$ and $\Omega(N/K, 3)$ do not depend on the choice of the free $\mathcal{O}_K[\Gamma]$-module F).

4.2 The theorems and conjectures

Having defined in §4.1 the invariants $\Omega(N/K, i)$, in this section we shall consider connections with other aspects of the arithmetic of N/K. In particular we shall discuss the conjectural relationship between $\Omega(N/K, i)$ for each $i = 1, 2$ and 3 and the root number class introduced in §1.

Conjecture 3 (Chinburg) If N/K is a finite Galois extension of global fields then

(i) $\Omega(N/K, 3) = W_{N/K}$.
(ii) $\Omega(N/K, 2) = W_{N/K}$.
(iii) $\Omega(N/K, 1) = 0$.

Note that, because of Theorem 2, any two of the conjectures 3(i), 3(ii) and 3(iii) imply the third.

Remark 4.18 Suppose χ is a complex valued character of Γ. If N/K is an extension of number fields recall that $\tilde{L}(s, \chi)$ denotes the extended Artin L-function of χ as discussed in §1; if N/K is an extension of global function fields we let $\tilde{L}(s, \chi)$ denote the usual Artin L-function of χ. The (Artin) root number of χ is a complex constant of absolute value 1 which is defined by the functional equation (1.6) (which is also valid in the global function field case). In §1 we discussed how, for tame Galois extensions of number fields, one can define an element $W_{N/K} \in Cl(\mathbf{Z}[\Gamma])$ using the root numbers associated to the symplectic characters χ of Γ. More generally in Fröhlich (1978) it is shown how to define a class $W_{N/K}$ coming from the symplectic root numbers of N/K without the tameness hypothesis. Fröhlich's definition of $W_{N/K}$ is recounted in Chinburg (1989a) and this definition applies equally well to function fields.

The following result connects $\Omega(N/K, 2)$ and Conjecture 3(ii) to the classical theory of the Galois structure of the ring of integers of tame Galois extensions of number fields as discussed in §1. The proof of this result given in Chinburg (1989c) is a modification of the proof given in Chinburg (1985) for number fields.

Theorem 4 (Chinburg (1989c)) Let N/K denote a finite tame Galois extension of global fields. Let $\Gamma = Gal(N/K)$. If F is any free rank one $\mathcal{O}_K[\Gamma]$-submodule of $\mathcal{O}_N$ then the quotient $\mathcal{O}_N/F$ is a finite $\mathbf{Z}[\Gamma]$-module of finite projective dimension and

$$\Omega(N/K,2) = (\mathcal{O}_N/F)_{\mathbf{Z}[\Gamma]} \,.$$

But if in particular N/K is a tame extension of number fields then $(F)_{\mathbf{Z}[\Gamma]} = 0$ and hence $\Omega(N/K,2) = (\mathcal{O}_N)_{\mathbf{Z}[\Gamma]}$. One can therefore regard $\Omega(N/K,2)$ as generalising the class $(\mathcal{O}_N)_{\mathbf{Z}[\Gamma]}$ to wildly ramified Galois extensions as well as to function fields. Conjecture 3(ii) thus conjecturally generalises Theorem 2 of §1.

To state what is known concerning Conjecture 3 we must recall some further definitions.

Let $D(\mathbf{Z}[\Gamma])$ be the kernel subgroup of $Cl(\mathbf{Z}[\Gamma])$ as introduced in §1. The argument which Fröhlich gives for number fields in Fröhlich (1983) (Proposition III 3.1(i)) implies that $W_{N/K} \in D(\mathbf{Z}[\Gamma])$ for all tame extensions N/K of global fields. A wildly ramified extension N/K of number fields for which $W_{N/K} \notin D(\mathbf{Z}[\Gamma])$ is constructed in Chinburg (1983a) (Proposition 7.1).

We will be concerned below with whether the equalities in Conjecture 3 hold modulo $D(\mathbf{Z}[\Gamma])$. An important complementary test of Conjecture 3 is provided by quaternion N/K, i.e., those for which $\Gamma = Gal(N/K)$ is isomorphic to the quaternion group H_8. For quaternion N/K, $Cl(\mathbf{Z}[\Gamma])$ has order two and equals $D(\mathbf{Z}[\Gamma])$. Quaternion N/K are the extensions of smallest degree in which $W_{N/K}$ can be non-trivial.

The following table lists results about $\Omega(N/K,2)$ and $\Omega(N/K,3)$ with later results appearing below earlier ones in each box. The results together with Theorem 2 give all that is known in each case about $\Omega(N/K,1)$. The Strong Stark Conjecture is stated in Conjecture 5 below.

We will now briefly discuss some of the ideas involved in the proofs of the results in Table I.

Let N/K be a tame Galois extension of number fields. By Theorem 4 the invariant $\Omega(N/K,2)$ equals the stable isomorphism class $(\mathcal{O}_N)_{\mathbf{Z}[\Gamma]}$. But as we have noted $W_{N/K} \in D(\mathbf{Z}[\Gamma])$ and hence the assertion

$$\Omega(N/K,2) \equiv W_{N/K} \text{ modulo } D(\mathbf{Z}[\Gamma])$$

Table I. Results on Galois structure invariants

INVARIANT	$\Omega(N/K,2)$	$\Omega(N/K,3)$
CONJECTURE	$= W_{N/K}$	$= W_{N/K}$
GENERAL RESULTS (number fields).	Fröhlich (1977): True modulo $D(\mathbf{Z}[\Gamma])$ if N/K is tame. Taylor (1981): True if N/K is tame.	Chinburg (1983a): Implied modulo $D(\mathbf{Z}[\Gamma])$ by the Strong Stark Conjecture.
GENERAL RESULTS (function fields).	Chinburg (1989c): True modulo $D(\mathbf{Z}[\Gamma])$ if N/K is tame.	Bae (1987): True modulo $D(\mathbf{Z}[\Gamma])$.
QUATERNION CASE (number fields).	Fröhlich (1977): True if N/K is tame. Kim (1989) (1990): True if $K = \mathbf{Q}$ and N is not totally ramified over 2.	Chinburg (1989a): True for some infinite families of quaternion $N/\mathbf{Q}$.
QUATERNION CASE (function fields).	Chinburg (1989c): True if N/K is tame.	

for tame extensions of number fields is equivalent to $(\mathcal{O}_N)_{\mathbf{Z}[\Gamma]} \in D(\mathbf{Z}[\Gamma])$, which is precisely the assertion of Corollary 7′ of §1. Similarly the claim that $\Omega(N/K,2) = W_{N/K}$ is, in this context, equivalent to Taylor's Theorem (Theorem 2 of §1).

In Chinburg (1989c) Fröhlich's methods and results are carried over to tame Galois extensions of function fields in the following way. In the function field case we always assume that S_∞ has the properties stated at the end of §4.1. We let $\mathcal{O}_{N,\infty}$ be the ring of elements of N which are regular at the places in S_∞ and define $\mathcal{O}_{K,\infty} = K \cap \mathcal{O}_{N,\infty}$. (Recall that, in this context, $\mathcal{O}_N$ denotes the ring of elements of N which are regular at all places outside S_∞, and that $\mathcal{O}_K = K \cap \mathcal{O}_N$). Let $K_0T(\mathbf{Z}[\Gamma])$ be the Grothendieck group of finite $\mathbf{Z}[\Gamma]$-modules of finite projective dimension. In Chinburg (1989c) it is shown that there is an F as in Lemma 4.5 such that $F = \beta\mathcal{O}_K[\Gamma]$ for some $\beta \in \mathcal{O}_N$ for which $\mathcal{O}_{N,\infty} \subset \beta\mathcal{O}_{K,\infty}[\Gamma]$. One then checks that the class

$$\Phi(N/K) = (\mathcal{O}_N/\beta\mathcal{O}_K[\Gamma]) - (\beta\mathcal{O}_{K,\infty}[\Gamma]/\mathcal{O}_{N,\infty})$$

in $K_0T(\mathbf{Z}[\Gamma])$ depends only on N/K, not on the choice of β or of S_∞ satisfying the above conditions. Further, the image of $\Phi(N/K)$ under the natural map of $K_0T(\mathbf{Z}[\Gamma])$ to $Cl(\mathbf{Z}[\Gamma])$ is $\Omega(N/K,2)$. Finally, one can describe $\Phi(N/K,2)$ in terms of Galois Gauss sums using Fröhlich's Hom-description of $K_0T(\mathbf{Z}[\Gamma])$. This leads to the tame function field results stated in Table I. To carry out the proofs, it is useful to rewrite $\Phi(N/K)$ in terms of Chern class (degree) invariants of the kind studied in Chapman (1989). (*Added 28 August 1990*: A sequel to Chinburg (1989c) will contain a proof that $\Omega(N/K,2) = W_{N/K}$ if N/K is a tame finite Galois extension of function fields.)

Kim proved the Theorem indicated in Table I by first obtaining a formula of the following kind for all finite Galois extensions N/K of number fields. (A similar formula holds also for function fields). Let $S_{wild}(K)$ be the set of finite places of K which are wildly ramified in the extension N/K. One can find a projective $\mathbf{Z}[\Gamma]$-submodule $\mathcal{O}'_N$ of $\mathcal{O}_N$ such that the index $[\mathcal{O}_N : \mathcal{O}'_N]_{\mathcal{O}_K}$ is finite and supported on $S_{wild}(K)$; such $\mathcal{O}'_N$ are specified by giving their localisations at each finite place w of K. For each such place w we let $v(w)$ be a place of N over w and let $\Gamma_{v(w)}$ be the decomposition group of $v(w)$. Induction of modules from $\Gamma_{v(w)}$ to Γ gives rise to a homomorphism

$$Ind_{\Gamma_{v(w)}}^{\Gamma} : Cl(\mathbf{Z}[\Gamma_{v(w)}]) \longrightarrow Cl(\mathbf{Z}[\Gamma]).$$

Kim proves that

$$\Omega(N/K,2) = (\mathcal{O}'_N)_{\mathbf{Z}[\Gamma]} + \sum_{w \in S_{wild}(K)} Ind_{\Gamma_{v(w)}}^{\Gamma} \Omega_w(\mathcal{O}'_N) \qquad (4.19)$$

where $\Omega_w(\mathcal{O}'_N)$ is a correction factor in $Cl(\mathbf{Z}[\Gamma_{v(w)}])$ which depends on w and $\mathcal{O}'_N$. If $Cl(\mathbf{Z}[\Gamma_{v(w)}]) = 0$ for $w \in S_{wild}(K)$ then Kim's formula simplifies to $\Omega(N/K,2) = (\mathcal{O}'_N)_{\mathbf{Z}[\Gamma]}$; the case in which $S_{wild}(K)$ is empty is just Theorem 4 for number fields. If $\Omega(N/K,2) = (\mathcal{O}'_N)_{\mathbf{Z}[\Gamma]}$ then one can try to show that $(\mathcal{O}'_N)_{\mathbf{Z}[\Gamma]} = W_{N/K}$ using Fröhlich's additive methods. This is the technique used in Kim (1989) to prove the result stated in Table I in case $K = \mathbf{Q}$ and $N/\mathbf{Q}$ is a quaternion extension having at least two places over 2. If there is just one place above 2 then the analysis is much more difficult (Kim (1990)) because the correction term $\Omega_w(\mathcal{O}'_N)$ must also be computed when w is the place of $\mathbf{Q}$ determined by the prime 2.

The Strong Stark Conjecture involved in the upper right of Table I is the multiplicative counterpart of Fröhlich's determination of the ideals generated by Galois Gauss sums in terms of resolvents (this is Theorem 5 of §1 – recall also Remark 3.18). This parallel was developed in Chinburg (1983a) and was an important motivation for the definition of $\Omega(N/K,3)$. We will now briefly recall Stark's conjecture and discuss its implications for $\Omega(N/K,3)$.

Let N/K be a finite Galois extension of global fields of group Γ. Let S be a finite Γ-stable set of places of N (containing the set $S_\infty(N)$ in case N is a number field). For any complex valued character χ of Γ we write $\tilde{L}_S(s,\chi)$

for the Artin L-function attached to χ and truncated by removing the Euler factors corresponding to places in S. We let $\tilde{L}_S^*(0,\chi)$ denote the first non-zero coefficient in the Taylor expansion at $s = 0$ of $\tilde{L}_S(s,\chi)$. For a fixed Γ-embedding

$$0 \longrightarrow X_S \overset{\imath}{\longrightarrow} U_S \tag{4.20}$$

(see (3.5)) we write $R(\chi,\imath)$ for the regulator associated to $\imath$ and to the character χ in Tate (1984). Thus $R(\chi,\imath)$ is a non-zero complex number. Recall however that this is not a regulator in the sense of §3 (but see the remarks following Remark 3.12). Define a complex number $A(\chi,\imath)$ by

$$A(\chi,\imath) = \frac{R(\chi,\imath)}{\tilde{L}_S^*(0,\chi)}.$$

Conjecture 5 (Stark) For each complex valued character χ of Γ and embedding $\imath$ as in (4.20) both of the following conditions are satisfied:

(a) (Tate's formulation, Tate (1984)) For each element α of the automorphism group $Aut(\mathbf{C}/\mathbf{Q})$ one has

$$A(\chi,\imath)^\alpha = A(\chi^\alpha,\imath).$$

In particular therefore $A(\chi,\imath) \in \mathbf{Q}(\chi)$.

(b) (The Strong Stark Conjecture, Chinburg (1983a)) The fractional $\mathcal{O}_{\mathbf{Q}(\chi)}$-ideal generated by $A(\chi,\imath)$ is an Euler characteristic ideal constructed from the Galois cohomologies of X_S and U_S.

A conjecture equivalent to (b) had been formulated earlier in Lichtenbaum (1975) in terms of étale cohomology groups. In Tate (1984) both (a) and (b) of Conjecture (6) are proved for rational valued characters χ. Results equivalent to Tate's were proved for this case in Bienenfeld and Lichtenbaum (1986) (for more details see also Lichtenbaum (1975) and Chinburg (1983b)). The Strong Stark Conjecture in the function field case was proved in Bae (1987) using the cohomological interpretation of L-functions; this implies Bae's result in Table I. For a discussion of the Strong Stark Conjecture for Dirichlet characters see Solomon (1987).

In Chinburg (1989a) it is shown that the Strong Stark Conjecture reduces the proof that $\Omega(N/K,3) = W_{N/K}$ to certain congruences for the (conjecturally algebraic) numbers $A(\chi,\imath)$. Because of Tate's result above, no conjectural assumptions are needed if all of the representations of Γ have rational character, e.g., if N/K is a quaternion extension.

In Chinburg (1989a) complex quaternion extensions N of $K = \mathbf{Q}$ are considered. The above congruences then concern the non-zero integer $\tilde{L}_{S_\infty(N)}(0, \chi(N/\mathbf{Q}))$ modulo powers of two, where $\chi(N/\mathbf{Q})$ denotes the two-dimensional irreducible symplectic character of $Gal(N/\mathbf{Q})$. Using Shintani's formulae these congruences are proved in Chinburg (1989a) for certain infinite families of quaternion $N/\mathbf{Q}$. This leads to the result stated in Table I that $\Omega(N/\mathbf{Q}, 3) = W_{N/\mathbf{Q}}$ for infinitely many such $N/\mathbf{Q}$. A sharper method for proving the required congruences results from the use of Hilbert modular forms and the q-expansion principle, as in Chinburg (1989b) and Schmidt (1987).

Finally we note an interesting consequence of Table I concerning the existence of certain distinguished subgroups of the unit group U_N of $\mathcal{O}_N$ in case N/K is an extension of number fields. For this we let $G_0(\mathbf{Z}[\Gamma])$ denote the Grothendieck group of the category $\mathcal{FG}_\Gamma$ of all finitely generated $\mathbf{Z}[\Gamma]$-modules. For each element X of $\mathcal{FG}_\Gamma$ we let (X) denote its class in $G_0(\mathbf{Z}[\Gamma])$. By means of the forgetful functor from the category $\mathcal{LF}_\Gamma$ of all finitely generated locally-free $\mathbf{Z}[\Gamma]$-modules to $\mathcal{FG}_\Gamma$ there is a natural map

$$f_\Gamma : Cl(\mathbf{Z}[\Gamma]) \longrightarrow G_0(\mathbf{Z}[\Gamma])_{\text{torsion}}$$

defined for each element X of $\mathcal{LF}_\Gamma$ by

$$(X)_{\mathbf{Z}[\Gamma]} \overset{f_\Gamma}{\longmapsto} (X) - rk(X) \cdot (\mathbf{Z}[\Gamma]).$$

In Curtis and Reiner (1987) it is shown that

$$f_\Gamma(D(\mathbf{Z}[\Gamma])) = 0. \tag{4.21}$$

Proposition 6 (Chinburg, Queyrut) Let N/K denote a finite Galois extension of number fields of Galois group Γ. The statement that $f_\Gamma(\Omega(N/K, 3)) = f_\Gamma(W_{N/K})$, which is implied by the Strong Stark conjecture, is equivalent to the existence of subgroups $\mathcal{E}$ of finite index in U_N for which both of the following conditions are satisfied:

(4.22) (i) There exists an isomorphism of $\mathbf{Z}[\Gamma]$-modules

$$\mathcal{E} \cong X_N .$$

(ii) In $G_0(\mathbf{Z}[\Gamma])$

$$(U_N/\mathcal{E}) = (Cl_N)$$

where here Cl_N is the ideal classgroup of N considered as an element of $\mathcal{FG}_\Gamma$.

Remark 4.23 The subgroups $\mathcal{E}$ thus predicted by this Proposition are not explicitly described. It is interesting therefore to ask whether the so-called 'special units' of Thaine, Rubin and of Kolyvagin can play any rôle here (see Rubin (1987) or Thaine (1988)). For example, suppose that Γ has prime order and that $K = \mathbf{Q}$. The homomorphism f_Γ is then an isomorphism by a theorem of Reiner. In Chinburg (1983a) (Theorem 3.3) it is shown that the Gras Conjecture concerning cyclotomic units implies that $\Omega(N/\mathbf{Q},3) = W_{N/\mathbf{Q}}$. For a precise statement of this conjecture see §5 of Gras and Gras (1977). By work of Greenberg (1977), the Gras Conjecture at primes $p \neq 2$ is a consequence of the Main Conjecture of Iwasawa theory over $\mathbf{Q}$ together with the vanishing of the cyclotomic μ-invariants associated to the abelian extensions of $\mathbf{Q}$. The Main Conjecture over $\mathbf{Q}$ was proved for $p \neq 2$ in Mazur and Wiles (1984), while the vanishing of the cyclotomic μ-invariants associated to abelian K over $\mathbf{Q}$ was proved in Ferrero and Washington (1979). More recently Wiles has given a proof of the Main Conjecture over $\mathbf{Q}$, as it concerns zeroes of distinguished polynomials, for $p = 2$ (see Wiles (1989)). In Chinburg (1989d), Chinburg indicates how, by combining this work of Wiles with that of Greenberg (1977) and of Gillard (1979), one can complete the proof of the Gras Conjecture for $p = 2$. We will now sketch this argument of Chinburg. (Note that in particular this will complete the proof that $\Omega(N/\mathbf{Q},3) = W_{N/\mathbf{Q}}$ if $N/\mathbf{Q}$ is a Galois extension of prime degree).

Suppose then that $N/\mathbf{Q}$ is a real abelian extension of odd degree, and that $p = 2$. We adopt the notation of §5 of Greenberg (1977), except that we here replace Greenberg's K with N and we define the cyclotomic unit group C_N as in Gillard (1979), rather than as in Greenberg (1977). The formulation of the Gras Conjecture on page 152 of Greenberg (1977) then agrees with that of Gras and Gras (1977) when $p = 2$. We let L (respectively M_0', respectively M_0) denote the maximal abelian pro-p-extension of N which is unramified (respectively unramified outside of p, respectively unramified outside of p and infinity) so that

$$N \subseteqq L \subset M_0' \subseteqq M_0 \ .$$

By considering complex conjugations, one sees that when $p = 2$, $Gal(M_0/M_0')$ is isomorphic to $(\mathbf{Z}/2\mathbf{Z})[\Delta]$ with $\Delta = Gal(N/\mathbf{Q})$. Greenberg's arguments show that in order to prove the Gras Conjecture it is in fact sufficient to prove that $e_\Psi(U_N'/\overline{C}_N')$ and $e_\Psi Gal(M_0'/N)$ have the same order, where here e_Ψ is the idempotent of $\mathbf{Z}_p[\Delta]$ corresponding to any non-trivial irreducible representation Ψ of Δ over $\mathbf{Q}_p$, U_N' is the product of the principal units in the completions of Δ above p, and $\overline{C}_N'$ is the intersection of U_N' with the closure of C_N. Because of Theorem 6.2 of Wiles (1989), the arguments on pages 154–5 and 150–2 of Greenberg (1977) express the order of $e_\Psi Gal(M_0/N)$ in terms of the values of p-adic L-functions at $s = 1$. But taking into account the fact that $Gal(M_0/M_0') = (\mathbf{Z}/2\mathbf{Z})[\Delta]$ when $p = 2$, this expresses the order of $e_\Psi Gal(M_0'/N)$ in terms of p-adic L-function values. The resulting expression now agrees with the formula for the order of $e_\Psi(U_N'/\overline{C}_N')$ as obtained in Lemma 5 of Gillard (1979) (see also the footnote to page 3 of Gillard (1979)), which therefore completes the proof.

To end this section we shall show how to prove Proposition 6. Firstly, using an isomorphism (3.5) it is always possible to find a subgroup $\mathcal{E}$ of finite index in U_N and satisfying condition (4.22)(i). For any such subgroup $\mathcal{E}$ one has

$$(\mathcal{E}) = (X_N). \tag{4.24}$$

By unpublished work of Queyrut (1983), the root number class $W_{N/K}$ lies in the kernel of the map f_Γ. Hence the defining sequence (4.11)(a) implies that the equality $f_\Gamma(\Omega(N/K,3)) = f_\Gamma(W_{N/K})$ is equivalent to the following equality in $G_0(\mathbf{Z}[\Gamma])$:

$$(U_S) - (X_S) = 0\,.$$

On the other hand, if $\mathcal{I}_S$ (respectively $\mathcal{P}_S$) denotes the group of $\mathcal{O}_N$-fractional ideals (respectively principal $\mathcal{O}_N$-fractional ideals) which have support only involving primes in S_f then one has natural exact sequences

$$0 \longrightarrow \mathcal{P}_S \longrightarrow \mathcal{I}_S \longrightarrow Cl_N \longrightarrow 0 \qquad (4.25)(i)$$

$$0 \longrightarrow U_N \longrightarrow U_S \longrightarrow \mathcal{P}_S \longrightarrow 0 \qquad (4.25)(ii)$$

$$0 \longrightarrow X_N \longrightarrow X_S \longrightarrow Div_{\mathbf{Z}}(S_f) \longrightarrow 0 \qquad (4.25)(iii)$$

and

$$0 \longrightarrow Div_{\mathbf{Z}}(S_f) \longrightarrow \mathcal{I}_S \longrightarrow 0 \qquad (4.25)(iv)$$

where the sequence (4.25)(i) is a consequence of the assumption (4.2). The relations in $G_0(\mathbf{Z}[\Gamma])$ produced by (4.24) and (4.25) give the following equalities in $G_0(\mathbf{Z}[\Gamma])$:

$$(U_S) - (X_S) = (U_N) - (X_N) - (Cl_N) \qquad (4.26)(i)$$

$$= (U_N/\mathcal{E}) - (Cl_N). \qquad (4.26)(ii)$$

But since the left hand side of (4.26)(i) is 0 if and only if

$$f_\Gamma(\Omega(N/K,3)) = f_\Gamma(W_{N/K})$$

the statement of Proposition 6 follows.

5 EXPLICIT GALOIS MODULES

Let N/F denote a finite abelian extension of number fields of Galois group Γ. In this final paragraph we return to consider the structure of $\mathcal{O}_N$ as a Galois module. As outlined in §1 there is a comprehensive theory for the case in which N/F is at most tamely ramified and $\mathcal{O}_N$ is considered as a $\mathbf{Z}[\Gamma]$-module. However we are now concerned with the case of wildly ramified extensions: in this case even today there are only partial results dealing with certain special classes of extensions. The best of these results is Leopoldt's theorem (to be described in §5.1) which, for any absolutely abelian extension, completely describes the structure of $\mathcal{O}_N$ as a module over its associated order $\mathcal{A}(N/\mathbf{Q})$ (recall that in general $\mathcal{A}(N/F)$ denotes the full set of elements of $F[\Gamma]$ that induce endomorphisms of $\mathcal{O}_N$). In this paragraph we shall describe a technique which, for example, provides a closely analogous result for a class of extensions arising from the torsion points of an elliptic curve with complex multiplication. It also gives a result dealing with rings of integers of ray class fields of an imaginary quadratic field, which may be considered as an integral contribution to the famous 'Jugendtraum' of Kronecker.

5.1 Leopoldt's theorem

Let $N/\mathbf{Q}$ be a finite abelian extension of Galois group Γ. For each $\chi \in \Gamma^{\dagger}$ let

$$e_\chi = \frac{1}{card(\Gamma)} \sum_{\gamma \in \Gamma} \chi(\gamma^{-1})\gamma$$

be the corresponding primitive idempotent in $\mathbf{Q}^c[\Gamma]$, and let $\mathcal{F}(\chi)$ denote the conductor of χ. Thus $\mathcal{F}(\chi)$ is a positive integer, and we can regard χ as a group homomorphism $(\mathbf{Z}/\mathcal{F}(\chi)\mathbf{Z})^* \longrightarrow (\mathbf{Q}^c)^*$. We decompose $\mathcal{F}(\chi)$ into 'tame' and 'wild' parts by setting

$$\mathcal{F}(\chi) = \mathcal{F}_t(\chi)\mathcal{F}_w(\chi)$$

where

$$\mathcal{F}_t(\chi) = \prod_{p \| \mathcal{F}(\chi)} p,$$

the product being taken over all rational primes p such that $p \mid \mathcal{F}(\chi)$ but $p^2 \nmid \mathcal{F}(\chi)$.

We define an equivalence relation on $\Gamma^{\dagger}$ with characters χ and ϕ equivalent if and only if $\mathcal{F}_w(\chi) = \mathcal{F}_w(\phi)$, and to each equivalence class Φ we associate the idempotent

$$e_\Phi = \sum_{\phi \in \Phi} e_\phi .$$

Since the $\Omega_{\mathbb{Q}}$-conjugates of any given character ϕ all have the same conductor (and hence lie in the same equivalence class), the idempotents e_Φ each belong to $\mathbb{Q}[\Gamma]$.

We also define the conductor and the kernel of an equivalence class Φ by

$$\mathcal{F}(\Phi) = l.c.m.\{\mathcal{F}(\chi) : \chi \in \Phi\}$$

and

$$ker(\Phi) = \bigcap_{\phi\in\Phi} ker(\phi).$$

To each character ϕ we now associate a Gauss sum as follows: let Φ be the equivalence class containing ϕ and set

$$\tau_\Phi(\phi) = \sum_{x\in(\mathbb{Z}/\mathcal{F}(\Phi)\mathbb{Z})^*} \phi(x)(\xi_{\mathcal{F}(\Phi)})^x ,$$

where, for any positive integer n, ξ_n denotes the primitive nth root of unity $e^{2\pi i/n}$. Although these Gauss sums are in general imprimitive , with $\mathcal{F}(\phi) < \mathcal{F}(\Phi)$, they do not all vanish since $(\mathcal{F}(\Phi)/\mathcal{F}(\phi), \mathcal{F}(\phi)) = 1$.

For each equivalence class Φ we define

$$T_\Phi = \frac{1}{[N^{ker(\Phi)} : \mathbb{Q}]}\sum_{\phi\in\Phi} \tau_\Phi(\phi).$$

We now have the notation to state

Theorem 1 (Leopoldt (1959))

$$\mathcal{A}(N/\mathbb{Q}) = \sum_\Phi e_\Phi \mathbb{Z}[\Gamma]$$

and

$$\mathcal{O}_N = (\sum_\Phi T_\Phi)\mathcal{A}(N/\mathbb{Q}) .$$

We note that the statement of this theorem comprises three elements:

(i) an explicit description of the associated order $\mathcal{A}(N/\mathbb{Q})$;
(ii) the assertion that $\mathcal{O}_N$ is a free $\mathcal{A}(N/\mathbb{Q})$-module;
(iii) an explicit generator of $\mathcal{O}_N$ over $\mathcal{A}(N/\mathbb{Q})$.

The construction of the generator involves the roots of unity $\xi_{\mathcal{F}(\Phi)}$, as well as the character values: thus to obtain the generator we have used the values of the transcendental function $e^{2\pi i x}$ at division points of its period. Below we will give a result (Theorem 3) which provides the same three elements in a rather different context, the generator being constructed from the values of a certain elliptic function at division points of its period lattice.

5.2 Hopf orders and the generalised Noether criterion

In general if we do not know that $\mathcal{O}_N$ is free over $\mathcal{A}(N/F)$ then we can at least ask whether it is locally-free. Over the integral group ring $\mathcal{O}_F[\Gamma]$ one has the following criterion:

Proposition 5.1 (Noether criterion) If N/F is a finite Galois extension of number fields (not necessarily abelian) with Galois group Γ, then $\mathcal{O}_N$ is locally-free over $\mathcal{O}_F[\Gamma]$ if and only if $trace_{N/F}(\mathcal{O}_N) = \mathcal{O}_F$.

Remark 5.2 This criterion is equivalent to the result of Theorem 1 of §1.

In the case that Γ is abelian, there is a generalisation of Proposition 5.1, due to Childs and Hurley, in which $\mathcal{O}_F[\Gamma]$ can be replaced by certain other $\mathcal{O}_F$-orders in $F[\Gamma]$, namely the Hopf orders.

Let

$$\Delta : F[\Gamma] \longrightarrow F[\Gamma] \otimes_F F[\Gamma]$$

be the F-linear map given by $\Delta(\gamma) = \gamma \otimes \gamma$ for each $\gamma \in \Gamma$. Then Δ is an algebra homomorphism and is called the comultiplication of $F[\Gamma]$. Now any $\mathcal{O}_F$-order $\mathcal{A}$ in $F[\Gamma]$ is necessarily $\mathcal{O}_F$-projective and so $\mathcal{A} \otimes_{\mathcal{O}_F} \mathcal{A}$ can be regarded as an $\mathcal{O}_F$-submodule of $F[\Gamma] \otimes_F F[\Gamma]$. We call $\mathcal{A}$ a Hopf order if $\Delta(\mathcal{A}) \subseteq \mathcal{A} \otimes_{\mathcal{O}_F} \mathcal{A}$. (This definition is somewhat simpler than the abstract definition of a Hopf algebra, since the augmentation map $\epsilon : \mathcal{A} \longrightarrow \mathcal{O}_F$ and the antipode map $S : \mathcal{A} \longrightarrow \mathcal{A}$, determined by $\epsilon(\gamma) = 1$ and $S(\gamma) = \gamma^{-1}$ for each $\gamma \in \Gamma$, are automatically inherited from the Hopf algebra structure of $F[\Gamma]$).

The condition that an order $\mathcal{A}$ in $F[\Gamma]$ be a Hopf order is a very strong one: if, for example, Γ has odd order then the only Hopf order in the rational group algebra $\mathbf{Q}[\Gamma]$ is the integral group ring $\mathbf{Z}[\Gamma]$. More Hopf orders are obtained however if we allow ramification in the base field at primes dividing the order of Γ. In particular, if F contains enough roots of unity, then the (unique) maximal order $\mathcal{M}$ of $F[\Gamma]$ is split (i.e., has a $\mathcal{O}_F$-basis of idempotents), and is therefore a Hopf order since Δ is an algebra homomorphism and $\mathcal{M} \otimes_{\mathcal{O}_F} \mathcal{M}$ is the unique maximal order in $F[\Gamma] \otimes_F F[\Gamma]$.

Let $\Sigma = \sum_{\gamma\in\Gamma}\gamma$ be the trace element of $F[\Gamma]$. To any Hopf order $\mathcal{A}$ in $F[\Gamma]$ we associate a fractional $\mathcal{O}_F$-ideal $i\mathcal{A}$ of $\mathcal{O}_F$ by the rule

$$\mathcal{A}^\Gamma \ (= \mathcal{A}\cap F\Sigma) \ = \ (i\mathcal{A})^{-1}\Sigma.$$

In fact $i\mathcal{A}$ is an integral ideal, and has the following property:

Proposition 5.3 (Generalised Noether Criterion) Let N/F be a finite abelian extension of number fields of Galois group Γ. Let $\mathcal{A}$ be any $\mathcal{O}_F$-order of $F[\Gamma]$ which is Hopf, and such that $\mathcal{O}_N$ is an $\mathcal{A}$-module. Then $\mathcal{O}_N$ is a locally-free $\mathcal{A}$-module if, and only if, $trace_{N/F}(\mathcal{O}_N) \ = \ i\mathcal{A}(N/F)$.

Proof We must show that, for each prime $\wp$ of F, the $\mathcal{O}_{F,\wp}$-order $\mathcal{O}_N\otimes_{\mathcal{O}_F}\mathcal{O}_{F,\wp}$ is free over $\mathcal{A}_\wp$ if and only if $(\mathcal{O}_N\otimes_{\mathcal{O}_F}\mathcal{O}_{F,\wp})\Sigma \ = \ (i\mathcal{A})_\wp$ (where here $\mathcal{O}_{F,\wp}$ denotes the completion of $\mathcal{O}_F$ at $\wp$).

We write $M = \mathcal{O}_N\otimes_{\mathcal{O}_F}\mathcal{O}_{F,\wp}$ and set $(i\mathcal{A})_\wp \ = \ t\mathcal{O}_{F,\wp}$. Thus

$$\mathcal{A}_\wp^\Gamma \ = \ t^{-1}\Sigma\mathcal{O}_{F,\wp} \ .$$

Now, if M is free over $\mathcal{A}_\wp$, say on the generator m, then

$$\mathcal{O}_{F,\wp} \ = \ M^\Gamma \ = \ m(\mathcal{A}_\wp^\Gamma) \ = \ m(t^{-1}\Sigma)\mathcal{O}_{F,\wp} \ = \ t^{-1}((M)\Sigma),$$

and so $(M)\Sigma \ = \ (i\mathcal{A})_\wp$.

Conversely suppose that $(M)\Sigma = (i\mathcal{A})_\wp$. Then $x.\Sigma = t$ for some $x\in M$. For $m\in M$ we define

$$\lambda(m) \ = \ t^{-1}\Sigma_{\gamma\in\Gamma}x.(m\gamma)\otimes\gamma^{-1} \ \in \ M\otimes_{\mathcal{O}_{F,\wp}}F_\wp[\Gamma].$$

In fact $\lambda(M)\subseteq M\otimes\mathcal{A}_\wp$ since $\lambda(m)$ is obtained by applying to $m\otimes 1\in M\otimes F_\wp[\Gamma]$ the element $(id\otimes S)\Delta(t^{-1}\Sigma)$ of $\mathcal{A}_\wp\otimes\mathcal{A}_\wp$ and then multiplying by $x\otimes 1$. Moreover $\lambda : M \longrightarrow M\otimes\mathcal{A}_\wp$ is a homomorphism of $\mathcal{A}_\wp$-modules, where $\mathcal{A}_\wp$ acts on $M\otimes\mathcal{A}_\wp$ by multiplication in the second factor, and is split by the homomorphism $m\otimes a\longmapsto ma$. It follows that M is projective over $\mathcal{A}_\wp$ and so, since we are working over the local ring $\mathcal{O}_{F,\wp}$, this implies that M is free over $\mathcal{A}_\wp$.

5.3 The map Ψ

We now use an elliptic curve E to generate Hopf orders. This will enable us to study the integral Galois module structure of fields arising from the division points of E.

Thus let E be an elliptic curve defined over a number field F with complex multiplication by the full ring of integers $\mathcal{O}_K$ of a quadratic imaginary field K. We assume that E has everywhere good reduction.

Fix a non-zero element $\pi \in \mathcal{O}_K$ and let $G = E[\pi]$ be the group of π-division points of E. Replacing F by a larger field if necessary we may assume that G is contained in the group of F-rational points $E(F)$ of E.

We consider the two F-algebras

$A = F[G]$ (the group algebra), and
$B = Map(G, F)$ (the algebra of functions $G \longmapsto F$ with pointwise addition and multiplication).

There is a non-degenerate F-linear pairing

$$< .,. >: B \times A \longrightarrow F$$

given by

$$< f, g >= f(g) \quad \text{for each } f \in B,\ g \in G.$$

Now let $\mathcal{G}$ be the group scheme over $\mathcal{O}_F$ determined by G: thus as a functor on the category of commutative F-algebras, $\mathcal{G}$ associates to each algebra R the group $\mathcal{G}(R)$ of π-division points in the group $E(R)$ of R-rational points of E. In particular, since $G \subseteqq E(F)$, we have $\mathcal{G}(F) = G$, and hence $\mathcal{G}$ has generic fibre $\mathcal{G}_{/F} = Spec(B)$, the constant group scheme of G over F. Since E has good reduction, $\mathcal{G}$ is finite and flat over $\mathcal{O}_F$. We may therefore identify $\mathcal{G}$ with $Spec(\mathcal{B})$ for some $\mathcal{O}_F$-order $\mathcal{B}$ in B. The fact that the scheme $\mathcal{G}$ is a group scheme means that $\mathcal{B}$ is a Hopf order. Let $\mathcal{A} \subset A$ be the $\mathcal{O}_F$-dual to $\mathcal{B}$ with respect to the pairing $< .,. >$. Thus $\mathcal{A}$ is a Hopf order in A, representing the Cartier dual $\mathcal{G}^D$ of the group scheme $\mathcal{G}$, and we have a non-degenerate $\mathcal{O}_F$-linear pairing

$$< .,. >: \mathcal{B} \times \mathcal{A} \longrightarrow \mathcal{O}_F\,.$$

Now let Q be an element of $E(F)$ and set

$$G_Q = \{Q' \in E(\mathbb{Q}^c) : \pi Q' = Q\}.$$

Our aim is essentially to study as a Galois module the ring of integers of the field $F(Q')$ obtained by adjoining to F the coordinates of Q' for any $Q' \in G_Q$. We have to make two adjustments, however, to the module to be studied. The first problem is that the degree of the extension $F(Q')/F$ depends on the point Q. In order to deal with all $Q \in E(F)$ uniformly, we replace $F(Q')$ by the Galois algebra

$$F_Q = Map_{\Omega_F}(G_Q, \mathbf{Q}^c).$$

Here Ω_F acts in the natural way on both G_Q and $\mathbf{Q}^c$ so that F_Q is a product of field extensions, one for each Galois orbit of points Q', and has F-dimension equal to the order of G.

Let $\mathcal{O}_Q$ denote the integral closure of $\mathcal{O}_F$ in F_Q. The group G acts on F_Q by translations on G_Q, and so we may consider $\mathcal{O}_Q$ as an $\mathcal{O}_F[G]$-module. The fields occuring in F_Q may however be wildly ramified over F, so that $\mathcal{O}_Q$ will not in general be locally-free over $\mathcal{O}_F[G]$. Thus we try to study $\mathcal{O}_Q$ as an $\mathcal{A}$-module, and here we need to make the second adjustment since $\mathcal{O}_Q$ does not necessarily admit $\mathcal{A}$. We define $\tilde{\mathcal{O}}_Q$ to be the largest $\mathcal{A}$-module in $\mathcal{O}_Q$. Thus

$$\tilde{\mathcal{O}}_Q = \{x \in \mathcal{O}_Q : x\mathcal{A} \subseteqq \mathcal{O}_Q\}$$

(and also

$$\tilde{\mathcal{O}}_Q \cong Hom_{\mathcal{O}_F[G]}(\mathcal{A}, \mathcal{O}_Q)$$

via the map taking $x \in \tilde{\mathcal{O}}_Q$ to the homomorphism $a \longmapsto xa$ for each $a \in \mathcal{A}$).

The module $\tilde{\mathcal{O}}_Q$ is clearly a lattice in F_Q, and, since $\mathcal{A}$ is a Hopf order, $\tilde{\mathcal{O}}_Q$ is in fact an order in F_Q: for if $x, y \in \tilde{\mathcal{O}}_Q$ and $a \in \mathcal{A}$ then using Sweedler's notational convention

$$\Delta(a) = \sum_{(a)} a_{(1)} \otimes a_{(2)} \in \mathcal{A} \otimes_{\mathcal{O}_F} \mathcal{A},$$

one has

$$(xy)a = \sum_{(a)} (xa_{(1)})(ya_{(2)}) \in \mathcal{O}_Q,$$

so that $xy \in \tilde{\mathcal{O}}_Q$; and $1 \in \tilde{\mathcal{O}}_Q$ since $\epsilon(\mathcal{A}) \subseteqq \mathcal{O}_F$.

Moreover, one has

Proposition 5.4 (Taylor (1988)) $\tilde{\mathcal{O}}_Q$ is locally-free over $\mathcal{A}$.

We can therefore consider the class $(\tilde{\mathcal{O}}_Q)$ of $\tilde{\mathcal{O}}_Q$ in the classgroup $Cl(\mathcal{A})$ of locally-free $\mathcal{A}$-modules, and can investigate how this class varies with Q. More precisely, we have a map

$$\Psi : E(F) \longrightarrow Cl(\mathcal{A})$$

given by

$$Q \longmapsto (\tilde{\mathcal{O}}_Q).$$

Proposition 5.5 (Taylor (1988)) Ψ is a homomorphism of groups.

Remark 5.6 The $\mathcal{O}_F$-orders $\tilde{\mathcal{O}}_Q$ represent principal homogeneous spaces for the group scheme $\mathcal{G}$ over $\mathcal{P} = Spec(\mathcal{O}_F)$, and so determine classes in the group $H^1(\mathcal{P}, \mathcal{G})$. Now by a spectral sequence argument (Waterhouse (1971) Theorem 2) one has $H^1(\mathcal{P}, \mathcal{G}) \cong Ext(\mathcal{G}^D, \mathcal{G}_m)$, where here $\mathcal{G}_m$ is the multiplicative group scheme over $\mathcal{G}$. Composing this isomorphism with the natural homomorphism

$$Ext(\mathcal{G}^D, \mathcal{G}_m) \longrightarrow H^1(\mathcal{G}^D, \mathcal{G}_m) = Pic(\mathcal{G}^D)$$

(Waterhouse (1971) §0), we obtain a homomorphism

$$H^1(\mathcal{P}, \mathcal{G}) \longrightarrow Pic(\mathcal{G}^D).$$

The map Ψ may be regarded as a special case of this homomorphism.

If we take $Q = \pi P$ for some $P \in E(F)$, then $G_Q \subseteq E(F)$ and so the class $(\mathcal{O}_Q)$ is trivial. Thus $\pi E(F) \subseteq ker(\Psi)$. Under suitable hypotheses we can say much more about $ker(\Psi)$.

Let w_K denote the number of roots of unity in K, and let $E(F)_{tor}$ be the torsion subgroup of $E(F)$. Then we have:

Theorem 2 (Srivastav and Taylor (1990)) If $(\pi, w_K) = 1$ then $E(F)_{tor} \subseteq ker(\Psi)$.

Remark 5.7 This is essentially a result about the rings of integers of the ray-class fields of K and so may be considered as an integral contribution to the famous 'Jugendtraum' of Kronecker.

Outline of proof Let $Q \in E(F)_{tor}$. We take an auxiliary prime l which splits in $K/\mathbf{Q}$, say $l.\mathcal{O}_K = \ell\bar{\ell}$. Let $F' = F(E[\ell^2])$, and let F'_Q and $\tilde{\mathcal{O}}'_Q$ be the analogues of F_Q and $\tilde{\mathcal{O}}_Q$ respectively, but defined here with respect to the basefield F'. Thus $\tilde{\mathcal{O}}'_Q$ is locally-free over $\mathcal{A}' = \mathcal{A} \otimes_{\mathcal{O}_F} \mathcal{O}_{F'}$ and we have a group homomorphism $\Psi' : E(F') \longrightarrow Cl(\mathcal{A}')$.

As in Theorem 3 below, one can construct a function f on E, defined over F', which has restriction f_Q to G_Q such that $\tilde{\mathcal{O}}'_F = f_Q \mathcal{A}'$. Thus $\Psi'(Q) = 1$.

To descend to F, we use the commutativity of the diagram

$$\begin{array}{ccc} E(F') & \xrightarrow{\Psi'} & Cl(\mathcal{A}') \\ \big\downarrow Tr & & \big\downarrow Res \\ E(F) & \xrightarrow{\Psi} & Cl(\mathcal{A}) \end{array}$$

where Tr is here the trace map with respect to the addition on E, and Res is the homomorphism of classgroups given by restriction of scalars. Since $Q \in E(F)$ one has $Tr(Q) = [F' : F]Q$ and hence $(\Psi(Q))^{[F':F]} = 1$.

By the theory of complex multiplication, $[F' : F] \mid l^2(l-1)^2$, and so, varying l, we have that $\Psi(Q)$ is annihilated by

$$\begin{aligned} h.c.f.\{l^2(l-1)^2 : l \text{ splits in } K\} &= h.c.f.\{(l-1)^2 : l \text{ splits in } K\} \\ &= (w_K)^2 , \end{aligned}$$

the last equality being by a standard lemma of class field theory. But we already know that $\Psi(\pi Q) = 1$, and so since $(\pi, w_K) = 1$ we deduce that $\Psi(Q) = 1$.

Remark 5.8(i) In general there appear to be many non-trivial classes of $Cl(\mathcal{A})$ in the image of Ψ.

Remark 5.8(ii) By means of this result, we obtain some implications of the conjecture of Birch and Swinnerton-Dyer for Galois module structure. Indeed, if the Hasse–Weil L-function $L(E/F, s)$ of E does not vanish at $s = 1$ then, conjecturally, $E(F)$ has rank 0 so that $E(F) = E(F)_{tor}$, and therefore $\Psi = 0$. Moreover, if we enlarge the field of definition of E to some extension N of F, and set $\mathcal{A}_N = \mathcal{A} \otimes_{\mathcal{O}_F} \mathcal{O}_N$, then we can still say something about the map $\Psi_N : E(N) \longrightarrow Cl(\mathcal{A}_N)$ even if $L(E/N, 1) = 0$; if $\tilde{\mathcal{O}}_{Q,N}$ denotes the analogue of $\tilde{\mathcal{O}}_Q$ but here defined with respect to the field N, then the commutativity of the diagram

$$\begin{array}{ccc} E(N) & \xrightarrow{\Psi} & Cl(\mathcal{A}_N) \\ \big\downarrow Tr & & \big\downarrow Res \\ E(F) & \xrightarrow{\Psi=0} & Cl(\mathcal{A}) \end{array}$$

implies that $(\tilde{\mathcal{O}}_{Q,N}) = 1$ in $Cl(\mathcal{A})$, so that $\tilde{\mathcal{O}}_{Q,N} \cong \mathcal{A}_N$ as an $\mathcal{A}$-module (although not necessarily as an $\mathcal{A}_N$-module). Thus, as with the tame theory of §1 we find that L-functions dominate the Galois module structure of rings of integers.

Remark 5.8(iii) We have seen that $< \pi E(F), E(F)_{tor} > \subseteqq ker(\Psi)$, but whether there is equality here remains an open question. Recent work of A. Agboola shows that equality holds in the analogous situation for abelian varieties over function fields.

5.4 Explicit results

We keep the notation and hypotheses of the previous section and impose some further assumptions.

We suppose that 2 splits in $K/\mathbf{Q}$, say $2\mathcal{O}_K = \ell.\bar{\ell}$. Let π be a prime element of $\mathcal{O}_K$ (possibly a rational prime) with $\pi \equiv \pm 1$ modulo 4, and set $q = \pi.\bar{\pi}$. We assume that $E[4] \subseteqq E(F)$ (in addition to the previous assumption that $E[\pi] \subseteqq E(F)$), and that the ramification index of π in F/K is $q-1$.

Fix an element Q of $E[\pi]$. We will give an explicit description of the Hopf order $\mathcal{A}$ and of the structure of $\tilde{\mathcal{O}}_Q$ as an $\mathcal{A}$-module.

Let S, T and R be points of $E(F)$ with annihilators precisely $\ell, \bar{\ell}$ and 4 respectively (so $E[2] = \{0, S, T, 2R\}$). Let D be a meromorphic elliptic function on E, defined over F, with divisor

$$(D) = (0) + (2R) - (S) - (T);$$

by the Abel–Jacobi theorem such a D exists and is unique up to multiplication by a non-zero constant in F. This function D was in fact introduced by Weber in his work on complex multiplication.

To describe the order $\mathcal{A}$ it is sufficient to specify its local completions $\mathcal{A}_\wp$ at each prime ideal $\wp$ of $\mathcal{O}_F$.

The result is the following

Theorem 3 (essentially from Taylor (1985)) The order $\mathcal{A}$ is given by

$$\mathcal{A}_\wp = \begin{cases} \mathcal{O}_{F,\wp}[G], & \text{if } \wp \nmid \pi, \\ \mathcal{O}_{F,\wp} + \sum_{i=0}^{q-2} \mathcal{O}_{F,\wp}\sigma_i, & \text{if } \wp \mid \pi, \end{cases}$$

where, for $0 \le i \le q-2$,

$$\sigma_i = \frac{1}{\pi} \sum_{g \in G} \left(\frac{D(g)}{D(R)} \right)^i g,$$

and

$$\tilde{\mathcal{O}}_Q = \frac{D(Q+R)}{D(Q'+R)} \mathcal{A},$$

where here Q' is any element of G_Q.

Here in fact F_Q is a field (not just a Galois algebra), and $\tilde{\mathcal{O}}_Q = \mathcal{O}_{F_Q}$ is its full ring of integers. Thus we have obtained, for the extension F_Q/F, the three elements of Theorem 1, namely an explicit description of the associated order $\mathcal{A}$, the freeness of $\mathcal{O}_Q$ over $\mathcal{A}$, and an explicit Galois generator.

5.5 Heegner points

In Theorems 2 and 3 we considered elliptic curves of a special type, namely those that admitted complex multiplication, and obtained a Galois module result for a field extension arising from an arbitrary F-rational point Q. In this section we apply similar methods to elliptic curves from another special family, namely those parameterised by a modular curve, and to certain special points Q, namely Heegner points.

We now take E to be an elliptic curve defined over $\mathbf{Q}$ (but not necessarily admitting complex multiplication), and we suppose that E is a Weil curve of Eisenstein type, with prime conductor N. Thus there is a map $\phi : X_0(N) \longrightarrow E$, defined over $\mathbf{Q}$, where $X_0(N)$ is the modular curve corresponding to the congruence subgroup $\Gamma_0(N)$ of $SL_2(\mathbf{Z})$. Set $m = (12, N-1)$ and suppose that $(\Delta(Nz)/\Delta(z))^{\frac{1}{m}}$ lies in the function field $\mathbf{Q}(E)$ (and not just in $\mathbf{Q}(X_0(N))$). The curve $X_0(N)$ has two cusps, at 0 and at ∞, and these map to the points 0 and S on E, where 0 is the identity of the group law, and S is a torsion point.

Let p be an odd prime number dividing the order of S. Then p also divides $N-1$. Let K be an imaginary quadratic field in which N splits, say $N\mathcal{O}_K = \mathcal{N}\overline{\mathcal{N}}$, and let $F = H_K(\mu_p)$, where H_K is the Hilbert class field of K and μ_p

is the group of pth roots of unity in $\mathbf{Q}^c$. Then $E[p] \subseteq E(F)$, and taking $\pi = p$, we obtain a group scheme $\mathcal{G}$ over $\mathcal{O}_F$ corresponding to $G = E[\pi]$ as before. In fact $\mathcal{G}$ splits as a product $\mathcal{G}_1 \times \mathcal{G}_2$ of its étale and connected parts, corresponding to a splitting $G_1 \times G_2$ of G, and the Hopf order $\mathcal{A}$ in $F[G]$ representing $\mathcal{G}^D$ is given by

$$\mathcal{A} = \mathcal{O}_F[G_1] \otimes \mathcal{M},$$

where $\mathcal{M}$ is the split maximal $\mathcal{O}_F$-order in $F[G_2]$. Thus we obtain a map

$$Cl(\mathcal{A}) \longrightarrow Cl(\mathcal{O}_F[G_1]) \times Cl(\mathcal{M});$$

and we then write

$$\xi_1 : Cl(\mathcal{A}) \longrightarrow Cl(\mathcal{O}_F[G_1]),$$

and

$$\xi_2 : Cl(\mathcal{A}) \longrightarrow Cl(\mathcal{M})$$

for the corresponding projection maps.

Now let P be a Heegner point on $X_0(N)$, given by a triple $(\mathcal{O}_K, \mathcal{N}, [\mathcal{P}])$ for some ideal class $[\mathcal{P}]$ of $\mathcal{O}_K$ (see §2 of Gross (1983)). Thus P corresponds to an isogeny of degree N from an elliptic curve with j-invariant $j(\mathcal{P})$ to one with j-invariant $j(\mathcal{P}.\overline{\mathcal{N}})$, both curves having complex multiplication by $\mathcal{O}_K$. Let $Q_{[\mathcal{P}]} = \phi(P) \in E(F)$. Then $\tilde{\mathcal{O}}_{Q_{[\mathcal{P}]}}$ is locally-free over $\mathcal{A}$ as before and so we obtain a class $\Psi(Q_{[\mathcal{P}]}) \in Cl(\mathcal{A})$.

The behaviour of this class is not yet fully understood, but we do have the following result:

Theorem 4 (Taylor)

(i) $\xi_2(\,\Psi(Q_{[\mathcal{P}]})\,) = 1$;

(ii) If $p \nmid d_K$, then $\xi_1(\,\Psi(Q_{[\mathcal{P}]}) - \Psi(Q_{[\mathcal{P}']})\,) = 1$ for all ideal classes $[\mathcal{P}]$ and $[\mathcal{P}']$ of $\mathcal{O}_K$. □

REFERENCES

S. Bae (1987)
'The conjecture of Lichtenbaum and Chinburg over function fields', Math. Ann., **285** 417–45.

E. Bayer-Fluckiger and H. W. Lenstra (Jr.) (1989)
'Forms in odd degree extensions and self-dual normal bases', to appear in Amer. J. Math.

M. Bienenfeld and S. Lichtenbaum (1986)
'Values of zeta- and L-functions at zero', to appear in Amer. J. Math.

D. Burns (1989)
'Factorisability and the arithmetic of wildly ramified Galois extensions', Séminaire de Théorie des Nombres de Bordeaux, **1** 59–66.

D. Burns (1990)
'Canonical factorisability and a variant of Martinet's conjecture', to appear in J. London Math. Soc.

Ph. Cassou-Noguès and M. J. Taylor (1983a)
'Constante de l'equation fonctionelle de la fonction L d'Artin d'une representation symplectique et modérée', Ann. Inst. Fourier, **33** 1–17.

Ph. Cassou-Noguès and M. J. Taylor (1983b)
'Local root numbers and Hermitian Galois module structure of rings of integers', Math. Ann., **263** 251–61.

Ph. Cassou-Noguès and M. J. Taylor (1989)
'The trace form and Swan modules', to appear.

R. Chapman (1989)
'L-functions and Galois module structure in real cyclotomic function fields', to appear.

T. Chinburg (1983a)
'On the Galois structure of algebraic integers and S-units', Invent. Math., **74** 321–49.

T. Chinburg (1983b)
'Derivatives of L-functions at $s = 0$ (after Stark, Tate, Bienenfeld and Lichtenbaum)', Compositio Mathematica, **48** 119–27.

T. Chinburg (1985)
'Exact sequences and Galois module structures', Annals of Maths., **121** 351–76.

T. Chinburg (1989a)
'The analytic theory of multiplicative Galois structure', Memoirs of the A.M.S. Vol. 77.

T. Chinburg (1989b)
'A quaternionic L-value congruence', J. of the Fac. of Sciences of the University of Tokyo, Sec. IA, **36** no. 3 765–87.

T. Chinburg (1989c)
'Galois structure invariants of global fields', to appear.

T. Chinburg (1989d)
Letter to D. Burns, December 1st, 1989.

C. Curtis and I. Reiner (1987)
Methods of Representation Theory, Vol.II. Wiley, New York.

B. Erez and J. Morales (1989)
'The hermitian structure of rings of integers in odd degree abelian extensions', preprint, University of Geneva.

B. Ferrero and L. Washington (1979)
'The Iwasawa invariant μ_p vanishes for abelian number fields', Ann. of Math., **109** 377–95.

A. Fröhlich (1976)
'Arithmetic and Galois module structure for tame extensions', J. reine u. angew. Math., **286/287** 380–440.

A. Fröhlich (1977)
'Galois module structure', in A. Fröhlich (ed.) *Algebraic Number Fields*, Proc. Durham Symp. 1975, Academic Press, London 133–91.

A. Fröhlich (1978)
'Some problems of Galois module structure for wild extensions', Proc. London Math. Soc., **27** 193–212.

A. Fröhlich (1983)
'Galois module structure of algebraic integers' Springer-Verlag, Heidelberg, New York, Tokyo.

A. Fröhlich (1984)
'Classgroups and Hermitian modules', Progress in Maths, Volume 84, Birkhäuser, Boston, Basel, Stuttgart.

A. Fröhlich (1988)
'Module defect and factorisability', Illinois J. Math., **32** 407–21.

A. Fröhlich (1989)
'L-values at zero and multiplicative Galois module structure (also Galois Gauss sums and additive Galois module structure)', J. reine u. angew. Math., **397** 42–99.

R. Gillard (1979)
'Unités cyclotomiques, unités semi-locales et $\mathbf{Z}_l$-extensions II', Ann. Inst. Fourier, **29** fasc. 4, 1–15.

G. Gras and M. N. Gras (1977)
'Calcul du nombre de classes et des unités des extensions abéliennes réeles de $\mathbf{Q}$', Bull. Sci. Math. de France, 2ᵉ série, **101** 97–129.

R. Greenberg (1977)
'On p-adic L-functions and cyclotomic fields II', Nagoya Math. J., **67** 139–58.

B. H. Gross (1984)
'Heegner points on $X_0(N)$', in R. A. Rankin (ed.) *Modular Forms*, Proc. Durham Symp. 1983, Ellis Horwood Ltd.

S. Kim (1989)
'A generalisation of Fröhlich's theorem to wildly ramified quaternion extensions of $\mathbf{Q}$', to appear in the Ill. J. of Math.

S. Kim (1990)
to appear.

H. W. Leopoldt (1959)
'Über die hauptordnung der ganzen elemente eines abelschen zahlkörpers', J. reine u. angew. Math., **201** 119–49.

S. Lichtenbaum (1975)
'Values of L-functions at s = 0', Astérisque, **24–5** 133–8.

J. Martinet (1977)
'Character theory and Artin L-functions', in A. Fröhlich (ed.) *Algebraic Number Fields*, Proc. Durham Symp. 1975, Academic Press London 1–87.

B. Mazur and A. Wiles (1984)
'Class fields of abelian extensions of $\mathbf{Q}$', Invent. Math., **76** 179–330.

J. Queyrut (1983)
Letter to T. Chinburg.

K. Rubin (1987)
'Global units and ideal class groups', Invent. Math., **89** 511–526.

T. Schmidt (1987)
'Quaternionic L-value congruences', Ph.D. thesis, Univ. of Pennsylvania.

D. Solomon (1987)
'Lichtenbaum's conjecture for Dirichlet characters', Ph.D. thesis, Brown University.

A. Srivastav and M. J. Taylor (1990)
'Elliptic curves with complex multiplication and Galois module structure', Invent. Math., **99** 165–84.

J. Tate (1950)
'Fourier analysis in number fields and Hecke's zeta-Functions', Ph.D. thesis, Princeton University. (Also in J. W. S. Cassels and A. Fröhlich (eds.) *Algebraic Number Theory*, Proc. Brighton Symp. 1965, Academic Press, London (1967)).

J. Tate (1966)
'The cohomology groups of tori in finite Galois extensions of number fields', Nagoya Math. J., **27** 709–19.

J. Tate (1984)
'Les conjectures de Stark sur les fonctions L d'Artin en $s = 0$', Notes d'un cours à Orsay rédigées par D. Bernardi et N. Schappacher, Progress in Maths, Vol 47, Birkhäuser, Boston–Basel–Stuttgart.

M. J. Taylor (1981)
'On Fröhlich's conjecture for rings of integers of tame extensions', Invent. Math., **63** 41–79.

M. J. Taylor (1985)
'Formal groups and the Galois module structure of local rings of integers', J. reine u. angew. Math., **358** 97–103.

M. J. Taylor (1988)
'Mordell–Weil groups and the Galois module structure of rings of integers', Illinois J. Math., **32** 428–52.

M. J. Taylor (1989)
'Rings of integers and trace forms for extensions of odd degree', Math. Zeit. **202** (3).

F. Thaine (1988)
'On the ideal class groups of real abelian number fields', Ann. of Math., **128** 1–18.

W. C. Waterhouse (1971)
'Principal homogeneous spaces and group scheme extensions', Trans. Amer. Math. Soc., **153** 181–9.

A. Wiles (1990)
'The Iwasawa conjecture for totally real fields', Ann. of Math, **131** 493–540.

Motivic p-adic L-functions

John Coates

Introduction. The connexions between special values of L-functions and arithmetic is an ancient and mysterious theme in number theory, which can be traced through the work of Dirichlet, Kummer, Minkowski, Siegel, Tamagawa, Weil, Birch and Swinnerton-Dyer, Iwasawa, ... Recently, Bloch and Kato [1], using ideas which rely heavily on the work of Fontaine, have succeeded in formulating a very general version of the classical Tamagawa number conjecture for linear algebraic groups for arbitrary motives over the rational field **Q**, which seems to contain as special cases all earlier conjectures about these questions. Needless to say, only a very modest amount of progress has been made so far towards proving the Bloch-Kato conjecture for specific motives over **Q** (essentially, the only cases where it can be established at present are for the Tate motives, and certain motives arising from elliptic curves with complex multiplication). In all the cases where proofs are known, the conjecture is established for each prime p separately, and the deepest part of the argument involves ideas from Iwasawa theory. Specifically, one must use a version for the motive of the so called 'main conjecture' of Iwasawa theory, which has now been completely proven for the above motives (apart from the troublesome primes 2 and 3 in the case of elliptic curves with complex multiplication), thanks to the beautiful work of Mazur, Wiles, Thaine, Kolyvagin and Rubin (see the article by Rubin in this volume). It does at least make sense to try to formulate the 'main conjecture' for arbitrary motives over **Q**, although one should have no illusions about the difficulty of proving it. The formulation of this 'main conjecture' involves, on the one hand, p-adic Iwasawa modules which are built out of the representations of the Galois group of **Q** given by the p-adic realisations of the motive (see the article by Greenberg in this volume for a discussion of the case when p is ordinary), and on the other hand, p-adic avatars of the complex L-function of the motive, which are built out of the critical special values of the complex L-function. The aim of the present article is to give a detailed conjectural description of these motivic p-adic L-functions, at least for primes p for which the motive has good ordinary reduction. Nearly everything which is contained in this paper is already given in the earlier articles by B. Perrin-Riou and myself ([3], [4], [10]). However, the assertions made about holomorphy in these earlier papers were too strong, and I have, I hope, corrected these here, as well as giving somewhat fuller versions of some of the crucial arguments about modifications of the Euler factors at both finite and infinite primes.

1. Notation and normalization. Let $\mathbf{Q}$ denote the field of rational numbers and $\mathbf{C}$ (resp. $\mathbf{R}$) the field of complex numbers (resp. real numbers). Throughout, p will signify an arbitrary prime number (we do not exclude p=2), and we write $\mathbf{Z}_p$, $\mathbf{Q}_p$, $\mathbf{C}_p$ for the ring of p-adic integers, the field of p-adic numbers, and the completion of an algebraic closure of the field of p-adic numbers. Let $\mathbf{U}$ denote the group of units of $\mathbf{Z}_p$. Let $\mathbf{A}$ denote the algebraic closure of $\mathbf{Q}$ in $\mathbf{C}$. We fix, once and for all, an embedding

$$j : \mathbf{A} \rightarrow \mathbf{C}_p \tag{1}$$

which we will often not make explicit in our formulae. Let $\mathbf{Q}^{ab}$ be the maximal abelian extension of $\mathbf{Q}$ in $\mathbf{A}$. If K/F is a Galois extension of fields, we write G(K/F) for the Galois group of K over F. For brevity, we put

$$G = G(\mathbf{A}/\mathbf{Q}) \quad , \quad G^{ab} = G(\mathbf{Q}^{ab}/\mathbf{Q}) \; .$$

We use the embedding (1) to identify complex and p-adic characters of finite order of G^{ab}. For each integer $m \geq 1$, let μ_m denote the group of m-th roots of unity in $\mathbf{A}$. Let Ξ be the group of all p-power roots of unity, and put

$$P = \mathbf{Q}(\Xi) \quad , \quad H = \mathbf{Q}(\Xi)^+ \quad , \quad J = G(H/\mathbf{Q}) \; . \tag{2}$$

(here the + denotes the maximal real subfield). We write

$$\psi : G(P/\mathbf{Q}) \rightarrow \mathbf{U} \tag{3}$$

for the isomorphism given by the action of this Galois group on W, i.e. $\zeta^{\sigma} = \zeta^{\psi(\sigma)}$ for all ζ in Ξ and σ in $\mathbf{Q}$. We also put

$$X = \mathrm{Hom}_{\mathrm{cont}}(\mathfrak{J}, \mathbf{C}_p^*) \ . \tag{4}$$

As far as the sign of the reciprocity law map is concerned, we must stress that we adopt throughout the geometric convention of [5], rather than the more classical arithmetic convention. Specifically, this convention is as follows. For each finite prime q, let Frob_q denote the arithmetic Frobenius, i.e. it operates on the algebraic closure of the field with q elements by sending x to x^q. Let C denote the idele class group of $\mathbf{Q}$. Let x_q be any idele whose q -th component is a local parameter at q, and all of whose other components are equal to 1. Then we choose the sign of the reciprocity map

$$r : C \to G^{ab} \tag{5}$$

such that $r(x_q)$ is an element of G^{ab} which acts on the algebraic closure of the residue field at q via the *inverse* of Frob_q. Let $\gamma : C \to \mathbf{C}^*$ be a continuous homomorphism. The complex L-function of γ is then defined, as usual, by the Euler product

$$L(\gamma, s) = \prod (1 - \gamma(x_q)/q^s)^{-1} \ , \tag{6}$$

where the product is taken over all finite primes q which are not ramified for γ, and x_q is as above. Similarly, if S is any finite set of primes of $\mathbf{Q}$, we write $L_S(\gamma, s)$ for the function obtained by omitting from (6) all Euler factors at primes which lie in S. Now let $\phi : G^{ab} \to A^*$ be any character of finite order. We define its associated idele class character

$$\phi_R : C \to A^* \tag{7}$$

via the formula $\phi_R = \phi \circ r$. Thus, if q is a finite prime which does not divide the conductor of ϕ, we have

$$\phi_R(x_q) = \phi(\mathrm{Frob}_q^{-1}) \ . \tag{8}$$

This last formula explains our choice of the sign of the reciprocity map (4), because it shows that the complex L-function (6) attached to ϕ_R coincides with the motivic L-function attached to ϕ (see §4).

2. p-adic pseudo-measures. The aim of this section is to give a slight generalization of the notion of a p-adic pseudo-measure which is given in [13]. Let **O** be the ring of integers of some finite extension of $\mathbf{Q}_p$, and let I be any profinite abelian group (in the rest of the paper, we take I = J). For simplicity, let X also denote the group of continuous homomorphisms from I to $\mathbf{C}_p{}^*$. The **O** - Iwasawa algebra $\mathfrak{I}$ of I is defined to be the projective limit of the group rings **O**[I/H], where H runs over the open subgroups of I. It is a compact algebra, which contains **O**[I] as a dense sub-algebra. The elements of $\mathfrak{I}$ are called integral measures on I (with values in **O**). This terminology is justified because, if μ is in $\mathfrak{I}$, and f is any continuous function from I to $\mathbf{C}_p$, we can define the integral

$$\int_I f \, d\mu$$

by passage to the limit from the case when f is locally constant. In this latter case, if H is an open subgroup of I such that f is locally constant modulo H, and if the image of μ in **O**[I/H] is equal to $\Sigma\mu(s)s$, then the value of the above integral is equal to $\Sigma\mu(s)f(s)$, where, in both sums, s runs over I/H. We shall need the following generalization of the notion of an integral measure on I, in order to take into account possible poles of our p-adic L-functions. Let $Q(\mathfrak{I})$ be the ring of quotients of $\mathfrak{I}$, i.e. the ring of all quotients α/β, where α and β belong to $\mathfrak{I}$ and β is not a divisor of 0. An element μ of $Q(\mathfrak{I})$ is said to be a measure if there exists a non-zero element d of **O** such that $d\mu$ belongs to $\mathfrak{I}$. We say that an element μ of $Q(\mathfrak{I})$ is a *pseudo-measure* if there exists a non-zero element d of **O**, a finite subset S of X, and non-negative integers $n(\xi)$ ($\xi \in S$), such that, for all choices of elements $\sigma(\xi)$ in I for ξ running over S, we have

$$d \prod_{(\xi \in S)} (\xi(\sigma(\xi)) - \sigma(\xi))^{n(\xi)} \mu \qquad (9)$$

belongs to the Iwasawa algebra $\mathfrak{I}$. It is clear that the pseudo-measures form a subring of $Q(\mathfrak{I})$. Suppose now that μ is a pseudo-measure. Let ϕ be any element of X which is distinct from all ξ in S. For each ξ in S, choose $\sigma(\xi)$ in I such that $\phi(\sigma(\xi)) \neq \xi(\sigma(\xi))$. We then define the integral of ϕ against μ by the formula

$$\int_I \phi \, d\mu = d^{-1} \prod_{(\xi \in S)} (\xi(\sigma(\xi)) - \phi(\sigma(\xi)))^{-n(\xi)} \int_I \phi \, d\lambda, \tag{10}$$

where λ denotes the integral measure (9). It is immediately verified that this definition is independent of all choices. Also, if λ given by (9) is an integral measure, we say that the pseudo-measure μ has a pole at each ξ in S of order $\leq n(\xi)$; the minimal value of $n(\xi)$ such that the expression (9) lies in $\mathfrak{I}$ is called the exact order of the pole of μ at ξ.

Finally, there is an important involution on the ring of pseudo-measures on I, which we denote by $\mu \to \mu^{\#}$. This involution is given on $\mathbf{O}[I]$ by the $\mathbf{O}$-linear map which sends σ to σ^{-1} for all σ in I, and it extends by continuity to $\mathfrak{I}$. It plainly extends to Q(I), and preserves the subring of pseudo-measures.

3. The cyclotomic theory. This section will be devoted to a brief account of the p-adic analogue of the Riemann zeta function. Recalling that X is given by (4), we write X_{alg} for the subgroup of X consisting of all characters of the form

$$\xi = \psi^n \chi \qquad (n \in \mathbf{Z}), \tag{11}$$

where χ is any character of finite order of $G(\mathbf{P}/\mathbf{Q})$, and ψ given by (3) is the p-adic cyclotomic character. Let τ_∞ denote the element of G given by complex conjugation. We are assuming that (11) is a character of the Galois group J, and this is clearly equivalent to the assertion that

$$\chi(\tau_\infty) = (-1)^n \quad . \tag{12}$$

The following is the basic existence theorem for the p-adic analogue of the Riemann zeta function. Fix an integer $m \leq 0$ and a character ϕ of finite order of $G(\mathbf{P}/\mathbf{Q})$, which satisfy

$$\phi(\tau_\infty) = (-1)^{m-1} \ . \tag{13}$$

The reason for this condition will become apparent later (in the notation and terminology explained later, we want the motive $\mathbf{Q}(m)$ twisted by ϕ to have weight ≥ 0 and to be critical at $s = 0$). Let $\mathbf{O}$ be the ring of integers of the field obtained by adjoining the values of ϕ to $\mathbf{Q}_p$, and let $\mathfrak{I}$ be the $\mathbf{O}$ - Iwasawa algebra of J. We remark that formula (15) below shows that, in the following theorem, the right hand side of (14) belongs to the field $\mathbf{A}$ of algebraic numbers, and so can be viewed as lying in $\mathbf{C}_p$ via the embedding (1).

Theorem 1. There exists a unique pseudo-measure $\mu = \mu(m,\phi)$ on the Galois group J satisfying :- (i). For all σ in J, $(\psi^{1-m}\phi^{-1}(\sigma) - \sigma)\mu$ belongs to the Iwasawa algebra $\mathfrak{I}$; (ii). If ξ given by (11) is any element of X_{alg} such that $m+n \leq 0$, then

$$\int_J \xi \, d\mu = L_T(\varpi_R, n+m) \, , \tag{14}$$

where $T = \{p\}$, and $\varpi = \phi\chi$. Moreover, μ has a pole of order 1 at $\psi^{1-m}\phi^{-1}$.

We sketch what is essentially Iwasawa's proof of the existence of μ. Put r=4 or r=p, according as p is even or odd, and put $r_k = rp^k$ for all $k \geq 0$. For each p-adic unit u, write $[u]_k$ for its class in the group of relatively prime classes of integers modulo r_k . The partial zeta function

$$\zeta(u, r_k ; s) = \Sigma \, w^{-s} \qquad (R(s) > 1) \quad ,$$

where the sum is over all positive integers in $[u]_k$, has an analytic continuation over the whole complex plane, apart from a simple pole at s=1. For each non-negative integer t, we have

$$\zeta(u, r_k ; -t) = - r_k^t \, B_{t+1} (\{u\}_k / r_k)/(t+1) \ , \tag{15}$$

where $\{u\}_k$ denotes the unique representative in $\mathbf{Z}$ of $[u]_k$,which lies between 0 and r_k ; here $B_{t+1}(x)$ denotes the (t+1)-th Bernoulli polynomial, which is defined by the expansion

$$ye^{yx}/(e^y - 1) = \Sigma_{(h\geq 0)} B_h(x) \, y^h/h! \quad .$$

In particular, we have

$$B_1(x) = x - 1/2\,,\ B_{t+1}(x) = x^{t+1} - (t+1)/2\, x^t + \dots\ . \tag{16}$$

For t fixed, let p^e denote the largest power of p occurring in the denominators of the coefficients of $B_{t+1}(x)/(t+1)$. One deduces immediately from (15) and (16) that, for all integers $k \geq 0$ and all p-adic units u, we have

$$\zeta(u, r_{k+e}\,; -t) \equiv t\, u^{t+1}/((t+1)r_{k+e}) + u^t\, \zeta(u, r_{k+e}\,; 0) \bmod r_k\ . \tag{17}$$

If v is also a p-adic unit, we define

$$\delta_t\,(u, v; r_k) = v^{t+1}\, \zeta(u, r_k\,; -t) - \zeta(uv, r_k\,; -t).$$

Then we claim that, for all integers $k \geq 0$, we have

$$\delta_t\,(u, v; r_k) \equiv (uv)^t\ \delta_0(u, v; r_k) \bmod r_k\ . \tag{18}$$

Note that (17) immediately implies the weaker version of (18), in which the first two r_k's appearing in (18) are replaced by r_{k+e}. But it is easy to see that this weaker congruence implies (18), when it is combined with the additional identity

$$\Sigma\, \zeta(z, r_h\,; s) = \zeta(u, r_k\,; s)\ , \tag{19}$$

where h is any integer $\geq k$, and z runs over any set of representatives in **U** of those classes modulo r_h which map to the class of u modulo r_k. Note that one obvious consequence of (18) is that $\delta_t(u, v; r_k)$ is integral at p for all $t \geq 0$, because this is plainly true for $t = 0$ from the explicit formula (16).

We can now construct the pseudo-measure μ. For each u in **U**, let $\sigma(u)$ denote the unique element of $G(P/\mathbf{Q})$ such that $\psi(\sigma(u)) = u$, and let $\tau(u)$ denote the restriction of $\sigma(u)$ to N. Let P_k be the field obtained by adjoining the group of r_k-th roots of unity to $\mathbf{Q}$, and let N_k be the maximal real subfield of P_k. We write $\sigma_k(u)$ (resp. $\tau_k(u)$) for the restriction of $\sigma(u)$ to P_k (resp. to N_k). Write V_k for any set of representatives in **U** of the group of relatively prime residue classes modulo r_k. We assume in what follows that k is so large that the conductor of ρ divides r_k. For each p-adic unit v, define the following element of the O - group ring of the Galois group of N_k over Q

$$\lambda_k(v) = \phi(\sigma(v))^{-1} \Sigma\, \delta_{-m}(u, v; r_k)\, \tau_k(u)^{-1}\, \phi(\sigma_k(u))^{-1} ,$$

where the sum is over all u in V_k. The identity (19) shows that, as k varies, the $\lambda_k(v)$ define an element $\lambda(v)$ of the Iwasawa algebra $\mathfrak{I}$. Put $\theta = \psi^{1-m}\phi^{-1}$. By virtue of (13), θ is a character of the Galois group J. If v is not of finite order in U, it is easy to see that $\theta(\tau(v)) - \tau(v)$ is not a zero divisor in $\mathfrak{I}$. For any such v, we define

$$\mu = \lambda(v).(\theta(\tau(v)) - \tau(v))^{-1}.$$

It is readily verified that μ is a pseudo-measure on J, which is independent of the choice of v of infinite order in U, and which satisfies assertion (i) of the above theorem. Assume now that k is so large that r_k is also divisible by the conductor of χ. To prove (ii), we note that, by definition, the integral of $\xi = \psi^n\chi$ against the measure $\lambda(v)$ is the p-adic limit as $k \to \infty$ of the expression

$$\phi(\sigma(v))^{-1} \Sigma_{(u)}\, \delta_{-m}(u, v; r_k)\, u^{-n}\, \varpi(\sigma(u))^{-1}.$$

Applying the congruence (18) for t = - n and t = - n - m, we deduce that this limit is equal to the limit as $k \to \infty$ of the expression

$$\phi(\sigma(v))^{-1} v^n\, \Sigma_{(u)}\, \delta_{-n-m}(u, v; r_k)\, \varpi(\sigma(u))^{-1} .$$

But, again using (19), we see that this last quantity has a value independent of k, which is given by

$$(\theta(\tau(v)) - \xi(\tau(v))\, L_T(\varpi_R, n+m) .$$

Assertion (ii) is now plain. We omit the proof of the final statement of the theorem, which is a well known consequence of the von-Staudt-Clausen theorem giving the exact power of p occuring in the denominators of the k-th Bernoulli numbers, where k runs over the positive even integers which are divisible by p-1.

4. Complex L-functions. Motives arise in nature as direct summands of the cohomology of a given dimension of a smooth projective algebraic variety defined over **Q**. However, we shall simply view motives in the naive sense, as being defined by a collection of realisations, satisfying certain axioms. Moreover, since we must consider the twists of our motives by arbitrary characters of finite order of G^{ab}, it is technically necessary to consider motives over **Q**, with coefficients in some finite extension K of **Q**. A detailed account of such motives and their realizations is given in §2 of [6], and we only briefly recall some of the key definitions here. Let $\Sigma(K)$ denote the set of embeddings of K in the complex field **C** . We identify the **C**-algebras $K\otimes\mathbf{C}$ (unless indicated to the contrary, all tensor products will be understood to be taken over **Q**) and $\mathbf{C}^{\Sigma(K)}$ via

$$K\otimes\mathbf{C} \cong \mathbf{C}^{\Sigma(K)} \ : u\otimes w \to (w.\,\sigma(u))_\sigma \ . \tag{20}$$

In addition, for each prime number l, we put

$$K_l = K\otimes\mathbf{Q}_l = \prod_{(\lambda\,|\,l)} K_\lambda \ ,$$

where λ runs over the primes of K dividing l, and K_λ denotes the completion of K at λ. By a homogeneous motive M over **Q**, with coefficients in K, of weight w(M) and dimension d(M), we mean a collection of Betti $H_B(M)$, de Rham $H_{DR}(M)$, and l-adic $H_l(M)$ (one for each prime l) realisations, which are, respectively, free modules over K, K, and K_l , all of the same rank d(M). Moreover, these realisations are endowed with the following additional structure :- (i). $H_B(M)$ admits an involution F_∞ ; (ii). The global Galois group G has a continuous action on $H_l(M)$ for each prime l, and there is an isomorphism

$$g_l : H_B(M)\otimes_K K_l \to \ H_l(M)$$

which transforms the involution F_∞ into the complex conjugation; (iii). There is a decreasing exhaustive filtration $\{F^kH_{DR}(M) : k \in \mathbf{Z}\}$ on the de Rham realisation; (iv). There is a Hodge decomposition into **C**-vector spaces

$$H_B(M)\otimes C = \oplus H^{i,j}(M) \quad , \tag{21}$$

where i,j run over a finite set of indices satisfying $i+j=w(M)$, and where F_∞ maps $H^{i,j}(M)$ to $H^{j,i}(M)$; (v). There is a $G_\infty = G(\mathbf{C}/\mathbf{R})$ - isomorphism of $\mathbf{C}$ - vector spaces (which also commutes with the action of K)

$$g_\infty : H_B(M)\otimes C \rightarrow H_{DR}(M)\otimes C \tag{22}$$

where complex conjugation acts on the space on the right via its action on $\mathbf{C}$, and on the space on the left via F_∞ on $H_B(M)$ and via its natural action on $\mathbf{C}$; (vi). Finally, for all $k \in \mathbf{Z}$, we have

$$g_\infty\,(\oplus_{(i\geq k)} H^{i,j}(M)) = F^k H_{DR}(M) \quad . \tag{23}$$

The first basic example of such a motive M is the Tate motive $\mathbf{Q}(m)$, for any m in $\mathbf{Z}$, which is of weight -2m and dimension 1. Let $V_l(\mu)$ be the tensor product over $\mathbf{Z}_l$ with $\mathbf{Q}_l$ of the projective limit of the Galois modules μ_{l^n} of l^n - th roots of unity, and let $V_l(\mu)^{\otimes m}$ be the m-th tensor power of $V_l(\mu)$. Then the realisations of $M = \mathbf{Q}(m)$ are given by

$$H_B(M) = K \ , \ H_{DR}(M) = K \ , \ H_l(M) = V_l(\mu)^{\otimes m} \otimes_{\mathbf{Q}_l} K_l \ .$$

The involution F_∞ is $(-1)^m$, and the action of G is the natural one. The Hodge decomposition is specified by taking $H^{-m,-m} = K\otimes C$, and the k-th term in the filtration of the de Rham cohomology is either K or 0, according as $k \leq -m$ or $k > -m$. The isomorphism (22) is given by $g_\infty(1\otimes 1) = 1\otimes(2\pi i)^m$.

If M is any such motive, we can construct the following motives from M :- (i). The twists M(n) for any n in $\mathbf{Z}$; the realisations of M(n) are the tensor products of the corresponding realisations of M and $\mathbf{Q}(n)$; (ii). The dual motive M^ ; the realisations of M^ are the dual vector spaces of the realisations of M.

We briefly recall the standard definitions and conjectures for the complex L-function attached to such a motive M. Put

$$\Gamma_{\mathbf{R}}(s) = \pi^{-s/2}\Gamma(s/2)\,,\ \Gamma_{\mathbf{C}}(s) = 2(2\pi)^{-s}\Gamma(s)\ .$$

For simplicity, we assume that, when $w(M)$ is even, F_∞ acts on $H^{k,k}$, where $k = w(M)/2$, via a scalar (this will be automatically implied by our assumption made later that M is critical at s=0). As is explained in §2 of [6], the fact that $H_B(M)\otimes\mathbf{C}$ is a free $K\otimes\mathbf{C}$ - module, when with the identification (20), yields a decomposition

$$H_B(M)\otimes\mathbf{C} = \oplus\, H_B(\sigma, M), \text{ where } H_B(\sigma, M) = H_B(M)\otimes_{(K,\sigma)}\mathbf{C}\,;$$

here σ runs over $\Sigma(K)$ and the tensor product on the right is taken by regarding $\mathbf{C}$ as a K-algebra via σ. Each $H_B(\sigma, M)$ admits a Hodge decomposition

$$H_B(\sigma, M) = \oplus\, H^{j,k}(\sigma, M)\ ,$$

and we let $h(j,k) = \mathbf{C}$ -dimension of $H^{j,k}(\sigma, M)$. This notation is justified, since it is shown in [6] that these dimensions are independent of $\sigma \in \Sigma(K)$. The Euler factor at ∞, which is also shown in [6] to be independent of the choice of σ, is then defined by

$$L_\infty(M, s) = L_\infty(\sigma, M, s) = \prod_{(U)} L_\infty(U, s)\ ,$$

where U runs over the direct summands of $H_B(\sigma, M)$ of either the form (i) $U = H^{j,k}(\sigma)\oplus H^{k,j}(\sigma)$ with $j < k$, or (ii) $U = H^{k,k}(\sigma)$, (where we have abbreviated $H^{j,k}(\sigma,M)$ by $H^{j,k}(\sigma)$) and $L_\infty(U, s)$ is given explicitly by :- (a). In case (i), $L_\infty(U, s) = \prod_{(j<k)} \Gamma_{\mathbf{C}}(s-j)^{h(j,k)}$; (b). In case (ii) when F_∞ acts on $H^{k,k}(\sigma)$ via $(-1)^k$, then $L_\infty(U,s) = \Gamma_{\mathbf{R}}(s-k)^{h(k,k)}$; (c). In case (ii) when F_∞ acts on $H^{k,k}(\sigma)$ via $(-1)^{k+1}$, then $L_\infty(U, s) = \Gamma_{\mathbf{R}}(s+1-k)^{h(k,k)}$. By contrast, the Euler factors at finite primes do depend on the choice of σ in $\Sigma(K)$. If q is a finite prime, let I_q denote the inertial subgroup in G of some fixed prime of A lying above q. Then the Euler factor at q is given by

$$L_q(\sigma, M, s) = (\sigma Z_q)(M, q^{-s}),$$

where

$$Z_q(M, X) = \det\left(1 - \mathrm{Frob}_q^{-1}.X \mid H_\lambda(M)^{I_q}\right)^{-1},$$

and where λ is any prime K not lying above q. We have imposed the standard hypothesis that $Z_q(M, X)$ is a rational function in X, with coefficients in K, which are independent of the choice of the prime λ. The complex L-function of M is then defined by the Euler product

$$\Lambda(\sigma, M, s) = \prod L_v(\sigma, M, s) ,$$

where v runs over all primes of **Q**, including $v = \infty$. We also write $L(\sigma, M, s)$ for this Euler product with the infinite Euler factor omitted. Note that we have

$$\Lambda(\sigma, M(n), s) = \Lambda(\sigma, M, s+n) \text{ for all } n \in \mathbf{Z} .$$

We assume that there exists a finite set of primes $S = S(M)$ such that (i) for each prime λ, and each q which is not in S and which does not lie below λ, the inertia group I_q operates trivially on $H_\lambda(M)$, and (ii) for q not in S, the reciprocal complex roots of $(\sigma Z_q)(M, X)^{-1}$ have absolute value equal to $q^{w(M)/2}$. Under additional hypotheses, one can then define the conductor of M and the global ε-factor $\varepsilon(\sigma, M, s)$ (see [14]). Here is the standard conjecture about the analytic continuation and functional equation of this L-function.

Conjecture A (Complex Version). $\Lambda(\sigma, M, s)$ has a meromorphic continuation over the whole complex plane to a function of order ≤ 1, and satisfies the functional equation

$$\Lambda(\sigma, M, s) = \varepsilon(\sigma, M, s)\Lambda(\sigma, M^\wedge(1), -s) \quad . \tag{24}$$

It is also conjectured that $\Lambda(\sigma, M, s)$ is entire if $w(M)$ is odd, and that the only possible pole which can occur, if $w(M)$ is even, is at $s = 1+w(M)/2$. In this latter case, the order of the pole is conjectured to be the K_λ - dimension of the subspace of $H_\lambda(M(w(M)/2))$ which is fixed by the global Galois group G, for any prime λ of K.

As is explained in [5] and [14], the global ε-factor $\varepsilon(\sigma, M, s)$ has a decomposition into local factors, which we shall see plays an important role in the non-archimedean theory. Let Θ denote the adele group of **Q**. Fix, once and for all, the Haar measure $dx = \prod dx_v$ on Θ, where dx_∞ is the usual Haar measure on $\mathbf{R}$, and, for each prime q, dx_q is the Haar measure on $\mathbf{Q}_q$ which gives $\mathbf{Z}_q$ volume 1. To define the local ε-factors, we must choose a complex character of the adele class group $\Theta/\mathbf{Q}$, and there are inescapably two natural choices. For the rest of this paper, ρ will denote i or -i, where i has its usual meaning as a complex number. Let η_ρ denote the character of $\Theta/\mathbf{Q}$ with components $\eta_{\rho,\infty}(x) = \exp(2\pi\rho x)$, and, for each finite prime q, $\eta_{\rho,q}(x) = \exp(-2\pi\rho x)$, where we have identified $\mathbf{Q}_q/\mathbf{Z}_q$ with the q-primary subgroup of $\mathbf{Q}/\mathbf{Z}$. For each place v of $\mathbf{Q}$, , let $\varepsilon_v(\sigma, M, \rho, s)$ denote Deligne's local ε-factor for the relative to the various choices just described (we have suppressed the the fixed measure dx_v in the notation, and we simply write ρ instead of the additive character η_ρ). Then we have

$$\varepsilon(\sigma, M, s) = \prod_v \varepsilon_v(\sigma, M, \rho, s) \quad , \tag{25}$$

where the product is taken over all primes v of $\mathbf{Q}$, including $v=\infty$. Note also that we have the fundamental relation

$$\varepsilon_v(\sigma, M, \rho, s)\, \varepsilon_v(\sigma, M^\wedge(1), -\rho, -s) = 1 \quad . \tag{26}$$

It is fundamental for the non-archimedean theory that we also consider the twists of our motive M by characters of finite order of G^{ab}, and we now briefly recall the definition of these twists . Let $\phi: G^{ab} \to A^*$ be a character of finite order, and assume that the values of φ lie in K. Following [6], §6, we can attach to φ a motive [φ] of dimension 1 and weight 0 over $\mathbf{Q}$, with coefficients in K. Let V(φ) be the vector space of dimension 1 over K, on which G acts via φ. We then define $H_B(\phi)$ to be the underlying space of V(φ), with the action of F_∞ given by $\phi(\tau_\infty)$, where τ_∞ is complex conjugation. The de Rham realisation is given by $H_{DR}(\phi) = (V(\phi)\otimes A)^G$, where the global Galois group G acts both on V(φ) via φ, and on A in the natural fashion (we endow the de Rham realisation with the trivial filtration for which F^k is 0 for $k > 0$, and the whole space for $k \leq 0$). The comparison isomorphism

$$g_{\phi,\infty} : H_B(\phi)\otimes\mathbf{C} \to H_{DR}(\phi)\otimes\mathbf{C} \tag{27}$$

is obtained by noting that $H_{DR}(\phi)$ provides a $\mathbf{Q}$ - structure for $H_B(\phi)\otimes A$, and then extending scalars from A to $\mathbf{C}$. For each finite prime λ of K, we take the λ-adic realisation to be $H_\lambda(\phi)$ to be a vector space of dimension 1 over the completion K_λ of K at λ, on which G acts via ϕ. For each embedding $\sigma: K \rightarrow \mathbf{C}$, we can apply the above motivic recipe for attaching a complex L-function $L(\sigma, \phi, s)$ to the motive $[\phi]$, and, in view of (8), we see that $L(\sigma, \phi, s)$ coincides with the L-function $L((\phi^\sigma)_R, s)$ defined by (6) - indeed, our sign of the reciprocity map was chosen to assure this. Now let M be a motive over $\mathbf{Q}$, with coefficients in K, as above. The twist $M(\phi)$ is then defined to be the motive over $\mathbf{Q}$, with coefficients in K, whose realisations are the tensor products over K of the realisations of M with the corresponding realisations of $[\phi]$.

5. Critical points and the period conjecture. Our goal in this section is to give a modified version of Deligne's period conjecture of [6], which seems essential for problems of p-adic interpolation. We shall be concerned with the following question. Let M be a fixed motive over $\mathbf{Q}$, with coefficients in some finite extension K of $\mathbf{Q}$, and consider twists of M of the form

$$W = M(n)(\phi)\ , \text{ with } \phi(\tau_\infty) = (-1)^n\ , \tag{28}$$

where n ranges over $\mathbf{Z}$, and ϕ over the characters of finite order of G^{ab} with values in K. How does the Deligne period $c^+(W)$ vary with n and ϕ? It turns out that the naive answer to this question is not precise enough for problems of p-adic interpolation, and our aim will be to use the properties of the complex L-function to give a finer answer, at least when both M and W are critical at $s = 0$.

We begin by briefly explaining the naive answer to the above question, which does not depend on any assumptions about M or W being critical at $s = 0$. In fact, the techniques of [6] reduce this to a problem of linear algebra (see [6] for the background material, which we do not repeat in detail here). We suppose always that K contains the values of ϕ. We assume that F_∞ acts on $H^{k,k}(M)$ by a scalar. In §2 of [6], Deligne attaches to W a period $c^+(W)$ in $(K\otimes\mathbf{C})^*$, which is well defined up to multiplication by an element of K^*. Let ρ denote a choice of either $+i$ or $-i$ in $\mathbf{C}$. Let $f(\phi)$ denote the conductor of ϕ, so that ϕ factors through the Galois group, which we denote by $\Delta(\phi)$, of the field generated over $\mathbf{Q}$ by the group of $f(\phi)$-th roots of unity. Following §6 of [6], we define the element $\delta_\rho(\phi) \in (K\otimes\mathbf{C})^*$ by

$$\delta_\rho(\phi) = \Sigma_{(\tau\in\Delta(\phi))}\, \phi^{-1}(\tau)\otimes(\exp(-2\pi\rho/f(\phi)))^\tau \quad . \tag{29}$$

If α = + or -, let $H_B(M)^\alpha$ denote the subspace of $H_B(M)$ on which F_∞ acts via the sign α, and let $d^\alpha(M)$ denote its K- dimension.

Lemma 2. Let W be the motive given by (28). Then, up to multiplication by an element of K^*, $c^+(W)$ coincides with

$$c^+(M)((2\pi i)^n\, \delta_\rho(\phi))^{d^+(M)}\,. \tag{30}$$

Proof. Let $T = M(n)$, and put $\varepsilon = \phi(\tau_\infty)$. Then (28) implies (see [6], p. 329) that

$$c^\varepsilon(T) = (2\pi i)^{nd^+(M)}\, c^+(M) \ . \tag{31}$$

Now $W = T(\phi)$, and (28) gives immediately

$$H_B(W)^+ = H_B(T)^\varepsilon \otimes_K H_B(\phi)\ ,\ H_{DR}(W)^+ = H_{DR}(T)^\varepsilon \otimes_K H_{DR}(\phi)\,.$$

By definition, $c^+(W)$ is the determinant of the comparison isomorphism

$$g_{R,\infty}{}^+ : H_B(W)^+ \otimes \mathbf{C} \to H_{DR}(W)^+ \otimes \mathbf{C}\ ,$$

computed relative to K-bases of the two sides (which each have cardinality equal to $d^+(M)$). Now a K-basis of the left hand side is given by $\{\alpha_i\otimes 1\otimes 1\}$, where $\{\alpha_i\}$ is a K-basis of $H_B(T)^\varepsilon$. Similarly, a K-basis of the right hand side is given by $\{\beta_i\otimes g\otimes 1\}$, where g is any non-zero element of $H_{DR}(\phi)$ and $\{\beta_i\}$ is a K-basis of $H_{DR}(T)^\varepsilon$. Thus $c^+(W)$ coincides, up to multiplication by an element of K^*, with coincides with $c^\varepsilon(T)g^{-d^+(M)}$. The assertion of the lemma now follows from (31), and the fact that, as remarked in §6 of [6], we can take $g=\delta_{-\rho}(\phi^{-1})/f(\phi)$, whence $g^{-1} = \delta_\rho(\phi)$.

Recall that an integer s = n is said to be *critical* for M if both the infinite Euler factors $L_\infty(\sigma,M, s)$ and $L_\infty(\sigma,M^\wedge(1), -s)$ are holomorphic at s = n. The following lemma (due to Bloch, Deligne, Scholl, ...) gives several useful equivalent forms of this definition. As before, we write $h(j,k)$ = C-dimension of $H^{j,k}(\sigma) = H^{j,k}(\sigma, M)$ (see (24) and (25)), where $\sigma \in \Sigma(K)$. We recall that both the infinite Euler factors, and these dimensions, are independent of the choice of σ.

Lemma 3. The following three assertions are equivalent for M:- (i). M is critical at $s = 0$; (ii). If $j < k$ and $h(j,k) \neq 0$, then $j < 0$ and $k \geq 0$, and, in addition, if $h(k,k) \neq 0$, then F_∞ acts on $H^{k,k}(\sigma)$ by $+1$ if $k < 0$ and by -1 if $k \geq 0$; (iii). The map

$$h_\infty : H_B(M)^+ \otimes \mathbf{R} \rightarrow (H_{DR}(M)/F^0H_{DR}(M))\otimes \mathbf{R} \quad , \tag{32}$$

induced from (22) is an isomorphism.

Proof. The equivalence of (i) and (ii) follows from the explicit formulae for the infinite Euler factors given above, and we do not give the details. Assume now that (ii) is valid. It follows that

$$d^+(M) = \Sigma_{(j<0)}\, h(j,k) \quad , \quad d^-(M) = \Sigma_{(j\geq 0)}\, h(j,k) \quad . \tag{33}$$

It follows from (23) and these formulae that the two sides of (32) have the same **R** - dimension, and so it suffices to prove (32) is injective. Again by (23), h_∞ will certainly be injective if

$$(H_B(M)^+\otimes \mathbf{C}) \cap (\oplus_{(\sigma)}\oplus_{(j\geq 0)} H^{j,k}(\sigma)) = 0 \ . \tag{34}$$

Let a **C** - basis of $H^{j,k}(\sigma)$ be given by $\{e_i(\sigma,j,k) : i = 1,\ldots,h(j,k)\}$. Then a **C** - basis of $H_B(W)^+\otimes \mathbf{C}$ is given by the set

$$\{e_i(\sigma,j,k) + F_\infty(e_i(\sigma,j,k)) : j < 0 \leq k \ , i = 1,\ldots,h_i(j,k),\ \sigma\in \Sigma(K)\},$$

together with the set $\{e_i(\sigma,k,k) : i = 1,\ldots,h_i(k,k),\ \sigma\in \Sigma(K)\}$ if $k = w(W)/2$ is < 0. Hence any non-zero element of $H_B(W)^+\otimes \mathbf{C}$ will have a non-zero projection on at least one of the subspaces $H^{j,k}(\sigma)$ with $j < 0$. This proves (34), and so also (iii). Conversely, assume (iii) holds. The equality of dimensions on the two sides of (32) shows that (33) is then valid. But, if $j < k$, then the space $H^{j,k} \oplus H^{k,j}$ contributes $h(j,k)$ to both $d^+(M)$ and $d^-(M)$, and so it follows from (33) that we must have $j < 0 \leq k$. If $H^{k,k} \neq 0$, we also conclude from (33) that F_∞ acts on $H^{k,k}$ by $+1$ if $k < 0$, and by -1 if $k \geq 0$. This establishes (ii), and completes the proof of the lemma.

Fix an embedding σ in $\Sigma(K)$. If v is any place of $\mathbf{Q}$, we define

$$R_v(\sigma, M, \rho, s) = L_v(\sigma, M, s)/(\varepsilon_v(\sigma, M, \rho, s)L_v(\sigma, M^\wedge(1), -s)). \tag{35}$$

As is already noted in [6] when v is non-archimedean (see Remark 5.2.1, p. 329), this ratio tends to be better behaved that the individual factors defining it. We shall exploit this fact in what follows. Clearly, we have

$$R_v(\sigma, M, \rho, s) = R_v(\sigma, M^\wedge(1), -\rho, -s)^{-1}.$$

It is therefore natural to ask whether one can define *canonical* new factors $E_v(\sigma, M, \rho, s)$ such that

$$R_v(\sigma, M, \rho, s) = E_v(\sigma, M, \rho, s)/E_v(\sigma, M^\wedge(1), -\rho, -s) \tag{36}$$

(of course, this last equation cannot characterize the factors $E_v(\sigma, M, \rho, s)$). In fact this is the case, as we shall subsequently explain. Note one immediate consequence of such a construction. Let S be any finite set of primes of $\mathbf{Q}$. Define the modified L-function

$$\Lambda_{(S)}(\sigma, M, \rho, s) = \Pi_{(v \in S)} E_v(\sigma, M, \rho, s) \,.\, \Pi L_v(\sigma, M, \rho, s),$$

where the latter product is over primes v not in S. Then we have the following modified form of the functional equation (24)

$$\Lambda_{(S)}(\sigma, M, \rho, s) = (\Pi_{(v \notin S)} \varepsilon_v(\sigma, M, \rho, s))\, \Lambda_{(S)}(\sigma, M^\wedge(1), -\rho, -s). \tag{37}$$

We now give the definition of the E_v - factors when $v = \infty$. For s in C, put

$$\rho^{-s} = \exp(-\rho\pi s/2), \quad \Gamma_{C,\rho}(s) = \rho^{-s}\Gamma_C(s).$$

We also recall that the Euler factor at ∞ is independent of the choice of the embedding σ. Similarly, the ε - factor at ∞ is independent of the choice of σ (see, for example, the explicit formulae given on p. 329 of [6]), so that we may drop the σ from our notation in this case. We then define

$$E_\infty(M, \rho, s) = \Pi\, E_\infty(U, \rho, s), \tag{38}$$

where U runs over the direct summands of the Hodge decomposition, and $E_\infty(U, \rho, s)$ is given explicitly by :-

(a) If $U = H^{j,k}(\sigma) \oplus H^{k,j}(\sigma)$ with $j<k$, then $E_\infty(U, \rho, s) = \Gamma_{C,\rho}(s - j)^{h(j,k)}$;

(b) If $U = H^{k,k}(\sigma)$ with $k \geq 0$, then $E_\infty(U, \rho, s) = 1$;

(c) If $U = H^{k,k}(\sigma)$ with $k<0$, then $E_\infty(U, \rho, s) = R_\infty(U, \rho , s)$.

From the table of values of the $\varepsilon_\infty(U, \rho, s)$ given on p. 329 of [6], we deduce easily that (36) is valid in case (a), and it is plainly true in cases (b) and (c).

If u and v are complex numbers, we write $u \sim v$ if there exists y in $\mathbf{Q}^*$ such that $u = yv$. The integer $r(M)$ defined by

$$r(M) = \Sigma_{(j<0)}\ jh(j,k)$$

plays an important role in what follows, thanks to the next crucial lemma.

Lemma 4. Assume that M is critical at $s = 0$. Then

$$E_\infty(M, \rho, 0) \sim (2\pi\rho)^{r(M)} , \tag{39}$$

where the rational number implicit in the $\sim$ is independent of the choice of ρ.

Corollary. Assume that M is critical at $s = 0$, and that W given by (28) is also critical at $s = 0$. Then

$$E_\infty(W, \rho, 0) \sim E_\infty(M, \rho, 0)(2\pi\rho)^{-nd^+(M)} . \tag{40}$$

To deduce the corollary, we first note that $h_M(j,k) = h_W(j\text{-}n, k\text{-}n)$, and that $d^+(M) = d^+(W)$, because of our hypothesis that $\phi(\tau_\infty) = (-1)^n$. As M is critical at $s = 0$, Lemma 3 shows that $d^+(M)$ is given by (33). On the other hand, since W is critical at $s = 0$, (ii) of Lemma 3 shows that $j\text{-}n < 0$ if and only if $j < 0$. Hence $r(W) = r(M) - nd^+(M)$, as required.

We now turn to the proof of Lemma 4. We prove the lemma by considering the three cases (a), (b), (c) for U above, and verifying (39) for each U, where $r(U) = jh(j,k)$ in case (a), $r(U) = 0$ in case (b), and $r(U) = kh(k,k)$ in case (c). This suffices since clearly $r(M) = \Sigma_{(U)} r(U)$. If s is in **Z** , we shall make use of the following classical properties of the Γ-function (see [6], p. 330) :-

$$\Gamma_{\mathbf{C}}(s) \sim (2\pi)^{-s} \ (s>0), \quad \Gamma_{\mathbf{R}}(s) \sim (2\pi)^{(1-s)/2} \ (s \text{ odd}),$$
$$\Gamma_{\mathbf{R}}(s) \sim (2\pi)^{-s/2} \ (s>0 \text{ and even}).$$

Suppose we are in case (a). Then

$$E_\infty(U, \rho, 0) = (\rho^j \, \Gamma_{\mathbf{C}}(-j))^{h(j,k)} \sim (2\pi\rho)^{r(U)} ,$$

as required. In case (b), (39) is plainly valid. Suppose finally that we are in case (c) (the one delicate case). Put $h = h(k,k)$. There are two possibilities, according as k is even or odd. (i). Assume k is even, so that F_∞ acts on U by $(-1)^k$. Then $\varepsilon_\infty(U, \rho, 0) = 1$ by the table on p. 329 of [6], and we have

$$L_\infty(U, 0) = \Gamma_{\mathbf{R}}(-k)^h \sim (2\pi)^{r(U)/2}, \ L_\infty(U\hat{\ }(1), 0) = \Gamma_{\mathbf{R}}(k+1)^h \sim (2\pi)^{r(U)/2},$$

whence (39) holds. (ii). Assume k is odd, so that F_∞ acts on U by $(-1)^{k+1}$. Then $\varepsilon_\infty(U, \rho, 0) = \rho^{kh}$ by the table on p. 329 of [6], and we have

$$L_\infty(U,0) = \Gamma_{\mathbf{R}}(1-k)^h \sim (2\pi)^{(k-1)h/2}, \ L_\infty(U\hat{\ }(1),0) = \Gamma_{\mathbf{R}}(j+2)^h \sim (2\pi)^{-(k+1)h/2},$$

whence (39) is again plain. The reader should also note that the unknown non-zero rational number implicit in (39) is independent of the choice of ρ.This completes the proof of Lemma 4.

We can now give the modified form of the period conjecture. Recall that we identify $K \otimes \mathbf{C}$ with $\mathbf{C}^{\Sigma(K)}$ via the isomorphism (20). Let $c^+(M) \in (K \otimes \mathbf{C})^*$ be the Deligne period of M as defined in §2 of [6], so that we can view $c^+(M)$ under the isomorphism (20) as a vector

$$c^+(M) = (c^+(\sigma, M))_{(\sigma \in \Sigma(K))} , \qquad (41)$$

whose components are well defined up to multiplication by a system of numbers $\sigma(\alpha)$ $(\sigma \in \Sigma(K))$, for any α in K^*. For each choice of $\rho = i$ or $-i$, we now put

$$\Omega_\rho(M) = (\Omega_\rho(\sigma, M)) = c^+(M)(2\pi\rho)^{r(M)} . \tag{42}$$

If ϕ is a character of finite order of G^{ab} with values in K, we also recall that $\delta_\rho(\phi)$ is given by (29), and its image under the isomorphism (20) is $\delta_\rho(\phi) = (\delta_\rho(\sigma, \phi))$, where

$$\delta_\rho(\sigma, \phi) = \Sigma_{(\tau \in \Delta(\phi))} (\phi^\sigma(\tau))^{-1}(\exp(-2\pi\rho/f(\phi)))^\tau \ . \tag{43}$$

Lemma 5. Assume that M is critical at s = 0, and that W given by (28) is also critical at s = 0. Then, for each $\sigma \in \Sigma(K)$, the quantity

$$\Lambda_{(\infty)}(\sigma, W, \rho)\ \Omega_\rho(\sigma, M)^{-1}\ \delta_\rho(\sigma, \phi)^{-d^+(M)} \tag{44}$$

does not depend on the choice of $\rho = i$ or $-i$.

Proof. It is plain that

$$\delta_{-\rho}(\sigma, \phi) = \phi^\sigma(\tau_\infty)\ \delta_\rho(\sigma, \phi). \tag{45}$$

The assertion of the lemma now follows from (28), Lemma 4, and the fact noted earlier that $r(W) = r(M) - nd^+(M)$.

Period Conjecture. Assume that M is critical at s = 0, and that W given by (28) is also critical at s = 0. Then there exists $\alpha \in K$ such that the expression (44) is of the form $\sigma(\alpha)$ for all $\sigma \in \Sigma(K)$.

Indeed, by Lemmas 3 and 4, we see that this conjecture is equivalent to Conjecture 2.8 of [6], applied to the motive W.

6. Modification of the Euler factor at p. We again let M be any motive over **Q** with coefficients in K. For this section, we drop the assumption that M is critical at s = 0, as it will not be needed. Let p be any prime number - the only restriction placed on p is given by Hypothesis I(p) below. Our aim in this section is to define a modification of the Euler factor at p, which is analogous to that already given for the Euler factor at ∞. Throughout, σ will denote any element of $\Sigma(K)$, and ρ will again denote i or -i.

Let G_p denote the absolute Galois group of $\mathbf{Q}_p$, and let I_p (resp. W_p) be the subgroup of G_p given by the inertial subgroup (resp. the Weil group). We fix an element Φ of G_p, whose image in G_p/I_p is the geometric Frobenius (i.e. the inverse of Frob_p). For each $s \in \mathbf{C}$, let

$$\omega_s : W_p \to \mathbf{C}^*$$

be the homomorphism which is trivial on I_p, and which satisfies $\omega_s(\Phi) = p^{-s}$. We also fix a prime number $l \neq p$, and a non-zero homomorphism $t_l : I_p \to \mathbf{Z}_l$. Let λ denote a prime of K above l, and K_λ the completion at λ. Now write W_p' for the Weil - Deligne group of $\mathbf{Q}_p$. Recall that the representations of W_p' are defined as follows (see [5], §8). Let V be a finite dimensional vector space over K_λ. Then a representation of W_p' in V is a pair $\Theta = (\gamma, N)$, where (i) $\gamma : W_p \to GL(V)$ is a homomorphism, whose kernel contains an open subgroup of I_p, and (ii) N is a nilpotent endomorphism of V such that

$$\gamma(\sigma)\, N\, \gamma(\sigma)^{-1} = \omega_1(\sigma)\, N \qquad \text{for all } \sigma \text{ in } W_p .$$

Given such a representation Θ, we can define the dual representation $\Theta^\wedge = (\gamma^\wedge, N^\wedge)$, where $\gamma^\wedge$ is the contragredient representation. Writing $I = I_p$ for brevity, we define

$$V_N = \mathrm{Ker}\,(N)\ , \quad Z_p(\Theta, X) = \det(\,1 - \gamma(\Phi)\, X \mid V_N^{\gamma(I)}\,)^{-1} .$$

Let $\sigma : K_\lambda \to \mathbf{C}$ denote a fixed extension of the embedding σ in $\Sigma(K)$. We then put $L_p(\sigma, \Theta, s) = (\sigma Z_p)(\Theta, p^{-s})$. Write $\varepsilon_p(\sigma, \gamma\otimes\omega_s, \rho)$ for Deligne's ε-factor attached to the representation $\gamma\otimes\omega_s$ of W_p on the complex vector space $V\otimes_{(K_\lambda, \sigma)}\mathbf{C}$, and define

$$\varepsilon_p(\sigma, \Theta, \rho\, , s) = \varepsilon_p(\sigma, \gamma\otimes\omega_s\, , \rho)\, \det(\, - \Phi.\, p^{-s} \mid V^{\gamma(I)}/V_N^{\gamma(I)}\,) .$$

By analogy with (35), we can now define $R_p(\sigma, \Theta, \rho, s)$ by simply replacing M throughout by Θ in the formula (35). As pointed out in Remark 5.2.1 of [6], the expression $R_p(\sigma, \Theta, \rho, s)$ is particularly well behaved. In particular, it does not change if we replace the representation $\Theta = (\gamma, N)$ by $\Theta' = (\gamma, 0)$. Finally, we recall (see [5], §8.5) that, for each representation $\Theta = (\gamma, N)$ of W_p', one can define a new representation $\Theta^{ss} = (\gamma^{ss}, N)$ called the Φ-semisimplification of Θ, which has the property that γ^{ss} is a semisimple representation of the ordinary Weil group W_p. Again, $R_p(\sigma, \Theta, \rho, s)$ does not change if we replace Θ by Θ^{ss}.

Now let us return to the λ-adic representation of W_p given by its natural action on $H_\lambda(M)$. By Grothendieck's theorem, this λ-adic representation gives rise to a unique representation $\Theta = (\gamma, N)$ of the Weil - Deligne group W_p' (see [5], §8).

Lemma 6. There exists a representation $\Theta' = (\gamma', N')$ of the Weil-Deligne group in $H_\lambda(M)$, which satisfies :- (i). $N' = 0$; (ii). If we extend scalars from K_λ to $\mathbf{C}$ via the embedding σ, then γ' is a semisimple complex representation of W_p ; and (iii). We have

$$R_p(\sigma, M, \rho, s) = R_p(\sigma, \Theta', \rho, s). \tag{46}$$

Proof. By the construction of the representation Θ via Grothendieck's theorem, we have that (46) is valid with Θ replaced by Θ'. On the other hand, it was remarked above that $R_p(\sigma, \Theta, \rho, s)$ does not change if we replace Θ by $\Theta_1 = (\gamma, 0)$, and subseqently Θ_1 by $\Theta' = \Theta_1{}^{ss}$. It is plain that this choice of Θ' satisfies the assertions of the lemma.

We can now define the factors $E_p(\sigma, M, \rho, s)$ satisfying (36). Let

$$\gamma' : W_p \to GL(Y), \quad \text{where } Y = H_\lambda(M) \otimes_{(K_\lambda, \sigma)} \mathbf{C} \tag{47}$$

be the semisimple complex representation of the Weil group given by Lemma 6. Let $Y = \oplus\, U$ be the decomposition of Y into irreducible complex representations of W_p. For each such representation U, we can define the expression $R_p(\sigma, U, \rho, s)$ by the formula (35) with M replaced by U, and, in view of (46), we have

$$R_p(\sigma, M, \rho, s) = \Pi_{(U)} R_p(\sigma, U, \rho, s). \tag{48}$$

Each U occurring in this decomposition is irreducible, and hence is known (see [5] , §4.10) to be of the form $\xi_U \otimes \omega_{s(U)}$, where s(U) is some complex number, and ξ_U is a complex representation of the Weil group such that $\xi_U(W_p)$ is a finite group. Consequently, the inverse roots of the polynomial det $(1 - \Phi. X \mid U)$ (note that we do not take the subspace of U fixed by I, but rather the whole of U) are all of the form a root of unity times one fixed root. Thus, assuming that these inverse roots are algebraic numbers, and viewing them as lying inside C_p via the embedding (1), we can unambiguously define $ord_p(U)$ to be $ord_p(\alpha)$ for any inverse root α of this polynomial; here ord_p denotes the order valuation of C_p , normalized so that $ord_p(p) = 1$. Note also that $ord_p(U)$ is independent of the choice of Φ, since the image of I in GL(U) is a finite group. Clearly, we have $ord_p(U\hat{}(1)) = - ord_p(U) - 1$, so that it is natural to impose the following hypothesis :-

Hypothesis I(p). For each U occurring in the above decomposition, we have $ord_p(U) \neq -1/2$.

For the rest of the paper, we assume that Hypothesis I(p) is valid for our motive M. We then define :-

(a). If $ord_p(U) > -1/2$, then $E_p(\sigma, U, \rho, s) = 1$;

(b). If $ord_p(U) < -1/2$, then $E_p(\sigma, U, \rho, s) = R_p(\sigma, U, \rho, s)$.

Note that the case (a) holds for U if and only if case (b) holds for $U\hat{}(1)$, because of Hypothesis I(p). Thus, putting

$$E_p(\sigma, M, \rho, s) = \Pi_{(U)} E_p(\sigma, U, \rho, s),$$

it follows from (48) that the equation (36) is valid, as required.

We now explicitly calculate the the E_p - factors in some simple cases. Let $d_p(\sigma,M)$ denote the number of inverse roots α of the polynomial

$$(\sigma Z_p(M, X))^{-1} = \sigma \det (1 - \Phi. X \mid H_\lambda(M)^I) \tag{49}$$

which satisfy $ord_p(\alpha) < 0$ (by hypothesis, the coefficients of this polynomial are algebraic numbers in C, and we view these as lying in

C_p via the embedding (1)). As usual, we say that M has good reduction at p if the inertia group $I = I_p$ acts trivially on $H_\lambda(M)$ for any prime λ of K not dividing p.

Lemma 7. Assume that M has good reduction at p. Let β (resp. α) run over all inverse roots, counted with multiplicity, of the polynomial (49) such that $\mathrm{ord}_p(\beta) > -1/2$ (resp. $\mathrm{ord}_p(\alpha) < -1/2$). Then

$$E_p(\sigma, M, \rho, s)/L_p(\sigma, M, s) = \Pi_{(\beta)}(1 - \beta p^{-s}).\ \Pi_{(\alpha)}(1 - \alpha^{-1}p^{s-1}).$$

Moreover, if ϕ is a non-trivial character of finite order of G(P/Q), we have

$$E_p(\sigma, M(\phi), \rho, s)/L_p(\sigma, M(\phi), s) = (\delta_p(\sigma, \phi)\, c(\phi)^{-s})^{-d_p(\sigma, M)}\, (\Pi_{(\alpha)}\ \alpha)^{-h(\phi)},$$

where $\delta_p(\sigma, \phi)$ is the Gauss sum given by (43), and $c(\phi) = p^{h(\phi)}$ is the conductor of ϕ.

Proof. The first assertion is immediate from the definitions because $\varepsilon_p(\sigma, U, \rho, s) = 1$, since U is an unramified representation of W_p . To prove the second, we note that our hypotheses that M has good reduction at p and that p actually divides the conductor of ϕ, imply easily that

$$E_p(\sigma, U(\phi), \rho, s)/L_p(\sigma, U(\phi), s)$$

is equal to 1 or $\varepsilon_p(\sigma, U(\phi), \rho, s)^{-1}$, according as $\mathrm{ord}_p(U)$ is $> -1/2$ or $< -1/2$. Let U be such that $\mathrm{ord}_p(U) < -1/2$. As U is an unramified representation of the Weil group, a standard formula (see (3.4.6) on p.15 of [14]) shows that

$$\varepsilon_p(\sigma, U(\phi), \rho, s) = \varepsilon_p(\sigma, \phi, \rho, s)^{\dim(U)}\ .\ (\det U)(\Phi^{h(\phi)})\ .$$

But $(\det U)(\Phi)$ = the product of the inverse roots of $\det(1 - \Phi X \mid U)$, and it is well known and readily verified that

$$\varepsilon_p(\sigma, \phi, \rho, s) = \delta_p(\sigma, \phi).\ c(\phi)^{-s}.$$

The second assertion now follows.

7. p - adic L-functions. We now take N to be any fixed motive over **Q** with coefficients in **Q** itself. Subsequently, we shall take the motive M considered in the earlier sections to be the extension of scalars of N to a variable finite extension K of **Q**. Our aim is to propose a definition of the p-adic L-function of N. While it is very probable that such p-adic analogues exist for all primes p, we can only make precise conjectures at present when N has good *ordinary* reduction at p. The definition of good reduction at p is given at the end of the previous section. The ordinarity hypothesis is the following condition on the p-adic realisation $V = H_p(N)$ of N as a representation for the local Galois group G_p of the algebraic closure of $\mathbf{Q}_p$ over $\mathbf{Q}_p$. There exists a decreasing filtration F^mV of V (with $F^mV = V$ (resp. 0) for m sufficiently small (resp. large)) of $\mathbf{Q}_p$-subspaces, which are stable under the action of G_p, such that, for all m in **Z**, G_p acts on $F^mV/F^{m+1}V$ via ψ^m; here ψ is the p-adic cyclotomic character (3). We suppose henceforth that N has good ordinary reduction at p. The same is then easily seen to be true for the motive N^(1).

We shall require two additional hypotheses, which are known in many cases, but which we must impose as axioms because of our naive definition of motives. It is well known (see, for example, [9], §6) that our hypothesis that V is ordinary at p implies that it is of Hodge-Tate type. We recall that this means the following. For each n in **Z**, let $C_p(n)$ be the 1-dimensional vector space over C_p, on which G_p acts via the normal action twisted by ψ^n. Then there is an isomorphism of G_p-modules

$$V\otimes C_p \cong \oplus_{(i \in \mathbf{Z})} C_p(-i)^{h(i)} \quad \text{, where } h(i) = \dim F^iV/F^{i+1}V; \qquad (50)$$

here the tensor product on the left is over $\mathbf{Q}_p$, and where G_p acts on this tensor product in the natural fashion, i.e. $\sigma(u\otimes v) = \sigma(u)\otimes\sigma(v)$ for all σ in G_p. The fist condition we impose is that the integers h(i) appearing in (50) are related to the complex Hodge numbers by

$$h(i) = h(i, w(N) - i) \text{ for all i in } \mathbf{Z}, \qquad (51)$$

where we recall that w(N) is the weight of the motive N. In fact, (51) has been proven by Faltings [7] when N is of the form $H^k(X)(n)$, for a smooth projective variety X over **Q** . As before, let

$$Z_p(N, X)^{-1} = \det(1 - \Phi . X \mid H_l(N)),$$

where l is any prime distinct from p. Let d = d(N), and let $\alpha_1, \dots, \alpha_d$ be the inverse roots in $\mathbf{C}_p$ of this polynomial, taken with multiplicity. Our second assumption is that, for each i in **Z**, the number of these inverse roots, counted with multiplicity, which satisfy $\mathrm{ord}_p(\alpha) = i$ is equal to the complex Hodge number h(i, w(N) - i). Note, in particular, that this implies that, in the ordinary case, the p-adic order of these inverse roots is an integer, and so Hypothesis I(p) is automatically valid. I understand that the second assumption is known to be true, by the work of Fontaine and Messing (see [8]), when N has good ordinary reduction and is of the form $H^k(X)(n)$, for a smooth projective variety X over **Q**. We assume from now on that these additional hypotheses are valid. This implies their validity for N^(1).

Lemma 8. Assume that N is critical at s = 0. Let α run over the inverse roots of the polynomial $Z_p(N, X)^{-1}$. The number of these α, counted with multiplicity, satisfying $\mathrm{ord}_p(\alpha) < 0$ is equal to $d^+(N)$. In other words, we have $d_p(N) = d^+(N)$.

Proof. This is plain from (33) and the second additional assumption made above.

Recall that X is the group of all continuous homomorphisms from the Galois group J = G(H/**Q**) to $\mathbf{C}_p^*$. We write X_{alg} for the subgroup of X consisting of all ξ of the form (11). For such a ξ, we take K to be the finite extension of **Q** generated by the values of χ, and let M be the motive over **Q** , which is given by extending the scalars of N to K in the obvious fashion. We then define

$$N(\xi) = M(n)(\chi). \tag{52}$$

Note that our fixed choice of the embedding (1), together with the fact that we take A to be the algebraic closure of **Q** in **C**, implies that K is actually given with a canonical embedding $\iota : K \to \mathbf{C}$. In the following, it is convenient and harmless to systematically omit the embedding ι from the notation, and simply write $\Lambda(N(\xi), s)$ instead of $\Lambda(\iota, N(\xi), s)$, etc.

We next consider a question which is important for the study the poles of both complex and p-adic L-functions. Let $Y_p(N)$ be the subspace of $H_p(N)$ given by

$$Y_p(N) = H_p(N)^{G(A/H)} \quad . \tag{53}$$

Since $G(A/H)$ is a normal subgroup of G, it is plain that $Y_p(N)$ is stable under the action of G, and so provides an abelian p-adic representation of G.

Lemma 9. Endowing $Y_p(N) \otimes \mathbf{C}_p$ with the linear action of G (i.e. $\sigma(u\otimes b) = \sigma(u)\otimes b$ for σ in G), it breaks up as a direct sum of G-modules

$$Y_p(N) \otimes \mathbf{C}_p \;\cong\; \oplus_{(\xi \in B(N))}\, \xi^{e(\xi)} \quad , \tag{54}$$

where $B(N)$ is some finite subset of X_{alg}, and the $e(\xi)$ are integers ≥ 1. Moreover, each $\xi \in B(N)$ is of the form $\xi = \psi^n\chi$, where $n = -\,w(M)/2$ and χ is a character of finite order of $G(P/\mathbf{Q})$.

Proof. Since the representation factors through the Galois group J, and the decomposition group of p in J is equal to J, it suffices to establish (54) as an isomorphism of G_p - modules. Now, viewed as a G_p - module, $Y_p(N)$ is of Hodge-Tate type, because it is easily seen that a sub-representation of a Hodge-Tate representation is again Hodge-Tate. Thus $Y_p(N)$ is an abelian p-adic Hodge-Tate representation of G_p. By a theorem of Tate ([11], §III - 7), this implies that $Y_p(N)$ is locally algebraic (note that in [11], it is necessary to assume that the restriction of the representation to the inertia group is semisimple, but it is pointed out in [12], §2 that this condition is automatically true). Since p is totally ramified in the fixed field of the kernel of the representation on $Y_p(N)$, it follows that, as a G_p-module, $Y_p(N)$ is a direct sum of simple locally algebraic abelian representations. Extending scalars to $\mathbf{C}_p$, we conclude that it is a direct sum of locally algebraic characters of G_p, which factor through J. But such characters are precisely the elements of X_{alg}, and so (54) follows. The final assertion is a consequence of the fact that, for any good prime $q \neq p$, the reciprocal complex roots of $\det(1 - \mathrm{Frob}_q^{-1}.X \mid Y_p(N))$ must have complex absolute value equal to $q^{w(N)/2}$, because of our hypothesis that N has weight w(N).

It is conjectured that the integers $e(\xi)$ occurring in the decomposition (54) are related to the poles of the complex L-functions by

$$e(\xi) = \text{order of pole of } L(M(\xi^{-1}), s) \text{ at } s = 1. \tag{55}$$

Moreover, for $\xi \in B(N)$, the function $L(M(\xi^{-1}), s)$ should be holomorphic at all points $s \neq 1$, and, for ξ in X_{alg} but not in B(N), this function should be entire.

Assume from now on that our motive N is critical at s=0. We then consider variable twists of N of the form $N(\xi)$, with ξ in X_{alg}, which are also critical at s = 0 (infinitely many such ξ clearly exist, since we can, in particular, take ξ to be any character of finite order of J). Let $c^+(N)$ be the Deligne period of N (it is well defined up to multiplication by a non-zero element of $\mathbb{Q}$). As earlier, let $r(N) = \Sigma_{(j<0)} jh(j,k)$. We assume the strong form of the Period Conjecture which is explained in §5. As always, let ρ denote i or -i.

Conjecture A (p-adic version). Assume that N is critical at s = 0, and let p be a good ordinary prime for N. For each choice of the Deligne period $c^+(N)$, there exists a unique pseudo-measure $\mu(c^+(N))$ on J as follows: for all ξ in X_{alg} such that (i) $N(\xi)$ is also critical at s = 0, and (ii) ξ^{-1} does not belong to B(N) and ξ does not belong to $B(N\hat{\ }(1))$, we have

$$\int_J \xi \, d\mu(c^+(N)) = \Lambda_{(S)}(N(\xi), \rho, 0)/(c^+(N)(2\pi\rho)^{r(N)}) \;, \tag{56}$$

where $S = \{\infty, p\}$, and $\Lambda_{(S)}(N(\xi), \rho, s)$ is the modified L-function defined in §5, for the standard embedding $\iota : K \to \mathbb{C}$.

Remarks.

1. Using Lemmas 5, 7, and 8, we see easily that the right hand side of (56) is independent of the choice of ρ, and hence so is the pseudo-measure $\mu(c^+(N))$.

2. Using the Period Conjecture of §5, and taking into account all the embeddings of K in **C** , one can show that the pseudo-measure $\mu(c^+(N))$ takes values in $\mathbf{Q}_p$.

3. The following conjecture about the possible poles of the pseudo-measure $\mu(c^+(N))$ should replace that proposed in our earlier papers [3], [4]. Our previous conjecture was too strong because it failed to take into account possible twisting by the characters of finite order of the Galois group J.

Holomorphy Conjecture (p-adic version). Let B(N) and B(N^(1)) be the subsets of X_{alg} occurring in the decomposition (54) for N and N^(1), respectively. Then there exists a non-zero b in $\mathbf{Z}_p$ such that

$$b\,\Pi_{(\xi \in B(N))}\,(\xi^{-1}(\sigma(\xi)) - \sigma(\xi))^{e(\xi)}\;\Pi_{(\eta \in B(N\wedge(1)))}\,(\eta(\sigma(\eta)) - \sigma(\eta))^{e(\eta)}\,\mu(c^+(M))$$

belongs to the Iwasawa algebra $\mathfrak{I} = Z_p[[\,J\,]]$, for all choices of $\sigma(\xi)$ and $\sigma(\eta)$ in J.

In parallel with (55), it seems reasonable to conjecture the even stronger assertion, that $\mu(c^+(N))$ will have poles of exact order $e(\xi)$ at ξ^{-1} for ξ in B(N), and of exact order $e(\eta)$ at each η in B(N^(1)).

4. If N = **Q**(m), where m is an odd negative integer, then N is critical at s = 0, and it is easily seen that Theorem 1 (with $\phi = 1$) shows that both Conjecture A and the Holomorphy Conjecture above do indeed hold for this motive. For further examples, see [4].

The pseudo-measure $\mu(c^+(N))$ satisfies a simple p-adic analogue of the functional equation of the complex L-function. Recall that $\mu \to \mu^\#$ is the involution of the ring of pseudo-measures on J, which is induced by sending σ in J to σ^{-1}. The conductor of N is an integral ideal of **Z** , which is prime to p because N has good reduction at p, and we write $\sigma(N)$ for its Artin symbol in J.

Put

$$\gamma(N) = \varepsilon(N,0)\,.\,(2\pi\rho)^{r(N)-r(N^\wedge(1))}(-1)^{r(N^\wedge(1))}\varepsilon_\infty(N,\rho,0)^{-1}.c^+(N)/c^+(N^\wedge(1)).$$

This number does depend on the choice of the periods $c^+(N)$, and $c^+(N^\wedge(1))$, but the arguments of §5 show that it is independent of the choice of ρ. Moreover, the arguments of [6], §5 prove that it lies in **Q**.

p - adic functional equation. We have

$$\mu(c^+(N)) = \gamma(N).\ \mu(c^+(N^\wedge(1))^{\#}\ .\ \sigma(M)^{\#}\ , \tag{57}$$

where $\sigma(M)$ denotes the Artin symbol of the conductor of M in the Galois group J.

Proof. It suffices to show that, for any ξ in X_{alg} satisfying the conditions set out in Conjecture A above, the integrals of ξ against both sides of (57) are equal. This follows immediately by combining (56), the modified functional equation (37), and the well known formula that, for $q \neq \infty$, p, we have

$$\varepsilon_q(N(\xi),\rho,0) = \varepsilon_q(N,\rho,0)\ \xi(\mathrm{Frob}_q^{-1})^{a(q)}\ ,$$

where $q^{a(q)}$ is the power of q in the conductor of M; this latter formula is valid because ξ is unramified at q.

In view of (57), we see that $\gamma(N)$ plays the role of a global p-adic ε-factor. We only make one observation here about its properties.

Lemma 10. Assume that w(N) is odd. Then

$$\gamma(N) = \varepsilon(N,0).\ (2\pi)^{(1+w(N))d^+(N)}\ .\ c^+(N)/c^+(N^\wedge(1)). \tag{58}$$

In particular, if $N = N^\wedge(1)$, and if we take $c^+(N) = c^+(N^\wedge(1))$, then

$$\gamma(N) = \varepsilon(N,0)\ . \tag{59}$$

Proof. We note that (59) follows immediately from (58), since $N = N^\wedge(1)$ implies that $w(N) = -1$. To prove (58), we observe that, because w(N) is odd, the only terms in the Hodge - decomposition (21) are the $H^{i,j}(N)$ with $i \neq j$, and hence the explicit formula for the infinite ε-factors, given in [6], §5, shows that

$\varepsilon_\infty(N, \rho, 0) = \rho^{b(N)}$, where $b(N) = \Sigma_{(j<0)} (w(N) - 2j + 1)h(j, w(N) - j)$.

In view of (33), we conclude that

$$\varepsilon_\infty(N, \rho, 0) = \rho^{(1 + w(N))d^+(N)} . (-1)^{r(N)} . \tag{60}$$

On the other hand, because w(N) is odd, it is readily verified that

$$r(N) - r(N^\wedge(1)) = (1 + w(N))d^+(N).$$

Since the right hand side of this last formula is even, (58) now follows from (60) and the definition of $\gamma(N)$.

We conclude by remarking that (59) is exactly what would be predicted by the main conjecture and algebraic arguments involving Iwasawa modules (see Proposition 1 of Greenberg's article in this volume).

References

1. Bloch, S., Kato, K., L-functions and Tamagawa number of motives, to appear.

2. Bloch, S., Kato, K., p-adic étale cohomology, Publ. Math. I.H.E.S. No. 63 (1986), 107-152.

3. Coates, J., Perrin-Riou, B., On p-adic L-functions attached to motives over **Q** , *Algebraic Number Theory*, Advanced Stud. Pure Math. 17 (1989), p. 23-54.

4. Coates, J., On p-adic L-functions, Séminaire Bourbaki, Exp. 701, Astérisque 177-178 (1989), p. 33-59.

5. Deligne, P., Les constantes des équations fonctionelles des fonctions L, Antwerp II, Springer L. N. 349 (1973), p. 501-595.

6. Deligne, P., Valeurs de fonctions L et périodes d'intégrales, Proc. Symp. Pure Math. A.M.S. 33 (1979), Vol. 2, p. 313-346.

7. Faltings, G., p-adic Hodge theory, Journal A.M.S. 1 (1988), p. 255-299.

8. Fontaine, J-M., Messing, W., p-adic periods and p-adic étale cohomology, Contemp. Math. A.M.S. 67 (1987), p. 179-207.

9. Greenberg, R., Iwasawa theory for p-adic representations, *Algebraic Number Theory,* Advanced Stud. Pure Math. 17 (1989), p. 97-137.

10. Perrin - Riou, B., Représentations p-adiques, périodes et fonctions L p-adiques, to appear.

11. Serre, J-P., *Abelian l-adic representations and elliptic curves,* Benjamin, New York, 1968.

12. Serre, J-P., Groupes algébriques associées aux modules de Hodge - Tate, Astérisque 65 (1979), p. 155-188.

13. Serre, J-P., Sur le residu de la fonction zêta p-adique d'un corps de nombres, C.R.A.S. Paris 287 (1978), p. 183-188.

14. Tate, J., Number Theoretic Background, Proc. Symp. Pure Math. A.M.S. 33 (1979), Vol. 2, p. 3-26.

Emmanuel College,
Cambridge CB2 3AP,
England.

The Beilinson conjectures

CHRISTOPHER DENINGER AND ANTHONY J. SCHOLL*

Introduction

The Beilinson conjectures describe the leading coefficients of L-series of varieties over number fields up to rational factors in terms of generalized regulators. We begin with a short but almost selfcontained introduction to this circle of ideas. This is possible by using Bloch's description of Beilinson's motivic cohomology and regulator map in terms of higher Chow groups and generalized cycle maps. Here we follow **[Bl3]** rather closely. We will then sketch how much of the known evidence in favour of these conjectures — to the left of the central point — can be obtained in a uniform way. The basic construction is Beilinson's Eisenstein symbol which will be explained in some detail. Finally in an appendix a map is constructed from higher Chow theory to a suitable Ext-group in the category of mixed motives as defined by Deligne and Jannsen. This smooths the way towards an interpretation of Beilinson's conjectures in terms of a Deligne conjecture for critical mixed motives **[Sc2]**. It also explains how work of Harder **[Ha2]** and Anderson fits into the picture.

For further preliminary reading on the Beilinson conjectures, one should consult the Bourbaki seminar of Soulé **[So1]**, the survey article by Ramakrishnan **[Ra2]** and the introductory article by Schneider **[Sch]**. For the full story see the book **[RSS]** and of course Beilinson's original paper **[Be1]**. Here one will also find the conjectures for the central and near-central points, which for brevity we have omitted here.

* Partially funded by NSF grant DMS-8610730

1. Motivic cohomology

Motivic cohomology is a kind of universal cohomology theory for algebraic varieties. There are two constructions both generalizing ideas from algebraic topology. The first one is due to Beilinson [**Be1**]. He defines motivic cohomology as a suitable graded piece of the γ-filtration on Quillen's algebraic K-groups tensored with $\mathbb{Q}$. This is analogous to the introduction of singular cohomology as a suitable graded piece of topological K-theory by Atiyah in [**At**] 3.2.7.

For smooth varieties there is a second more elementary construction which is due to Bloch [**Bl2,3,4**]. It is modeled on singular cohomology: instead of continuous maps from the n-simplex to a topological space one considers algebraic correspondences from the algebraic n-simplex $\Delta_n \cong \mathbb{A}^n$ to the variety. We proceed with the details:

Let k be a field and set for $n \geq 0$

$$\Delta_n = \operatorname{Spec} k[T_0, \dots, T_n]/(\Sigma T_i - 1).$$

There are face maps

$$\partial_i : \Delta_n \hookrightarrow \Delta_{n+1} \quad \text{for} \quad 0 \leq i \leq n+1 \tag{1.1}$$

which in coordinates are given by

$$\partial_i(t_0, \dots, t_n) = (t_0, \dots, t_{i-1}, 0, t_i \dots, t_n).$$

Let X be an equidimensional scheme over k. A face of $X \times \Delta_m$ is the image of some $X \times \Delta_{m'}$, $m' < m$ under a composition of face maps induced by (1.1)

$$\partial_i : X \times \Delta_n \hookrightarrow X \times \Delta_{n+1}.$$

We denote by $z^q(X,n)$ the free abelian group generated by the irreducible codimension q subvarieties of $X \times \Delta_n$ meeting all faces properly. Here subvarieties $Y_1, Y_2 \subset X \times \Delta_n$ of codimensions c_1, c_2 are said to meet properly if every irreducible component of $Y_1 \cap Y_2$ has codimension $\geq c_1 + c_2$ on X. Observe that $z^q(X,n)$ is a subgroup of correspondences from Δ_n to X. Using the differential

$$d = \sum_{i=0}^{n+1} (-1)^i \partial_i^* : z^q(X, n+1) \to z^q(X,n)$$

one obtains a complex of abelian groups $z^q(X, \cdot)$. If X is a smooth quasiprojective variety over k setting

$$\Gamma_X(q)^{\cdot} = z^q(X, 2q - \cdot)$$

we define:

$$H^p_{\mathcal{M}}(X,\Lambda(q)) = H^p(\Gamma_X(q)^{\cdot} \otimes \Lambda) \tag{1.2}$$

for any ring Λ. By one of the main results of Bloch these groups coincide for $\Lambda = \mathbb{Q}$ with the groups $H^p_{\mathcal{M}}(X,\mathbb{Q}(q))$ defined by Beilinson using algebraic K-theory. Using either definition the following formal properties can be proved:

(1.3) Theorem. *(1) $H^{\cdot}_{\mathcal{M}}(\ \ ,\mathbb{Q}(*))$ is a contravariant functor from the category of smooth quasiprojective varieties over k into the category of bigraded $\mathbb{Q}$-vector spaces. For proper maps $f: X \to Y$ of pure codimension $c = \dim Y - \dim X$ we also have covariant functoriality with a shift of degrees*

$$f_*: H^{\cdot}_{\mathcal{M}}(X,\mathbb{Q}(*)) \to H^{\cdot+2c}_{\mathcal{M}}(Y,\mathbb{Q}(*+c)).$$

(2) There is a cup product which is contravariant functorial, associative and graded commutative with respect to $\cdot$.

(3) There are functorial isomorphisms compatible with the product structure

$$H^{2p}_{\mathcal{M}}(X,\mathbb{Q}(p)) = CH^p(X) \otimes \mathbb{Q}.$$

(4) $H^1_{\mathcal{M}}(X,\mathbb{Q}(1)) = \Gamma(X,\mathcal{O}^*) \otimes \mathbb{Q}$ functorially.

(5) Let $i: Y \hookrightarrow X$ be a closed immersion (of smooth varieties) of codimension c with open complement $j: U = X - Y \hookrightarrow X$. Then there is a functorial long exact localization sequence

$$\cdots \to H^{\cdot-2c}_{\mathcal{M}}(Y,\mathbb{Q}(*-c)) \xrightarrow{i_*} H^{\cdot}_{\mathcal{M}}(X,\mathbb{Q}(*)) \xrightarrow{j^*} H^{\cdot}_{\mathcal{M}}(U,\mathbb{Q}(*))$$
$$\to H^{\cdot+1-2c}_{\mathcal{M}}(Y,\mathbb{Q}(*-c)) \to \cdots.$$

(6) If $\pi: X' \to X$ is a finite galois covering with group G we have $\pi_*\pi^* = |G|\ id$ and $\pi^*\pi_* = \sum_{\sigma\in G}\sigma^*$. In particular

$$\pi^*: H^{\cdot}_{\mathcal{M}}(X,\mathbb{Q}(*)) \xrightarrow{\sim} H^{\cdot}_{\mathcal{M}}(X',\mathbb{Q}(*))^G$$

is an isomorphism, i.e. $H^{\cdot}_{\mathcal{M}}(\ \ ,\mathbb{Q}(*))$ has galois descent.

For zero dimensional X over $\mathbb{Q}$ the motivic cohomology groups are known by the work of Borel [**Bo1,Bo2**] on algebraic K-theory of number fields. A proof of the following result which does not make use of algebraic K-theory seems to be out of reach.

(1.4) Theorem. *Let k be a number field, $X = \operatorname{Spec} k$. Then*

(1)
$$\begin{aligned} \dim_{\mathbb{Q}} H^1_{\mathcal{M}}(X, \mathbb{Q}(q)) &= r_1 + r_2 && \text{if } q > 1 \text{ is odd} \\ &= r_2 && \text{if } q > 1 \text{ is even} \end{aligned}$$

where r_1, r_2 denote the numbers of real resp. complex places of k.

(2)
$$H^p_{\mathcal{M}}(X, \mathbb{Q}(q)) = 0 \quad \text{for} \quad p \neq 1.$$

Observe that for X as in the theorem we have:

$$H^1_{\mathcal{M}}(X, \mathbb{Q}(1)) = k^* \otimes_{\mathbb{Z}} \mathbb{Q}.$$

In view of the class number formula which involves a regulator formed with the units of k we see that for arithmetic purposes the groups $H^p_{\mathcal{M}}(X, \mathbb{Q}(q))$ may have to be replaced by smaller ones:

If X is a variety over $\mathbb{Q}$ we set:

(1.6) $H^p_{\mathcal{M}}(X, \mathbb{Q}(q))_{\mathbb{Z}} = H^p_{\mathcal{M}}(X, \mathbb{Q}(q))$ for $q > p$.

(1.7) $H^p_{\mathcal{M}}(X, \mathbb{Q}(q))_{\mathbb{Z}} = \operatorname{Im}(H^p_{\mathcal{M}}(\mathcal{X}, \mathbb{Q}(q)) \to H^p_{\mathcal{M}}(X, \mathbb{Q}(q))$ for $q \leq p$.

Here $\mathcal{X}$ is a proper regular model of X over $\operatorname{Spec} \mathbb{Z}$ which is supposed to exist. The groups $H^p_{\mathcal{M}}(\mathcal{X}, \mathbb{Q}(q))$ are either defined by the above construction which also works over $\operatorname{Spec} \mathbb{Z}$ or by using the K-theory of $\mathcal{X}$. The "motivic cohomology groups of an integral model" $H^p_{\mathcal{M}}(X, \mathbb{Q}(q))_{\mathbb{Z}}$ are independent of $\mathcal{X}$. It is a conjecture that (1.6) holds if the definition in (1.7) is extended to $q > p$.

2. Deligne cohomology and regulator map

The definition of Deligne cohomology which is about to follow may seem rather unmotivated at first. We refer to (2.9) below where a conceptual interpretation of these groups as Ext's in a category of mixed Hodge structures is described.

(2.1) For a subring Λ of $\mathbb{C}$ we set $\Lambda(q)=(2\pi i)^q\Lambda\subset\mathbb{C}$. Let X be a smooth projective variety over $\mathbb{C}$ and consider the following complex of sheaves on the analytic manifold X_{an}:

$$\mathbb{R}(q)_{\mathcal{D}}=(R(q)\to\mathcal{O}\to\cdots\to\Omega^{q-1})$$

in degrees 0 to q. We set

$$H^p_{\mathcal{D}}(X,\mathbb{R}(q))=H^p(X_{\mathrm{an}},\mathbb{R}(q)_{\mathcal{D}}).$$

Apart from the Deligne cohomology groups we need the singular (Betti) cohomology groups of X_{an}

$$H^p_B(X,\Lambda(q))=H^p_{\mathrm{sing}}(X_{\mathrm{an}},\Lambda(q))$$

and the de Rham groups

$$H^p_{DR}(X)=H^p_{\mathrm{Zar}}(X,\Omega^{\cdot}_{X/\mathbb{C}})\cong H^p(X_{\mathrm{an}},\Omega^{\cdot}).$$

(2.2) If X is smooth projective over $\mathbb{R}$ there is an antiholomorphic involution F_∞ on $(X_{\mathbb{C}})_{\mathrm{an}}$, the infinite Frobenius. We set

$$H^p_{\mathcal{D}}(X,\mathbb{R}(q))=H^p_{\mathcal{D}}(X_{\mathbb{C}},\mathbb{R}(q))^+$$

where the superscript $+$ denotes the fixed module under

$$\overline{F}^*_\infty=F^*_\infty\circ(\text{complex conjugation on the coefficients}).$$

The groups $H^p_B(X,\Lambda(q))$ are defined similarly if $1/2\in\Lambda$. Observe that under the comparison isomorphism

$$H^p_{DR}(X_{\mathbb{C}}/\mathbb{C})\xrightarrow{\sim}H^p_B(X_{\mathbb{C}},\mathbb{C})$$

the de Rham conjugation corresponds to $\overline{F}^*_\infty$ and hence

$$H^p_{DR}(X)\xrightarrow{\sim}H^p_B(X,\mathbb{C}).$$

(2.3) Recall that if $u : \mathcal{A}^{\cdot} \to \mathcal{B}^{\cdot}$ is a morphism of complexes of sheaves the cone of u is the complex

$$\mathrm{Cone}(\mathcal{A}^{\cdot} \xrightarrow{u} \mathcal{B}^{\cdot}) = \mathcal{A}^{\cdot}[1] \oplus \mathcal{B}^{\cdot}$$

with the differentials

$$\begin{aligned} \mathcal{A}^{q+1} \oplus \mathcal{B}^{q} &\to \mathcal{A}^{q+2} \oplus \mathcal{B}^{q+1} \\ (a,b) &\mapsto (-d(a), u(a) + d(b)). \end{aligned}$$

There are quasi-isomorphisms on X_{an}

$$\mathrm{Cone}(\Omega^{\geq q} \oplus \mathbb{R}(q) \to \Omega^{\cdot})[-1] \xrightarrow{\sim} \mathbb{R}(q)_{\mathcal{D}}$$

where $\Omega^{\geq q} = (0 \to \cdots \to 0 \to \Omega^{q} \to \Omega^{q+1} \to \cdots)$ and u is the difference of the obvious embeddings. For a smooth projective variety X over $\mathbb{R}$ or $\mathbb{C}$ we thus obtain a long exact sequence

(2.3.1) $$\begin{aligned} \cdots \to H^{p}_{\mathcal{D}}(X, \mathbb{R}(q)) \to F^{q} H^{p}_{DR}(X) \oplus H^{p}_{B}(X, \mathbb{R}(q)) \\ \to H^{p}_{DR}(X) \to H^{p+1}_{\mathcal{D}}(X, \mathbb{R}(q)) \to \cdots. \end{aligned}$$

Recall here the definition of the Hodge filtration (for $X/\mathbb{C}$ say):

$$F^{q} H^{p}_{DR}(X) = \mathrm{Im}(H^{p}_{\mathrm{Zar}}(X, \Omega^{\geq q}_{X/\mathbb{C}}) \to H^{p}_{\mathrm{Zar}}(X, \Omega^{\cdot}_{X/\mathbb{C}}))$$

and observe that by GAGA and the degeneration of the Hodge spectral sequence we have

$$F^{q} H^{p}_{DR}(X) \cong H^{p}_{\mathrm{Zar}}(X, \Omega^{\geq q}_{X/\mathbb{C}}) \cong H^{p}(X_{\mathrm{an}}, \Omega^{\geq q}).$$

Now assume that X is a variety over $\mathbb{R}$. Using Hodge theory we obtain for $q > \frac{p}{2} + 1$ exact sequences

$(\mathcal{B})$ $$0 \to F^{q} H^{p}_{DR}(X) \to H^{p}_{B}(X, \mathbb{R}(q-1)) \to H^{p+1}_{\mathcal{D}}(X, \mathbb{R}(q)) \to 0$$

$(\mathcal{D})$ $$0 \to H^{p}_{B}(X, \mathbb{R}(q)) \to H^{p}_{DR}(X)/F^{q} \to H^{p+1}_{\mathcal{D}}(X, \mathbb{R}(q)) \to 0.$$

For a smooth projective variety X over $\mathbb{Q}$ these define $\mathbb{Q}$-structures

(2.3.2) $$\begin{aligned} \mathcal{B}_{p,q} &= \det(H^{p}_{B}(X_{\mathbb{R}}, \mathbb{Q}(q-1))) \otimes \det(F^{q} H^{p}_{DR}(X/\mathbb{Q}))^{\vee} \\ \mathcal{D}_{p,q} &= \det(H^{p}_{DR}(X/\mathbb{Q})/F^{q}) \otimes \det(H^{p}_{B}(X_{\mathbb{R}}, \mathbb{Q}(q))^{\vee} \end{aligned}$$

on $\det H^{p+1}_{\mathcal{D}}(X_{\mathbb{R}}, \mathbb{R}(q))$. Here $\det W$ denotes the highest exterior power of a finite dimensional vector space and $^{\vee}$ is the dual.

(2.4) For smooth quasiprojective varieties X over $\mathbb{C}$ the above definition of Deligne cohomology leads to vector spaces which are in general infinite dimensional. A more sophisticated definition imposing growth conditions at infinity remedies this defect.

By resolution of singularities there exists an open immersion

$$j : X \hookrightarrow \overline{X}$$

of X into a smooth, projective variety $\overline{X}$ over $\mathbb{C}$ such that the complement $D = \overline{X} - X$ is a divisor with only normal crossings. Consider the natural maps of complexes of sheaves $\Omega^{\geq q}_{\overline{X}}\langle D\rangle \to j_*\Omega^{\cdot}_X$ on $\overline{X}_{\text{an}}$ and $\mathbb{R}(q) \to \Omega^{\cdot}_X$ on X_{an}. Choose injective resolutions

$$\mathbb{R}(q) \xrightarrow{\sim} I^{\cdot} \quad \text{and} \quad \Omega^{\cdot}_X \xrightarrow{\sim} J^{\cdot}$$

and set

$$Rj_*\mathbb{R}(q) = j_*I^{\cdot} \quad \text{and} \quad Rj_*\Omega^{\cdot}_X = j_*J^{\cdot}.$$

We get induced maps on $\overline{X}_{\text{an}}$

$$\Omega^{\geq q}_{\overline{X}}\langle D\rangle \to Rj_*\Omega^{\cdot}_X \quad \text{and} \quad Rj_*\mathbb{R}(q) \to Rj_*\Omega^{\cdot}_X$$

and using the difference of these we can form

$$\mathbb{R}(q)_{\mathcal{D}} = \operatorname{Cone}(\Omega^{\geq q}_{\overline{X}}\langle D\rangle \oplus Rj_*\mathbb{R}(q) \to Rj_*\Omega^{\cdot}_X)[-1].$$

The Deligne cohomology groups

$$H^p_{\mathcal{D}}(X, \mathbb{R}(q)) = H^p(\overline{X}_{\text{an}}, \mathbb{R}(q)_{\mathcal{D}})$$

are independent of the choice of resolutions and compactification. As before we can define $\mathcal{D}$-cohomology of varieties over $\mathbb{R}$.

For X over $\mathbb{R}$ or $\mathbb{C}$ there is still a long exact sequence (2.3.1) where now $F^qH^p_{DR}(X)$ is the Deligne Hodge filtration on $H^p_{DR}(X)$. Observe that by the degeneration of the logarithmic Hodge spectral sequence

$$F^qH^p_{DR}(X) \cong H^p(\overline{X}_{\text{an}}, \Omega^{\geq q}_{\overline{X}}\langle D\rangle).$$

Assertions (1), (2), (5), (6) of theorem (1.3) have their counterparts for Deligne cohomology. The analogue of assertion (4) is the formula

(2.4.1) $$H^1_{\mathcal{D}}(X, \mathbb{R}(1)) = \{g \in H^0(X_{\text{an}}, \mathcal{O}/\mathbb{R}(1)) \mid dg \in H^0(\overline{X}_{\text{an}}, \Omega^1_{\overline{X}}\langle D\rangle)\}$$

which follows immediately from the definition. The typical element of this group should be thought of as an $\mathbb{R}$-linear combination of logarithms of regular invertible functions on X.

(2.5) In the proofs of the Beilinson conjectures a more explicit description of $\mathcal{D}$-cohomology in terms of $\mathcal{C}^\infty$-differential forms is used. Let $\mathcal{A}^{\cdot}$ be the de Rham complex of real valued $\mathcal{C}^\infty$-forms and let $\pi_k : \mathbb{C} \to \mathbb{R}(k)$, $\pi_k(z) = \frac{1}{2}(z+(-1)^k\bar{z})$ be the natural projection. There is a quasi isomorphism u

$$\begin{array}{l} \mathbb{R}(q)_{\mathcal{D}} = \mathrm{Cone}(\Omega_{\overline{X}}^{\geq q}\langle D\rangle \oplus Rj_*\mathbb{R}(q) \to Rj_*\Omega_X^{\cdot})[-1] \\ \quad {\scriptstyle u}\downarrow \\ \widetilde{\mathbb{R}(q)}_{\mathcal{D}} := \mathrm{Cone}(\Omega_{\overline{X}}^{\geq q}\langle D\rangle \to j_*\mathcal{A}_X^{\cdot} \otimes \mathbb{R}(q-1))[-1] \end{array}$$

on $\overline{X}_{\mathrm{an}}$ induced by the projection

$$\Omega_{\overline{X}}^{\geq q}\langle D\rangle \oplus Rj_*\mathbb{R}(q) \to \Omega_{\overline{X}}^{\geq q}\langle D\rangle$$

and by the composition:

$$Rj_*\Omega_X^{\cdot} = j_*J^{\cdot} \to j_*\mathcal{A}_X^{\cdot} \otimes \mathbb{C} \xrightarrow{\pi_{q-1}} j_*\mathcal{A}_X^{\cdot} \otimes \mathbb{R}(q-1).$$

In particular

$$H^p_{\mathcal{D}}(X,\mathbb{R}(q)) \cong H^p_{\mathcal{D}}(\overline{X}_{\mathrm{an}},\widetilde{\mathbb{R}(q)}_{\mathcal{D}}).$$

For $p=q$ we obtain by a straightforward computation

$$\text{(2.5.1)} \quad H^p_{\mathcal{D}}(X,\mathbb{R}(p)) \cong \frac{\left\{ \begin{array}{c} \varphi \in H^0(X_{\mathrm{an}},\mathcal{A}^{p-1}\otimes\mathbb{R}(p-1)) | d\varphi = \pi_{p-1}(\omega), \\ \omega \in H^0(\overline{X}_{\mathrm{an}},\Omega^p_{\overline{X}}\langle D\rangle) \end{array} \right\}}{dH^0(X_{\mathrm{an}},\mathcal{A}^{p-2}\otimes\mathbb{R}(p-1))}.$$

In case $p=1$ we find

$$H^1_{\mathcal{D}}(X,\mathbb{R}(1)) \cong \{\varphi \in H^0(X_{\mathrm{an}},\mathcal{A}^0) \,|\, d\varphi = \pi_0(\omega), \omega \in H^0(\overline{X}_{\mathrm{an}},\Omega^1_{\overline{X}}\langle D\rangle)\}.$$

Under this isomorphism the section g of (2.4.1) is mapped to $\varphi = \pi_0(g)$ and $\omega = dg$.

Using (2.5.1) as an identification the cup product of classes $[\varphi_a]$, $[\varphi_b]$ in $H^a_{\mathcal{D}}(X,\mathbb{R}(a))$ resp. $H^b_{\mathcal{D}}(X,\mathbb{R}(b))$ with associated forms ω_a, ω_b is represented by

$$\varphi_a \cup \varphi_b := \varphi_a \wedge \pi_b\omega_b + (-1)^a \pi_a\omega_a \wedge \varphi_b.$$

One checks that $\omega_a \wedge \omega_b$ is associated to $\varphi_a \cup \varphi_b$.

We also note that in this description the boundary map in (2.4.1) is given by

$$\begin{aligned}\partial: H^p_{\mathcal{D}}(X,\mathbb{R}(p)) &\to F^p H^p_{DR}(X) \oplus H^p_B(X,\mathbb{R}(p))\\ \varphi &\mapsto (\omega,[\omega]).\end{aligned}$$

Observe that $\pi_{p-1}[\omega]=0$ and hence $[\omega] \in H^p_B(X,\mathbb{R}(p))$.

(2.6) The final ingredient in the formulation of the Beilinson conjectures is the regulator map. This is a co- and contravariant functorial homomorphism

$$r_{\mathcal{D}}: H^{\cdot}_{\mathcal{M}}(X,\mathbb{Q}(*)) \to H^{\cdot}_{\mathcal{D}}(X,\mathbb{R}(*))$$

for smooth quasiprojective varieties X over $\mathbb{R}$ or $\mathbb{C}$ which commutes with cup products. If motivic cohomology is described in terms of K-theory $r_{\mathcal{D}}$ is a generalized Chern character. In the description of $H^{\cdot}_{\mathcal{M}}$ given in section 1 $r_{\mathcal{D}}$ becomes a generalized cycle map (see (2.8)). There is a commutative diagram:

$$(2.6.1)\qquad \begin{array}{ccc} H^1_{\mathcal{M}}(X,\mathbb{Q}(1)) & \xrightarrow{r_{\mathcal{D}}} & H^1_{\mathcal{D}}(X,\mathbb{R}(1)) \\ \| & \overset{\log}{\nearrow} & \downarrow \wr \pi_0 \\ \mathcal{O}^*(X)\otimes\mathbb{Q} & \xrightarrow[\log||]{} & \left\{\varphi \in \Gamma(X,\mathcal{A}^0) \,\middle|\, \begin{array}{l} d\varphi = \pi_0(\omega),\ \omega \text{ with} \\ \text{log-sing. at infinity}\end{array}\right\} \end{array}$$

If X is a smooth quasiprojective variety over $\mathbb{Q}$ the regulator map is defined by composition:

$$r_{\mathcal{D}}: H^{\cdot}_{\mathcal{M}}(X,\mathbb{Q}(*)) \xrightarrow{\text{res}} H^{\cdot}_{\mathcal{M}}(X_{\mathbb{R}},\mathbb{Q}(*)) \xrightarrow{r_{\mathcal{D}}} H^{\cdot}_{\mathcal{D}}(X_{\mathbb{R}},\mathbb{R}(*)).$$

(2.7) The formal properties of motivic and Deligne cohomology and of the regulator map which we have mentioned up to now are sufficient for an understanding of the proofs of Beilinson's conjectures in the cases sketched in section 4. For Bloch's actual construction of the regulator map as a generalized cycle class map in (2.8) however more properties of Deligne cohomology are required. We list them briefly:

(2.7.1) There are relative $\mathcal{D}$-cohomology groups $H^p_{\mathcal{D},Y}(X,\mathbb{R}(q))$ for smooth, quasi-projective X over $\mathbb{R}$ and $\mathbb{C}$ and arbitrary closed subschemes Y of X. These fit into a co- and contravariant functorial long exact sequence

$$\to H^p_{\mathcal{D},Y}(X,\mathbb{R}(q)) \to H^p_{\mathcal{D}}(X,\mathbb{R}(q)) \to H^p_{\mathcal{D}}(X-Y,\mathbb{R}(q)) \to H^{p+1}_{\mathcal{D},Y}(X,\mathbb{R}(q)) \to \cdots$$

(2.7.2) If $Y \subset X$ has pure codimension q there is a contravariant functorial cycle class $[Y]$ in $H^{2q}_{\mathcal{D},Y}(X,\mathbb{R}(q))$. Moreover weak purity holds: in other words, $H^p_{\mathcal{D},Y}(X,\mathbb{R}(q))=0$ for $p<2q$.

(2.7.3) Homotopy: $H^p_{\mathcal{D}}(X\times\mathbb{A}^1,\mathbb{R}(q))=H^p_{\mathcal{D}}(X,\mathbb{R}(q))$.

(2.7.4) For $Y \subset X$ of pure codimension there are complexes of $\mathbb{R}$-vector spaces $D^{\cdot}_Y(X,q)$ which are contravariant functorial with respect to cartesian diagrams

$$\begin{array}{ccc} Y' & \hookrightarrow & X' \\ \downarrow & & \downarrow \\ Y & \hookrightarrow & X \end{array}$$

and such that

$$H^p_{\mathcal{D},Y}(X,\mathbb{R}(q))=H^p(D^{\cdot}_Y(X,q)) \quad \text{functorially.}$$

Remarks. (1) The relative $\mathcal{D}$-cohomology groups are defined by

$$H^p_{\mathcal{D},Y}(X,\mathbb{R}(q))=H^p(X,\mathrm{Cone}(\mathbb{R}(q)_{\mathcal{D},X}\xrightarrow{\mathrm{res}}\mathbb{R}(q)_{\mathcal{D},X-Y})[-1])$$

where $\mathbb{R}(q)_{\mathcal{D},X}$ and $\mathbb{R}(q)_{\mathcal{D},X-Y}$ are the Deligne complexes on X and $X-Y$ computed with respect to compatible compactifications. The long exact sequence is then an immediate consequence.

(2) For the complexes $D^{\cdot}(X,q)=D^{\cdot}_{\phi}(X,q)$ we can choose:

$$D^{\cdot}(X,q)=\varinjlim s\check{C}(\mathcal{U},\mathbb{R}(q)_{\mathcal{D}})$$

the limit over all coverings $\mathcal{U}$ of X_{an} of the associated simple complex to the Čech complex with coefficients in $\mathbb{R}(q)$. Moreover

$$D^{\cdot}_Y(X,q):=\mathrm{Cone}(D^{\cdot}(X,q)\xrightarrow{\mathrm{res}}D^{\cdot}(X-Y,q))[-1].$$

(2.8) We now proceed to the construction **[Bl3]** of the regulator map for smooth quasiprojective varieties X over $\mathbb{R}$ or $\mathbb{C}$. Consider the cohomological double complex

$$D^{\cdot}(X^*,q)=D^{\cdot}(X\times\Delta_{-*},q)$$

non-zero for $\cdot\geq 0$, $*\leq 0$ with $*$-differential:

$$d=\sum_{i=0}^{-a}(-1)^i\partial_i^*:D^b(X^a,q)\to D^b(X^{a+1},q).$$

Similarly another double complex is defined

$$D^{\cdot}_{\mathrm{supp}}(X^*,q) = \varinjlim_{Z\in z^q(X,-*)} D^{\cdot}_{\mathrm{Supp}\,Z}(X_{-*},q).$$

For technical reasons we truncate these complexes (non-trivially) in large negative $*$-degree:

$$\mathbb{D}^{\cdot}_{(\mathrm{supp})}(X^*,q) = \tau_{*\geq -N} D^{\cdot}_{(\mathrm{supp})}(X^*,q)$$

where $N >> 0$ is an even integer.

Consider the spectral sequence

$$E_1^{a,b} = H^b(\mathbb{D}^{\cdot}(X^a,q)) \Rightarrow H^{a+b}(s\mathbb{D}^{\cdot}(X^*,q))$$

where s denotes the associated simple complex of a double complex. Because of the homotopy axiom

$$E_1^{a,b} = H^b_{\mathcal{D}}(X,\mathbb{R}(q)) \quad \text{for} \quad -N \leq a \leq 0,\ b \geq 0$$

and $E_1^{a,b} = 0$ for all other a,b. Moreover $d_1^{a,b} = 0$ except for a even, $-N \leq a < 0$ and $b \geq 0$ in which case $d_1^{a,b} = \mathrm{id}$. Hence we obtain isomorphisms

$$H^p(s\mathbb{D}^{\cdot}(X^*,q)) = H^p_{\mathcal{D}}(X,\mathbb{R}(q)).$$

In the spectral sequence

$$E_1^{a,b} = H^b(\mathbb{D}^{\cdot}_{\mathrm{supp}}(X^a,q)) \Rightarrow E^{a+b} = H^{a+b}(s\mathbb{D}^{\cdot}_{\mathrm{supp}}(X^*,q))$$

we have

$$E_1^{a,b} = \varinjlim_{Z\in z^q(X,-a)} H^b_{\mathcal{D},\mathrm{Supp}\,Z}(X^a,\mathbb{R}(q))$$

for $-N \leq a \leq 0$, $b \geq 0$ and $E_1^{a,b} = 0$ otherwise. The cycle map induces a natural map of complexes

$$\Gamma_X(q)^* \to E_1^{*-2q,2q}$$

and hence for all p a map

$$H^p_{\mathcal{M}}(X,\mathbb{Z}(q)) \to E_2^{p-2q,2q}.$$

Due to weak purity the groups $E_1^{a,b}$ are zero for $b < 2q$ and all $r \geq 1$. Hence there are natural maps

$$E_2^{a,2q} \twoheadrightarrow E_\infty^{a,2q} \hookrightarrow E^{a+2q}.$$

Choosing $-N < p-2q$ the regulator map $r_{\mathcal{D}}$ is defined by composition:

$$\begin{array}{ccc} H^p_{\mathcal{M}}(X,\mathbb{Z}(q)) & \longrightarrow & H^p(s\mathbf{D}^{\cdot}_{\mathrm{supp}}(X^*,q)) \\ {\scriptstyle r_{\mathcal{D}}}\downarrow & & \downarrow{\scriptstyle \mathrm{nat.}} \\ H^p_{\mathcal{D}}(X,\mathbb{R}(q)) & = & H^p(s\mathbf{D}^{\cdot}(X^*,q)). \end{array}$$

It is independent of N. Similarly a regulator (or cycle) map into continuous étale cohomology [**Ja1**] can be constructed.

(2.9) We now sketch how the notions introduced fit into the philosophy of motives. More details will be given in the appendix.

Assume X is smooth, projective over $\mathbb{R}$. Let $MH_{\mathbb{R}}$ be the abelian category of $\mathbb{R}$-mixed Hodge structures with the action of a real Frobenius. According to Beilinson ([**Be3**], see also [**Ca**]) there is a natural isomorphism for $p+1<2q$

$$H^{p+1}_{\mathcal{D}}(X,\mathbb{R}(q)) \xrightarrow{\sim} \mathrm{Ext}^1_{MH_{\mathbb{R}}}(\mathbb{R}(0), H^p_B(X)(q)).$$

One would like to give a similar interpretation to the motivic cohomology groups as Ext-groups in a suitable abelian category of "mixed motives". The ultimate definition of such a category remains to be found. However via realizations (ℓ-adic, Betti, ...) working definitions have been found for $MM_{\mathbb{Q}}$ and $MM_{\mathbb{Z}}$ the categories of mixed motives over $\mathbb{Q}$ resp. $\mathbb{Z}$, see [**De3,Ja2,Sc2**]. It is shown in the appendix that for smooth, projective varieties X over $\mathbb{Q}$ there are natural maps (conjecturally isomorphisms) for $p+1<2q$

$$H^{p+1}_{\mathcal{M}}(X,\mathbb{Q}(q)) \rightarrow \mathrm{Ext}^1_{MM_{\mathbb{Q}}}(\mathbb{Q}(0), H^p(X)(q))$$

and one hopes that the image of $H^{p+1}_{\mathcal{M}}(X,\mathbb{Q}(q))_{\mathbb{Z}}$ is precisely $\mathrm{Ext}^1_{MM_{\mathbb{Z}}}(\mathbb{Q}(0), H^p(X)(q))$. Moreover there is a commutative diagram

$$\begin{array}{ccc} H^{p+1}_{\mathcal{M}}(X,\mathbb{Q}(q)) & \xrightarrow{r_{\mathcal{D}}} & H^{p+1}_{\mathcal{D}}(X_{\mathbb{R}},\mathbb{R}(q)) \\ \downarrow & & \downarrow{\scriptstyle \wr} \\ \mathrm{Ext}^1_{MM_{\mathbb{Q}}}(\mathbb{Q}(0), H^p(X)(q)) & \xrightarrow{H_B} & \mathrm{Ext}^1_{MH_{\mathbb{R}}}(\mathbb{R}(0), H^p_B(X_{\mathbb{R}})(q)) \end{array}$$

where H_B maps a mixed motive to the Betti realization over $\mathbb{R}$ endowed with its mixed $\mathbb{R}$-Hodge structure.

3. The conjectures

Recall the definition of the i-th L-series of a smooth projective variety X over $\mathbb{Q}$ by the following Euler product:

$$L(H^i(X),s)=\prod_p P_p(H^i(X),p^{-s}).$$

Here we have set

$$P_p(H^i(X),t)=\det\left(1-Fr_p t \mid H^i_{\acute{e}t}(X_{\overline{\mathbb{Q}}_p},\mathbb{Q}_\ell)^{I_p}\right)$$

where ℓ is a prime different from p, I_p is the inertia group in $G_{\mathbb{Q}_p}$ and Fr_p is the inverse of a Frobenius element in $G_{\mathbb{Q}_p}$. For primes p where X has good reduction the polynomial $P_p(H^i(X),t)$ has coefficients in $\mathbb{Q}$ independent of ℓ. The product of the $P_p(H^i(X),p^{-s})$ extended over the good primes converges absolutely in the usual topology for $\mathrm{Re}\, s > \frac{i}{2}+1$. Conjectures **[Se]**:

- The polynomials $P_p(H^i(X),t)$ lie in $\mathbb{Q}[t]$ for *all* p, and are independent of ℓ, and nonvanishing for $|t|<p^{-1-i/2}$.
- The Euler product has a meromorphic continuation to the whole plane.
- There is a functional equation relating $L(H^i(X),s)$ and $L(H^i(X),i+1-s)$ as in **[Se]**.

Concerning the special values of these L-functions there is the following conjecture.

(3.1) Conjecture. *Assume* $n>\frac{i}{2}+1$. *Then:*
(3.1.1) $r_{\mathcal{D}}\otimes\mathbb{R}: H^{i+1}_{\mathcal{M}}(X,\mathbb{Q}(n))_{\mathbb{Z}}\otimes\mathbb{R}\to H^{i+1}_{\mathcal{D}}(X_{\mathbb{R}},\mathbb{R}(n))$ *is an isomorphism.*
(3.1.2) $r_{\mathcal{D}}(\det H^{i+1}_{\mathcal{M}}(X,\mathbb{Q}(n))_{\mathbb{Z}}) = L(H^i(X),n)\mathcal{D}_{i,n}$ *in* $\det H^{i+1}_{\mathcal{D}}(X_{\mathbb{R}},\mathbb{R}(n))$ *with* $\mathcal{D}_{i,n}$ *as defined in (2.3.2).*

If the above hypothesis on the L-function of $H^i(X)$ are satisfied assertion (3.1.2) is equivalent to:

$$r_{\mathcal{D}}(\det H^{i+1}_{\mathcal{M}}(X,\mathbb{Q}(n))_{\mathbb{Z}})=L(H^i(X),i+1-n)^*\mathcal{B}_{i,n} \tag{3.1.3}$$

in $\det H^{i+1}_{\mathcal{D}}(X_{\mathbb{R}},\mathbb{R}(n))$ where $L(H^i(X),k)^*$ denotes the leading coefficient at $s=k$ in the Taylor development of the L-series **[Ja3]**.

The following result on the order of vanishing follows from a straightforward calculation and the expected functional equation**[Sch]**:

$$\begin{aligned}\mathrm{ord}_{s=i+1-n}L(H^i(X),s)&=\dim H^{i+1}_{\mathcal{D}}(X_{\mathbb{R}},\mathbb{R}(n))\\&=\dim H^{i+1}_{\mathcal{M}}(X,\mathbb{Q}(n))_{\mathbb{Z}}\quad\text{assuming (3.1.1).}\end{aligned}\tag{3.1.4}$$

Observe that the conjectures determine the special values of the L-series up to a non-vanishing rational number. Equation (3.1.3) is the original proposal by Beilinson. The version (3.1.2) is a reformulation due to Deligne. It requires less information about the L-series to make sense.

For the remaining values of n: the right central point $n=\frac{i}{2}+1$ and the central point $n=\frac{i+1}{2}$ the conjectures have to be modified ([**Bel**], Conjecture 3.7 *et seq.*). Since we don't deal with examples for these cases we skip the formulation. A uniform approach is possible in the framework of mixed motives: The Beilinson conjectures are seen to be equivalent to a Deligne conjecture for critical mixed motives [**Sc2**]. An integral refinement of the conjectures has been proposed by Bloch and Kato [**BlK**] using their philosophy of Tamagawa measures for motives. Essentially the only case where (3.1.1) is known is for X the spectrum of a number field. In this case the result is due to Borel with a different definition of the regulator map. For a comparison of the regulator maps see [**Bel,Rap**]. For a proof of Borel's result the K-theoretical approach to motivic cohomology is essential.

In a number of cases to be treated in section 5 and 6 the following weakened version of the conjectures can be proved.

(3.2) Conjecture. *Assume $n>\frac{i}{2}+1$. Then (3.1.1) and (3.1.2) (or (3.1.3)) hold with $H^{i+1}_{\mathcal{M}}(X,\mathbb{Q}(n))_{\mathbb{Z}}$ replaced by a suitable $\mathbb{Q}$-subspace.*

Thus motivic cohomology as we have defined it should at least be large enough so that a sensible regulator can be formed having the expected relation to the L-values.

(3.3) Generalization to Chow motives. For some well known L-series the above framework is too restrictive. For example the Dirichlet L-functions of algebraic number theory are not covered. This is remedied by extending the above notions and conjectures to the category of Chow motives, which should be thought of as generalised varieties. Fix a number field $T/\mathbb{Q}$ – the field of coefficients. Let $\mathcal{V}_k$ be the category of smooth projective varieties over a field k. Consider the category $\mathcal{C}_k(T)$ with objects TX for each object X in $\mathcal{V}_k$ and morphisms

$$\mathrm{Hom}(TX,TY)=CH^{\dim Y}(X\times_k Y)\otimes T.$$

For $a:TX_1\to TX_2$ and $b:TX_2\to TX_3$ composition is defined by intersecting cycles:

$$b\circ a=p_{13*}(p_{12}^*a\cdot p_{23}^*b)$$

where $p_{ij}:X_1\times X_2\times X_3\to X_i\times X_j$ are the projections. Sending X to TX and a morphism f to its graph $\tilde{f}$ defines a covariant functor from

$\mathcal{V}_k \to \mathcal{C}_k(T)$. The category of effective Chow motives $\mathcal{M}_k^+(T)$ is obtained by adding images of projectors to $\mathcal{C}_k(T)$. Objects are pairs $M = (TX, p)$ where $p \in \mathrm{End}(TX)$, $p^2 = p$ and morphisms are the obvious ones. Setting

$$H_?(M) = p^*(H_?(X) \otimes T)$$

the cohomologies and conjecture (3.1) factorize over $\mathcal{M}_{\mathbb{Q}}^+(T)$. They determine special values of $T \otimes \mathbb{C}$-valued L-series $L(H^i(M), s)$ up to numbers in T^*. See **[Be1,Ja3,Kl,Ma]** for more details.

Remarks. The category $\mathcal{M}_{\mathbb{Q}}^+(T)$ is not abelian. If instead of Chow theory one considers cycles modulo homological equivalence one obtains what is essentially Grothendieck's category of (effective) motives. Standard conjectures on algebraic cycles would imply that it is an abelian semisimple category. Nowadays these motives are called pure in contrast to more general "mixed" motives which should come e.g. from the H^i of singular varieties. The category of these mixed motives $MM_{\mathbb{Q}}$ was already alluded to in section (2.9). As yet there is no Grothendieck style definition for $MM_{\mathbb{Q}}$ using cycles but only a definition via realizations.

As an example of a Chow motive let us construct the motive M_χ of a Dirichlet character χ of a number field k: Via class field theory we may view χ as a one-dimensional representation of the absolute galois group G_k of k with values in a number field T:

$$\chi : G_k \to T^*.$$

We may assume that T is generated over $\mathbb{Q}$ by the values of χ. Choose a finite abelian extension F/k such that χ factorizes over $G = \mathrm{Gal}(F/k)$ and set

$$M_\chi = e_\chi(T\mathrm{Spec}(F))$$

in $\mathcal{M}_k^+(T)$ where e_χ is the idempotent:

$$e_\chi = \frac{1}{|G|} \sum_{\sigma \in G} \chi(\sigma)\tilde{\sigma}^{-1} \quad \text{in} \quad T[G].$$

Observe that M_χ is independent of the choice of F.

We end this section with a short discussion of known cases for the conjectures. (3.1) is known for $X = \mathrm{Spec}\, F$, $F/\mathbb{Q}$ a number field **[Bo2]** and for the motives M_χ attached to Dirichlet characters of $k = \mathbb{Q}$ or $k = K$ an imaginary quadratic field **[Be1,Den2]**. In section 5 and 6 we will deduce the evidence for the weak conjecture (3.2) from the theory of Beilinson's

Eisenstein symbol map (section 4). The logical dependencies in our approach are depicted in a diagram:

		modular curves/$\mathbb{Q}$; Shimura curves/$\mathbb{Q}$ **[Ra1]**
		$\uparrow$
		modular curves **[Be1,Be2,SS1]**
		$\uparrow$
Eisenstein symbol	$\rightarrow$	modular forms of weight ≥ 2 **[Sc2]**
$\downarrow$	$\searrow$	
Dirichlet characters of imaginary quadratic fields **[Den2]**		algebraic Hecke characters of imaginary quadratic fields **[Den1,Den2]**
$\downarrow$		$\downarrow$
Dirichlet characters of $\mathbb{Q}$ **[Be1,Den2,N]**		CM elliptic curves over number fields of Shimura type **[Bl1,Den1]**

4. Kuga-Sato varieties and the Eisenstein symbol

The Eisenstein symbol is a certain "universal" construction of elements of motivic cohomology of an elliptic curve, or more generally self-products of an elliptic curve. It has its origins in the work of Bloch **[Bl1]** on K_2 of elliptic curves but was constructed in generality by Beilinson **[Be2]**. For a constant elliptic curve a slightly refined construction is made in **[Den1]**.

4.1 We first introduce the modular and Kuga-Sato varieties. In what follows n will be an integer ≥ 3. Let M_n be the modular curve of level n, parameterising elliptic curves E together with level n structure $(\mathbb{Z}/n)^2 \xrightarrow{\sim} E[n]$. Thus the set of complex points $M_n(\mathbb{C})$ is the disjoint union of $\phi(n)$ copies of $\Gamma(n)\backslash\mathcal{H}$, the quotient of the upper half-plane by the principal congruence subgroup $\Gamma(n) \subset SL_2(\mathbb{Z})$.

The assumption $n \geq 3$ assures that there is a universal family of elliptic curves:

$$\pi : X_n \rightarrow M_n.$$

Write $\overline{M}_n$ for the usual compactification of M_n, and $M_n^\infty = \overline{M}_n - M_n$ for the cusps of $\overline{M}_n$ (a sum of copies of $\mathrm{Spec}\mathbb{Q}(\zeta_n)$). Then we can consider the minimal (regular) model of X_n over $\overline{M}_n$:

$$\bar{\pi} : \overline{X}_n \rightarrow \overline{M}_n$$

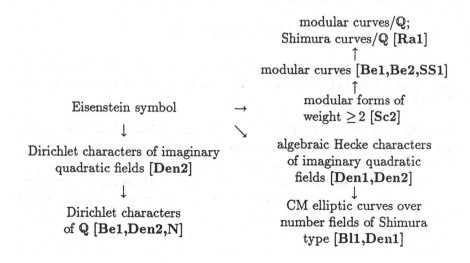

whose restriction to M_n is just π. For each cusp $s \in M_n^\infty$, the fibre $\bar{\pi}^{-1}(s)$ is a Néron polygon with n sides. Write $\hat{X}_n \subset \overline{X}_n$ for the connected component of the smooth part (Néron model) of $\overline{X}_n$. Then $\bar{\pi}^{-1}(s) \cap \hat{X}_n$ is (non-canonically) isomorphic to the multiplicative group $\mathbb{G}_m$.

(4.2) For $l \geq 0$ write X_n^l, $\overline{X}_n^l$, $\hat{X}_n^l$ for the l-fold fibre product of X_n (resp. $\overline{X}_n$, $\hat{X}_n$) over $\overline{M}_n$. The variety $\overline{X}_n^l$ has singularities for $l \geq 2$; we shall consider these in 5.2 below. Since X_n^l is a group scheme over M_n, in addition to the obvious projections

$$p_i : X_n^l \to X_n \quad \text{for } 1 \leq i \leq l$$

onto the factors, there is a further projection

$$p_0 = -p_1 - \cdots - p_l : X_n^l \to X_n.$$

These $(l+1)$ projections p_i allow us to regard X_n^l as a closed subscheme of X_n^{l+1}. This gives an action of the symmetric group $\mathfrak{S}_{l+1}$ on X_n^l, permuting the projections $p_0, \ldots p_l$. The same construction works also for $\hat{X}_n^l$.

(4.3) From the localisation sequence (1.3.5) for the pair $(\hat{X}_n^l, X_n^l)$ we have:

$$\text{(4.3.1)} \quad \begin{aligned} &H_{\mathcal{M}}^{l+1}(\hat{X}_n^l, \mathbb{Q}(l+1)) \to H_{\mathcal{M}}^{l+1}(X_n^l, \mathbb{Q}(l+1)) \\ &\qquad \to H_{\mathcal{M}}^l(M_n^\infty \times \mathbb{G}_m^l, \mathbb{Q}(l)) \to H_{\mathcal{M}}^{l+2}(\hat{X}_n^l, \mathbb{Q}(l+1)). \end{aligned}$$

Consider the eigenspaces for the sign character sgn_{l+1} of $\mathfrak{S}_{l+1}$. Under the involution $\sigma : x \mapsto x^{-1}$ of $\mathbb{G}_m$, the motivic cohomology (1.3.4) of $\mathbb{G}_m$ decomposes:—

$$H_{\mathcal{M}}^1(\mathbb{G}_m/k, \mathbb{Q}(1)) = k[x, x^{-1}]^* \otimes \mathbb{Q} = \underbrace{k^* \otimes \mathbb{Q}}_{\sigma=+1} \oplus \underbrace{\mathbb{Q}.x}_{\sigma=-1}.$$

Using this it is not hard to see that

$$\text{(4.3.2)} \qquad H_{\mathcal{M}}^l(M_n^\infty \times \mathbb{G}_m^l, \mathbb{Q}(l))_{\mathrm{sgn}_{l+1}} = H_{\mathcal{M}}^0(M_n^\infty, \mathbb{Q}(0)) = \mathbb{Q}[M_n^\infty].$$

Here $\mathbb{Q}[M_n^\infty]$ denotes the set of $\mathbb{Q}$-valued functions on the closed points of M_n^∞. By composing (4.3.1) and (4.3.2) we therefore obtain a "residue map" in motivic cohomology:

$$\text{(4.3.3)} \qquad H_{\mathcal{M}}^{l+1}(X_n^l, \mathbb{Q}(l+1))_{\mathrm{sgn}_{l+1}} \xrightarrow{\mathrm{Res}_{\mathcal{M}}^l} \mathbb{Q}[M_n^\infty].$$

Beilinson's key result is then:—

(4.4) Theorem ([Be2], Theorem 3.1.7). $\mathrm{Res}^l_{\mathcal{M}}$ *is surjective for* $l \geq 1$.

(4.5) Remarks. This theorem can be viewed as a generalisation of the theorem of Manin and Drinfeld, which is the case $l=0$. For then $X^0_n = M_n$ and (4.3.1) comes from the exact sequence

$$\begin{array}{ccccccccc} 0 & \longrightarrow & \mathcal{O}^*(\overline{M}_n) & \longrightarrow & \mathcal{O}^*(M_n) & \xrightarrow{\mathrm{Div}} & \mathbb{Z}[M^\infty_n] & \xrightarrow{c} & \mathrm{Pic}\overline{M}_n \\ & & \| & & & & & & \\ & & \mathbb{Q}(\zeta_n)^* & & & & & & \end{array}$$

by tensorising with $\mathbb{Q}$. Here $\mathrm{Res}^0_{\mathcal{M}} = \mathrm{Div}$ is the divisor map, and c maps a divisor supported on the cusps to its class in $\mathrm{Pic}\overline{M}_n$. According to the Manin-Drinfeld theorem, the divisors of degree zero

$$\mathbb{Z}[M^\infty_n]^0 \stackrel{\mathrm{def}}{=} \ker\left\{\mathbb{Z}[M^\infty_n] \xrightarrow{\deg} \mathbb{Z}\right\}$$

are torsion in $\mathrm{Pic}\,\overline{M}_n$, or equivalently

$$\mathrm{Res}^0_{\mathcal{M}} : H^1_{\mathcal{M}}(M_n, \mathbb{Q}(1)) \twoheadrightarrow \mathbb{Q}[M^\infty_n]^0.$$

For $l>0$ the picture should be even better. Firstly, there is no restriction to divisors of degree zero. Secondly, the general Beilinson conjectures would imply that $\mathrm{Res}^l_{\mathcal{M}}$ is actually an isomorphism. (To see this one examines carefully the exact sequence (4.3.1).) This makes Beilinson's proof of (4.4) philosophically reasonable—he constructs a totally explicit left inverse to $\mathrm{Res}^l_{\mathcal{M}}$, the *Eisenstein symbol* map

$$\mathcal{E}^l_{\mathcal{M}} : \mathbb{Q}[M^\infty_n] \to H^{l+1}_{\mathcal{M}}(X^l_n, \mathbb{Q}(l+1))_{\mathrm{sgn}_{l+1}}$$

whose construction we now describe.

(4.6) Let $U_n \subset X_n$ be the complement of the n^2 sections of order dividing n, and write

$$U^{l\prime}_n = \bigcap_{0\leq i\leq l} p_i^{-1}(U_n) \subset X^l_n.$$

We first construct symbols on $U^{l\prime}_n$ as follows. Start with any invertible functions $g_0, \ldots\, g_l \in \mathcal{O}^*(U_n)$. Then

(4.6.1) $$p_0^*(g_0) \cup \cdots \cup p_l^*(g_l) \in H^{l+1}_{\mathcal{M}}(U^{l\prime}_n, \mathbb{Q}(l+1)).$$

To get an element of $H^{l+1}_{\mathcal{M}}(X^l_n, \mathbb{Q}(l+1))$ we apply three projectors:

—$U^{l\prime}_n$ is stable under the symmetric group $\mathfrak{S}_{l+1}$, and we take the sgn-eigenspace;

—The group of sections of finite order $(\mathbb{Z}/n)^{2l}$ acts on $U^{l\prime}_n$ by translations, and we project onto the subspace of invariants;

—For an integer $m \geq 1$, there is a multiplication map

$$[m^{-1}]: H^{\cdot}_{\mathcal{M}}(U^{l\prime}_n, \mathbb{Q}(*)) \to H^{\cdot}_{\mathcal{M}}(U^{l\prime}_n, \mathbb{Q}(*))$$

defined as follows: consider the diagram

$$\begin{array}{ccc} U^{l\prime}_n & \xleftarrow{\ j\ } & U^{l\prime}_{mn} \\ & & \downarrow{\scriptstyle [\times m]} \\ & & U^{l\prime}_n \end{array}$$

Here j denotes the inclusion map, and the multiplication $[\times m]$ is a Galois étale covering with group $(\mathbb{Z}/m)^{2l}$. By (1.3.6) we have

$$\begin{array}{ccc} H^{\cdot}_{\mathcal{M}}(U^{l\prime}_n, \mathbb{Q}(*)) & \xrightarrow{\ j^*\ } & H^{\cdot}_{\mathcal{M}}(U^{l\prime}_{mn}, \mathbb{Q}(*))_{(\mathbb{Z}/m)^{2l}} \\ & {\scriptstyle [m^{-1}]} \searrow & \wr \uparrow {\scriptstyle [\times m]^*} \\ & & H^{\cdot}_{\mathcal{M}}(U^{l\prime}_{mn}, \mathbb{Q}(*)) \end{array}$$

whence there is a map $[m^{-1}]$ as indicated. Denote by a subscript l the maximal quotient of $H^{\cdot}_{\mathcal{M}}(U^{l\prime}_n, \mathbb{Q}(*))$ on which $[m^{-1}]$ is multiplication by m^{-l}, for every $m \geq 1$. (In fact it suffices to consider only one $m > 1$.)

(4.7) Theorem. *The restriction from X^l_n to $U^{l\prime}_n$ induces an isomorphism*

$$H^{\cdot}_{\mathcal{M}}(X^l_n, \mathbb{Q}(*))_{\mathrm{sgn}_{l+1}} \xrightarrow{\sim} H^{\cdot}_{\mathcal{M}}(U^{l\prime}_n, \mathbb{Q}(*))_{\mathrm{sgn}_{l+1}, (\mathbb{Z}/n)^{2l}, l}.$$

Applying this to the elements (4.6.1) projected to the right hand group gives a map

$$\bigotimes^{l+1} \mathcal{O}^*(U_n) \otimes \mathbb{Q} \to H^{l+1}_{\mathcal{M}}(X^l_n, \mathbb{Q}(l+1))_{\mathrm{sgn}_{l+1}}. \tag{4.7.1}$$

(4.8) Lemma. *The divisor map $\mathcal{O}^*(U_n) \otimes \mathbb{Q} \to \mathbb{Q}[(\mathbb{Z}/n)^2]^0$ is surjective.*

Proof. Let $s : M_n \to X_n$ be a section of order dividing n, and let $e : M_n \to X_n$ be the unit section. We have to show that $\mathcal{O}(s - e)$ is torsion in

PicX_n. It certainly is torsion in the relative Picard group Pic(X_n/M_n), so for some $N \geq 1$ and some line bundle $\mathcal{L}$ on M_n we have $\mathcal{O}(s-e)^{\otimes N} \simeq \pi^*\mathcal{L}$. Hence $\mathcal{L} = e^*\pi^*\mathcal{L} \simeq e^*\mathcal{O}(s-e)^{\otimes N} = e^*\mathcal{O}(-e)^{\otimes N} = \mathcal{N}_e^{\otimes N}$ where $\mathcal{N}_e$ is the normal bundle of the unit section. Hence $\mathcal{L} \simeq \underline{\omega}_{X_n/M_n}^{\otimes(-N)}$, and $\underline{\omega}^{\otimes 12}$ is trivial (a nowhere-vanishing section being the discriminant Δ).

We now have a diagram:

$$\begin{array}{ccc}
\bigotimes^{l+1} \mathbb{Q}[(\mathbb{Z}/n)^2]^0_{\mathrm{sgn}_{l+1},(\mathbb{Z}/n)^{2l}} & \xleftarrow{\ \mathrm{Div}\ } & \bigotimes^{l+1} \mathcal{O}^*(U_n) \\
\uparrow \wr\, \vartheta & \searrow\ (\text{dotted}) & \downarrow \\
\mathbb{Q}[(\mathbb{Z}/n)^2]^0 & \xrightarrow[E^l_{\mathcal{M}}]{} & H^{l+1}_{\mathcal{M}}(X^l_n, \mathbb{Q}(l+1))_{\mathrm{sgn}_{l+1}}
\end{array}$$

It is not hard to show that the map (4.7.1) factors through the dotted arrow as shown. The isomorphism ϑ is given by

$$\beta \mapsto \beta \otimes \alpha \otimes \cdots \otimes \alpha, \qquad \alpha = n^2(0) - \sum_{x \in (\mathbb{Z}/n)^2} (x).$$

This defines a composite map $E^l_{\mathcal{M}}$ as indicated, which is "almost" the Eisenstein symbol.

(4.9) At this point we want to describe the composite of the map $E^l_{\mathcal{M}}$ just constructed with the regulator map. Let us restrict attention to the component of $M_n(\mathbb{C})$ containing the cusp at infinity, and write τ for the variable on the complex upper half-plane. The corresponding component of $X^l_n(\mathbb{C})$ then can be described as the quotient

$$\Gamma(n)\backslash \mathcal{H} \times \mathbb{C}^l/\mathbb{Z}^{2l}$$

where the actions of $\Gamma(n)$ and $\mathbb{Z}^{2l}$ are given by:

$$\begin{pmatrix} a & b \\ c & d \end{pmatrix} : (\tau, z_1, \ldots, z_l) \mapsto \left(\frac{a\tau+b}{c\tau+d}, \frac{z_1}{c\tau+d}, \ldots, \frac{z_l}{c\tau+d}\right)$$

$$(u_1, v_1, \ldots, u_l, v_l) : (\tau, z_1, \ldots, z_l) \mapsto (\tau, z_1 + u_1\tau + v_1, \ldots, z_l + u_l\tau + v_l).$$

Let $\beta \in \mathbb{Q}[(\mathbb{Z}/n)^2]^0$. In terms of the description (2.5.1) of Deligne cohomology by differential forms, $r_{\mathcal{D}} E^l_{\mathcal{M}}(\beta) \in H^{l+1}_{\mathcal{D}}(X^l_n/\mathbb{R}, \mathbb{R}(l+1))$ is represented by

(4.9.1) $$\phi = \sum_{j=0}^{l} \sum_{c_1, c_2 \in \mathbb{Z}}{}' \frac{\psi_\beta(c_1, c_2)\mathrm{Im}(\tau)}{(c_1\tau + c_2)^{j+1}(c_1\bar{\tau} + c_2)^{l-j+1}} (d\bar{z}_1 \wedge \cdots$$
$$\cdots \wedge d\bar{z}_j \wedge dz_{j+1} \wedge \cdots \wedge dz_l)_{\mathrm{sgn}_l} + (d\tau,\ d\bar{\tau} \text{ term})$$

where for $c=(c_1,c_2)\in(\mathbb{Z}/n)^2$

$$\psi_\beta(c)=\sum_{d\in(\mathbb{Z}/n)^2}\beta(d)e^{2\pi i(c_1d_2-c_2d_1)/n}.$$

(The omitted terms in (4.9.1) involving $d\tau$, $d\bar{\tau}$ vanish in the applications of §§5, 6.) See 4.12 below for remarks concerning the proof of this formula.

(4.10) To pass from $E^l_{\mathcal{M}}$ to $\mathcal{E}^l_{\mathcal{M}}$ we first recall that the set of closed points of M_n^∞ is canonically isomorphic to

$$GL_2(\mathbb{Z}/n)/\begin{pmatrix}* & * \\ 0 & \pm 1\end{pmatrix}.$$

The definition of the residue map (4.3.2) involves choosing for each $s\in M_n^\infty$ an isomorphism of the fibre of $\hat{X}$ at s with $\mathbb{G}_m$ (see 4.1 above), and the two such isomorphisms are interchanged by $-1\in GL_2(\mathbb{Z}/n)$. If we replace $H^0_{\mathcal{M}}(M_n^\infty,\mathbb{Q}(0))$ by the (non-canonically) isomorphic space $V^{(-)^l}$ defined as

$$V^{\pm}=\left\{f{:}\,GL_2(\mathbb{Z}/n)\to\mathbb{Q}\;\middle|\;f\left(g\begin{pmatrix}* & * \\ 0 & 1\end{pmatrix}\right)=f(g)=\pm f(-g)\right\}$$

then the map $\mathrm{Res}^l_{\mathcal{M}}$ becomes $GL_2(\mathbb{Z}/n)$-equivariant.

(4.11) Now consider the family of maps

$$\lambda^l_n:\mathbb{Q}[(\mathbb{Z}/n)^2]^0\to V^{(-)^l}$$
$$(\lambda^l_n\phi)(g)=\sum_{x,y=0}^{n-1}\phi(g.{}^t(x,y))B_{l+2}\left(\frac{y}{n}\right)$$

(where B_{l+2} are Bernoulli polynomials). It is fairly elementary to prove that λ^l_n is surjective. (These maps are the finite level analogues of the horospherical map τ of **[Be2]**, paragraph following 3.1.6.) One now proves that (up to a non-zero constant factor) the diagram

(4.11.1)
$$\begin{array}{ccc} & & \mathbb{Q}[(\mathbb{Z}/n)^2]^0 \\ & \swarrow^{E^l_{\mathcal{M}}} & \downarrow{\scriptstyle\lambda^l_n} \\ H^{l+1}_{\mathcal{M}}(X^l_n,\mathbb{Q}(l+1))_{\mathrm{sgn}_{l+1}} & \xrightarrow[\mathrm{Res}^l_{\mathcal{M}}]{} & V^{(-)^l} \end{array}$$

is commutative, and that $E^l_{\mathcal{M}}$ factors through λ^l_n. Thus there is a map

$$\mathcal{E}^l_{\mathcal{M}}{:}\,V^{(-)^l}\to H^{l+1}_{\mathcal{M}}(X^l_n,\mathbb{Q}((l+1))_{\mathrm{sgn}_{l+1}}$$

satisfying

$$\mathcal{E}^l_{\mathcal{M}} \circ \lambda^l_n = E^l_{\mathcal{M}} \quad \text{and} \quad \mathrm{Res}^l_{\mathcal{M}} \circ \mathcal{E}^l_{\mathcal{M}} = \mathrm{id}.$$

This proves Theorem 4.4.

(4.12) We finally say some words about the commutativity of (4.11.1), on which the theorem rests. Beilinson's original proof uses the fact (from Borel's theorem) that the regulator map

$$r_{\mathcal{D}}: H^0_{\mathcal{M}}(M^{\infty}_n, \mathbb{Q}(0)) \to H^0_{\mathcal{D}}(M^{\infty}_n/\mathbb{R}, \mathbb{R}(0))$$

is injective. From this we see that one need only check the commutativity of the analogue of (4.11.1) in Deligne cohomology. To do this Beilinson explicitly calculates $r_{\mathcal{D}} \circ E^l_{\mathcal{M}}$, by integrating along the fibres of the projection $X^l_n(\mathbb{C}) \to M_n(\mathbb{C})$—see [**Be2**] §3.3 for details. (The resulting formula we gave as (4.9.1) above.)

An alternative proof [**SS2**] is by direct computation of $\mathrm{Res}^l_{\mathcal{M}} \circ E^l_{\mathcal{M}}$ using the Néron model of X^l_n. In this approach, the formula (4.9.1) is obtained as a *consequence* of the commutativity of (4.11.1). In fact, the analogue of $\mathrm{Res}^l_{\mathcal{M}}$ in Deligne cohomology is an *isomorphism*

$$H^{l+1}_{\mathcal{D}}(X^l_n/\mathbb{R}, \mathbb{R}(l+1))_{\mathrm{sgn}_{l+1}} \xrightarrow{\sim} H^0_{\mathcal{D}}(M^{\infty}_n/\mathbb{R}, \mathbb{R}(0))$$

(by consideration of the Hodge numbers) whose inverse is given by real analytic Eisenstein series.

5. L-functions of modular forms.

In this section we sketch how, mildly generalising the results of Beilinson, the Eisenstein symbol can be used to exhibit a relation between special values of L-functions of cusp forms of weight ≥ 2 and higher regulators.

5.1 Let $k \geq 0$ be an integer, and f a classical cusp form of weight $k+2$, which we assume to be a newform on some $\Gamma_0(N)$ with character χ_f. For simplicity we shall assume that the field generated by the Fourier coefficients of f is $\mathbb{Q}$.

As is well known **[Del]**, attached to f is a strictly compatible system of ℓ-adic representations $\{V_\ell(f)\}$, whose associated L-function is the Hecke L-series $L(f,s)$. Moreover $V_\ell(f)$ is a subspace of the parabolic cohomology

$$H^1_{\text{ét}}(\overline{M}_n \otimes \overline{\mathbb{Q}}, \phi_* \mathrm{Sym}^k R^1 \pi_* \mathbb{Q}_\ell) \tag{5.1.1}$$

for suitable n. (Recall that ϕ denotes the inclusion $M_n \hookrightarrow \overline{M}_n$.) In Lemma 7 of **[Del]** a canonical resolution of singularities of $\overline{X}^k_n$ is constructed, which we denote by $\overline{\overline{X}}^k_n$, and it is shown that $V_\ell(f)$ is a constituent of $H^{k+1}_{\text{ét}}(\overline{\overline{X}}^k_n \otimes \overline{\mathbb{Q}}, \mathbb{Q}_\ell)$.

5.2 Theorem [Sc1]. *There exists a projector Π_f in the ring of algebraic correspondences on $\overline{\overline{X}}^k_n$ modulo homological equivalence such that for every prime ℓ*

$$V_\ell(f) = \Pi_f \left[H^{k+1}_{\text{ét}}(\overline{\overline{X}}^k_n \otimes \overline{\mathbb{Q}}, \mathbb{Q}_\ell) \right].$$

Remarks. (1) In fact Π_f annihilates H^i for $i \neq k+1$.

(2) The pair $V(f) = \left[\overline{\overline{X}}^k_n, \Pi_f\right]$ is a *motive* in the sense of Grothendieck (cf. 3.3 above); by the above remark and the theorem, the ℓ-adic representations of $V(f)$ are $\{V_\ell(f)\}$. The Betti realisation of $V(f)$ is given by the singular parabolic cohomology groups (Eichler-Shimura). It has Hodge type $(k+1,0)+(0,k+1)$ and the $(k+1,0)$ part is spanned by the differential form on $\overline{\overline{X}}^k_n$

$$\omega_f = 2\pi i f(\tau) d\tau \wedge dz_1 \wedge \cdots \wedge dz_k.$$

(3) A construction of $V(f)$ as a motive defined by absolute Hodge cycles was given by Jannsen (**[Ja2]**, §1; see also **[Scha]**, V.1.1).

(4) For the purposes of testing Beilinson's conjectures, one would like $V(f)$ to be a Chow motive (3.3). In general this seems rather difficult to establish. However, one can consider in place of $V_\ell(f)$ the whole parabolic

cohomology group (5.1.1) of level n. There is then a Chow motive with this group for its ℓ-adic realisation. (See step (i) below.)

(5) One may also consider, for p prime to the level of f, the p-adic realisation $V_p(f)$, which is a crystalline representation of $\mathrm{Gal}(\overline{\mathbb{Q}}_p/\mathbb{Q}_p)$ **[Fa,FM]**. A consequence of 5.2 is that the characteristic polynomial of Frobenius on the associated filtered module is the Hecke polynomial $t^2 - a_p t + \chi_f(p)p^{k+1}$.

(5.3) Sketch of the construction. For $k = 0$ the theorem amounts to the decomposition of the Jacobian of $\overline{M}_n$ under the action of the Hecke algebra, and is classical. In this case the problem (4) does not arise. In the case $k > 0$ there are two steps:

(i) The use of automorphisms: acting on X_n^k one has the following groups of automorphisms and characters:

—$(\mathbb{Z}/n)^{2k}$, the translations by sections of finite order;

—μ_2^k, inversions in the components of the fibres;

—$\mathfrak{S}_k$, the symmetric group.

These generate a group Γ of automorphisms of X_n^k, and this extends to a group of automorphisms of $\overline{\overline{X}}{}_n^k$. There is a unique character of Γ which restricts to the trivial character on $(\mathbb{Z}/n)^{2k}$, the product character on μ_2^k, and the sign character of $\mathfrak{S}_k$. This defines a projector Π in the group algebra $\mathbb{Q}[\mathrm{Aut}\,\overline{\overline{X}}{}_n^k]$. By explicit calculation of the cohomology of the boundary of $\overline{\overline{X}}{}_n^k$ one shows that Π cuts out the parabolic cohomology (5.1.1).

(ii) To pass to the individual $V(f)$'s one projects using an idempotent in the Hecke algebra (which is semisimple as an algebra of correspondences modulo homological equivalence).

(5.4) The integers $s = 1,\ldots,k+1$ are critical for $L(f,s)$. At these points the Beilinson conjectures reduce to the conjunction of Deligne's conjecture (already proved in **[De2]**) and the vanishing of $\Pi\left[H_{\mathcal{M}}^{k+2}(\overline{\overline{X}}{}_n^k, \mathbb{Q}(r))_{\mathbb{Z}}\right]$ for $1 \leq r \leq k+1$, $r \neq k/2$ (for which there is at present no evidence). At $s = -l \leq 0$ the L-function has a simple zero, and the conjectures predict a relation between $L'(f,-l)$ and a regulator coming from $H_{\mathcal{M}}^{k+2}(\overline{\overline{X}}{}_n^k, \mathbb{Q}(k+l+2))$.

The target for this regulator is the Deligne cohomology group

$$H_{\mathcal{D}}^{k+2}(\overline{\overline{X}}{}_n^k/\mathbb{R}, \mathbb{R}(k+l+2)) = H_B^{k+1}(\overline{\overline{X}}{}_n^k, \mathbb{R}(k+l+1))^+$$

and its Π_f-component is the space $(H_B(V(f)) \otimes_{\mathbb{Q}} \mathbb{R}(k+l+1))^+$, which is one-dimensional.

(5.5) Theorem. *There is a subspace* $\mathcal{P}_n \subset H^{k+2}_{\mathcal{M}}(\overline{\overline{X}}{}^k_n, \mathbb{Q}(k+l+2))$ *such that*

$$\Pi_f[r_{\mathcal{D}}(\mathcal{P}_n)] = L'(f,-l).(H_B(V(f))\otimes\mathbb{Q}(k+l+1))^+.$$

(5.6) Remarks. (1) For $k=0$ (the case of modular curves) this was proved by Beilinson [**Be1,Be2,SS1**]. The main ideas for the general case can already be found there. The case $k=1$, $l=0$ was also considered by Ramakrishnan (unpublished). Full details for the general case will appear in [**Sc3**].

(2) Recall that for the correct formalism of Beilinson's conjecture it is necessary to consider "motivic cohomology over $\mathbb{Z}$" (cf. 1.7 above). Although in general we cannot prove that $\mathcal{P}_n \subset H^{k+2}_{\mathcal{M}}(\overline{\overline{X}}{}^k_n, \mathbb{Q}(k+l+2))_{\mathbb{Z}}$, we have the following:

(i) Standard conjectures on the K-theory of varieties over finite fields would imply that $H^{k+2}_{\mathcal{M}}(\overline{\overline{X}}{}^k_n, \mathbb{Q}(k+l+2))_{\mathbb{Z}} = H^{k+2}_{\mathcal{M}}(\overline{\overline{X}}{}^k_n, \mathbb{Q}(k+l+2))$ except in the case $k=l=0$.

(ii) For curves these conjectures are known [**Ha1**]. Thus for $k=0$ the only obstruction to integrality occurs when $l=0$; in this case it is known (see [**SS1**], §7) that $\mathcal{P}_n \subset H^2_{\mathcal{M}}(\overline{M}_n, \mathbb{Q}(2))_{\mathbb{Z}}$.

(iii) For $k>0$ one can at least show that $\mathcal{P}_n$ contains enough elements which are integral away from primes dividing n, using a modification of a trick of Soulé [**So1**].

(5.7) Construction of $\mathcal{P}_n$. Consider the diagram

$$\begin{array}{ccccc} X^k_n & \xleftarrow{\;p\;} & X^{k+l}_n & \xrightarrow{\;q\;} & X^l_n \\ \downarrow{\scriptstyle r} & & & & \\ S = \mathrm{Spec}\,\mathbb{Q}(\zeta_n) & & & & \end{array}$$

where p, q are the projections onto the first k and last l factors of the fibre product, respectively. We define two subspaces

$$\mathcal{U}_n, \mathcal{V}_n \subset H^{k+2}_{\mathcal{M}}(X^k_n, \mathbb{Q}(k+l+2))(\Pi)$$

(where the projector Π is as in (5.3.1) above) as follows:

$$\mathcal{U}_n = p_*\left(q^* H^{l+1}_{\mathcal{M}}(X^l_n, \mathbb{Q}(l+1)) \cup H^{k+l+1}_{\mathcal{M}}(X^{k+l}_n, \mathbb{Q}(k+l+1))\right)$$
$$\mathcal{V}_n = r^* H^1_{\mathcal{M}}(S, \mathbb{Q}(l+1)) \cup H^{k+1}_{\mathcal{M}}(X^k_n, \mathbb{Q}(k+1)).$$

(Note that the Eisenstein symbol and Borel's theorem give a plentiful supply of elements of $\mathcal{U}_n$ and $\mathcal{V}_n$.) Let σ be the restriction

$$H^{k+2}_{\mathcal{M}}(\overline{\overline{X}}{}^k_n, \mathbb{Q}(k+l+2))(\Pi) \xrightarrow{\sigma} H^{k+2}_{\mathcal{M}}(X^k_n, \mathbb{Q}(k+l+2))(\Pi)$$

(which is in fact an inclusion) and write

$$\mathcal{Q}_n = \sigma^{-1}(\mathcal{U}_n + \mathcal{V}_n).$$

We then define (cf. 4.10 above)

$$\mathcal{P}_n = \bigcup_{n|n'} \rho^l_{n,n'*}(\mathcal{Q}_{n'}).$$

(5.8) Calculation of the regulator. At this point we should observe that the assumption that the Fourier coefficients of f are rational simplifies the calculation somewhat; in particular, we need not distinguish between f and its complex conjugate. There is a nondegenerate pairing (Poincaré duality)

$$<,>: H_B(V(f)) \times H_B(V(f)) \to \mathbb{Q}(-k-1)$$

and one has to prove that

(5.8.1) $$<r_{\mathcal{D}}(\mathcal{P}_n), \omega_f> = L'(f,-l).c^+(V(f)(k+l+1))\cdot\mathbb{Q}.$$

Here c^+ is Deligne's period **[De2]**. To calculate the left hand side we pull back to $X^{k+l}_{n'}$ for suitable n', and use the description (2.5) of the cup-product in Deligne cohomology. One obtains an integral of the form

(5.8.2) $$\frac{1}{(2\pi i)^{k+l}} \int_{X^{k+l}_{n'}} \mathcal{E}^{k+l}_{\mathcal{D}} \wedge \overline{q^* E_{l+2}} \wedge p^* \omega_f.$$

In this expression $\mathcal{E}^{k+l}_{\mathcal{D}}$ is the image of an Eisenstein symbol in Deligne cohomology, and E_{l+2} is a (variable) weight $k+2$ holomorphic Eisenstein series. This is a standard Rankin-Selberg integral and can be calculated explicitly. The Eisenstein series E_{l+2} is a linear combination of Eisenstein series E_χ, for various Dirichlet characters χ with $\chi(-1) = (-1)^l$, and the integral becomes a linear combination of terms which, up to a finite number of Euler factors, are of the form

$$L(f, k+l+2).L(f \otimes \chi, k+1).L(\chi.\chi_f, k+l+2)^{-1}.$$

At this stage one applies Shimura's algebraicity results on the twisted L-functions $L(f \otimes \chi, k+1)$ (which are critical values) and the functional equation for $L(f,s)$. In this way it can be shown that the left hand side of (5.8.1) is contained in the right hand side. The final step is to prove the equality—that is, to find suitable Eisenstein symbols for which the integral (5.8.2) is non-zero. For this one has to analyse the bad Euler factors carefully, and it is essential to work adelically. See **[Be2]**, §4 or **[SS1]**, §§2,4,5,6 for further details.

6. L-functions of algebraic Hecke characters

In this section we describe a construction involving the Eisenstein symbol which will give us elements in the motivic cohomology of motives attached to Hecke characters of imaginary quadratic number fields. The regulators of these elements have the expected relation to special values of Hecke L-series. As a corollary one obtains results on Beilinson's conjectures for CM elliptic curves of Shimura type and for Dirichlet characters of $\mathbb{Q}$ and of imaginary number field. Full details are contained in **[Den1,2]**.

(6.1) Consider an algebraic Hecke character $\epsilon: I_K/K^* \to \mathbb{C}^*$ of weight w of an imaginary quadratic field K. We wish to understand the special values $L(\epsilon, n)$ for $n > \frac{w}{2}+1$ of the corresponding L-series in terms of Beilinson's conjectures. In fact one can treat the L-values of all conjugates of ϵ simultaneously. Thus it is better to take a slightly different point of view and to look at the associated CM character

$$\phi: I_K \to T^*.$$

Here T/K is a number field and there exist integers a, b with $a+b=w$ such that

$$\phi(x) = x^a \bar{x}^b \quad \text{for all } x \text{ in } K^* \subset I_K.$$

From ϕ we obtain an L-series taking values in $T \otimes \mathbb{C} = \mathbb{C}^{\mathrm{Hom}(T,\mathbb{C})}$ by setting $L(\phi, s) = (L(\phi_\sigma, s))_\sigma$ where ϕ_σ is the Hecke character associated to ϕ via the embedding σ of T.

For critical n Beilinson's conjectures reduce to the Deligne conjecture, which for $L(\phi, n)$ is proved in **[GS1,2]** and in much greater generality in **[Bla]**.

For non-critical n we first have to find a Chow motive (3.3) with coefficients in T whose L-series equals $L(\phi, s)$. Note that if ϕ is a Dirichlet character χ of K—i.e. if $a=b=0$—we can take the motive M_χ constructed in (3.3). For the general case one needs the theory of CM elliptic curves of Shimura type **[GS1]**. These are elliptic curves E with CM by $\mathcal{O}_K$ which are defined over an abelian extension F of K such that the extension $F(E_{\mathrm{tors}})/K$ is abelian as well. One checks that $e_0 = [E \times 0]$, $e_2 = [0 \times E]$ and $e_1 = 1 - e_0 - e_2$ are pairwise orthogonal projectors of the motive $\mathbb{Q}E$ in $\mathcal{M}_F^+(\mathbb{Q})$. The motive $h_1(E) = e_1(\mathbb{Q}E)$ in $\mathcal{M}_F^+(\mathbb{Q})$, viewed as a motive in $\mathcal{M}_K^+(\mathbb{Q})$, will be called M.

(6.2) Proposition. *For $w \geq 1$ and possibly after enlarging the field T there exists an elliptic curve as above such that $M^{\otimes w}$ contains a direct factor M_ϕ with $\mathrm{End}(M_\phi) = T$ and $L(H^w(M_\phi), s) = L(\phi, s)$.*

In the last equation M_ϕ is viewed as a motive in $\mathcal{M}_K^+(T)$ via **[De2]** 2.1.

Note that it is sufficient to treat Hecke characters of positive weight since multiplication of ϕ by the norm just results in a shift by one of s in the L-series. For the same reason we may assume that $a, b \geq 0$.

(6.3) Theorem. *Assume that $w \geq 1$, $n > \frac{w}{2} + 1$ and in addition that n is non-critical for M_ϕ, i.e. $n >$ Max(a, b). Then the L-series $L(\bar{\phi}, s)$ has a first order zero for $s = -l := w + 1 - n$ and there is an element ξ in $H_{\mathcal{M}}^{w+1}(M_\phi, \mathbb{Q}(n))$ such that*

$$r_{\mathcal{D}}(\xi) \equiv L'(\bar{\phi}, -l)\eta \quad \text{mod } T^*$$

in the free rank one $T \otimes \mathbb{R}$-module

$$H_{\mathcal{D}}^{w+1}(M_{\phi\mathbb{R}}, \mathbb{R}(n)) = H_B^w(M_{\phi\mathbb{R}}, \mathbb{R}(n-1)).$$

Here η is a T-generator of $H_B^w(M_{\phi\mathbb{R}}, \mathbb{Q}(n-1))$.

Remarks. (1) In general the conjectures involve the motivic cohomology of an integral model. However since $E \to \text{Spec} K$ has potential good reduction one can show that

$$H_{\mathcal{M}}^{w+1}(M_\phi, \mathbb{Q}(n)) = H_{\mathcal{M}}^{w+1}(M_\phi, \mathbb{Q}(n))_{\mathbb{Z}}$$

for $n \neq \frac{w}{2} + 1$, using **[So1]** 3.1.3, Corollary 2.

(2) In **[Den1]** a refined version of (6.3) is proved for $w = 1$ where one considers motivic cohomology with almost integral coefficients. This was possible by a careful reexamination of the entire (slightly modified) construction of Beilinson's Eisenstein symbol specialised to a constant elliptic curve.

(6.4) Construction of ξ and calculation of $r_{\mathcal{D}}(\xi)$. For simplicity we shall assume that $l \geq 0$. For the finitely many negative l in the theorem a slightly different construction is required. Set $k = w + 2l > 0$ and fix some integer $N \geq 1$. For a choice of a square root of the discriminant d_K of K consider the map

$$\delta = (\text{id}, \sqrt{d_K}) : E \to E^2 = E \times_F E$$

and let pr : $E^{l+w} = E^l \times_F E^w \to E^w$ be the projection. Choose a Galois extension F' of F such that the N-torsion points of $E' = E \otimes_F F'$ are rational over F'. The choice of a level N structure $\alpha : (\mathbb{Z}/N)^2 \xrightarrow{\sim} E'_N$ on E' determines a commutative diagram

$$\begin{array}{ccc} E' & \xrightarrow{i_\alpha} & X_N \\ \downarrow & & \downarrow \\ \mathrm{Spec} F' & \longrightarrow & M_N \end{array}$$

Using (1.3)(6) we find a canonical map $\mathcal{E}_{\mathcal{M}}$ independent of α which makes the following diagram commute:

$$\begin{array}{ccc} \mathbb{Q}[(\mathbb{Z}/N)^2]^0 & \xrightarrow{E^k_{\mathcal{M}}} & H^{k+1}_{\mathcal{M}}(X^k_N, \mathbb{Q}(k+1))_{\mathrm{sgn}_{k+1}} \\ \Vert & & \downarrow i^*_\alpha \\ \mathbb{Q}[E'_N]^0 & & H^{k+1}_{\mathcal{M}}(E'^k, \mathbb{Q}(k+1))_{\mathrm{sgn}_{k+1}} \\ \uparrow & & \uparrow \\ \mathbb{Q}[E_N]^0 & \xrightarrow{\mathcal{E}_{\mathcal{M}}} & H^{k+1}_{\mathcal{M}}(E^k, \mathbb{Q}(k+1))_{\mathrm{sgn}_{k+1}} \end{array}$$

Now consider the following composition $\mathcal{K}_{\mathcal{M}}$ of maps:

$$\begin{array}{ccc} H^{k+1}_{\mathcal{M}}(E^k, \mathbb{Q}(k+1))_{\mathrm{sgn}_{k+1}} & \xrightarrow{(\delta^l \times \mathrm{id})^*} & H^{k+1}_{\mathcal{M}}(E^{l+w}, \mathbb{Q}(k+1)) \\ \Big| & & \downarrow \mathrm{pr}_* \\ \Big| & & H^{w+1}_{\mathcal{M}}(E^w, \mathbb{Q}(n)) \\ \Big| \mathcal{K}_{\mathcal{M}} & & \downarrow \\ \Big| & & H^{w+1}_{\mathcal{M}}(h_1(E)^{\otimes w}, \mathbb{Q}(n)) \\ \downarrow & & \downarrow \\ H^{w+1}_{\mathcal{M}}(M_\phi, \mathbb{Q}(n)) & \longleftarrow & H^{w+1}_{\mathcal{M}}(M^{\otimes w}, \mathbb{Q}(n)) \end{array}$$

For $l < 0$ the map $\mathcal{K}_{\mathcal{M}}$ is defined differently **[Den2]** §2. The required element ξ is obtained in the form $\xi = \mathcal{K}_{\mathcal{M}} \mathcal{E}_{\mathcal{M}}(\beta)$ for suitable N and divisor β in $\mathbb{Q}[E_N]^0$. To prove that it has the right properties we must first of all calculate explicitly the analogous maps $\mathcal{E}_{\mathcal{D}}$ and $\mathcal{K}_{\mathcal{D}}$ in Deligne cohomology. For $\mathcal{K}_{\mathcal{D}}$ this is easy. For $\mathcal{E}_{\mathcal{D}}$ we can use formula (4.9.1) for $E^k_{\mathcal{D}}$ specialised to the value of τ corresponding to our elliptic curve E. Note that in order to derive (4.9.1) Beilinson makes essential use of the compactification $\overline{M}_N$ of M_N—see **[Be2]**, §3.3. In **[Den1]** a different method for the calculation of $\mathcal{E}_{\mathcal{D}}$ is described which only uses analysis on E itself.

Looking at (4.9.1) we see that $\mathcal{E}_{\mathcal{D}}(\beta)$ is a certain linear combination of Eisenstein-Kronecker series. Hence it comes as no surprise that for suitable β the element $\mathcal{K}_{\mathcal{D}}\mathcal{E}_{\mathcal{D}}(\beta)$ is related to $L'(\bar{\phi},-l)$ as specified in the theorem.

(6.5) Corollary. *(1) Let E/F be a CM elliptic curve of Shimura type as above. Then for $n \geq 2$ the weak Beilinson conjecture (3.2) holds for $L(H^1(E),n)$.*

(2) Assume that F is Galois over $\mathbb{Q}$ and let F^+ be a real subfield of F, i.e. $F^+ = F^\sigma \cap \mathbb{R}$ for some embedding σ of F into $\mathbb{C}$. Then for any elliptic curve E^+/F^+ whose base change to F is of Shimura type the analogue of (1) holds.

Remark. (2) generalises the case of CM elliptic curves over $\mathbb{Q}$ at $n = 2$ treated by Bloch **[Bl1]** and Beilinson **[Be1]**; see also **[DW]**.

(6.6) Dirichlet characters. Given a character

$$\chi : G_K \to T^*$$

we can attach to it the motive M_χ of (3.3) and the twist $M_\phi(1)$ of a motive M_ϕ as in (6.2) for $\phi = \chi N_{K\otimes\mathbb{R}/\mathbb{R}}$. Possibly after extension of scalars both motives have the same L-function and should in fact be equal. The Beilinson conjectures for $M_\phi(1)$ follow from the theorem. For M_χ one can prove them directly using the map

$$\mathcal{K}_{\mathcal{M}}\mathcal{E}_{\mathcal{M}} : \mathbb{Q}[E_N]^0 \to H^1_{\mathcal{M}}(M_\chi, \mathbb{Q}(l+1)), \quad l > 0$$

where E is a CM elliptic curve of Shimura type over an abelian extension of K trivialising χ and $\mathcal{K}_{\mathcal{M}}$ is defined by composition:

$$\begin{array}{ccc} H^{2l+1}_{\mathcal{M}}(E^{2l}, \mathbb{Q}(2l+1))_{\mathrm{sgn}_{l+1}} & \xrightarrow{(\delta^l)^*} & H^{2l+1}_{\mathcal{M}}(E^l, \mathbb{Q}(2l+1)) \\ \downarrow \mathcal{K}_{\mathcal{M}} & & \downarrow \mathrm{pr}_* \\ H^1_{\mathcal{M}}(M_\chi, \mathbb{Q}(l+1)) & \xleftarrow{e_\chi} & H^1_{\mathcal{M}}(\mathrm{Spec} F, \mathbb{Q}(l+1)). \end{array}$$

By a very simple argument **[Den2]** (3.6) one can use the theory over K to prove the Beilinson conjectures for Dirichlet characters of $\mathbb{Q}$ as well. The complete results are these:

(6.7) Theorem. *For $k=\mathbb{Q}$ or K consider a character*

$$\chi: G_k \to T^*$$

and let $L(\chi,s) = \big(L(\sigma\chi,s)\big)_\sigma$ be its $T\otimes\mathbb{C}$-valued L-series. For $l>0$ the map

$$r_{\mathcal{D}}\otimes\mathbb{R}: H^1_{\mathcal{M}}(M_\chi,\mathbb{Q}(l+1))_{\mathbb{Z}}\otimes\mathbb{R} \to H^1_{\mathcal{D}}(M_{\chi\mathbb{R}},\mathbb{R}(l+1))$$

is an isomorphism of free $T\otimes\mathbb{R}$-modules. For $k=\mathbb{Q}$ and $\chi(c)=(-1)^l$ or $k=K$ their rank equals one. In this case we have

$$c_{M_\chi} \equiv L'(\chi,-l) \quad \mathrm{mod}\ T^*$$

where $c_{M_\chi}\in(T\otimes\mathbb{R})^*/T^*$ denotes the regulator.

Remarks. (1) That $r_{\mathcal{D}}\otimes\mathbb{R}$ is an isomorphism follows from the work of Borel [**Bo1**] and Beilinson [**Be1**], app. to §2; see also [**Rap**].

(2) For $k=\mathbb{Q}$ a different proof of the theorem is given in [**Be1**] §7, see also [**N,E**].

Appendix: motivic cohomology and extensions

In this appendix, we outline without proof the construction of extensions of motives attached to elements of motivic cohomology. Details should appear in a future paper by the second author. The underlying idea is certainly not new, aand is implicit in the constructions of [**Bl2**]. To motivate it, we consider first the case of ordinary Chow theory (i.e., $H^{2q}_{\mathcal{M}}(X,\mathbb{Q}(q))$). The corresponding extensions appear first in a paper of Deligne ([**De2**], 4.3). So let X be smooth and projective over $\mathbb{Q}$, and let y be a cycle of codimension q, homologous to zero. Write Y for the support of y. Then there is an exact sequence of mixed motives (in the sense of [**Ja2**], Chap.1):

$$0\to h^{2q-1}(X)\to h^{2q-1}(X-Y)\to h^{2q}_Y(X)\xrightarrow{\gamma} h^{2q}(X)\cdots.$$

The cycle class gives a map $cl(y):\mathbb{Q}(-q)\to h^{2q}_Y(X)$, and by hypothesis $\gamma\circ cl(y)=0$. Hence by pullback we obtain an extension

$$0\to h^{2q-1}(X)\to E_y\to\mathbb{Q}(-q)\to 0.$$

Theorem [Ja2]. *The class of the extension E_y depends only on the rational equivalence class of y. The following diagram commutes:*

$$\begin{array}{ccc}\ker\{CH^q(X)\to H^{2q}(\overline{X},\mathbb{Q}_\ell(q))\} & \xrightarrow{y\mapsto E_y} & \mathrm{Ext}^1_{MM_{\mathbb{Q}}}(\mathbb{Q}(-q),h^{2q-1}(X)) \\ \downarrow \text{cycle} & & \downarrow \ell\text{-adic realisation} \\ \ker\{H^{2q}(X,\mathbb{Q}_\ell(q))\to H^{2q}(\overline{X},\mathbb{Q}_\ell(q))\} & \xrightarrow{H\text{–}S} & H^1(\overline{\mathbb{Q}}/\mathbb{Q},H^{2q-1}(\overline{X},\mathbb{Q}_\ell(q)))\end{array}$$

Here H–S denotes the edge homomorphism in the Hochschild-Serre spectral sequence (in continuous étale cohomology [**Ja1**])

$$E_2^{ab}=H^a(\overline{\mathbb{Q}}/\mathbb{Q},H^b(\overline{X},\mathbb{Q}_\ell)(q))\Rightarrow H^{a+b}(X,\mathbb{Q}_\ell(q)).$$

There is a similar statement for Deligne cohomology (cf. 2.9 above).

We now imitate this construction for higher cycles. In an attempt to make the notation tidier we write Δ^n_X for $\Delta_n\times X$, and $\partial\Delta^n_X$ for the union of the codimension one faces of Δ^n_X. By the normalisation theorem, any element of $H^{2q-n}_{\mathcal{M}}(X,\mathbb{Q}(q))=CH^q(X,n)\otimes\mathbb{Q}$ may be represented by a cycle $y\in z^q(X,n)$ with $\partial_i^*(y)=0$ for $0\le i\le n$. Choosing such a representative y, write $Y=\mathrm{supp}(y)$, $\partial Y=Y\cap\partial\Delta^n_X$, $U=\Delta^n_X-Y$, and $\partial U=U\cap\partial\Delta^n_X$. We consider the motive $h^{2q-1}(U,\partial U)$ which fits into a long exact sequence

$$\text{(A.1)}\quad h^{2q-2}(U)\to h^{2q-2}(\partial U)\to h^{2q-1}(U,\partial U)\to h^{2q-1}(U)\to h^{2q-1}(\partial U).$$

By purity we have $h^{2q-2}(U)=h^{2q-2}(\Delta^n_X)\equiv h^{2q-2}(X)$. It is also easy to deduce that $h^{2p-2}(\partial U)=h^{2p-2}(\partial\Delta^n_X)$ by considering the spectral sequence expressing the cohomology of $\partial\Delta^n_X$ in terms of that of its faces.

Lemma. *There is a decomposition*

$$\text{(A.2)}\qquad h^i(\partial\Delta^n_X)\xrightarrow{\sim}h^i(X)\oplus h^{i-n+1}(X).$$

(In fact this decomposition is given by the 1- and sgn-eigenspaces for the action of the symmetric group of degree n.) Thus the sequence (A.1) becomes

$$0\to h^{2q-n-1}(X)\to h^{2q-1}(U,\partial U)\to h^{2q-1}(U)\to h^{2q-1}(\partial U).$$

This fits into a bigger diagram:

$$
\begin{array}{ccccccccc}
 & & & & 0 & & & & \\
 & & & & \downarrow & & & & \\
 & & & & h^{2q-n-1}(X) & & & & \\
 & & & & \downarrow & & & & \\
 & & 0 & & h^{2q-1}(U,\partial U) & \longrightarrow & \ker\beta & & \\
 & & \downarrow & & \downarrow & & \downarrow & & \\
0 & \longrightarrow & h^{2q-1}(\Delta^n_X) & \longrightarrow & h^{2q-1}(U) & \longrightarrow & h^{2q}_Y(\Delta^n_X)^0 & \longrightarrow & 0\\
 & & \downarrow & & \downarrow & & \downarrow{\scriptstyle\beta} & & \\
0 & \longrightarrow & h^{2q-1}(\partial\Delta^n_X) & \longrightarrow & h^{2q-1}(\partial U) & \longrightarrow & h^{2q}_{\partial Y}(\partial\Delta^n_X)^0 & \longrightarrow & 0
\end{array}
$$

Here we have written

$$
\begin{aligned}
h^{2q}_Y(\Delta^n_X)^0 &= \ker\{h^{2q}_Y(\Delta^n_X) \to h^{2q}(\Delta^n_X)\}\\
h^{2q}_{\partial Y}(\partial\Delta^n_X)^0 &= \ker\{h^{2q}_{\partial Y}(\partial\Delta^n_X) \to h^{2q}(\partial\Delta^n_X)\}.
\end{aligned}
$$

The cycle class of y gives a map $\mathbb{Q}(-q) \to \ker\beta$. From the snake lemma and (A.2) we have a long exact sequence

(A.3) $$0 \to h^{2q-n-1}(X) \to h^{2q-1}(U,\partial U) \to \ker\beta \to h^{2q-n}(X).$$

Since $n > 0$ the composite map $\mathbb{Q}(-q) \to \ker\beta \to h^{2q-n}(X)$ is zero (by weights), hence by pullback we obtain an extension

$$0 \to h^{2q-n-1}(X) \to E_y \to \mathbb{Q}(-q) \to 0.$$

Theorem. *The class of the extension E_y depends only on the class of y in $H^{2q-n}_{\mathcal{M}}(X,\mathbb{Q}(q))$. The following diagram commutes:*

$$
\begin{array}{ccc}
H^{2q-n}_{\mathcal{M}}(X,\mathbb{Q}(q)) & \xrightarrow{\;y\mapsto E_y\;} & \mathrm{Ext}^1_{MM_{\mathbb{Q}}}(\mathbb{Q}(-q), h^{2q-n-1}(X)) \\
\big\downarrow{\scriptstyle \text{cycle}} & & \big\downarrow{\scriptstyle \ell\text{-adic realisation}} \\
H^{2q-n}(X,\mathbb{Q}_\ell(q)) & \xrightarrow{\text{Hochschild-Serre}} & H^1(\overline{\mathbb{Q}}/\mathbb{Q}, H^{2q-n-1}(\overline{X},\mathbb{Q}_\ell(q)))
\end{array}
$$

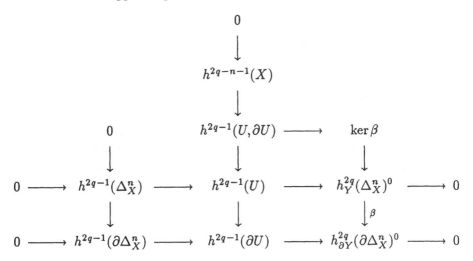

The analogous statement 2.9 for Deligne cohomology also holds.

Remark. In this construction, we have in the interest of clarity freely used "relative" and "local" motives $h^i_\bullet(-)$, $h^i(-,\bullet)$. Lest this trouble the reader, we point out that the extension E_y really belongs to the category $MM_{\mathbb{Q}}$ of mixed motives generated by $h^i(V)$ for quasi-projective varieties $V/\mathbb{Q}$ (see **[Ja2]**, Appendix C2). Indeed, the "relative" motive $h^{2q-1}(U, \partial U)$ can be constructed as part of the motive of a suitable singular variety (a mapping cylinder) and the motive $\ker \beta$ is simply the Tate twist of an Artin motive. Therefore the objects in the exact sequence (A.3) are all motives in $MM_{\mathbb{Q}}$. To construct the arrows we need only work in the various realisations, and there the relative and local cohomology groups are available.

References

[At] M. F. Atiyah; *K*-theory. Benjamin, 1969

[Be1] A. A. Beilinson; Higher regulators and values of *L*-functions. J. Soviet Math. **30** (1985), 2036–2070

[Be2] A. A. Beilinson; Higher regulators of modular curves. Applications of algebraic *K*-theory to algebraic geometry and number theory (Contemporary Mathematics **5** (1986)), 1–34

[Be3] A. A. Beilinson; Notes on absolute Hodge cohomology. Applications of algebraic *K*-theory to algebraic geometry and number theory (Contemporary Mathematics **55** (1986)), 35–68

[Bla] D. Blasius; On the critical values of Hecke *L*-series. Ann. Math. **124** (1986), 23–63

[Bl1] S. Bloch; Lectures on algebraic cycles. Duke Univ. Math. series **4** (1980)

[Bl2] S. Bloch; Algebraic cycles and higher *K*-theory. Advances in Math. **61** (1986) 267–304

[Bl3] S. Bloch; Algebraic cycles and the Beilinson conjectures. Contemporary Mathematics **58** (1), 1986, 65–79

[Bl4] S. Bloch; Algebraic cycles and higher *K*-theory: correction. Preprint, University of Chicago (1989)

[BlK] S. Bloch, K. Kato; *L*-functions and Tamagawa numbers of motives. Preprint, 1989

[**Bo1**] A. Borel; Stable real cohomology of arithmetic groups. Ann. Sci. ENS **7** (1974) 235–272

[**Bo2**] A. Borel; Cohomologie de SL_n et valeurs de fonctions zêta. Ann. Scuola Normale Superiore **7** (1974) 613–616

[**Ca**] J. Carlson; Extensions of mixed Hodge structures. Journées de géométrie algébrique d'Angers 1979, 107–127 (Sijthoff & Noordhoff, 1980)

[**De1**] P. Deligne; Formes modulaires et représentations l-adiques. Sém. Bourbaki, exposé 355. Lect. notes in mathematics **179,** 139–172 (Springer, 1969)

[**De2**] P. Deligne; Valeurs de fonctions L et périodes d'intégrales. Proc. Symp. Pure Math. AMS **33** (1979), 313–346

[**De3**] P. Deligne; Le groupe fondamentale de la droite projective moins trois points. Galois groups over **Q** (ed. Y. Ihara, K. Ribet, J.-P. Serre). MSRI publications **16,** 1989

[**Den1**] C. Deninger; Higher regulators and Hecke L-series of imaginary quadratic fields I. Inventiones math. **96** (1989) 1–69

[**Den2**] C. Deninger; Higher regulators and Hecke L-series of imaginary quadratic fields II. Ann. Math. (to appear)

[**DW**] C. Deninger, K. Wingberg; On the Beilinson conjectures for elliptic curves with complex multiplication. In [**RSS**].

[**E**] H. Esnault; On the Loday symbol in the Deligne-Beilinson cohomology. K-theory **3** (1989), 1–28

[**Fa**] G. Faltings; p-adic representations and crystalline cohomology. To appear

[**FM**] J.-M. Fontaine, W. Messing; p-adic periods and p-adic étale cohomology. Current trends in arithmetical algebraic geometry, ed. K. Ribet (Contemporary Mathematics **67** (1987)), 179–207

[**GS1**] C. Goldstein, N. Schappacher; Séries d'Eisenstein et fonctions L de courbes elliptiques à multiplication complexe. J. Reine angew. Math. **327** (1981), 184–218

[**GS2**] C. Goldstein, N. Schappacher; Conjecture de Deligne et Γ-hypothèse de Lichtenbaum sur les corps quadratiques imaginaires. C.R.A.S. **296,** Sér. I (1983), 615–618

[**Ha1**] G. Harder; Die Kohomologie S-arithmetischer Gruppen über Funktionenkörpern. Inventiones math. **42** (1977), 135–175

[Ha2] G. Harder; Arithmetische Eigenschaften von Eisensteinklassen, die modulare Konstruktion von gemoschten Motiven und von Erweiterungen endlicher Galoismoduln. Preprint, 1989

[Ja1] U. Jannsen; Continuous étale cohomology. Math. Annalen **280** (1988) 207–245

[Ja2] U. Jannsen; Mixed motives and algebraic K-theory. Lecture notes in math. **1400** (1990)

[Ja3] U. Jannsen; Deligne homology, Hodge-$\mathcal{D}$-conjecture and motives. In **[RSS]**

[Kl] S. Kleiman; Motives. Algebraic Geometry, Oslo 1970 (ed. F. Oort). Walters-Noordhoff, Groningen 1972, 53–82

[Ma] Yu. I. Manin; Correspondences, motives and monoidal correspondences. Math. USSR Sbornik **6** (1968) 439–470

[N] J. Neukirch; The Beilinson conjecture for algebraic number fields. In **[RSS]**

[Ra1] D. Ramakrishnan; Higher regulators of quaternionic Shimura curves and values of L-functions. Applications of algebraic K-theory to algebraic geometry and number theory (Contemporary Mathematics **5** (1986)), 377–387

[Ra2] D. Ramakrishnan; Regulators, algebraic cycles, and values of L-functions. Algebraic K-theory and algebraic number theory (Contemporary Mathematics **83** (1989)), 183–310

[Rap] M. Rapoport; Comparison of the regulators of Beilinson and of Borel. In **[RSS]**

[RSS] M. Rapoport, N. Schappacher, P. Schneider (ed.); Beilinson's conjectures on special values of L-functions. (Academic Press, 1988)

[Scha] N. Schappacher; Periods of Hecke characters. Lecture notes in mathematics **1301** (1988)

[SS1] N. Schappacher, A. J. Scholl; Beilinson's theorem on modular curves. In **[RSS]**

[SS2] N. Schappacher, A. J. Scholl; The boundary of the Eisenstein symbol. Preprint, 1990

[Sch] P. Schneider; Introduction to the Beilinson conjectures. In **[RSS]**

[Sc1] A. J. Scholl; Motives for modular forms. Inventiones math. **100** (1990), 419–430

[Sc2] A. J. Scholl; Remarks on special values of L-functions. This volume.

[Sc3] A. J. Scholl; Higher regulators and special values of L-functions of modular forms. In preparation

[Se] J.-P. Serre; Facteurs locaux des fonctions zêta des variétés algébriques (définitions et conjectures). Séminaire Delange-Pisot-Poitou 1969/ 70, exposé 19

[So1] C. Soulé; Groupes de Chow et K-théorie de variétés sur un corps fini. Math. Annalen **268** (1984), 237–253

[So2] C. Soulé; Régulateurs. Séminaire Borbaki, exposé 644. Astérisque **133/134** (1986), 237–253

Iwasawa theory for motives

RALPH GREENBERG

Let $V = \{V_\ell\}$ be a compatible system of ℓ-adic representations of $G_{\mathbf{Q}} = \mathrm{Gal}(\overline{\mathbf{Q}}/\mathbf{Q})$. We will think of V_ℓ as the ℓ-adic *homology* of a motive M defined over $\mathbf{Q}$. The corresponding L-function is defined by an Euler product

$$L_V(s) = \prod_q E_q(q^{-s})^{-1}$$

where

$$E_q(T) = \det(I - \mathrm{Frob}_{\overline{q}} T|_{(V_\ell)_{I_{\overline{q}}}})$$

for any $\ell \neq q$. Here $\mathrm{Frob}_{\overline{q}}$ is an *arithmetic* Frobenius for some place $\overline{q}$ of $\overline{\mathbf{Q}}$ lying over q, $I_{\overline{q}}$ denotes the corresponding inertia subgroup of $G_{\mathbf{Q}}$, and $(V_\ell)_{I_{\overline{q}}}$ is the maximal *quotient* space of V_ℓ on which $I_{\overline{q}}$ acts trivially. We are interested in the values of $L_V(s)$ where s is an integer, but, by taking a Tate twist of M if necessary, we can consider just the value $L_V(1)$. This turns out to be a convenient normalization. The conjectural functional equation for $L_V(s)$ (stated more precisely in [4] and [23]) can be put in the form

$$A_V^s \Gamma_V(s) L_V(s) = w_v A_{V^*}^{2-s} \Gamma_{V^*}(2-s) L_{V^*}(2-s)$$

where $V^* = \{V_\ell^*\}$ is the compatible system of ℓ-adic representations defined by $V_\ell^* = \mathrm{Hom}_{\mathbf{Q}_\ell}(V_\ell, \mathbf{Q}_\ell(1))$. A_V and A_{V^*} are certain positive constants. $\Gamma_V(s)$ and $\Gamma_{V^*}(s)$ are certain products of Γ-functions.

We will assume throughout this paper that $L_V(1)$ is a critical value in the sense of Deligne. In the above notation, this means that neither $\Gamma_V(s)$ nor $\Gamma_{V^*}(s)$ has a pole at $s = 1$. Then $L_{V^*}(1)$ is also a critical value. To describe 'Iwasawa Theory' for the motive M, we choose a prime p such that V_p is 'ordinary' in a sense we will explain later. This assumption on p seems to be essential in both the analytic side of the theory (the existence of p-adic L-functions with certain properties) as well as the algebraic side (the definition of a natural 'Selmer group'). Then we can formulate a 'Main Conjecture' which gives a link between the zeros of the p-adic L-function and the structure of the

Selmer group. The most natural and general context at present where one can give such a conjecture seems to be when V_p is in an analytic family of ordinary p-adic representations as considered for example in Hida [7] and in Mazur and Wiles [16].

In this paper, we will consider just the simplest case: an analytic family of twists $\{V_p \otimes \varphi\}$. The twists that we allow can be described as follows. Consider the canonical isomorphism

$$\mathcal{N} : \mathrm{Gal}\,(\mathbf{Q}(\mu_{p^\infty})/\mathbf{Q}) \xrightarrow{\sim} \mathbf{Z}_p^\times,$$

where μ_{p^∞} denotes the p-power roots of unity. Now $\mathbf{Z}_p^\times = \mu_{p-1} \times (1+p\mathbf{Z}_p)$ for odd p. (For $p=2$, $\mathbf{Z}_2^\times = \mu_2 \times (1+4\mathbf{Z}_2)$.) We let $\kappa :\ \mathrm{Gal}\,(\mathbf{Q}(\mu_{p^\infty})/\mathbf{Q}) \to 1+p\mathbf{Z}_p$ (or $1+4\mathbf{Z}_2$ if $p=2$) denote the projection. We can write $\mathcal{N} = \omega\kappa$ where ω can be thought of as a $\mathbf{Z}_p$-valued Dirichlet character of conductor p (or 4 if $p=2$). The homomorphism κ induces an isomorphism $\kappa : \Gamma \to 1+p\mathbf{Z}_p$ (or $1+4\mathbf{Z}_2$) where $\Gamma = \mathrm{Gal}\,(\mathbf{Q}_\infty/\mathbf{Q})$ and $\mathbf{Q}_\infty$ is the so-called cyclotomic $\mathbf{Z}_p$-extension of $\mathbf{Q}$. We have $\mathbf{Q}_\infty = \cup_{n\geq 1}\mathbf{Q}_n$ where $\mathbf{Q}_n$ is a cyclic extension of $\mathbf{Q}$ of degree p^n.

Let $\mathbf{C}_p$ denote the completion of $\overline{\mathbf{Q}}_p$ with its usual absolute value. We will consider twists by the elements of $\mathrm{Hom}_{\mathrm{cont}}(\Gamma, \mathbf{C}_p^\times)$. Each such element φ defines a 1-dimensional representation W_φ of $G_\mathbf{Q}$ over $\mathbf{C}_p$ (which factors through Γ). We then define $V_p \otimes \varphi$ as the $\mathbf{C}_p$-linear representation of $G_\mathbf{Q}$ on $V_p \otimes_{\mathbf{Q}_p} W_\varphi$. If one fixes a topological generator γ_0 of Γ, then φ is determined by the value $\varphi(\gamma_0)$ which lies in the open unit disc U in $\mathbf{C}_p$ centered at 1. One can think of the family of twists $\{V_p \otimes \varphi\}$ as parametrized by the value $\varphi(\gamma_0)$ in U. Fix a basis for V_p such that the $\mathbf{Z}_p$-lattice T_p spanned by this basis is invariant under the action of $G_\mathbf{Q}$. If φ_1 and φ_2 are such that $|\varphi_1(\gamma_0) - \varphi_2(\gamma_0)| < \varepsilon$, then each matrix entry function for $V_p \otimes \varphi_1$ will differ from the corresponding one for $V_p \otimes \varphi_2$ by less than ε (uniformly as functions on $G_\mathbf{Q}$). As an example, let $\varphi = \kappa^n$, where $n \in \mathbf{Z}$. Then $\varphi(\gamma_0) = u_0^n$ where $u_0 = \kappa(\gamma_0)$ is a topological generator of $1+p\mathbf{Z}_p$ (or $1+4\mathbf{Z}_2$ for $p=2$). Two such φ's, $\varphi_1 = \kappa^{n_1}$ and $\varphi_2 = \kappa^{n_2}$ will be close (as above) if $n_1 \equiv n_2$ modulo a high power of p.

1 THE ANALYTIC THEORY

We will briefly describe three fundamental properties of the p-adic L-function (whose existence in general is conjectural). A much more detailed description of the crucial interpolation factors c_φ is given in Coates article in this volume. We can think of the p-adic L-function as a $\mathbf{C}_p$-valued function defined on the analytic family of p-adic representations $\{V_p \otimes \varphi\}$. We will write it as $L_p(\varphi, V)$. It is defined for all but finitely many $\varphi \in \mathrm{Hom}_{\mathrm{cont}}(\Gamma, \mathbf{C}_p^\times)$.

It will be necessary to fix embeddings $\sigma_p : \overline{\mathbf{Q}} \to \mathbf{C}_p$ and $\sigma_\infty : \overline{\mathbf{Q}} \to \mathbf{C}$. A character φ of Γ of finite order can then be regarded as either $\mathbf{C}_p$-valued or $\mathbf{C}$-valued. For such φ, the embedding σ_p determines a prime λ_p of the field $E = \mathbf{Q}$ (values of φ) lying over p. The twist $V_p \otimes \varphi$ can be thought of as part (corresponding to λ_p) of a compatible system of λ-adic representations of $G_{\mathbf{Q}}$, λ running over the primes of E. The embedding σ_∞ then gives us the $\mathbf{C}$-valued L-function $L_V(s, \varphi)$. If $L_V(1)$ is a critical value, then so are all of the values $L_V(1, \varphi)$ since the Dirichlet characters φ are even and the conjectural functional equation for $L_V(s, \varphi)$ (which relates this function to $L_{V^*}(2-s, \varphi^{-1})$) will have the same Γ-factors as that for $L_V(s)$.

The p-adic L-function $L_p(\varphi, V)$ should have the following properties:

(A) $L_p(\varphi, V) = \varphi(\theta_V)$ for all but finitely many $\varphi \in \mathrm{Hom}_{\mathrm{cont}}(\Gamma, \mathbf{C}_p^\times)$, where θ_V is an element of the field of fractions of the Iwasawa algebra Λ.

(B) If φ has finite order, then $L_p(\varphi, V) = \sigma_p \circ \sigma_\infty^{-1}(c_\varphi L_V(1, \varphi))$.

(C) $L_p(\varphi, V) = u_\varphi L_p(\varphi^{-1}, V^*)$ for $\varphi \in \mathrm{Hom}_{\mathrm{cont}}(\Gamma, \mathbf{C}_p^\times)$, where u_φ is a unit in $\mathbf{C}_p$.

In property (A), the Iwasawa algebra Λ is the completed group algebra for Γ over $\mathbf{Z}_p$. It can be described as $\Lambda = \varprojlim \mathbf{Z}_p[\mathrm{Gal}(\mathbf{Q}_n/\mathbf{Q})]$, where the inverse limit is defined by the maps induced from the restriction homomorphisms $\mathrm{Gal}(\mathbf{Q}_{n+1}/\mathbf{Q}) \to \mathrm{Gal}(\mathbf{Q}_n/\mathbf{Q})$. One can also think of Λ as a formal power series ring: $\Lambda = \mathbf{Z}_p[[\gamma_0 - 1]]$. If $\varphi \in \mathrm{Hom}_{\mathrm{cont}}(\Gamma, \mathbf{C}_p^\times)$, then φ can be uniquely extended to a continuous $\mathbf{Z}_p$-algebra homomorphism $\varphi : \Lambda \to \mathbf{C}_p$. If $\lambda \in \Lambda$, $\lambda = f(\gamma_0 - 1)$ say, then $\varphi(\lambda) = f(\varphi(\gamma_0) - 1)$, which is an analytic function of the parameter $\varphi(\gamma_0)$ in the disc U. If $\theta = \lambda\delta^{-1}$, then $\varphi(\theta) = \varphi(\lambda)\varphi(\delta)^{-1}$ is meromorphic on this disc. One shows easily that it will have only finitely many zeros and poles in this disc (assuming $\theta \neq 0$). If one considers φ of the form κ^s, where $s \in \mathbf{Z}_p$, then $\varphi(\theta)$ will be a meromorphic function in the variable s.

The interpolation factors c_φ (described in detail in [1]) which occur in (B) all involve Deligne's period Ω_V (together with other more elementary quantities). The definition of Ω_V given in [4] depends on choosing lattices in the Betti and De Rham cohomology of the motive M and so is defined only up to a nonzero rational factor. For this reason, the p-adic L-function is only well-defined up to such a factor. Conjecturally, $L_V(1)/\Omega_V$ is rational. Then all of the values $c_\varphi L_V(1, \varphi)$, where φ is of finite order, will turn out to be algebraic complex numbers. Applying $\sigma_p \circ \sigma_\infty^{-1}$ then gives us values in $\mathbf{C}_p$. Once the properly chosen interpolation factors c_φ are specified, properties (A) and (B) determine uniquely the function $L_p(\varphi, V)$.

It is useful also to consider φ of the form $\varphi = \kappa^n = \mathcal{N}^n\omega^{-n}$, where $n \in \mathbf{Z}$. The representation $V_p \otimes \varphi = V_p(n) \otimes \omega^{-n}$, where $V_p(n)$ is the Tate twist, can again be regarded as part of a compatible system of λ-adic representations, λ running over the primes of $E = \mathbf{Q}$ (values of ω). The value at $s = 1$ of the corresponding complex L-function is $L_{V(n)}(1, \omega^{-n}) = L_V(1-n, \omega^{-n})$. Some of these values for $n \neq 0$ may also turn out to be critical. For such n, we should have $L_p(\kappa^n, V) = \sigma_p \circ \sigma_\infty^{-1}(c_n L_V(1-n, \omega^{-n}))$ for a suitable interpolation factor c_n. This is also discussed more precisely in Coates article [1].

The p-adic functional equation in (C) should be a direct consequence of the complex functional equation which relates $L_V(1,\varphi)$ to $L_{V^*}(1,\varphi^{-1})$ for all φ of finite order, together with properties (A) and (B). One can give the following alternative statement. The automorphism $\gamma \to \gamma^{-1}$ determines an involution of the $\mathbf{Z}_p$-algebra Λ and its field of fractions which we denote by $\theta \to \theta^\iota$. If $\varphi \in \mathrm{Hom}_{\mathrm{cont}}(\Gamma, \mathbf{C}_p^\times)$, then $\varphi(\theta^\iota) = \varphi^{-1}(\theta)$. Thus (C) implies that $\varphi(\theta_V(\theta_{V^*}^\iota)^{-1}) = u_\varphi$, a unit of $\mathbf{C}_p^\times$ for all φ. One sees from this that $\theta_V = u\theta_{V^*}^\iota$ where $u \in \Lambda^\times$.

2 THE ALGEBRAIC THEORY

Let $G_{\mathbf{Q}_p} = \mathrm{Gal}(\overline{\mathbf{Q}}_p/\mathbf{Q}_p)$. We identify $G_{\mathbf{Q}_p}$ with the decomposition subgroup in $G_{\mathbf{Q}}$ for some prime $\overline{\mathfrak{p}}$ of $\overline{\mathbf{Q}}$ lying over p. We assume that the p-adic representation of $G_{\mathbf{Q}_p}$ on V_p is ordinary in the following sense. There should be a filtration F^iV_p (where $i \in \mathbf{Z}$) of subspaces invariant under $G_{\mathbf{Q}_p}$ such that

(1) $F^{i+1}V_p \subseteq F^iV_p$. $F^iV_p = V_p$ for $i << 0$, $F^iV_p = 0$ for $i >> 0$.
(2) The inertia group $I_{\mathbf{Q}_p}$ acts on $\mathrm{gr}^i(V_p) = F^iV_p/F^{i+1}V_p$ by $\mathcal{N}^i$, where $\mathcal{N}$ is the p-power cyclotomic character.

The simplest examples are the 1-dimensional p-adic representations $V_p = Q_p(n)$ on which $G_{\mathbf{Q}}$ acts by $\mathcal{N}^n$, where $n \in \mathbf{Z}$. As another example, let E be an elliptic curve defined over $\mathbf{Q}$ with good, ordinary reduction at p. Let $\overline{E}$ be the reduced elliptic curve. One has a surjective homomorphism of the Tate modules $T_p(E) \to T_p(\overline{E})$, regarded as $G_{\mathbf{Q}_p}$-modules. Let $T_p^1(E)$ be the kernel. $T_p(E)$ has $\mathbf{Z}_p$-rank 2; $T_p(\overline{E})$ and $T_p^1(E)$ both have $\mathbf{Z}_p$-rank 1. On $V_p = V_p(E) = T_p(E) \otimes \mathbf{Q}_p$, one gets a filtration: $F^0V_p = V_p$, $F^1V_p = T_p^1(E) \otimes \mathbf{Q}_p$, $F^2V_p = 0$. $I_{\mathbf{Q}_p}$ acts trivially on $\mathrm{gr}^0(V_p)$ (i.e., by $\mathcal{N}^0$) and by $\mathcal{N}^1$ on $\mathrm{gr}^1(V_p)$ (since the determinant of the Galois action on $T_p(E)$ is $\mathcal{N}$).

Let T_p be a $G_{\mathbf{Q}}$-invariant lattice in V_p. Let $A = V_p/T_p$. Then A is a $G_{\mathbf{Q}}$-module which is isomorphic to $(\mathbf{Q}_p/\mathbf{Z}_p)^d$ as a group, where $d = \dim_{\mathbf{Q}_p}(V_p)$.

(We mention two examples: $\mu_{p^\infty} = \mathbf{Q}_p(1)/\mathbf{Z}_p(1)$. $E_{p^\infty} = V_p(E)/T_p(E)$ for an elliptic curve $E/\mathbf{Q}$.) We prefer to denote F^1V_p by F^+V_p. Let F^+A denote its image in A. As before, $\Gamma = \mathrm{Gal}(\mathbf{Q}_\infty/\mathbf{Q})$. The 'Selmer group' $S_A(\mathbf{Q}_\infty)$ is the subgroup of $H^1(\mathrm{Gal}(\overline{\mathbf{Q}}/\mathbf{Q}_\infty), A) = H^1(\mathbf{Q}_\infty, A)$ defined by certain local triviality conditions. For each place λ of $\mathbf{Q}_\infty$, let I_λ denote the inertia group for some place of $\overline{\mathbf{Q}}$ lying over λ. (The choice won't matter. For $\lambda = \pi$, the unique prime of $\mathbf{Q}_\infty$ over p, the choice determines the filtration of V_p and hence F^+A.) Here is the definition of our Selmer group.

$$S_A(\mathbf{Q}_\infty) = \ker\left(H^1(\mathbf{Q}_\infty, A) \to H^1(I_\pi, A/F^+A) \times \prod_{\lambda \neq \pi} H^1(I_\lambda, A)\right).$$

Let Σ be a finite set of primes containing p and ∞ and all primes ramified in the representation of $G_{\mathbf{Q}}$ on V_p. Let $\mathbf{Q}_\Sigma$ be the maximal extension of $\mathbf{Q}$ unramified outside Σ. Hence V_p is a representation space for $\mathrm{Gal}(\mathbf{Q}_\Sigma/\mathbf{Q})$. Also $\mathbf{Q}_\infty \subseteq \mathbf{Q}_\Sigma$ since only the prime p is ramified in $\mathbf{Q}_\infty/\mathbf{Q}$. Let Σ_∞ be the set of primes of $\mathbf{Q}_\infty$ lying over those in Σ. If $\lambda \notin \Sigma_\infty$, then I_λ acts trivially on A and so $H^1(I_\lambda, A) = \mathrm{Hom}(I_\lambda, A)$. From this one easily derives the following alternative description:

$$S_A(\mathbf{Q}_\infty) = \ker\left(H^1(\mathbf{Q}_\Sigma/\mathbf{Q}_\infty, A) \to H^1(I_\pi, A/F^+A) \times \prod_{\substack{\lambda \in \Sigma_\infty \\ \lambda \neq \pi}} H^1(I_\lambda, A)\right)$$

Now Γ acts naturally on $S_A(\mathbf{Q}_\infty)$ and so we can regard $S_A(\mathbf{Q}_\infty)$ as well as its Pontryagin dual $X = S_A(\mathbf{Q}_\infty)\hat{} = \mathrm{Hom}(S_A(\mathbf{Q}_\infty), \mathbf{Q}_p/\mathbf{Z}_p)$ as Λ-modules. X is a compact Λ-module and can be shown to be finitely generated. (See [5].) The following conjecture is crucial.

Conjecture 1 $S_A(\mathbf{Q}_\infty)\hat{}$ is a torsion Λ-module.

We would then say that $S_A(\mathbf{Q}_\infty)$ is Λ-cotorsion. If X is any finitely generated torsion Λ-module, then it is well-known that X is pseudo-isomorphic to a Λ-module of the form $\oplus_{i=1}^t \Lambda/(\lambda_i)$, where $\lambda_i \in \Lambda, \lambda_i \neq 0$. (This means that there exists a Λ-homomorphism with finite kernel and cokernel.) The $\mathbf{Q}_p$-vector space $X \otimes_{\mathbf{Z}_p} \mathbf{Q}_p$ would then be finite dimensional. The characteristic ideal of the Λ-module X is defined as $(\lambda_X) = \prod_{i=1}^t (\lambda_i)$. The generator λ_X is only defined up to a factor in $\Lambda^\times$. One possible choice would be $\lambda_X = p^\mu \prod_\varepsilon (\gamma_0 - \varepsilon)$, where ε runs over the eigenvalues (counted with their multiplicities) of γ_0 acting on $X \otimes_{\mathbf{Z}_p} \mathbf{Q}_p$. (The value of μ is related to the structure of the $\mathbf{Z}_p$-torsion subgroup of X as a Λ-module.) Notice that if $\varphi \in \mathrm{Hom}_{\mathrm{cont}}(\Gamma, \mathbf{C}_p^\times)$, then $\varphi(\lambda_X) = 0$ if and only if $\varphi(\gamma_0)$ is one of the eigenvalues ε.

Some motivation for the above conjecture comes from studying the structure of the global and local H^1's occurring in the definition of $S_A(\mathbf{Q}_\infty)$. This is discussed fully in [5], Sections 3, 4, and 5. The conjecture is equivalent to the assertion that, as a group, $S_A(\mathbb{Q}_\infty) \cong (\mathbb{Q}_p/\mathbb{Z}_p)^e \times C$ for some e, where C has bounded exponent. We include some examples at the end of this paper where we can actually verify this.

We will also need the Λ-modules $H^0(\mathbb{Q}_\infty, A) = A^{G_{\mathbb{Q}_\infty}}$ and $H^0(\mathbb{Q}_\infty, A^*) = (A^*)^{G_{\mathbb{Q}_\infty}}$ which we write more briefly as $A(\mathbb{Q}_\infty)$ and $A^*(\mathbb{Q}_\infty)$. Here we define $A^* = V_p^*/T_p^*$ where $V_p^* = \mathrm{Hom}\,(V_p, \mathbb{Q}_p(1))$ as before and $T_p^* = \mathrm{Hom}\,(T_p, \mathbb{Z}_p(1)) \subseteq V_p^*$. Both $A(\mathbb{Q}_\infty)$ and $A^*(\mathbb{Q}_\infty)$ are obviously Λ-cotorsion modules.

3 THE MAIN CONJECTURE

We will now describe the conjectural relationship between the analytic and algebraic theories. Define $\theta_A^{\mathrm{alg}} = \lambda_A \delta_A^{-1}$, where $\lambda_A = \lambda_{S_A(\mathbb{Q}_\infty)^\frown}$ and $\delta_A = (\lambda_{A(\mathbb{Q}_\infty)^\frown})(\lambda^{\iota}_{A^*(\mathbb{Q}_\infty)^\frown})$. Of course, θ_A^{alg} is well-defined only up to a factor in $\Lambda^\times$. We will now denote the θ_V occurring in the definition of the p-adic L-function by θ_V^{anal}. As we mentioned before, it is only well-defined up to a factor in $\mathbb{Q}^\times$.

Main conjecture $\theta_V^{\mathrm{anal}} = \theta_A^{\mathrm{alg}} \cdot \beta$, where $\beta \in \mathbb{Q}^\times \cdot \Lambda^\times$.

If $\varphi \in \mathrm{Hom}_{\mathrm{cont}}(\Gamma, \mathbb{C}_p^\times)$ and $\beta \in \mathbb{Q}^\times \cdot \Lambda^\times$, then $\varphi(\beta) \neq 0$. Thus the above conjecture gives an algebraic interpretation of the zeros and poles of the p-adic L-function $L_p(\varphi, V)$. Also $A(\mathbb{Q}_\infty)$ is infinite precisely when $\dim(V_p^{G_{\mathbb{Q}_\infty}}) \geq 1$, i.e., when V_p contains a $G_{\mathbb{Q}}$-subrepresentation factoring through Γ. Thus $L_p(\varphi, V)$ should have no poles when V_p and V_p^* have no such subrepresentations, which of course is usually the case.

The truth of the conjecture as formulated above is independent of the choice of lattice T_p. A change in T_p will not affect the ideal (δ_A) and will affect the characteristic ideal $(\lambda_{S_A(\mathbb{Q}_\infty)^\frown})$ only up to multiplication by a power of p (which is computed by Perrin-Riou in [17]). The factor β compensates for this ambiguity. However, the choice of lattice T_p in V_p should somehow determine $\mathbb{Z}$-lattices in the Betti and DeRham cohomology of M well-enough so that one can specify Ω_V up to a rational factor with numerator and denominator prime to p. It seems that this should be possible to do under very general assumptions by making use of a recent theorem of Fontaine and Lafaille. One could then hopefully make a more precise definition of θ_V^{anal} (which one might denote by θ_A^{anal}) so that the factor β in the above conjecture can be taken in $\Lambda^\times$. Since then $\varphi(\beta)$ would be a unit in $\mathbb{C}_p$, we would have a precise algebraic interpretation of the p-adic valuation of $\varphi(\theta_A^{\mathrm{anal}}) = L_p(\varphi, V)$.

The p-adic functional equation implies that $(\theta_V^{\text{anal}})^\iota$ and $\theta_{V^*}^{\text{anal}}$ should differ just by a factor in $\Lambda^\times$. On the algebraic side, obviously $(\delta_A)^\iota = \delta_{A^*}$. The compatibility of the main conjecture with the functional equation follows from the theorem below, which is proved in [5] by using Tate's local and global duality theorems.

Theorem 1 Assume that $S_A(\mathbf{Q}_\infty)$ is Λ-cotorsion. Then so is $S_{A^*}(\mathbf{Q}_\infty)$ and we have $(\lambda_{S_{A^*}(\mathbf{Q}_\infty)^\frown}) = (\lambda^\iota_{S_A(\mathbf{Q}_\infty)^\frown})$.

We want to make an observation about the special case where $V_p^* \cong V_p$ as $G_{\mathbf{Q}}$-representation spaces. Then both A and A^* are of the form $V_p/(\text{a lattice})$, although they are not necessarily isomorphic. (One might say they are isogenous.) We will assume that Conjecture 1 is valid for A. Then, as we mentioned above, $\lambda_A/\lambda_{A^*} = p^e u$, where $e \in \mathbf{Z}$, $u \in \Lambda^\times$. By Theorem 1, $\lambda_{A^*}/\lambda_A^\iota \in \Lambda^\times$. Hence $\lambda_A/\lambda_A^\iota = p^e u'$, where $u' \in \Lambda^\times$. But since ι is an automorphism of Λ, it is clear that $e = 0$. Now let $\Delta = \{id, \iota\}$. Then $\iota \to \lambda_A/\lambda_A^\iota$ defines an element σ_A in $H^1(\Delta, \Lambda^\times)$. Assume that p is odd. We then have $\Lambda^\times \cong \mathbf{Z}_p^\times \times (1 + (\gamma_0 - 1)\Lambda) \cong \mu_{p-1} \times$ (a pro-p group). Δ acts trivially on the constants $\mathbf{Z}_p^\times$ and hence on μ_{p-1}. Thus $H^1(\Delta, \Lambda^\times) \cong H^1(\Delta, \mu_{p-1}) = \text{Hom}(\Delta, \mu_{p-1}) = \text{Hom}(\Delta, \pm 1)$. Let σ'_A be the image of σ_A in the last group and let $\varepsilon_A = \sigma'_A(\iota)$. Since σ_A is equivalent to the cocycle σ'_A, one can change λ_A by a factor in $\Lambda^\times$ obtaining another generator (which we will still call λ_A) of the characteristic ideal of $S_A(\mathbf{Q}_\infty)^\frown$ with the property that $\lambda_A^\iota = \varepsilon_A \lambda_A$ where $\varepsilon_A = \pm 1$. Since p is odd, $\gamma_0 = \gamma_1^2$ for some $\gamma_1 \in \Gamma$. Let $\lambda_0 = \gamma_1 - \gamma_1^\iota$. Then $\lambda_0^\iota = -\lambda_0$ and $(\gamma_0 - 1) = (\lambda_0)$ as ideals in Λ. One easily sees that $\varepsilon_A = (-1)^{m_0}$ where m_0 is the highest power of the irreducible element $\gamma_0 - 1$ (or λ_0) which divides λ_A. If φ_0 is the trivial character of Γ, one can think of m_0 as the multiplicity of φ_0 as a zero of the function $\varphi \to \varphi(\lambda_A)$ (regarding this for example as an analytic function of the parameter $\varphi(\gamma_0)$, $\varphi \in \text{Hom}_{\text{cont}}(\Gamma, \mathbf{C}_p^\times)$). Let $n = \dim_{\mathbf{Q}_p}(X \otimes_{\mathbf{Z}_p} \mathbf{Q}_p)$ where $X = S_A(\mathbf{Q}_\infty)^\frown$. Then n is the total number of zeros (counting multiplicities) of the above function. Also, since $\lambda_A^\iota = \pm\lambda_A$, if φ is a zero, then so is φ^{-1} (and with the same multiplicity). Also $\varphi = \varphi^{-1}$ only for the trivial character $\varphi = \varphi_0$. This proves the following proposition.

Proposition 1 Let p be an odd prime. Assume that $V_p^* \cong V_p$ and that $S_A(\mathbf{Q}_\infty)$ is Λ-cotorsion. The characteristic ideal for $X = S_A(\mathbf{Q}_\infty)^\frown$ has a generator λ_A such that $\lambda_A^\iota = \varepsilon_A \lambda_A$, where $\varepsilon_A = (-1)^n$, $n = \text{rank}_{\mathbf{Z}_p}(X)$.

If $p = 2$, then the situation is slightly more complicated since there is then also one nontrivial φ (of order 2) such that $\varphi = \varphi^{-1}$. One finds that $H^1(\Delta, \Lambda^\times)$

has order 4. The cocycle σ_A reflects the parities of the multiplicities of both φ's as zeros of $\varphi \to \varphi(\lambda_A)$.

On the analytic side one can also define a cocycle σ_V^{anal} which sends ι to $\theta_V^{\text{anal}}/(\theta_V^{\text{anal}})^{\iota}$, still assuming that $V_p \cong V_p^*$. For odd p, one obtains just as above a sign $\varepsilon_V^{\text{anal}}$. By the main conjecture, one should have $\varepsilon_V^{\text{anal}} = \varepsilon_A$. The sign in the complex functional equation should often be the same as $\varepsilon_V^{\text{anal}}$, but not always. The existence of trivial zeros of p-adic L-functions has a bearing on this question. (See [9], [14] and also Example 3 below.)

We want to mention one more theorem which shows that the main conjecture is compatible with exact sequences. More precisely, assume that we have an exact sequence of $G_{\mathbf{Q}}$-representation spaces over $\mathbf{Q}_p$:

$$0 \to V_p' \to V_p \to V_p'' \to 0.$$

We assume that V_p and hence V_p' and V_p'' are ordinary p-adic representations and that they come from compatible systems of ℓ-adic representations V, V', and V'', say. If T_p is a $G_{\mathbf{Q}}$-invariant lattice in V_p, then one can easily choose such lattices $T_p' \subset V_p'$, $T_p'' \subset V_p''$ so that we get an exact sequence

$$0 \to A' \to A \to A'' \to 0$$

where of course $A = V_p/T_p$, $A' = V_p'/T_p'$, and $A'' = V_p''/T_p''$. Assume that $L_V(1)$ is a critical value and that $S_A(\mathbf{Q}_\infty)$ is Λ-cotorsion. Then the same statement will be true for V' and V''. In defining the p-adic L-functions, we take $\Omega_V = \Omega_{V'}\Omega_{V''}$. In [6], we intend to give a proof of the following result.

Theorem 2 The main conjecture for any two of A, A', A'' implies the main conjecture for the third.

4 EXAMPLES

Example 1 Our first example is a reformulation of the classical main conjecture of Iwasawa. Let n be a positive, even integer. We consider first the compatible system $V = \{\mathbf{Q}_\ell(n)\}$ of 1-dimensional ℓ-adic representations. ($\mathbf{Q}_\ell(n)$ is the ℓ-adic homology of the motive denoted by $\mathbf{Q}(-n)$ in [4].) We have $L_V(s) = \prod_q (1 - q^n q^{-s})^{-1} = \zeta(s-n)$. Hence $L_V(1) = \zeta(1-n)$, which is a critical value. Let φ be a character of Γ of finite order. In the interpolation property (B) which characterizes the p-adic L-function, one takes $c_{\varphi_0} = 1 - p^{n-1}$, $L_V(1, \varphi_0) = \zeta(1-n)$ for the trivial character $\varphi = \varphi_0$ and $c_\varphi = 1$, $L_V(1, \varphi) = L(1-n, \varphi)$ (a Dirichlet L-function!) if $\varphi \neq \varphi_0$. This is essentially the p-adic L-function constructed by Kubota and Leopoldt.

Now $V_p = \mathbf{Q}_p(n)$ is ordinary for every prime p. For simplicity, we will assume that p is odd. We can take $\Sigma = \{p, \infty\}$. Since $n > 0$, we have $F^+V_p = V_p$ and so $A/F^+A = 0$, where $A = \mathbf{Q}_p(n)/\mathbf{Z}_p(n) = \mathbf{Q}_p/\mathbf{Z}_p(n)$. The second description of the Selmer group becomes quite simple in this case: $S_A(\mathbf{Q}_\infty) = H^1(\mathbf{Q}_\Sigma/\mathbf{Q}_\infty, A)$. Let $K_\infty = \mathbf{Q}(\mu_{p^\infty})$, $\Delta = \mathrm{Gal}(K_\infty/\mathbf{Q}_\infty)$, and $X_\infty = \mathrm{Gal}(M_\infty/K_\infty)$ where M_∞ denotes the maximal abelian pro-p extension of K_∞ contained in $\mathbf{Q}_\Sigma$. Since $(|\Delta|, p) = 1$ and since G_{K_∞} acts trivially on A, we get isomorphisms

$$H^1(\mathbf{Q}_\Sigma/\mathbf{Q}_\infty, A) \xrightarrow{\sim} H^1(\mathbf{Q}_\Sigma/K_\infty, A)^\Delta \cong \mathrm{Hom}_\Delta(X_\infty, A) = \mathrm{Hom}(X_\infty^{\omega^n}, \mathbf{Q}_p/\mathbf{Z}_p(n))$$

where $X_\infty^{\omega^n}$ denotes the subgroup (a direct summand) of X_∞ on which Δ acts by ω^n. Thus $S_A(\mathbf{Q}_\infty)$ is essentially the Pontryagin dual of the Galois group $X_\infty^{\omega^n}$ (but with a twisting by κ^{-n} for the action of Γ). Iwasawa proved that X_∞^+ (the subgroup of X_∞ on which complex conjugation in Δ acts by $+1$) is a torsion Λ-module. Since n is even, it follows that $S_A(\mathbf{Q}_\infty)$ is in fact Λ-cotorsion. If $n \not\equiv 0 \pmod{p-1}$, then both $A(\mathbf{Q}_\infty)$ and $A^*(\mathbf{Q}_\infty)$ are zero. The main conjecture would then relate the characteristic ideal of the Λ-module $X_\infty^{\omega^n}$ to a Kubota–Leopoldt p-adic L-function. If one unravels the twisting, one finds that this is one of the two forms of Iwasawa's classical main conjecture. (The factor β will be in $\Lambda^\times$.) It is the version proved by Rubin using Kolyvagin's Euler system formed from cyclotomic units. (See [19].)

If $n \equiv 0 \pmod{p-1}$, it is rather simple to verify the main conjecture. We then have $A(\mathbf{Q}_\infty) = A$ and $A^*(\mathbf{Q}_\infty) = 0$. Also let M_∞^0 denote the maximal abelian pro-p extension of $\mathbf{Q}_\infty$ contained in $\mathbf{Q}_\Sigma$. Then $\mathrm{Gal}(M_\infty^0/\mathbf{Q}_\infty) = X_\infty^{\omega^n}$ since $\omega^n = \omega^0$. But one verifies easily that $M_\infty^0 = \mathbf{Q}_\infty$ by using the fact that $\mathbf{Q}_1$ is the only cyclic extension of $\mathbf{Q}$ of degree p unramified outside Σ. Hence $X_\infty^{\omega^n} = 0$. Thus one can take $\theta_A^{\mathrm{alg}} = 1/(\gamma_0 - \kappa^{-n}(\gamma_0))$. The known properties of the Kubota–Leopoldt p-adic L-function for the character ω^0 imply that θ_V differs from θ_A^{alg} by a factor $\beta \in \Lambda^\times$. In particular, $L_p(\varphi, V)$ has a pole at $\varphi = \kappa^{-n}$. As a concrete illustration, let $\varphi = \varphi_0$. Then $L_p(\varphi_0, V) = (1-p^{n-1})\zeta(1-n) = -(1-p^{n-1})B_n/n = u/p^{1+v_p(n)}$ by the well-known Clausen–von Staudt theorem about the denominator of the Bernoulli number B_n when $(p-1)|n$. Here $v_p(n)$ is the p-adic valuation of n and $u \in \mathbf{Z}_p^\times$. But it is easy to see that $\varphi_0(\theta_A^{\mathrm{alg}}) = 1/(1 - \kappa(\gamma_0)^{-n})$ is also of the form $u'/p^{1+v_p(n)}$, where $u' \in \mathbf{Z}_p^\times$.

We will consider now $V^* = \{\mathbf{Q}_\ell(1-n)\}$. Then we have $L_{V^*}(1) = \zeta(n)$. The p-adic L-function is also a form of the Kubota–Leopoldt p-adic L-function, but the interpolation factors are slightly more complicated. One can choose them so that the p-adic functional equation in (C) takes the simple form $L_p(\varphi, V) = L_p(\varphi^{-1}, V^*)$. Now this time we have $A^* = \mathbf{Q}_p/\mathbf{Z}_p(1-n)$ and $F^+A^* = 0$. Thus $S_{A^*}(\mathbf{Q}_\infty) \subseteq \mathrm{Hom}_\Delta(X_\infty, \mathbf{Q}_p/\mathbf{Z}_p(1-n))$ and the local triviality condition at π requires that a cocycle (or homomorphism in this case) become trivial when restricted to the inertia group in X_∞ for any prime over π. Let $Y_\infty = \mathrm{Gal}(L_\infty/K_\infty)$, where L_∞ is the maximal abelian unramified pro-p extension of K_∞. We have

$$\begin{aligned} S_{A^*}(\mathbf{Q}_\infty) &= \mathrm{Hom}_\Delta(Y_\infty, \mathbf{Q}_p/\mathbf{Z}_p(1-n)) \\ &= \mathrm{Hom}(Y_\infty^{\omega^{1-n}}, \mathbf{Q}_p/\mathbf{Z}_p(1-n)). \end{aligned}$$

Again we get essentially the Pontryagin dual of a Galois group $Y_\infty^{\omega^{1-n}}$ (slightly twisted). Iwasawa proved that Y_∞ is Λ-torsion and so Conjecture 1 is true in this case. The main conjecture is the version of Iwasawa's main conjecture which was proved by Mazur and Wiles in [15].

We would like to give a hint about how the equivalence of the two versions of Iwasawa's main conjecture can be proved. In this case, it is a theorem of Iwasawa. Let $K_0 = \mathbf{Q}(\mu_p)$. We consider for simplicity just an extension L of K_0 such that L/K_0 is unramified, $L/\mathbf{Q}$ is Galois, and $\mathrm{Gal}(L/K_0) \cong \mu_p^{1-n}$ as a Δ-module. (Here μ_p^{1-n} is a cyclic group of order p on which Δ acts by ω^{1-n}. Such an isomorphism gives an element of order p in $S_{A^*}(\mathbf{Q}_\infty)$ fixed by the action of Γ.) By class field theory one finds that there exists a nontrivial ideal class c of K_0 such that $c^p = 1$ and $\delta(c) = c^{\omega^{1-n}(\delta)}$ for $\delta \in \Delta$. If $\mathfrak{a} \in c$, then $\mathfrak{a}^p$ is principal. One can choose a generator α such that the subgroup H of $K_0^\times/(K_0^\times)^p$ generated by α is contained in $(K_0^\times/(K_0^\times)^p)^{\omega^{1-n}}$. Thus $H \cong \mu_p^{1-n}$ as a Δ-module. Let $M = K_0(\sqrt[p]{\alpha})$. Then $M/\mathbf{Q}$ is Galois and, by Kummer theory, $\mathrm{Gal}(M/K_0) \cong \mathrm{Hom}(H, \mu_p) \cong \mu_p^n$ as a Δ-module. Such an isomorphism gives an element of order p in $S_A(\mathbf{Q}_\infty)$, again fixed by Γ.

Example 2 Let E be an elliptic curve defined over $\mathbf{Q}$ with good, ordinary reduction at p. Consider the compatible system of ℓ-adic representations $V = \{V_\ell(E)\}$, where $V_\ell(E) = T_\ell(E) \otimes \mathbf{Q}_\ell$, $T_\ell(E)$ denoting the Tate module for E for the prime ℓ. The corresponding L-function $L_V(s)$ is the Hasse–Weil L-function for E over $\mathbf{Q}$. Assume that E is a modular elliptic curve. Then $L_V(s)$ is the Mellin transform of a modular form f_E of weight 2. The value $L_V(1)$ is a critical value. If Ω_E is the real period, then $L_V(1)/\Omega_E$ is rational. Also, as we discussed earlier, $V_p(E)$ is an ordinary p-adic representation. The Euler factor for p in $L_V(s)$ is of the form $(1-\alpha_p p^{-s})(1-\beta_p p^{-s})$, where $\alpha_p + \beta_p = a_p$, $\alpha_p\beta_p = p$. Here a_p is the pth Fourier coefficient of f_E and, since E has ordinary reduction at p, we have $p \nmid a_p$. Identifying the numbers α_p, β_p with their images under the embedding σ_p, one of them (say β_p) is a p-adic unit.

Mazur and Swinnerton-Dyer [13] have constructed a p-adic L-function having the properties (A), (B), and (C). We want to state here just the interpolation property for $\varphi = \varphi_0$:

$$L_p(\varphi_0, V) = (1 - \alpha_p p^{-1})(1 - \alpha_p p^{-1}) L_V(1)/\Omega_E.$$

Note $L_p(\varphi_0, V) = 0$ if and only if $L_V(1) = 0$ because $\alpha_p p^{-1} = \beta_p^{-1} \neq 1$. (The complex absolute value of α_p and β_p is $\sqrt{p}$.)

In [12], Mazur formulated a conjecture relating the p-adic L-function to essentially the classical Selmer group for E over $\mathbf{Q}_\infty$. Now in this case we can take $A = V_p(E)/T_p(E) \cong E_{p^\infty}$ and $F^+A = E^1_{p^\infty}$, where $E^1_{p^\infty} = \ker(E_{p^\infty} \to \overline{E}_{p^\infty})$, as discussed previously. Also, one can verify that $A(\mathbf{Q}_\infty)$ and $A^*(\mathbf{Q}_\infty)$ are both finite. (Actually, $A \cong A^*$ by the Weil pairing.) The following result shows the equivalence of Mazur's conjecture and the main conjecture stated in this paper.

Proposition 2 $S_A(\mathbf{Q}_\infty)$ is isomorphic to the p-primary subgroup of the classical Selmer group $S_E^{\text{class}}(\mathbf{Q}_\infty)$ as Λ-modules.

This is a special case of a much more general result for arbitrary abelian varieties defined over number fields which will be proved in [2]. We will sketch here a simple argument that applies to this case. For each prime λ of $\mathbf{Q}_\infty$, let D_λ denote a decomposition group in $\operatorname{Gal}(\overline{\mathbf{Q}}/\mathbf{Q}_\infty)$ for a prime of $\overline{\mathbf{Q}}$ lying over λ and let I_λ denote the corresponding inertia group. If ℓ is a prime, $\ell \neq p$, then ℓ is unramified in $\mathbf{Q}_\infty/\mathbf{Q}$ and its decomposition group in $\Gamma = \operatorname{Gal}(\mathbf{Q}_\infty/\mathbf{Q})$ is nontrivial (and hence isomorphic to $\mathbf{Z}_p$). It follows that $D_\lambda/I_\lambda \cong \prod_{q\neq p} \mathbf{Z}_q$, a projective limit of finite groups of order prime to p. Therefore, in the definition of $S_A(\mathbf{Q}_\infty)$, one can replace I_λ by D_λ when $\lambda \neq \pi$. For $\lambda = \pi$, if we replace I_π by D_π, we obtain a subgroup of $S_A(\mathbf{Q}_\infty)$ which we denote by $S_A^{\text{str}}(\mathbf{Q}_\infty)$ (the strict Selmer group). The quotient $S_A(\mathbf{Q}_\infty)/S_A^{\text{str}}(\mathbf{Q}_\infty)$ can be identified with a subgroup of $H^1(D_\pi/I_\pi.A/F^+A) = H^1(D_\pi/I_\pi, \overline{E}_{p^\infty}) = \overline{E}_{p^\infty}/(\operatorname{Frob}_p - 1)\overline{E}_{p^\infty}$ which is 0 because Frob_p acts nontrivially on the group $\overline{E}_{p^\infty}$. (In fact, Frob_p acts by multiplication by β_p.) Thus we have $S_A^{\text{str}}(\mathbf{Q}_\infty) = S_A(\mathbf{Q}_\infty)$.

Both $S_A(\mathbf{Q}_\infty)$ and the p-primary subgroup of $S_E^{\text{class}}(\mathbf{Q}_\infty)$ are contained in $H^1(\mathbf{Q}_\infty, E_{p^\infty})$. If λ is any prime of $\mathbf{Q}_\infty$, let $(\mathbf{Q}_\infty)_\lambda$ denote the union of the completions of the $\mathbf{Q}_n$'s under λ. It is an extension of $\mathbf{Q}_\ell$, where $\lambda|\ell$, and we have $\operatorname{Gal}((\mathbf{Q}_\infty)_\lambda/\mathbf{Q}_\ell) \cong \mathbf{Z}_p$. The decomposition group D_λ can be identified with $\operatorname{Gal}(\overline{\mathbf{Q}}_\ell/(\mathbf{Q}_\infty)_\lambda)$. In defining $S_E^{\text{class}}(\mathbf{Q}_\infty)_{p\text{-prim}}$, it is enough to determine $\ker(H^1(D_\lambda, E_{p^\infty}) \to H^1(D_\lambda, E(\overline{\mathbf{Q}}_\ell)))$. To prove the proposition, we must show that this kernel is zero for $\ell \neq p$ and that it is the same as $\ker(H^1(D_\pi, E_{p^\infty}) \to H^1(D_\pi, \overline{E}_{p^\infty}))$ for $\lambda = \pi$.

For $\ell \neq p$, it is quite easy to see that the above kernel is zero. If $F/\mathbf{Q}_\ell$ is any finite extension, then $E(F)$ is an ℓ-adic Lie group. One then has a canonical decomposition $E(F) = (E(F)_{p\text{-prim}}) \times M(F)$, where we define $M(F) = p^a E(F)$ for $a \gg 0$. From this, we obtain a decomposition $E(\overline{\mathbf{Q}}_\ell) = E_{p^\infty} \times M$ as $G_{\mathbf{Q}_\ell}$-modules, where $M = \cup_F M(F)$, F running over all finite extensions of $\mathbf{Q}_\ell$. The map $H^1(D, E_{p^\infty}) \to H^1(D, E_{p^\infty} \times M)$ is injective for any subgroup D of $G_{\mathbf{Q}_\ell}$.

Let $\lambda = \pi$. We have the Kummer exact sequence.

$$0 \to E((\mathbf{Q}_\infty)_\pi) \otimes (\mathbf{Q}_p/\mathbf{Z}_p) \xrightarrow{\delta} H^1(D_\pi, E_{p^\infty}) \to H^1(D_\pi, E(\overline{\mathbf{Q}}_p))$$

and also the exact sequence

$$H^1(D_\pi, F^+E_{p^\infty}) \xrightarrow{\varepsilon} H^1(D_\pi, E_{p^\infty}) \to H^1(D_\pi, \overline{E}_{p^\infty}).$$

$Im(\delta)$ is the kernel occurring in the definition of $S_E^{\text{class}}(\mathbf{Q}_\infty)_{p\text{-prim}}$. $Im(\varepsilon)$ is the kernel occurring in the definition of $S_A^{\text{str}}(\mathbf{Q}_\infty) = S_A(\mathbf{Q}_\infty)$. Let $P \in E((\mathbf{Q}_\infty)_\pi)$. For $t \geq 1$, let $Q \in E(\overline{\mathbf{Q}}_p)$ be such that $p^t Q = P$. Then $\delta(P \otimes (1/p^t))$ is the cocycle σ defined by $\sigma(g) = g(Q) - Q$ for any $g \in D_\pi$. For $g \in I_\pi$, it is obvious that $\sigma(g) \in E^1_{p^\infty} = F^+E_{p^\infty}$. Hence σ is mapped to a cocycle $\overline{\sigma} \in H^1(D_\pi, \overline{E}_{p^\infty})$ such that $\overline{\sigma}|_{I_\pi}$ is trivial. But, as noted before, this means that $\overline{\sigma}$ is trivial. Hence $Im(\delta) \subseteq Im(\varepsilon)$. Now $\Gamma = \text{Gal}((\mathbf{Q}_\infty)_\pi/\mathbf{Q}_p)$. Both $Im(\delta)$ and $Im(\varepsilon)$ are Λ-submodules of $H^1(D_\pi, E_{p^\infty})$. We have a surjective Λ-module homomorphism $\Phi : Im(\varepsilon)\hat{} \to Im(\delta)\hat{}$. If one applies Corollary 1 of Proposition 1 in [5] to $F^+E_{p^\infty}$, one finds that $H^1(D_\pi, F^+E_{p^\infty})\hat{}$ is a torsion-free Λ-module of rank 1. The same is true for $Im(\varepsilon)\hat{}$, since ε has finite kernel. Thus either Φ is an isomorphism or $Im(\delta)\hat{}$ is a torsion Λ-module. But $Im(\delta)\hat{}$ has Λ-rank ≥ 1 as one can see by considering the maps $E((\mathbf{Q}_n)_\pi) \otimes (\mathbf{Q}_p/\mathbf{Z}_p) \to (E((\mathbf{Q}_\infty)_\pi) \otimes (\mathbf{Q}_p/\mathbf{Z}_p))^{\Gamma_n}$ for $n \geq 1$, where $\Gamma_n = \text{Gal}((\mathbf{Q}_\infty)_\pi/(\mathbf{Q}_n)_\pi)$. The first group is isomorphic to $(\mathbf{Q}_p/\mathbf{Z}_p)^{p^n}$ and the kernel is easily verified to be finite (using the fact that $E_{p^\infty}((\mathbf{Q}_\infty)_\pi)$ is finite). These remarks show that $Im(\delta)\hat{}$ has Λ-rank 1 and that $Im(\delta) = Im(\varepsilon)$. Proposition 2 follows.

Both Conjecture 1 and the main conjecture have been proven when E is an elliptic curve/$\mathbf{Q}$ with complex multiplication. Conjecture 1 is a consequence of a theorem of Rohrlich that we will state below together with a theorem of Rubin [20], giving an annihilator for a certain Selmer group. This result of Rubin is a weak version of the so-called 'two variable main conjecture' which

Rubin has now proven completely [21]. The main conjecture for $S_A(\mathbf{Q}_\infty)$, where $A = E_{p^\infty}$, follows from this.

If E doesn't have complex multiplication, then very little is known. One can prove that the natural map $S_E^{\text{class}}(\mathbf{Q}) \to S_E^{\text{class}}(\mathbf{Q}_\infty)^\Gamma$ has finite kernel and cokernel. In particular, if both the Mordell–Weil group $E(\mathbf{Q})$ and the Tate–Shafarevich group $Ш_E(\mathbf{Q})$ are finite, then it follows that $S_A(\mathbf{Q}_\infty)^\Gamma$ is finite, where still $A = E_{p^\infty}$. This easily implies that $S_A(\mathbf{Q}_\infty)$ is Λ-cotorsion. More generally, $S_A(\mathbf{Q}_\infty)$ is Λ-cotorsion if and only if both $\operatorname{rank}_{\mathbf{Z}}(E(\mathbf{Q}_n))$ and $\operatorname{corank}_{\mathbf{Z}_p}(Ш_E(\mathbf{Q}_n)_{p\text{-prim}})$ are bounded as $n \to \infty$. (The $\mathbf{Z}_p$-corank is just the $\mathbf{Z}_p$-rank of the Pontryagin dual.) Of course, conjecturally $Ш_E(\mathbf{Q}_n)$ is finite for all n (and hence has $\mathbf{Z}_p$-corank 0). A theorem of Rohrlich [18] states that $L_V(1,\varphi) \neq 0$ for all but finitely many characters φ of Γ of finite order. (Equivalently, $\theta_V \neq 0$.) It follows that the order of vanishing at $s = 1$ of the Hasse–Weil L-function for E over $\mathbf{Q}_n$ is bounded as $n \to \infty$. The Birch and Swinnerton-Dyer conjecture then implies that $\operatorname{rank}_{\mathbf{Z}}(E(\mathbf{Q}_n))$ is bounded. Thus it is certainly very reasonable to believe that $S_A(\mathbf{Q}_\infty)$ is Λ-cotorsion. The main conjecture would then imply that $\varphi_0(\theta_V) = 0$ precisely when $\varphi_0(\lambda_A) = 0$. Now $\varphi_0(\theta_V) = L_p(\varphi_0, V)$ vanishes if and only if $L_V(1) = 0$. On the algebraic side, $\varphi_0(\lambda_A) = 0$ is equivalent to the assertion that $S_A(\mathbf{Q}_\infty)^\Gamma$ is infinite. The main conjecture thus implies the following statement: $L_V(1) = 0$ if and only if either $E(\mathbf{Q})$ is infinite or the p-primary subgroup of $Ш_E(\mathbf{Q})$ is infinite for all primes p where E has good, ordinary reduction.

Example 3 Consider the elliptic curve $E = X_0(11)$. Let $p = 11$ throughout this example. E has split, multiplicative reduction at p. Let V be as in Example 2. The p-adic L-function for such E has been constructed by Mazur, Tate, and Teitelbaum in [14]. The interpolation property at $\varphi = \varphi_0$ takes the form

$$L_p(\varphi_0, V) = (1 - \beta_p^{-1}) L_V(1)/\Omega_E.$$

$L_V(1) \neq 0$ but nevertheless, $L_p(\varphi_0, V) = 0$ since now $\beta_p = 1$. (This is an example of a so-called 'trivial zero' for a p-adic L-function, i.e., a zero arising from the vanishing of an Euler factor contribution to the interpolation factor c_φ.) We want to explain why $\lambda_{S_{E_{p^\infty}}(\mathbf{Q}_\infty)^\wedge}$ also vanishes at $\varphi = \varphi_0$. This can be verified in a very general situation (see [5]) but here we can take advantage of some facts that simplify the situation.

Using a result of Kolyvagin, we can verify that $S_E^{\text{class}}(\mathbf{Q})_{p\text{-prim}} = 0$ for $p = 11$. (This is predicted by the Birch and Swinnerton-Dyer conjecture.) The inflation-restriction exact sequence implies that $H^1(\mathbf{Q}, E_{p^\infty}) \xrightarrow{\sim} H^1(\mathbf{Q}_\infty, E_{p^\infty})^\Gamma$.

(Here one needs the fact that $E_{p^\infty}(\mathbf{Q}_\infty) = 0$. This follows from $E_{p^\infty}(\mathbf{Q}_\infty)^\Gamma = E(\mathbf{Q})_{p\text{-prim}} = 0$.) One can then prove by a rather standard type of argument that $S_E^{\text{class}}(\mathbf{Q})_{p\text{-prim}} \xrightarrow{\sim} S_E^{\text{class}}(\mathbf{Q}_\infty)^\Gamma_{p\text{-prim}}$. The crucial point is that the restriction maps $H^1(\mathbf{Q}_\ell, E(\overline{\mathbf{Q}}_\ell)) \to H^1((\mathbf{Q}_n)_\lambda, E(\overline{\mathbf{Q}}_\ell))$ are injective for all n, where ℓ is any prime, λ any prime of $\mathbf{Q}_n$ over ℓ. For $\ell = p$, this follows from the Tate parametrization $E(\overline{\mathbf{Q}}_p) = \overline{\mathbf{Q}}_p^\times / q_E^{\mathbf{Z}}$, where q_E turns out to have the form $q_E = pu$, $u \in \mathbf{Z}_p^\times$ but $\notin (\mathbf{Z}_p^\times)^p$. (Its approximate value is given in [14].) From this one sees that the norm map $E((\mathbf{Q}_n)_\pi) \to E(\mathbf{Q}_p)$ is surjective for all $n \geq 1$. Tate duality then gives the injectivity of the above restriction map. For $\ell \neq p$, the surjectivity of the norm map $E((\mathbf{Q}_n)_\lambda) \to E(\mathbf{Q}_\ell)$ follows from the fact that E has good reduction at ℓ and that $(\mathbf{Q}_n)_\lambda/\mathbf{Q}_\ell$ is unramified. We again get the injectivity we want. Since $S_E^{\text{class}}(\mathbf{Q}_\infty)^\Gamma_{p\text{-prim}} = 0$, it follows that $S_E^{\text{class}}(\mathbf{Q}_\infty)_{p\text{-prim}} = 0$. (Let $X = S_E^{\text{class}}(\mathbf{Q}_\infty)\hat{}_{p\text{-prim}}$. Then X is a finitely generated Λ-module such that $X/(\gamma_0 - 1)X = 0$. This implies that $X = 0$.)

The Tate parametrization gives us an exact sequence:

$$0 \to \mu_{p^\infty} \to E_{p^\infty} \to \mathbf{Q}_p/\mathbf{Z}_p \to 0$$

of $G_{\mathbf{Q}_p}$-modules, where $G_{\mathbf{Q}_p}$ acts trivially on $\mathbf{Q}_p/\mathbf{Z}_p$. Thus $V_p = T_p(E) \otimes \mathbf{Q}_p$ is clearly ordinary. Let $A = E_{p^\infty}$. We have $F^+A = \mu_{p^\infty}$. Thus we can define both $S_A(\mathbf{Q}_\infty)$ and $S_A^{\text{str}}(\mathbf{Q}_\infty)$, but this time they could be different because $H^1(D_\pi/I_\pi, A/F^+A) = H^1(D_\pi/I_\pi, \mathbf{Q}_p/\mathbf{Z}_p) = \mathbf{Q}_p/\mathbf{Z}_p$. Thus $S_A(\mathbf{Q}_\infty)/S_A^{\text{str}}(\mathbf{Q}_\infty)$ is isomorphic to a subgroup of $\mathbf{Q}_p/\mathbf{Z}_p$. The analogue of Proposition 2 is that $S_E^{\text{class}}(\mathbf{Q}_\infty) = S_A^{\text{str}}(\mathbf{Q}_\infty)$. This is also a special case of the general result proved in [2], but one can give a simple, direct proof using Tate's parametrization and Kummer theory for $\overline{\mathbf{Q}}_p^\times$.

It follows that $S_A^{\text{str}}(\mathbf{Q}_\infty) = 0$. We will show that $S_A(\mathbf{Q}_\infty) = \mathbf{Q}_p/\mathbf{Z}_p$ and that Γ acts trivially. Hence we see that φ_0 is a zero of $\lambda_{S_A(\mathbf{Q}_\infty)\hat{}}$. (In fact, it's the only zero. It should be possible to computationally verify that φ_0 is also the only zero of $L_p(\varphi, V)$, but we haven't done this.) Let $\Sigma = \{p, \infty\}$. Since E has good reduction at all primes $\ell \neq p$, V_p can be viewed as a representation space for $\mathrm{Gal}(\mathbf{Q}_\Sigma/\mathbf{Q})$. Let $h_i = \mathrm{corank}_{\mathbf{Z}_p}(H^i(\mathrm{Gal}(\mathbf{Q}_\Sigma/\mathbf{Q}), A))$ for $i = 0, 1, 2$. Tate's calculation of the global Euler characteristic for finite $\mathrm{Gal}(\mathbf{Q}_\Sigma/\mathbf{Q})$-modules can be used to show that $h_0 - h_1 + h_2 = -\dim(V_p^-) = -1$, where V_p^- is the (-1)-eigenspace for $\mathrm{Gal}(\mathbf{C}/\mathbf{R})$. Thus $h_1 = 1 + h_0 + h_2 \geq 1$. Let C denote any subgroup of $H^1(\mathbf{Q}_\Sigma/\mathbf{Q}, A)$ such that $C \cong \mathbf{Q}_p/\mathbf{Z}_p$. (It turns out that $h_1 = 1$ and so C is uniquely determined.) We will also let C denote its image under the isomorphism $H^1(\mathbf{Q}_\Sigma/\mathbf{Q}, A) \xrightarrow{\sim} H^1(\mathbf{Q}_\Sigma/\mathbf{Q}_\infty, A)^\Gamma$. We will prove that C is contained in $S_A(\mathbf{Q}_\infty)^\Gamma$. Our assertion about the precise structure of

$S_A(\mathbf{Q}_\infty)$ clearly follows from this. Consider the following sequence of maps

$$C \subseteq H^1(\mathbf{Q}_\Sigma/\mathbf{Q}, A) \to H^1(G_{\mathbf{Q}_p}, A) \to H^1(G_{\mathbf{Q}_p}, A/F^+A)$$
$$= H^1(G_{\mathbf{Q}_p}, \mathbf{Q}_p/\mathbf{Z}_p) = \mathrm{Hom}(\mathrm{Gal}(K_\infty/\mathbf{Q}_p), \mathbf{Q}_p/\mathbf{Z}_p),$$

where K_∞ is the maximal abelian pro-p extension of $\mathbf{Q}_p$. Local class field theory implies that $\mathrm{Gal}(K_\infty/\mathbf{Q}_p) \cong \mathbf{Z}_p^2$ and that K_∞ is the compositum of the $\mathbf{Z}_p$-extension $(\mathbf{Q}_\infty)_\pi$ of $\mathbf{Q}_p$ and the maximal unramified $\mathbf{Z}_p$-extension of $\mathbf{Q}_p$. Hence it is clear that $I_\pi \subseteq \mathrm{Gal}(\overline{\mathbf{Q}}_p/K_\infty)$. Thus the image of C in $H^1(I_\pi, A/F^+A)$ is trivial. This is enough to prove that $C \subset S_A(\mathbf{Q}_\infty)$.

Example 4 Let f_{12} denote the unique, normalized cusp form of weight 12, level 1. Let $V = V(f_{12}) = \{V_\ell\}$ denote the corresponding compatible system of ℓ-adic representations. Then $L_V(s) = \sum_{n=1}^{\infty} \tau(n)n^{-s}$ where $\tau(n)$ is Ramanujan's τ-function. The Euler factor for p is $(1-\alpha_p p^{-s})(1-\beta_p p^{-s})$ where $\alpha_p + \beta_p = \tau(p)$, $\alpha_p\beta_p = p^{11}$. If $p \nmid \tau(p)$, then, as in Example 2, one of the numbers α_p, β_p (say β_p again) will be a p-adic unit (under the fixed embedding σ_p). With this assumption, Manin [10] has constructed a p-adic L-function $L_p(\varphi, V)$. The interpolation property for $\varphi = \varphi_0$ is

$$L_p(\varphi_0, V) = (1-\alpha_p p^{-1})(1-\alpha_p p^{-11})L_V(1)/\Omega_V$$

where Ω_V is chosen so that $L_V(1)/\Omega_V \in \mathbf{Q}$. We would like to formulate a more precise main conjecture (where $\beta \in \Lambda^\times$ rather than just $\in \mathbf{Q}^\times \cdot \Lambda^\times$). Unfortunately in general we don't know how to do this. In this example, we suspect that the simple choice $\Omega_V = L_V(1)$ will work for all primes p such that V_p is ordinary except for $p = 691$. When $p = 691$, there are two distinct $G_{\mathbf{Q}}$-invariant lattices (up to homothety) and two corresponding choices of Ω_V, which we believe should be $L_V(1)$ and $pL_V(1)$. The first choice will make $p|\theta_V^{\mathrm{anal}}$ in Λ. The second will make θ_V^{anal} invertible in Λ.

Mazur and Wiles have proved that V_p is ordinary if $p \nmid \tau(p)$. (See [24] for a more general result.) The set of primes p such that $p|\tau(p)$ (e.g., $p = 2, 3, 5, 7, 2411$) should conjecturally be infinite although very sparse. Let T_p be a $G_{\mathbf{Q}}$-invariant lattice in V_p and let $A = V_p/T_p$. If $p \nmid \tau(p)$, then we suspect that (for this specific example) the Selmer group $S_A(\mathbf{Q}_\infty)$ should be trivial if $\beta_p \not\equiv 1 (\bmod p\mathbf{Z}_p)$. (This certainly will be false in general for other cusp forms.) Note that $\beta_p \equiv \tau(p) \pmod{p\mathbf{Z}_p}$. Conjecturally there should be infinitely many many p such that $\tau(p) \equiv 1 (\bmod p)$. We will consider three such $p: p = 11, 23$, and 691.

For $p = 11$, we will show that $S_A(\mathbf{Q}_\infty) \cong \mathbf{Q}_p/\mathbf{Z}_p$ as a group and that Γ acts by κ^5. Thus κ^{-5} is the only zero of $\lambda_{\widehat{S_A(\mathbf{Q}_\infty)}}$. On the analytic side, one has (by

Manin [11]) that $L_p(\kappa^{-5}, V) = cL_{V(-5)}(1, w^5) = cL_V(6, w^5)$ for some explicit factor c. Now w^5 is an odd quadratic Dirichlet character (of conductor 11) and we find that $L_V(6, w^5)$ is forced to vanish because of the sign in the functional equation for $L_V(s, w^5)$. (The functional equation relates the values at s and $12-s$.) Thus κ^{-5} is also a zero of θ_V^{anal}.

To prove our result about $S_A(\mathbf{Q}_\infty)$, we first point out that $W_p = V_p \otimes \kappa^{-5}$ satisfies $W_p^* \cong W_p$ as $G_{\mathbf{Q}}$-representation spaces. To see this, let ℓ be any prime, $\ell \neq p$. Let α_ℓ, β_ℓ be the eigenvalues of $\mathrm{Frob}_{\bar{\ell}}$ acting on V_p, where $\bar{\ell}$ is any prime of $\overline{\mathbf{Q}}$ over ℓ. (All $\ell \neq p$ are unramified in the representation on V_p since f_{12} has level 1.) Then $\alpha_\ell\beta_\ell = \ell^{11}$. The eigenvalues γ_ℓ, δ_ℓ of $\mathrm{Frob}_{\bar{\ell}}$ acting on W_p satisfy $\gamma_\ell\delta_\ell = \ell^{11}\kappa(\ell)^{-10} = \ell$ since $\kappa(\ell)^{10} = w(\ell)^{-10}\ell^{10} = \ell^{10}$. On W_p^*, the eigenvalues of $\mathrm{Frob}_{\bar{\ell}}$ are $\ell\gamma_\ell^{-1} = \delta_\ell$ and $\ell\delta_\ell^{-1} = \gamma_\ell$. Thus, by the Tchebotarev density theorem, the representations of $G_{\mathbf{Q}}$ on W_p and W_p^* have the same trace. Since V_p and hence W_p are known to be irreducible, we must have $W_p^* \cong W_p$. Let $B = A \otimes \kappa^{-5} = W_p/R_p$, where $R_p = T_p \otimes \kappa^{-5}$. Proposition 1 implies that we can choose a generator λ_B of the characteristic ideal of $S_B(\mathbf{Q}_\infty)^\wedge$ such that $\varphi(\lambda_B) = \varepsilon\varphi^{-1}(\lambda_B)$ for all $\varphi \in \mathrm{Hom}_{\mathrm{cont}}(\Gamma, \mathbf{C}_p^\times)$ where $\varepsilon = (-1)^n$, $n = \mathrm{corank}_{\mathbf{Z}_p}(S_B(\mathbf{Q}_\infty))$. But $A \cong B$ as $G_{\mathbf{Q}_\infty}$-modules and so it is obvious that $S_A(\mathbf{Q}_\infty) = S_B(\mathbf{Q}_\infty)$ as groups. Once we show that they are isomorphic to $\mathbf{Q}_p/\mathbf{Z}_p$, it follows that $n = 1$ and that $\varphi_0(\lambda_B) = 0$. It is then clear that Γ acts trivially on $S_B(\mathbf{Q}_\infty)$. Now one can verify that for the action of Γ, we have $S_B(\mathbf{Q}_\infty) = S_A(\mathbf{Q}_\infty) \otimes \kappa^{-5}$ and hence Γ does act on $S_A(\mathbf{Q}_\infty)$ by κ^5.

It remains to verify that $S_A(\mathbf{Q}_\infty) \cong \mathbf{Q}_p/\mathbf{Z}_p$ as a group. Now we have a well-known congruence $f_{12} \equiv f_2 (\bmod 11)$ where $f_2 = f_E$ is the weight 2, level 11 cusp form corresponding to the elliptic curve $E = X_0(11)$ discussed in Example 3. As a consequence of this congruence together with the fact that E_p is irreducible as a representation space for $G_{\mathbf{Q}}$ over $\mathbf{Z}/p\mathbf{Z}$, we see that $A[p] \cong E_p$ as $G_{\mathbf{Q}}$-modules. Here $A[p]$ denotes the elements of order dividing p in A; E_p is the p-torsion on E.

Now Mazur and Wiles prove that for any p such that $p \nmid \tau(p)$, there is a filtration for the action of $G_{\mathbf{Q}_p}$ on V_p, $0 \subset U_p \subset V_p$, where $\dim(U_p) = 1$, $I_{\mathbf{Q}_p}$ acts trivially on V_p/U_p and by $\mathcal{N}^{11}$ on U_p. (See [24].) Thus $F^+V_p = U_p$. Let $p = 11$. We let $F^+A[p]$ denote $F^+A \cap A[p]$ (and similarly define F^+E_p). This defines a filtration of the $G_{\mathbf{Q}_p}$-modules $A[p]$ and E_p and it is clear that an isomorphism $A[p] \xrightarrow{\sim} E_p$ must send $F^+A[p]$ to F^+E_p. (The inertia group $I_{\mathbf{Q}_p}$ acts by ω^{11} and ω^1 respectively. We have $\omega^{11} = \omega^1$. They are nontrivial while the action of $I_{\mathbf{Q}_p}$ on $A[p]/F^+A[p]$ and E_p/F^+E_p is trivial.) We will use the notation $S_{A[p]}(\mathbf{Q}_\infty)$ for the subgroup of $H^1(\mathbf{Q}_\infty, A[p])$ defined by local triviality conditions analogous to those defining $S_A(\mathbf{Q}_\infty)$, using $F^+A[p]$ in place of F^+A. Similarly, we define $S_{E_p}(\mathbf{Q}_\infty)$. The isomorphism $A[p] \cong E_p$ shows that $S_{A[p]}(\mathbf{Q}_\infty) \cong S_{E_p}(\mathbf{Q}_\infty)$.

The exact sequence $0 \to A[p] \to A \xrightarrow{p} A \to 0$ induces a surjective homomorphism $H^1(\mathbf{Q}_\infty, A[p]) \to H^1(\mathbf{Q}_\infty, A)[p]$. It is also injective, since $H^0(\mathbf{Q}_\infty, A) = A(\mathbf{Q}_\infty) = 0$ in this case. (This is true because $A[p](\mathbf{Q}_\infty) = E_p(\mathbf{Q}_\infty) = 0$ as noted in Example 3.) It follows that we have an injective map $S_{A[p]}(\mathbf{Q}_\infty) \to S_A(\mathbf{Q}_\infty)[p]$. We will show that this map is an isomorphism. To prove surjectivity, we must verify that an element of $H^1(\mathbf{Q}_\infty, A[p])$ which satisfies the local triviality conditions defining $S_A(\mathbf{Q}_\infty)$ already satisfies those defining $S_{A[p]}(\mathbf{Q}_\infty)$. It is enough to verify the injectivity of the maps $H^1(I_\lambda, A[p]) \to H^1(I_\lambda, A)$ for all primes $\lambda \neq \pi$ and of the map $H^1(I_\pi, A[p]/F^+A[p]) \to H^1(I_\pi, A/F^+A)$. For $\lambda \neq \pi$, the kernel of the above map is the cokernel of the map $H^0(I_\lambda, A) \xrightarrow{p} H^0(I_\lambda, A)$. This cokernel is zero since $H^0(I_\lambda, A) = A$ is divisible. As for π, since $A[p]/F^+A[p] = (A/F^+A)[p]$ and $H^0(I_\pi, A/F^+A) = A/F^+A$ is again divisible, we see that the kernel of the corresponding map is also zero. All of these considerations apply to E_p and E_{p^∞}. We conclude that

$$S_A(\mathbf{Q}_\infty)[p] = S_{A[p]}(\mathbf{Q}_\infty) = S_{E_p}(\mathbf{Q}_\infty) = S_{E_{p^\infty}}(\mathbf{Q}_\infty)[p].$$

In Example 3, we proved that $S_{E_{p^\infty}}(\mathbf{Q}_\infty) \cong \mathbf{Q}_p/\mathbf{Z}_p$. Hence $S_A(\mathbf{Q}_\infty)[p]$ is cyclic. It follows that either $S_A(\mathbf{Q}_\infty)$ is a finite cyclic p-group or is isomorphic to $\mathbf{Q}_p/\mathbf{Z}_p$. But Proposition 10 of [5] shows that $S_A(\mathbf{Q}_\infty)^\wedge$ has no nontrivial finite Λ-submodules and hence is certainly not finite. (One must check the proof of this proposition to see that it applies to $p = 11$.) It follows that $S_A(\mathbf{Q}_\infty) \cong \mathbf{Q}_p/\mathbf{Z}_p$ as we stated.

Now we consider $p = 23$. We will also show that $S_A(\mathbf{Q}_\infty) \cong \mathbf{Q}_p/\mathbf{Z}_p$ as a group. It follows that Γ acts on $S_A(\mathbf{Q}_\infty)$ by $\varphi = \kappa^{s_0}$ for some $s_0 \in \mathbf{Z}_p$. The main conjecture states that κ^{s_0} is the only zero of $L_p(\varphi, V)$, but we can prove nothing in this direction.

Let K be the Hilbert class field of $\mathbf{Q}(\sqrt{-23})$. Then $\Delta = \mathrm{Gal}(K/\mathbf{Q})$ is isomorphic to the symmetric group S_3. Let ρ denote the unique 2-dimensional complex irreducible representation of Δ (which can be realized over $\mathbf{Z}$). This Artin representation corresponds to a cusp form f_1 of weight 1, level 23 with coefficients in $\mathbf{Z}$. It is known that the action of $G_{\mathbf{Q}}$ on $A[p]$ for $p = 23$ factors through Δ and is given by $\overline{\rho}$, the reduction of ρ mod 23. This is equivalent to a congruence $f_{12} \equiv f_1 (\bmod 23)$. Let W_p be a 2-dimensional $\mathbf{Q}_p$-vector space on which Δ acts by ρ, R_p a $G_{\mathbf{Q}}$-invariant lattice, and $B = W_p/R_p$. Then $A[p] \cong B[p]$ as $G_{\mathbf{Q}}$-modules.

Let $\mathfrak{P}$ be any one of the three prime of K lying over p. The corresponding decomposition (and inertia) subgroup $\Delta_{\mathfrak{P}}$ of Δ can be identified with $\mathrm{Gal}(\mathbf{Q}_p(\sqrt{-23})/\mathbf{Q}_p)$ since $K_{\mathfrak{P}} = \mathbf{Q}_p(\sqrt{-23})$. As a $\Delta_{\mathfrak{P}}$-representation space, we have $W_p = \oplus_\chi W_p^\chi$, where $\chi \in \widehat{\Delta}_{\mathfrak{P}} = \{\chi_0, \chi_1\}$, χ_0 denoting the trivial character. Each W_p^χ has dimension 1. Now W_p is not ordinary in the sense defined earlier, but nevertheless it has two possible filtrations as a $G_{\mathbf{Q}_p}$-representation space. For our purpose here, we will consider the filtration $0 \subset W_p^{\chi_1} \subset W_p$ and we will denote $W_p^{\chi_1}$ by F^+W_p. Let F^+B denote its image in B. Thus we can define Selmer groups $S_B(\mathbf{Q}_\infty)$ and $S_B^{\mathrm{str}}(\mathbf{Q}_\infty)$ just as before. The inertia group $I_{\mathbf{Q}_p}$ acts on F^+V_p by $\mathcal{N}^{11}$ and hence on $F^+A[p] = (F^+A) \cap A[p]$ by ω^{11}, which is a character of $G_{\mathbf{Q}_p}$ of order 2. One easily sees that $\omega^{11} = \chi_1$ and therefore the isomorphism $A[p] \tilde{\to} B[p]$ must send $F^+A[p]$ to $F^+B[p]$. The argument that we described for the case $p = 11$ applies here without change. We find that $S_A(\mathbf{Q}_\infty)[p] = S_B(\mathbf{Q}_\infty)[p]$. It is enough therefore to prove that $S_B(\mathbf{Q}_\infty) \cong \mathbf{Q}_p/\mathbf{Z}_p$.

We will actually show that $S_B(\mathbf{Q}_\infty) \cong \mathbf{Q}_p/\mathbf{Z}_p$ and Γ acts trivially. The argument is similar to the one in Example 3. We have $B/F^+B \cong \mathbf{Q}_p/\mathbf{Z}_p$ on which $G_{\mathbf{Q}_p}$ acts trivially. Let $\Sigma = \{p, \infty\}$. B is a $\mathrm{Gal}(\mathbf{Q}_\Sigma/\mathbf{Q})$-module and $\mathrm{corank}_{\mathbf{Z}_p}(H^1(\mathbf{Q}_\Sigma/\mathbf{Q}, B)) \geq 1$. We can pick a subgroup $C \cong \mathbf{Q}_p/\mathbf{Z}_p$ in $H^1(\mathbf{Q}_\Sigma/\mathbf{Q}, B)$. Just as previously, we find that $S_B(\mathbf{Q}_\infty)/S_B^{\mathrm{str}}(\mathbf{Q}_\infty)$ is isomorphic to a subgroup of $\mathbf{Q}_p/\mathbf{Z}_p$ and that $C|_{\mathbf{Q}_\infty} \cong \mathbf{Q}_p/\mathbf{Z}_p$ is contained in $S_B(\mathbf{Q}_\infty)$. We will show that $S_B^{\mathrm{str}}(\mathbf{Q}_\infty) = 0$ by proving the equivalent statement that $S_B^{\mathrm{str}}(\mathbf{Q}_\infty)^\Gamma = 0$. The assertion that $S_B(\mathbf{Q}_\infty) \cong \mathbf{Q}_p/\mathbf{Z}_p$ (with a trivial action of Γ) follows.

We will consider the subgroup $S_B(\mathbf{Q})$ of $H^1(\mathbf{Q}, B)$ defined by local triviality conditions analogous to before, except of course involving inertia group in $G_{\mathbf{Q}}$. We have

$$S_B(\mathbf{Q}) = \ker(H^1(\mathbf{Q}_\Sigma/\mathbf{Q}, B) \to H^1(I_{\mathbf{Q}_p}, B/F^+B)).$$

Since $(|\Delta|, p) = 1$, we have an isomorphism

$$H^1(\mathbf{Q}_\Sigma/\mathbf{Q}, B) \to H^1(\mathbf{Q}_\Sigma/K, B)^\Delta = \mathrm{Hom}_\Delta(\mathrm{Gal}(M/K), B).$$

Here M denotes the maximal abelian pro-p extension of K contained in Q_Σ. Let $U = \prod_{\mathfrak{P}|p} U_{\mathfrak{P}}$, where $U_{\mathfrak{P}}$ is the unit group in the completion $K_{\mathfrak{P}}$. Let E be the unit group of K, which we identify with its image in U. Let $\overline{E}$

denote its closure in U. One finds that $U/\overline{E} \cong \mathbf{Z}_p^4 \times T$, where T is finite of order prime to p. (One can verify this by a calculation with the roots of $X^3 - X + 1$, which are in E.) Also, the class number of K is not divisible by 23. Class field theory then implies that $\operatorname{Gal}(M/K) \cong \mathbf{Z}_p^4$. Each irreducible representation of Δ occurs with multiplicity 1 in $\operatorname{Gal}(M/K)$ and so one finds that $\operatorname{Gal}(M/K)^\rho = \operatorname{Gal}(M_\rho/K) \cong \mathbf{Z}_p^2$ for a certain field M_ρ, $K \subset M_\rho \subset M$. Hence we have $H^1(\mathbf{Q}_\Sigma/\mathbf{Q}, B) = \operatorname{Hom}_\Delta(\operatorname{Gal}(M_\rho/K), B) \cong \mathbf{Q}_p/\mathbf{Z}_p$. (The $\mathbf{Z}_p$-corank is therefore exactly 1. The subgroup C referred to before must be all of $H^1(\mathbf{Q}_\Sigma/\mathbf{Q}, B)$. This fact will simplify our discussion.)

Let $\sigma \in H^1(\mathbf{Q}_\Sigma/\mathbf{Q}, B)$ have order p. Then $\sigma|_K$ is a surjective Δ-homomorphism: $\operatorname{Gal}(M/K) \to B[p]$. Let $\widetilde{B} = B[p], \widetilde{U}_\mathfrak{P} = U_\mathfrak{P}/U_\mathfrak{P}^p$, and $\widetilde{U} = \prod_{\mathfrak{P}|p} \widetilde{U}_\mathfrak{P}$. The image $\widetilde{E}$ of E in $\widetilde{U}$ is nonzero because otherwise $U/\overline{E}$ would have nontrivial p-torsion. By class field theory, we obtain a surjective Δ-homomorphism $\varphi : \widetilde{U} \to \widetilde{B}$ such that $\widetilde{E} \subseteq \ker(\varphi)$. Now Δ acts on $\widetilde{B}$ by $\overline{\rho}$. Regarding $\widetilde{U}$ as a representation space for Δ over $\mathbf{Z}/p\mathbf{Z}$ (completely reducible since $p \nmid |\Delta|$), $\overline{\rho}$ occurs with multiplicity 2 and $\widetilde{E}$ is a Δ-subspace of $\widetilde{U}^{\overline{\rho}}$. Since $\dim_{\mathbf{Z}/p\mathbf{Z}}(\widetilde{E}) = 2$, we clearly get an isomorphism $\widetilde{U}^{\overline{\rho}}/\widetilde{E} \xrightarrow{\sim} \widetilde{B}$ induced from φ. Let $\mathfrak{P}|p$. Then $\Delta_\mathfrak{P}$ acts on $\widetilde{U}_\mathfrak{P}$ and the χ_0 and χ_1 components each have dimension 1. Let $\widetilde{U}_0 = \prod_{\mathfrak{P}|p} \widetilde{U}_\mathfrak{P}^{\chi_0}$. This is a Δ-subspace of $\widetilde{U}$ in which $\overline{\rho}$ occurs with multiplicity 1. We have a decomposition $\widetilde{B} = \widetilde{B}^{\chi_0} \times \widetilde{B}^{\chi_1}$ as a Δ-module. The local triviality condition (for σ) defining $S_B(\mathbf{Q})$ is equivalent to requiring that $\varphi(\widetilde{U}_\mathfrak{P}) \subseteq F^+\widetilde{B} = \widetilde{B}^{\chi_1}$, that is $\varphi(\widetilde{U}_\mathfrak{P}^{\chi_0}) = 0$, for each $\mathfrak{P}|p$. Hence, if $S_B(\mathbf{Q}) \neq 0$, we must have $\widetilde{U}_0 \subset \ker(\varphi)$. This implies that $\widetilde{U}_0^{\overline{\rho}} = \widetilde{E}$. Thus $\widetilde{E}$ is in the kernel of the projection map $\widetilde{U} \to \widetilde{U}_\mathfrak{P} \to \widetilde{U}_\mathfrak{P}^{\chi_1}$ for each $\mathfrak{P}$. If $\varepsilon \in E$, one can write $\varepsilon^2 = \varepsilon_0\varepsilon_1$ where $\Delta_\mathfrak{P}$ acts on ε_i by χ_i. We would then have $\varepsilon_1 \in U_\mathfrak{P}^p$. But, again by a calculation with the roots of $X^3 - X + 1$, one finds that this is not the case. Hence $S_B(\mathbf{Q}) = 0$.

Let $C = H^1(\mathbf{Q}_\Sigma/\mathbf{Q}, B)$. We have a map

$$C \to H^1(G_{\mathbf{Q}_p}, B/F^+B) = \operatorname{Hom}(G_{\mathbf{Q}_p}, \mathbf{Q}_p/\mathbf{Z}_p).$$

This map must be injective because $S_B(\mathbf{Q}) = 0$. We denote its image also by C. Since $C \cong \mathbf{Q}_p/\mathbf{Z}_p$, we can write $C = \operatorname{Hom}(\operatorname{Gal}(K_\infty/\mathbf{Q}_p), \mathbf{Q}_p/\mathbf{Z}_p)$ for a certain $\mathbf{Z}_p$-extension K_∞ of $\mathbf{Q}_p$. K_∞ is determined by its group of universal norms $\operatorname{UnivNorm}(K_\infty/\mathbf{Q}_p) = \cap_n N_{K_n/\mathbf{Q}_p}(K_n^\times)$, where K_n is the nth layer in K_∞. If $K_\infty \neq K_\infty^{\text{unr}}$, the unramified $\mathbf{Z}_p$-extension of $\mathbf{Q}_p$, then $\operatorname{UnivNorm}(K_\infty/\mathbf{Q}_p)$ is of the form $\mu_{p-1} \cdot q^{\mathbf{Z}}$ where $q = p^a u$, $a = p^t, t \geq 0$, and $u \in 1 + p\mathbf{Z}_p$. The fact that $S_B(\mathbf{Q}) = 0$ means that $K_\infty \cap K_\infty^{\text{unr}} = \mathbf{Q}_p$ and this implies that $q = pu$. Now $C = H^1(\mathbf{Q}_\Sigma/\mathbf{Q}, B) \xrightarrow{\sim} H^1(\mathbf{Q}_\Sigma/\mathbf{Q}_\infty, B)^\Gamma$. Let K_∞^{cycl}

denote the cyclotomic $\mathbf{Z}_p$-extension of $\mathbf{Q}_p$, $K_\infty^{\text{cycl}} = (\mathbf{Q}_\infty)_\pi$. Then $S_B^{\text{str}}(\mathbf{Q}_\infty)^\Gamma = 0$ if and only if $K_\infty \cap K_\infty^{\text{cycl}} = \mathbf{Q}_p$ or equivalently, $u \notin (1+p\mathbf{Z}_p)^p = 1+p^2\mathbf{Z}_p$. K_∞ has the following more concrete description. Fix a prime $\mathfrak{P}$ of K lying over p and let K_0 be the decomposition subfield $K^{\Delta_{\mathfrak{P}}}$. Then $\text{Gal}(M_\rho/K)^{\chi_0}$ has $\mathbf{Z}_p$-rank 1 and hence corresponds to a $\mathbf{Z}_p$-extension of K_0. Completing at $\mathfrak{P}$ gives a $\mathbf{Z}_p$-extension of $\mathbf{Q}_p$ which one easily sees is K_∞. It is an interesting exercise in class field theory to compute $q = pu$ and we will just give the result here. Let ε_0 be a fundamental unit of K_0, π_0 a generator of the prime ideal of K_0 lying below $\mathfrak{P}$. Let g be a generator of $\text{Gal}(K/\mathbf{Q}(\sqrt{-23}))$. Put $\varepsilon_1 = g(\varepsilon_0)/g^2(\varepsilon_0)$ and $\eta_1 = g(\pi_0)/g^2(\pi_0)$. Regarding these numbers in $K_{\mathfrak{P}}$, it turns out that $2\log_p(u) = 3\log_p(\pi_0) - 3b\log_p(\varepsilon_0)$ where $b = \log_p(\eta_1)/\log_p(\varepsilon_1)$. This determines u and one finds by calculation that $u \not\equiv 1(\bmod p^2\mathbf{Z}_p)$.

We consider the q occurring above as almost an analogue of the Tate period q_E occurring in Example 3. Let E be an elliptic curve/$\mathbf{Q}$ with split multiplicative reduction at p. Let $A = E_{p^\infty}$. Then $G_{\mathbf{Q}_p}$ acts trivially on $A/F^+A \cong \mathbf{Q}_p/\mathbf{Z}_p$. Let Σ be a finite set of primes containing ∞ and all primes where E has bad reduction. Assume that $S_E^{\text{class}}(\mathbf{Q})_{p\text{-prim}}$ is finite. Then one can show that $H^1(\mathbf{Q}_\Sigma/\mathbf{Q}, A)$ has $\mathbf{Z}_p$-corank 1 and hence contains a unique $C \cong \mathbf{Q}_p/\mathbf{Z}_p$. The image of C in $H^1(G_{\mathbf{Q}_p}, A/F^+A) = \text{Hom}(G_{\mathbf{Q}_p}, \mathbf{Q}_p/\mathbf{Z}_p)$ will then be nontrivial (and so isomorphic to $\mathbf{Q}_p/\mathbf{Z}_p$ again) and therefore will determine a $\mathbf{Z}_p$-extension K_∞ of $\mathbf{Q}_p$. One can show that $q_E \in \text{UnivNorm}(K_\infty/\mathbf{Q}_p)$. If q is a generator of $\text{UnivNorm}(K_\infty/\mathbf{Q}_p)$ mod μ_{p-1} as described earlier, then $q_E = q^e\varepsilon$ for some $e \geq 1$ and $\varepsilon \in \mu_{p-1}$. Either q or q_E would determine the $\mathbf{Z}_p$-extension K_∞.

We will now give some examples where $\lambda_{S_A(\mathbf{Q}_\infty)^\frown}$ is divisible by p in Λ (i.e., the μ-invariant is positive). Such examples will arise whenever p is an irregular prime, $p|B_k$ say, where $0 < k \leq p-3$, k even. As is well-known, there will then be a congruence mod p between an Eisenstein series and a cusp form (an eigenform for the Hecke operators), both of level 1, weight k. We will consider the cases where the cusp form has coefficients in $\mathbf{Q}$: $f_{12}, p = 691$; $f_{16}, p = 3617$; $f_{18}, p = 43867$; $f_{20}, p = 283$ or 617; $f_{22}, p = 131$ or 593; and $f_{26}, p = 657931$. (In each case f_k is the unique, normalized cusp form of level 1, weight k.) One will have a congruence $a_n^{(k)} \equiv \sigma_{k-1}(n)(\bmod p)$ where $a_n^{(k)}$ is the nth Fourier coefficient of f_k, $\sigma_{k-1}(n) = \sum_{d|n} d^{k-1}$, and p is one of the above primes (which are the primes dividing B_k). Let $V = V(f_k) = \{V_\ell\}$ be the compatible system of ℓ-adic representations corresponding to f_k. We have $a_p^{(k)} \equiv 1 \not\equiv 0(\bmod p)$ and so (by a theorem of Mazur and Wiles), the p-adic representation V_p is ordinary. Let T_p be a $G_{\mathbf{Q}}$-invariant lattice and $A = V_p/T_p$. Then the above congruence implies that $A[p]$ is actually reducible as a representation of $G_{\mathbf{Q}}$ over $\mathbf{Z}/p\mathbf{Z}$ and has composition factors μ_p^{k-1} and $\mathbf{Z}/p\mathbf{Z}(=\mu_p^0)$. We can choose the lattice T_p so that $\mu_p^{k-1} \subset A[p]$. Now $I_{\mathbf{Q}_p}$ acts on F^+V_p by $\mathcal{N}^{k-1}$ and on V_p/F^+V_p trivially. Hence it is clear that $\mu_p^{k-1} \subset F^+A[p] = A[p] \cap F^+A$. Let $W_p = \mathbf{Q}_p(k-1)$, as in Example 1 except that now $k-1$ is positive and *odd*. Thus $F^+W_p = W_p$. We let $R_p = \mathbf{Z}_p(k-1)$ and $B = W_p/R_p = \mu_{p^\infty}^{k-1}$. Therefore, we have $B[p] \subset A[p]$ as $G_{\mathbf{Q}}$-modules, $F^+B[p] \subset F^+A[p]$ as $G_{\mathbf{Q}_p}$-modules, and we obtain (in essentially the same way as earlier for $p = 11$ and 23) homomorphisms

$$S_B(\mathbf{Q}_\infty)[p] = S_{B[p]}(\mathbf{Q}_\infty) \to S_{A[p]}(\mathbf{Q}_\infty) \to S_A(\mathbf{Q}_\infty)[p].$$

The kernels of the above maps are finite. Referring back to our first example, we have that $S_B(\mathbf{Q}_\infty) = \mathrm{Hom}\,(X_\infty^{\omega^{k-1}}, \mathbf{Q}_p/\mathbf{Z}_p(k-1))$. But Iwasawa [8] proved that X_∞^- (the subgroup of X_∞ on which complex conjugation in Δ acts by -1) is a Λ-module of rank $(p-1)/2$ and, more precisely, each component $X_\infty^{\omega^n}$, where n is odd, has Λ-rank 1. Thus $S_B(\mathbf{Q}_\infty)$ is not Λ-cotorsion. As we explain in [5] in a much more general context, this is related to the fact that $\zeta(s)$ has a simple zero at $s = 1-(k-1) = -k$ forced by a pole of the Γ-factor in its functional equation. Thus $S_{B[p]}(Q_\infty)^\wedge$ is a $\Lambda/p\Lambda$-module of rank ≥ 1. (This rank turns out to in fact be 1.) Consequently, we see that $S_A(\mathbf{Q}_\infty)[p]^\wedge$ also has $\Lambda/p\Lambda$-rank ≥ 1. This implies that p divides a generator of the characteristic ideal of $S_A(\mathbf{Q}_\infty)^\wedge$.

We believe that in all of the above cases $S_A(\mathbf{Q}_\infty)$ has exponent p and its dual is a free $\Lambda/p\Lambda$-module of rank 1. In order to make the main conjecture valid in the precise form (where $\beta \in \Lambda^\times$), one should take $\Omega_V = L_V(1)$ when T_p is the lattice chosen above. One can give an interesting criterion for $S_A(\mathbf{Q}_\infty)$ to have the above structure in terms of the ideal class group $Cl(F)$ of a certain field F. Let $K_0 = \mathbf{Q}(\mu_p)$, $\Delta = \mathrm{Gal}\,(K_0/\mathbf{Q})$. Let F be the cyclic extension of K_0 of degree p which is Galois over $\mathbf{Q}$ and such that Δ acts on $\mathrm{Gal}\,(F/K_0)$ by ω^{k-1}, i.e., $\mathrm{Gal}\,(F/K_0) \cong \mu_p^{k-1}$. (In the notation of [5], this field is $L_{\omega^{k-1}}$.) Let $C = Cl(F)/Cl(F)^p$. Let g be a generator of $\mathrm{Gal}\,(F/K_0)$. Then $\mathrm{Gal}\,(F/\mathbf{Q})$ acts on C and Δ acts on each of the subquotients C/C^{g-1}, $C^{g-1}/C^{(g-1)^2}$, etc. The criterion is that the ω^0-component of Δ acting on $C^{g-1}/C^{(g-1)^2}$ is trivial. We will not try to explain this here. If the class number h_{K_0} of K_0 is not divisible by p^2, then the criterion is more simply that $C^{g-1} = 0$. This turns out to imply that $\mathrm{Gal}\,(F/K_0)$ acts trivially on $Cl(F)_{p\text{-prim}}$ and that $p^2 \nmid h_F$. Unfortunately, it would seem rather difficult to verify these criteria for any of our cases. In [5], we gave the simpler criterion for the case $k = 12, p = 691$, mistakenly assuming that $p^2 \nmid h_{K_0}$. (We are grateful to David Penman for pointing this error out to us.) The assumption that $p^2 \nmid h_{K_0}$ is however true for example when $p = 131$, 283, and 593.

REFERENCES

1. J. Coates, 'p-Adic L-functions for motives', this volume.

2. J. Coates and R. Greenberg, 'Selmer groups for abelian varieties', to appear.

3. J. Coates and B. Perrin-Riou, 'On p-adic L-functions attached to motives over $\mathbf{Q}$', Advanced Studies in Pure Mathematics, **17**, (1989), p. 23–54.

4. P. D. Deligne, 'Valeurs de fonctions L et périodes d'intégrales', Proc. Symp. Pure Math., **33** (1979), p. 313–46.

5. R. Greenberg, 'Iwasawa theory for p-adic representations', Advanced Studies in Pure Mathematics, **17** (1989), p. 97–137.

6. R. Greenberg, 'Iwasawa theory for p-adic representations II', to appear.

7. H. Hida, 'Galois representations into $\mathbf{Z}_p[[X]]$ attached to ordinary cusp forms', Invent, Math., **85** (1986), 545–613.

8. K. Iwasawa, 'On $\mathbf{Z}_\ell$-extensions of algebraic number fields', Ann. of Math., **98** (1973), p. 246–326.

9. J. Jones, 'Iwasawa theory at multiplicative primes', thesis (1987), Harvard University.

10. J. Manin, 'Periods of parabolic forms and p-adic Hecke series', Math. Sbornik, **92** (1973), p. 371–93.

11. J. Manin, 'The values of p-adic Hecke series at integer points of the critical strip', Math. Sbornik, **93** (1974), p. 631–7.

12. B. Mazur, 'Rational points of abelian varieties with values in towers of number fields', Invent. Math., **18** (1972), p. 183–266.

13. B. Mazur and P. Swinnerton-Dyer, 'Arithmetic of Weil curves', Invent. Math., **25** (1974), p. 1–61.

14. B. Mazur, J. Tate, and J. Teitelbaum, 'On p-adic analogues of the conjectures of Birch and Swinnerton-Dyer', Invent. Math., **84** (1986), p. 1–48.

15. B. Mazur and A. Wiles, 'Class fields of abelian extensions of $\mathbf{Q}$', Invent. Math., **76** (1984), p. 179–330.

16. B. Mazur and A. Wiles, 'On p-adic analytic families of Galois representations', Compositio Math., **59** (1986), p. 231–64.

17. B. Perrin-Riou, 'Variation de la fonction L p-adique par isogénie', Advanced Studies in Pure Mathematics, **17** (1989), p. 347–58.

18. D. Rohrlich, 'On L-functions of elliptic curves and cyclotomic towers', Invent. Math., **75** (1984), p. 409–23.

19. K. Rubin, 'The main conjecture. Appendix to: *Cyclotomic Fields* by S. Lang, Grad. Texts in Math., Springer-Verlag (1989).

20. K. Rubin, 'On the main conjecture of Iwasawa theory for imaginary quadratic fields', Invent. Math., **93** (1988), p. 701–13.

21. K. Rubin, 'The "main conjecture" of Iwasawa theory for imaginary quadratic fields', to appear in Invent. Math.

22. P. Schneider, 'Motive Iwasawa theory', Advanced Studies in Pure Mathematics, **17** (1989), p. 421–56.

23. J. P. Serre, 'Facteurs locaux des fonctions zeta des variétés algébriques (définitions et conjectures)', Sem. Delange—Pisot–Poitou, (1969/70), exp. 19.

24. A. Wiles, 'On ordinary λ-adic representations associated to modular forms', Invent. Math., **94** (1988), p. 529–73.

Kolyvagin's work on modular elliptic curves

BENEDICT H. GROSS

1. Let $X_0(N)$ be the modular curve over $\mathbf{Q}$ which classifies elliptic curves with a cyclic N-isogeny. Let $K = \mathbf{Q}(\sqrt{-D})$ be an imaginary quadratic field of discriminant $-D$, where all prime factors of N are split. For simplicity, we assume that $D \neq 3, 4$, so the integers $\mathcal{O}$ of K have unit group $\mathcal{O}^\times = \langle \pm 1 \rangle$. Choose an ideal $\mathcal{N}$ of $\mathcal{O}$ with $\mathcal{O}/\mathcal{N} \simeq \mathbf{Z}/N\mathbf{Z}$.

We consider K, and all other number fields in this paper, as subfields of $\mathbf{C}$. Then the complex tori $\mathbf{C}/\mathcal{O}$ and $\mathbf{C}/\mathcal{N}^{-1}$ define elliptic curves related by a cyclic N-isogeny, hence a complex point x_1 of $X_0(N)$. The theory of complex multiplication shows that the point x_1 is rational over K_1, the Hilbert class field of K.

Let E be a modular elliptic curve of conductor N over $\mathbf{Q}$, and fix a parametrization $\varphi : X_0(N) \longrightarrow E$ which maps the cusp ∞ of $X_0(N)$ to the origin of E. Once φ has been chosen, there is a unique invariant differential ω on E over $\mathbf{Q}$ such that $\varphi^*(\omega)$ is the differential $\Sigma a_n q^n dq/q$ associated to a normalized ($a_1 = 1$) newform on $X_0(N)$. Write $\omega_0 = c\omega$, where ω_0 is a Néron differential on E. It is known that c is an integer, and we may assume that $c \geq 1$.

Let $y_1 = \varphi(x_1)$ in $E(K_1)$, and define the point $y_K = \mathrm{Tr}_{K_1/K}(y_1)$ in $E(K)$. This point is obtained by adding y_1 to its conjugates, using the group law on E. If $\mathcal{N}'$ is another ideal with $\mathcal{O}/\mathcal{N}' \simeq \mathbf{Z}/N\mathbf{Z}$, and y'_K is the corresponding point in $E(K)$, we have $y'_K = \pm y_K +$ (torsion). Hence the canonical height $\hat{h}(y_K)$ is well-defined, independent of the choice of $\mathcal{N}$. Zagier and I proved the limit formula [GZ; Ch. I, (6.5)]:

$$L'(E/K,1) = \frac{\iint_{E(\mathbf{C})} \omega \wedge \overline{i\omega}}{\sqrt{D}} \cdot \hat{h}(y_K). \tag{1.1}$$

In particular, the point y_K has infinite order if and only if $L'(E/K,1) \neq 0$.

By comparing (1.1) with the conjecture of Birch and Swinnerton-Dyer for $L(E/K, s)$, Zagier and I were led to the following [GZ; Ch. V, 2.2].

Conjecture 1.2 Assume that $\hat{h}(y_K) \neq 0$, or equivalently, that the point y_K has infinite order in $E(K)$. Then

(1) the group $E(K)$ has rank 1, so the index $I_K = [E(K) : \mathbf{Z}y_K]$ is finite,
(2) the Tate–Shafarevich group $Ш(E/K)$ is finite; its order is given by

$$\#Ш(E/K) = (I_K/c \cdot \Pi_{p|N} m_p)^2$$

where $m_p = (E(\mathbf{Q}_p) : E^0(\mathbf{Q}_p))$.

In (2), note that both the index I_K and the integer c depend on the parametrization φ, but that the ratio I_K/c is independent of the parametrization chosen. Since c and the local factors m_p are integers, the formula in (2) predicts that the order of $Ш(E/K)$ should always divide $(I_K)^2$. This implies, by the existence of the Cassels pairing, that the group $Ш(E/K)$ should always be annihilated by I_K.

Kolyvagin has proved a great part of Conjecture 1.2. His main result is the following [K1, Thm. A].

Theorem 1.3 (Kolyvagin) Assume that the point y_K has infinite order in $E(K)$. Then

(1) the group $E(K)$ has rank 1,
(2) the group $Ш(E/K)$ is finite, of order dividing $t_{E/K} \cdot (I_K)^2$.

In part (2) of this theorem, $t_{E/K}$ is an integer ≥ 1, whose prime factors depend only on the curve E: they consist of 2 and the odd primes p where the Galois group of the extension $\mathbf{Q}(E_p)$ is smaller than expected.

In many cases, Theorem 1.3 reduces the conjecture of Birch and Swinnerton-Dyer to a finite amount of computation. For example, let $E = X_0(37)/w_{37}$ be the curve $y^2 + y = x^3 - x$, and let φ be the modular parametrization of degree 2. Then $c = 1$ and $m_{37} = 1$ in part (2) of Conjecture 1.2, so we expect that $\#Ш(E/K) = (I_K)^2$ when y_K has infinite order. Kolyvagin shows that $t_{E/K}$ is a power of 2 in this case, and that $t_{E/K} = 1$ when I_K is odd. To prove the full conjecture of Birch and Swinnerton-Dyer for E over K, one must construct non-trivial elements in $Ш(E/K)$ when $I_K > 1$. (Kolyvagin's method suggests such a construction – see §11). We remark that in this case the point y_K lies in $E(\mathbf{Q})$, which is infinite cyclic and generated by $P = (0, 0)$.

Writing $y_K = m_K \cdot P$ we find $I_K = \pm m_K$; the integers m_K appear as Fourier coefficients of a modular form of weight 3/2 for $\Gamma_0(4 \cdot 37)$ [Z; §5].

2. We will not prove all of Theorem 1.3, but will sketch the proof of a slightly weaker result to illustrate Kolyvagin's main argument. In all that follows, we assume that the curve E does not have complex multiplication over $\mathbf{C}$. (This excludes only thirteen j-invariants.) Then Serre has shown that the extension $\mathbf{Q}(E_p)$ generated by the p-division points of E has Galois group isomorphic to $GL_2(\mathbf{Z}/p\mathbf{Z})$ over $\mathbf{Q}$ for all sufficiently large primes p [S; Thm. 2]. In fact, if E is semi-stable (i.e., if N is square-free), the Galois group of $\mathbf{Q}(E_p)/\mathbf{Q}$ is isomorphic to $GL_2(\mathbf{Z}/p\mathbf{Z})$ for all $p \geq 11$ [Ma; Thm. 4].

The first (crucial) observation is the following. If y_K has infinite order in $E(K)$, one does not know *a priori* that the index $[E(K) : \mathbf{Z}y_K]$ is finite. However, since the group $E(K)$ is finitely generated, the point y_K is not infinitely divisible in $E(K)$. In other words, there are only finitely many integers n such that $y_K = nP$ with $P \in E(K)$.

Proposition 2.1 Let p be an odd prime such that the extension $\mathbf{Q}(E_p)$ has Galois group $GL_2(\mathbf{Z}/p\mathbf{Z})$, and assume that p does not divide y_K in $E(K)$. Then

(1) the group $E(K)$ has rank 1,
(2) the p-torsion subgroup $Ш(E/K)_p$ is trivial.

When y_K has infinite order in $E(K)$, Proposition 2.1 applies for almost all primes p. Our hypotheses imply that p does not divide the index $I_K = [E(K) : \mathbf{Z}y_K]$, so the conclusion is consistent with part (2) of Conjecture 1.2. Kolyvagin obtains Theorem 1.3 by refining the argument for primes p which divide y_K, using the fact that p^n does not divide y_K for large n. The p-primary component of $Ш(E/K)$ is bounded using his techniques on ideal class groups (see [R2]). When the Galois group of $\mathbf{Q}(E_p)$ is strictly contained in $GL_2(\mathbf{Z}/p\mathbf{Z})$, he uses Serre's result that the Galois group of $\mathbf{Q}(E_{p^n})$ has bounded index in $GL_2(\mathbf{Z}/p^n\mathbf{Z})$ for $n \to \infty$.

In fact, what we will prove involves the Selmer group $\mathrm{Sel}(E/K)_p$ at p, which sits in an exact sequence of $\mathbf{Z}/p\mathbf{Z}$-vector spaces

$$(2.2) \qquad 0 \longrightarrow E(K)/pE(K) \xrightarrow{\delta} \mathrm{Sel}(E/K)_p \longrightarrow Ш(E/K)_p \longrightarrow 0.$$

By our hypothesis on $\mathbf{Q}(E_p)$, the group $E(K)$ contains no p-torsion and the dimension of $E(K)/pE(K)$ over $\mathbf{Z}/p\mathbf{Z}$ is equal to the rank of $E(K)$.

Proposition 2.3 Let p be an odd prime such that the extension $\mathbf{Q}(E_p)$ has Galois group $GL_2(\mathbf{Z}/p\mathbf{Z})$, and assume that p does not divide y_K in $E(K)$. Then the group $\mathrm{Sel}(E/K)_p$ is cyclic, generated by δy_K.

The proof of Proposition 2.3 (following Kolyvagin) has three steps. The first is the construction of certain cohomology classes $c(n) \in H^1(K, E_p)$ from Heegner points of conductor n for K, and the study of their amazing properties. The second is the use of Tate duality to obtain information on the local components of elements in the Selmer group $\mathrm{Sel}_p(E/K)$ from the classes $c(n)$. The third is the use of the Čebotarev density theorem to convert information on the local components of the Selmer group to an upper bound on its order. Proposition 2.1 is an immediate corollary of Proposition 2.3, using (2.2).

3. We begin with a construction of the cohomology classes $c(n)$, or rather, with a description of the properties of Heegner points on which the construction depends.

Let $n \geq 1$ be an integer which is prime to N, and let $\mathcal{O}_n = \mathbf{Z} + n\mathcal{O}_K$ be the order of index n in $\mathcal{O}_K$. The ideal $\mathcal{N}_n = \mathcal{N} \cap \mathcal{O}_n$ is an invertible $\mathcal{O}_n$-module with $\mathcal{O}_n/\mathcal{N}_n \simeq \mathbf{Z}/N\mathbf{Z}$. Consequently the elliptic curve $\mathbf{C}/\mathcal{O}_n$ (with its cyclic N-isogeny to $\mathbf{C}/\mathcal{N}_n^{-1}$) defines a complex point x_n on $X_0(N)$. The theory of complex multiplication shows that the point x_n is rational over K_n, the ring class field of conductor n over K. We have a field diagram with Galois groups marked:

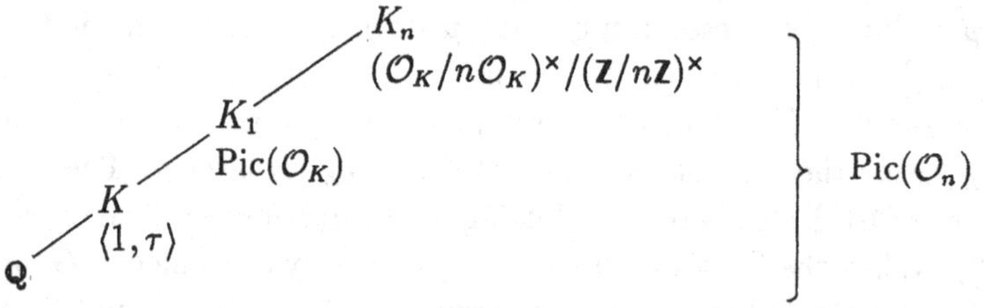

Here τ is complex conjugation, which lifts to an involution of K_n and acts on $\mathrm{Gal}(K_n/K)$ by: $\tau\sigma\tau^{-1} = \sigma^{-1}$.

We will only consider the points x_n on $X_0(N)$, and their images $y_n = \varphi(x_n)$ in $E(K_n)$, when the integer n is square-free. We insist that every prime factor ℓ of n satisfies:

(3.1) $\qquad \ell$ does not divide $N \cdot D \cdot p$.

This hypothesis implies that the prime ℓ is unramified in the extension $K(E_p)$. We let Frob(ℓ) be the conjugacy class in Gal($K(E_p)/\mathbf{Q}$) containing the Frobenius substitutions of the prime factors of ℓ, and further insist that

$$\text{(3.2)} \qquad \mathrm{Frob}(\ell) = \mathrm{Frob}(\infty)$$

as conjugacy classes in Gal($K(E_p)/\mathbf{Q}$). Here Frob(∞) is the conjugacy class of complex conjugation τ. There are an infinite number of primes ℓ satisfying (3.2), by Čebotarev's density theorem.

A simple implication of (3.2) is that Frob(ℓ) $= \tau$ in Gal($K/\mathbf{Q}$). Hence the prime (ℓ) remains inert in K; we let λ denote its unique prime factor. The implication Frob(ℓ) = Frob(∞) in Gal($\mathbf{Q}(E_p)/\mathbf{Q}$) is equivalent to the congruences:

$$\text{(3.3)} \qquad a_\ell \equiv \ell + 1 \equiv 0 \qquad (\mathrm{mod}\ p),$$

where $\ell + 1 - a_\ell$ is the number of points on the reduction $\tilde{E}$ over the finite field $F_\ell = \mathbf{Z}/\ell\mathbf{Z}$. Indeed, the characteristic polynomial of Frob(ℓ) acting on E_p is known to be $x^2 - a_\ell x + \ell$, whereas the characteristic polynomial of Frob(∞) $= \tau$ is known to be $x^2 - 1 = (x-1)(x+1)$.

Let F_λ denote the residue field of K at λ, which has ℓ^2 elements. By (3.2) the prime λ splits completely in the extension $K(E_p)$. Hence $\tilde{E}(F_\lambda)_p \simeq (\mathbf{Z}/p\mathbf{Z})^2$; in fact we have:

$$\text{(3.4)} \qquad \tilde{E}(F_\lambda)_p^{\pm} \simeq \mathbf{Z}/p\mathbf{Z}$$

where $\pm$ denote the eigenspaces for the automorphism group $\langle 1, \tau \rangle$. Indeed $\tilde{E}(F_\lambda)^+$ has order $\ell+1-a_\ell$, and $\tilde{E}(F_\lambda)^-$ has order $\ell+1+a_\ell$; both are divisible by p by (3.3).

We recall that n is square-free. Write $n = \prod \ell$ and let G_n be the Galois group of the extension K_n/K_1. Then $G_n \simeq \prod G_\ell$ where, for each $\ell | n$, G_ℓ is the subgroup fixing the subfield $K_{n/\ell}$. The subgroups $G_\ell \simeq F_\lambda^\times / F_\ell^\times$ are cyclic of order $\ell + 1$. Let σ_ℓ be a fixed generator of G_ℓ; the augmentation ideal of the group ring $\mathbf{Z}[G_\ell]$ is principal and generated by $(\sigma_\ell - 1)$. Let Tr_ℓ be the element $\sum_{G_\ell} \sigma$ in $\mathbf{Z}[G_\ell]$, and let D_ℓ be a solution of

$$\text{(3.5)} \qquad (\sigma_\ell - 1) \cdot D_\ell = \ell + 1 - \mathrm{Tr}_\ell$$

in $\mathbf{Z}[G_\ell]$. Then D_ℓ is well-defined up to addition of elements in the subgroup $\mathbf{Z} \cdot \mathrm{Tr}_\ell$ (Kolyvagin uses the solution $D_\ell^0 = \sum_{i=1}^{\ell} i \cdot \sigma_\ell^i = -\sum_{i=1}^{\ell+1} (\sigma_\ell^i - 1)/(\sigma_\ell - 1)$

but this has little advantage over the others). Finally, define $D_n = \prod D_\ell$ in $\mathbf{Z}[G_n]$.

Proposition 3.6 The point $D_n y_n$ in $E(K_n)$ gives a class $[D_n y_n]$ in $(E(K_n)/pE(K_n))$ which is fixed by G_n.

Proof It suffices to show that $[D_n y_n]$ is fixed by σ_ℓ, for all primes $\ell | n$, as these elements generate G_n. Hence we must prove that $(\sigma_\ell - 1)D_n y_n$ lies in $pE(K_n)$.

Write $n = \ell \cdot m$. By (3.5) we have $(\sigma_\ell - 1)D_n = (\sigma_\ell - 1)D_\ell \cdot D_m = (\ell + 1 - \mathrm{Tr}_\ell)D_m$ in $\mathbf{Z}[G_n]$. Hence

$$(\sigma_\ell - 1)D_n y_n = (\ell + 1)D_m y_n - D_m(\mathrm{Tr}_\ell y_n).$$

Since $\ell + 1 \equiv 0 \pmod p$ by (3.3), it suffices to show that $\mathrm{Tr}_\ell y_n$ lies in $pE(K_m)$. This follows from part (1) of the following proposition, and the congruence $a_\ell \equiv 0 \pmod p$ of (3.3).

Proposition 3.7 Let $n = \ell \cdot m$. Then

(1) $\mathrm{Tr}_\ell y_n = a_\ell \cdot y_m$ in $E(K_m)$.
(2) Each prime factor λ_n of ℓ in K_n divides a unique prime λ_m of K_m, and we have the congruence $y_n \equiv \mathrm{Frob}(\lambda_m)(y_m) \pmod{\lambda_n}$.

Proof This follows from the corresponding facts about the points x_n and x_m on $X_0(N)$ over K_n. If T_ℓ denotes the Hecke correspondence, which is self-dual of bidegree $\ell + 1$, we have: $\mathrm{Tr}_\ell x_n = T_\ell(x_m)$ as an equality of divisors of degree $\ell + 1$ on $X_0(N)$ over K_m [G; §6]. Since $\varphi(T_\ell d) = a_\ell \cdot \varphi(d)$ for any divisor d on $X_0(N)$, this proves (1).

To prove (2), we note that by class-field theory, the prime λ is split completely in K_m/K (as it is principal, and generated by an integer ℓ prime to m). The factors λ_m of λ in K_m are totally ramified in $K_n : \lambda_m = (\lambda_n)^{\ell+1}$. In particular, the residue field F_{λ_n} has ℓ^2 elements and is canonically isomorphic to F_λ. We have the congruence: $x_n \equiv \mathrm{Frob}(\lambda_m)(x_m)$ on $X_0(N)$ over F_{λ_n}. Indeed, the points in the divisor $T_\ell(x_m)$ are the conjugates of x_n over K_m; these are all congruent to $x_n \pmod{\lambda_n}$ as λ_m is totally ramified in K_n/K_m. The Eichler–Shimura congruence relation $T_\ell \equiv Fr_\ell + Fr_\ell^t \pmod \ell$ shows that at least one point in the divisor $T_\ell x_m$ is congruent to $\mathrm{Frob}(\lambda_m)(x_m) \pmod{\lambda_n}$. Hence all points in the divisor are congruent to $\mathrm{Frob}(\lambda_m)(x_m)$; this also follows from the fact that the residue field has ℓ^2 elements, so $\alpha^\ell \equiv \alpha^{1/\ell}$.

The two properties of Heegner points in Proposition 3.7 show that the collection $\{y_n\}$ forms an 'Euler system', in the language of Kolyvagin [K1; §1]. In the next section, we show how they may be used to construct cohomology classes $c(n)$ in $H^1(K, E_p)$. We observe that since $\mathrm{Tr}_\ell y_n = a_\ell y_m$ lies in $pE(K_n)$, the class $[D_n y_n]$ in $E(K_n)/pE(K_n)$ is independent of the choice of solutions D_ℓ of (3.5). It depends on the choice of generators σ_ℓ of G_ℓ only up to scaling by $(\mathbf{Z}/p\mathbf{Z})^\times$.

4. We retain the notation $n = \prod \ell$ with ℓ satisfying (3.1) and (3.2). Let $\mathcal{G}_n$ be the Galois group of K_n over K; this sits in an exact sequence $0 \to G_n \to \mathcal{G}_n \to \mathrm{Gal}(K_1/K) \to 0$. Let S be a set of coset representatives for G_n in $\mathcal{G}_n$, and define

$$P_n = \sum_{\sigma \in S} \sigma(D_n y_n) \qquad \text{in } E(K_n). \tag{4.1}$$

By Proposition 3.6, the class $[P_n]$ in $E(K_n)/pE(K_n)$ is fixed by $\mathcal{G}_n$. We use the same set S to define P_m for any $m|n$; note that $P_1 = \sum_{\sigma \in S} \sigma y_1 = \mathrm{Tr}_{K_1/K}(y_1) = y_K$. The class $[P_n]$ is independent of the choice of S, and depends on the choice of generators σ_ℓ of G_ℓ, for $\ell|n$, only up to scaling by $(\mathbf{Z}/p\mathbf{Z})^\times$.

The exact sequence $0 \to E_p \to E \xrightarrow{p} E \to 0$ of group schemes over $\mathbf{Q}$ gives, on taking cohomology (Galois = étale) over K and K_n, a commutative diagram

$$\begin{array}{ccccccccc}
 & & & & & & 0 & & \\
 & & & & & & \downarrow & & \\
 & & & & & & H^1(K_n/K, E)_p & & \\
 & & & & & & \downarrow{\scriptstyle \mathrm{Inf}} & & \\
0 & \longrightarrow & E(K)/pE(K) & \xrightarrow{\delta} & H^1(K, E_p) & \longrightarrow & H^1(K, E)_p & \longrightarrow & 0 \\
 & & \downarrow & & {\scriptstyle \mathrm{Res}}\downarrow \wr & & {\scriptstyle \mathrm{Res}}\downarrow & & \\
0 & \longrightarrow & (E(K_n)/pE(K_n))^{\mathcal{G}_n} & \xrightarrow{\delta_n} & H^1(K_n, E_p)^{\mathcal{G}_n} & \longrightarrow & H^1(K_n, E)_p^{\mathcal{G}_n}. & &
\end{array} \tag{4.2}$$

Both rows of (4.2), and the right column, are exact. The restriction from $H^1(K, E_p)$ to $H^1(K_n, E_p)^{\mathcal{G}_n}$ is an isomorphism, as its kernel is $H^1(K_n/K, E_p(K_n))$ via inflation and its cokernel injects into $H^2(K_n/K, E_p(K_n))$ via transgression in the Hochschild–Serre spectral sequence. These cohomology groups are both trivial by the following.

Lemma 4.3 The curve E has no p-torsion rational over K_n.

Proof If not, either $E_p(K_n) = \mathbb{Z}/p\mathbb{Z}$ or $E_p(K_n) = (\mathbb{Z}/p\mathbb{Z})^2$. The first implies that E_p has a cyclic subgroup scheme over $\mathbb{Q}$, as K_n is Galois over $\mathbb{Q}$. Hence the Galois group of $\mathbb{Q}(E_p)$ is contained in a Borel subgroup of $GL_2(\mathbb{Z}/p\mathbb{Z})$. If $E_p(K_n) = (\mathbb{Z}/p\mathbb{Z})^2$, then $\mathbb{Q}(E_p)$ is a subfield of K_n and we have a surjective homomorphism $\mathcal{G}_n \twoheadrightarrow GL_2(\mathbb{Z}/p\mathbb{Z})$. This is impossible: when $p > 2$, $GL_2(\mathbb{Z}/p\mathbb{Z})$ is not a quotient of a group of 'dihedral' type.

We now define Kolyvagin's cohomology classes. Let $c(n)$ be the unique class in $H^1(K, E_p)$ such that

$$\text{(4.4)} \qquad \operatorname{Res} c(n) = \delta_n[P_n] \quad \text{in } H^1(K_n, E_p)^{\mathcal{G}_n}.$$

Let $d(n)$ be the image of $c(n)$ in $H^1(K, E)_p$. Since Res $d(n) = 0$ by the commutativity of (4.2) and the exactness of the bottom row, there is a unique class $\tilde{d}(n)$ in $H^1(K_n/K, E)_p = H^1(\mathcal{G}_n, E(K_n))_p$ such that

$$\text{(4.5)} \qquad \operatorname{Inf} \tilde{d}(n) = d(n) \quad \text{in } H^1(K, E)_p.$$

W. McCallum has observed that the class $c(n)$ is represented by the 1-cocycle

$$\text{(4.6)} \qquad f(\sigma) = \sigma\Big(\frac{1}{p}\, P_n\Big) - \frac{1}{p}\, P_n - \frac{(\sigma - 1)P_n}{p}$$

on $\mathrm{Gal}(\overline{K}/K)$. Here $\frac{1}{p}\, P_n$ is a fixed p^{th} root of P_n in $E(\overline{K})$, and $\frac{(\sigma-1)P_n}{p}$ is the unique p^{th} root of $(\sigma - 1)P_n$ in $E(K_n)$, which exists by Lemma 4.3. The class $\tilde{d}(n)$ is represented by the 1-cocycle

$$\tilde{f}(\sigma) = -\frac{(\sigma - 1)P_n}{p}$$

on $\mathcal{G}_n$.

Proposition 4.7 (1) The class $c(n)$ is trivial in $H^1(K, E_p)$ if and only if $P_n \in pE(K_n)$.
(2) The class $d(n)$ is trivial in $H^1(K, E)_p$, and the class $\tilde{d}(n)$ is trivial in $H^1(K_n/K, E)_p$, if and only if $P_n \in pE(K_n) + E(K)$.

Proof This follows from their definitions and the diagram (4.2).

Note The class $c(1)$ is trivial if and only if $P_1 = y_K$ is divisible by p in $E(K)$, and the classes $d(1)$ and $\tilde{d}(1)$ are always globally trivial.

5. We now discuss the action of $\mathrm{Gal}(K/\mathbf{Q}) = \langle 1, \tau \rangle$ on the cohomology classes $c(n)$ in $H^1(K, E_p)$. Since p is odd, we have a direct sum decomposition into eigenspaces for τ:

$$(5.1) \qquad H^1(K, E_p) = H^1(K, E_p)^+ \oplus H^1(K, E_p)^-.$$

We will see that the class $c(n)$ lies in one of these eigenspaces, whose sign depends both on E and the number of primes ℓ dividing n.

Let $\epsilon = \pm 1$ be the eigenvalue of the Fricke involution w_N on the eigenform $f = \Sigma a_n q^n$ associated to the modular curve E:

$$(5.2) \qquad f|w_N = \epsilon \cdot f.$$

Then the L-function of E over $\mathbf{Q}$ satisfies a functional equation with sign $= -\epsilon$.

Complex conjugation τ acts on the Galois extension K_n, and hence on the point y_n in $E(K_n)$.

Proposition 5.3 We have $y_n^\tau = \epsilon \cdot y_n^{\sigma'} +$ (torsion) in $E(K_n)$, for some $\sigma' \in \mathcal{G}_n$.

Proof This follows from the identity [G, §5]

$$x_n^\tau = w_N(x_n^{\sigma'})$$

for some σ' in $\mathcal{G}_n$. Hence

$$(x_n - \infty)^\tau = w_N(x_n - \infty)^{\sigma'} + (w_N\infty - \infty).$$

Since $w_N\infty$ is the cusp 0 of $X_0(N)$, and the class of $(0 - \infty)$ is torsion in the Jacobian, this gives the claim on the curve E.

Proposition 5.4 (1) The class $[P_n]$ lies in the $\epsilon_n = \epsilon \cdot (-1)^{f_n}$ eigenspace for τ in $(E(K_n)/pE(K_n))^{\mathcal{G}_n}$, where $f_n = \#\{\ell : \ell | n\}$.
(2) The class $c(n)$ lies in the ϵ_n-eigenspace for τ in $H^1(K, E_p)$, and the class $d(n)$ lies in the ϵ_n-eigenspace for τ in $H^1(K, E)_p$.

Proof Recall that $P_n = \sum_{\sigma \in S} \sigma D_n y_n$ in $E(K_n)$, where S is a set of coset representatives for G_n in $\mathcal{G}_n$. For any $\sigma \in \mathcal{G}_n$ we have the commutation relation $\tau\sigma = \sigma^{-1}\tau$. Hence $\tau P_n = \sum_S \sigma^{-1}\tau D_n y_n$.

But $D_n = \prod_{\ell | n} D_\ell$ in $\mathbf{Z}[G_n]$, where $D_\ell \in \mathbf{Z}[G_\ell]$ is a solution (well-defined up to $m\ \mathrm{Tr}_\ell$) of $(\sigma_\ell - 1)D_\ell = \ell + 1 - \mathrm{Tr}_\ell$. Applying τ on the right and left of this identity, we find

$$\begin{aligned}(\sigma_\ell - 1)D_\ell\tau &= \tau(\sigma_\ell - 1)D_\ell \\ &= (\sigma_\ell^{-1} - 1)\tau D_\ell \\ &= -\sigma_\ell^{-1}(\sigma_\ell - 1)\tau D_\ell.\end{aligned}$$

Hence $\tau D_\ell = -\sigma_\ell D_\ell \tau + k\mathrm{Tr}_\ell$ for some $k \in \mathbf{Z}$, as $\tau D_\ell + \sigma_\ell D_\ell \tau$ is annihilated by $(\sigma_\ell - 1)$. (For $D_\ell^0 = \sum_1^\ell i\sigma_\ell^i$, one has $k = \ell$). Since $\mathrm{Tr}_\ell y_n = a_\ell y_{n/\ell} \equiv 0$ in $pE(K_n)$, we have

$$\tau P_n \equiv (-1)^{f_n} \cdot \prod_{\ell | n} \sigma_\ell \cdot \sum_S \sigma^{-1} \cdot D_n(\tau y_n) \pmod{pE(K_n)}.$$

But $\tau y_n = \epsilon \cdot \sigma'(y_n) +$ torsion, by Proposition 5.3, for some σ' in $\mathcal{G}_n$. But Lemma 4.3 shows that $E(K_n)_p = 0$. Hence

$$\tau P_n \equiv \epsilon_n \cdot \prod_{\ell | n} \sigma_\ell \cdot \sigma' \cdot \sum_S \sigma^{-1} D_n y_n \pmod{pE(K_n)}.$$

The sum $\sum \sigma^{-1} D_n y_n$ is $\equiv P_n$, as $[D_n y_n]$ is fixed by G_n and $\{\sigma^{-1}\}$ is another set of coset representatives for G_n in $\mathcal{G}_n$. Since $[P_n]$ is fixed by $\mathcal{G}_n$, we have

$$\tau P_n \equiv \epsilon_n \cdot P_n \qquad \mod pE(K_n)$$

which proves (1). The statements in (2) are an immediate corollary, as all the maps in the diagram (4.2) commute with the action of $\mathrm{Gal}(K/\mathbf{Q}) = \langle 1, \tau \rangle$.

Since $d(n) \in H^1(K, E)_p^{\epsilon_n}$, we may refine Proposition 4.7, part (2).

Corollary 5.5 The class $d(n)$ is trivial in $H^1(K, E)_p^{\epsilon_n}$ if and only if $P_n \in pE(K_n) + E(K)^{\epsilon_n}$.

6. Recall that

$$(6.1) \qquad Ш(E/K)_p = \ker(H^1(K, E)_p \longrightarrow \coprod_v H^1(K_v, E)_p)$$

where the sum is taken over all places v of K. The Selmer group $\mathrm{Sel}(E/K)_p$ is, by definition, the largest subgroup of $H^1(K, E_p)$ which maps to $Ш(E/K)_p$ in $H^1(K, E)_p$. We now wish to decide if the class $c(n)$ is in the Selmer group, i.e., if the class $d(n)$ is locally trivial at all places of K. We note that $\delta_n[P_n]$ is in the Selmer group of E over K_n, and is fixed by $\mathcal{G}_n$, but restriction does not necessarily induce an isomorphism: $\mathrm{Sel}(E/K)_p \to \mathrm{Sel}(E/K_n)_p^{\mathcal{G}_n}$.

Proposition 6.2 (1) The class $d(n)_v$ is locally trivial in $H^1(K_v, E)_p$ at the archimedean place $v = \infty$, and at all finite places v of K which do not divide n.
(2) If $n = \ell m$ and λ is the unique prime of K dividing ℓ, the class $d(n)_\lambda$ is locally trivial in $H^1(K_\lambda, E)_p$ if and only if $P_m \in pE(K_{\lambda_m}) = pE(K_\lambda)$ for one (and hence all) places λ_m of K_m dividing λ.

Proof If $v = \infty, K_v = \mathbf{C}$ is algebraically closed and the Galois cohomology of E is trivial. If $v \nmid n$ then the class $d(n)$ is inflated from the class $\widetilde{d(n)}$ of an extension K_n/K which is unramified at v. Hence $d(n)_v$ lies in the subgroup $H^1(K_v^{un}/K_v, E)$, where K_v^{un} is the maximal unramified extension. This group is trivial when E has good reduction at v [M; Ch. I, §3], so $d(n)_v = 0$ for $v \nmid N$.

If $v|N$ the curve E has bad reduction: let E^0 be the connected component of the Néron model and $\phi = E/E^0$ the group of components. Then $H^1(K_v^{un}/K_v, E^0) = 0$, so $H^1(K_v^{un}/K_v, E)$ injects into $H^1(F_v, \phi)$ [M; Ch. I, Prop. 3.8]. But the class $d(n)_v$ is represented by a cocycle with values in a subgroup E' with $(E' : E^0)$ prime to p. Indeed, let J be the Jacobian of $X_0(N)$. Then for any place w dividing v in K_n, the class of the Heegner divisor $(x_n) - (\infty)$ in $J(K_{n,w})$ lies, up to translation by the rational torsion point $(0) - (\infty)$, in J^0 [GZ; III, 3.1]. Hence y_n is, up to translation by rational torsion on E, in E^0. Since $E(\mathbf{Q})_p = 0$ by assumption, the points y_n (and hence $D_n y_n$ and P_n) lie in a subgroup E' whose image in ϕ has order prime to p. Since $d(n)_v$ is killed by p, we have $d(n)_v = 0$.

(2) We recall that the prime λ splits completely in K_m, and each factor λ_m is totally ramified, of degree $\ell + 1$, in K_n. The localization $d(n)_\lambda$ is represented by the cocycle $\sigma \longmapsto \frac{(\sigma-1)P_n}{p}$ on $\mathrm{Gal}(K_{\lambda_n}/K_{\lambda_m}) \simeq G_\ell$ with values in $E(K_{\lambda_n})$. Since $\ell \nmid N$ the curve E has good reduction at λ; let E^1 denote the subgroup of points reducing to the identity. Since E^1 is a pro-ℓ-group and $\ell \neq p$, the cohomology group $H^1(G_\ell, E^1(K_{\lambda_n}))_p = 0$. Hence $d(n)_\lambda$ is trivial if and only if it has trivial image in

$$H^1(G_\ell, \widetilde{E}(F_{\lambda_n}))_p = \mathrm{Hom}(G_\ell, \widetilde{E}(F_\lambda)_p),$$

where $\widetilde{E} = E/E^1$ is the reduced curve. The image of $d(n)_\lambda$ is represented by the cocycle $\sigma \longmapsto$ reduction of $-\dfrac{(\sigma - 1)P_n}{p}$. Since G_ℓ is cyclic, generated by σ_ℓ, we see that the local class $d(n)_\lambda$ is trivial if and only if the point $Q_n = \dfrac{(\sigma_\ell - 1)P_n}{p}$ has trivial reduction (mod λ_n). Since σ_ℓ acts trivially on $\widetilde{E}(F_{\lambda_n}) = \widetilde{E}(F_\lambda)$, the reduction $\widetilde{Q}_n$ is contained in $\widetilde{E}(F_\lambda)_p$.

Since $P_n = \sum_S \sigma D_m \cdot D_\ell \cdot y_n$ and $(\sigma_\ell - 1)D_\ell = \ell + 1 - \mathrm{Tr}_\ell$, we have

$$Q_n = \sum_S \sigma D_m \left(\frac{\ell+1}{p} y_n - \frac{a_\ell}{p} y_m \right)$$

by Proposition 3.7, part (1). By part (2) of that proposition, we have the congruence:

$$\frac{\ell+1}{p} y_n - \frac{a_\ell}{p} y_m \equiv \frac{(\ell+1)\mathrm{Frob}(\lambda_m) - a_\ell}{p} y_m \qquad (\mathrm{mod}\ \lambda_n)$$

at all places λ_n dividing λ in K_n. For any $\sigma \in \mathcal{G}_n$ we conjugate this congruence $(\mathrm{mod}\ \sigma^{-1}\lambda_n)$ by σ to obtain

$$\sigma \left(\frac{\ell+1}{p} y_n - \frac{a_\ell}{p} y_m \right) \equiv \sigma \left(\frac{(\ell+1)(\mathrm{Frob}\ \sigma^{-1}\lambda_m) - a_\ell}{p} \right) y_m \quad (\mathrm{mod}\ \lambda_n).$$

But $\sigma \cdot \mathrm{Frob}(\sigma^{-1}\lambda_m) = \mathrm{Frob}(\lambda_m) \cdot \sigma$, so we obtain

$$\sigma \left(\frac{\ell+1}{p} y_n - \frac{a_\ell}{p} y_m \right) \equiv \left(\frac{(\ell+1)(\mathrm{Frob}\ \lambda_m) - a_\ell}{p} \right) \sigma y_m \quad (\mathrm{mod}\ \lambda_n).$$

Hence

$$Q_n \equiv \frac{(\ell+1)(\mathrm{Frob}\ \lambda_m) - a_\ell}{p} P_m \quad (\mathrm{mod}\ \lambda_n).$$

The reduction $\tilde{P}_m$ lies in the ϵ_m-eigenspace for $\mathrm{Frob}(\ell)$ on $\tilde{E}(F_\lambda)/p\tilde{E}(F_\lambda)$. Since $(\ell+1)\mathrm{Frob}(\ell) - a_\ell$ annihilates $\tilde{E}(F_\lambda)$, and the ϵ_m-eigenspace of p-torsion is cyclic, we see that $\tilde{Q}_n = 0$ if and only if $\tilde{P}_m \in p\tilde{E}(F_\lambda)$. Since E^1 is p-divisible, this is equivalent to the divisibility $P_m \in pE(K_{\lambda_m})$.

Note We have seen that the class $d(1)$ is always globally trivial, hence is locally trivial at all places of K. This is in accord with Proposition 6.2, part (1). For a more interesting example, assume $n = \ell$ is prime. Then, by Proposition 4.7, the class $d(\ell)$ is globally trivial if and only if $P_\ell \in pE(K_\ell) + E(K)$. By Proposition 6.2, the class $d(\ell)$ is locally trivial at all places $v \neq \lambda$ of K, and is locally trivial at λ if and only if $P_1 = y_K \in pE(K_\lambda)$.

7. We now review the relevant results of Tate local duality [T, §2], [M, Ch. I] which will be used in the proof of Proposition 2.3. In this section, we let K_λ be a local field, with ring of integers $\mathcal{O}_\lambda$ and finite residue field F_λ of characteristic ℓ. We let E be an elliptic curve over K_λ, with good reduction over $\mathcal{O}_\lambda$.

Let p be a prime, with $p \neq \ell$. Then E_p is a finite étale group scheme of rank p^2 over $\mathcal{O}_\lambda$. The Kummer sequence $0 \to E_p \to E \xrightarrow{p} E \to 0$ induces an isomorphism

$$E(K_\lambda)/pE(K_\lambda) \overset{\delta}{\underset{\sim}{\to}} H^1(\mathcal{O}_\lambda, E_p), \tag{7.1}$$

as $H^1(\mathcal{O}_\lambda, E) = 0$. Since the subgroup $E^1(K_\lambda)$ is ℓ-divisible, the group $E(K_\lambda)/pE(K_\lambda)$ is isomorphic to $\widetilde{E}(F_\lambda)/p\widetilde{E}(F_\lambda)$, so has dimension ≤ 2 over $\mathbf{Z}/p\mathbf{Z}$, with equality holding if all the p-torsion on E is rational over K_λ.

The Weil pairing $\{,\} : E_p \times E_p \longrightarrow \mu_p$ of finite group schemes over K_λ induces a cup-product pairing in Galois (= étale, or flat) cohomology:

$$H^1(K_\lambda, E_p) \times H^1(K_\lambda, E_p) \longrightarrow H^2(K_\lambda, \mu_p). \tag{7.2}$$

The invariant map of local class field theory gives a canonical isomorphism $H^2(K_\lambda, \mu_p) = \mathrm{Br}(K_\lambda)_p \underset{\sim}{\to} \frac{1}{p}\mathbf{Z}/\mathbf{Z} = \mathbf{Z}/p\mathbf{Z}$, and Tate's local duality theorem states that the resulting pairing of $\mathbf{Z}/p\mathbf{Z}$-vector spaces

$$\langle,\rangle : H^1(K_\lambda, E_p) \times H^1(K_\lambda, E_p) \longrightarrow \mathbf{Z}/p\mathbf{Z} \tag{7.3}$$

is alternating and non-degenerate (see [M, Ch. I, Corollary 2.3]).

The Kummer sequence $0 \to E_p \to E \xrightarrow{p} E \to 0$ gives a short exact sequence in cohomology:

$$0 \longrightarrow E(K_\lambda)/pE(K_\lambda) \longrightarrow H^1(K_\lambda, E_p) \longrightarrow H^1(K_\lambda, E)_p \to 0. \tag{7.4}$$

The subspace $E(K_\lambda)/pE(K_\lambda) \underset{\sim}{\to} H^1(\mathcal{O}_\lambda, E_p)$ is isotropic for the pairing $\langle,\rangle$ induced by cup-product, as $H^2(\mathcal{O}_\lambda, \mu_p) = 0$.

Proposition 7.5 The pairing $\langle,\rangle$ of (7.3) induces a non-degenerate pairing of $\mathbf{Z}/p\mathbf{Z}$-vector spaces (of dimension ≤ 2)

$$\langle,\rangle : E(K_\lambda)/pE(K_\lambda) \times H^1(K_\lambda, E)_p \longrightarrow \mathbf{Z}/p\mathbf{Z}.$$

Proof It suffices to check that the subspace $H^1(\mathcal{O}_\lambda, E_p)$ is maximal isotropic, or equivalently, that $\dim\ H^1(K_\lambda, E)_p = \dim\ E(K_\lambda)_p$. This is a general fact, due to Tate [T, §2] (see [M, Ch. I, Thm. 2.6]); we give a proof using tame local class field theory. Let K_λ^{un} be the completion of the maximal unramified extension of K_λ; since $H^1(K_\lambda^{un}/K_\lambda, E) = H^1(\mathcal{O}_\lambda, E) = 0$, restriction induces an isomorphism $H^1(K_\lambda, E)_p \underset{\sim}{\to} H^1(K_\lambda^{un}, E)_p^{\mathrm{Frob}(\lambda)}$. The latter group is isomorphic to $H^1(K_\lambda^{un}, E_p)^{\mathrm{Frob}(\lambda)}$, using the Kummer sequence and the fact that

$E(K_\lambda^{un})$ is p-divisible. Since the residue field of K_λ^{un} is algebraically closed, $H^1(K_\lambda^{un}, E_p)^{\mathrm{Frob}(\lambda)} = \mathrm{Hom}(\mathrm{Gal}(\overline{K}_\lambda/K_\lambda^{un}), E_p)^{\mathrm{Frob}(\lambda)}$. But the homomorphism of $\mathrm{Gal}(\overline{K}_\lambda/K_\lambda^{un})$ to E_p must kill the wild inertia subgroup (as $\ell \neq p$), and factor through the maximal pro-p quotient of the tame inertia group. This quotient is isomorphic to $\mathbf{Z}_p(1) = T_p\mathbf{G}_m$ as a $\mathrm{Frob}(\lambda)$-module, so $H^1(K_\lambda, E)_p$ is isomorphic to $\mathrm{Hom}(\mu_p, E_p)^{\mathrm{Frob}(\lambda)}$. The latter space has the same dimension as $\widetilde{E}(F_\lambda)_p \simeq E(K_\lambda)_p$, by the Weil pairing.

We henceforth assume that the p-torsion on E is rational over K_λ, so the $\mathbf{Z}/p\mathbf{Z}$-vector spaces in Proposition 7.5 each have dimension $= 2$. In this case there is an elegant formula for the pairing $\langle , \rangle$, which is due to Kolyvagin and gives an independent proof of its non-degeneracy. To $c_1 \in E(K_\lambda)/pE(K_\lambda)$ we associate the point $e_1 = \left(\frac{1}{p}\, c_1\right)^{\mathrm{Frob}(\lambda)-1}$ in $E(K_\lambda)_p$. To $c_2 \in H^1(K_\lambda, E_p)$ we associate a homomorphism $\phi_2 : \mu_p \to E_p(K_\lambda)$ as above, using tame local class field theory. Fix a primitive p^{th} root ζ of 1 in K_λ^*, and let $\phi_2(\zeta) = e_2$ in $E(K_\lambda)_p$. Then

$$\zeta^{\langle c_1, c_2\rangle} = \{e_1, e_2\}, \tag{7.6}$$

where $\{,\}$ is the Weil pairing on E_p. A proof of (7.6) may be found in the appendix of [W].

8. We now apply Proposition 7.5 in the specific local situation which arises in the study of Heegner points: K is an imaginary quadratic extension of $\mathbf{Q}$ and K_λ is the completion of K at an inert prime $\lambda = (\ell)$. The curve E is defined over $\mathbf{Q}$, so $\mathrm{Gal}(K/\mathbf{Q}) = \mathrm{Gal}(K_\lambda/\mathbf{Q}_\ell) = \langle 1, \tau\rangle$ acts on the $\mathbf{Z}/p\mathbf{Z}$-vector spaces $E(K_\lambda)/pE(K_\lambda)$ and $H^1(K_\lambda, E)_p$.

We assume, as usual, that p is odd and that ℓ satisfies the congruences $\ell+1 \equiv a_\ell \equiv 0 \pmod p$ of (3.3). Then the eigenspaces $E(K_\lambda)_p^{\pm}$ for τ each have dimension 1 over $\mathbf{Z}/p\mathbf{Z}$.

Proposition 8.1 (1) The eigenspaces $(E(K_\lambda)/pE(K_\lambda))^{\pm}$ and $H^1(K_\lambda, E)_p^{\pm}$ for $\mathrm{Gal}(K_\lambda/\mathbf{Q}_\ell)$ each have dimension 1 over $\mathbf{Z}/p\mathbf{Z}$.
(2) The pairing $\langle , \rangle$ of (2.3) induces non-degenerate pairings of $\mathbf{Z}/p\mathbf{Z}$-vector spaces

$$\langle , \rangle^{\pm} : (E(K_\lambda)/pE(K_\lambda))^{\pm} \times H^1(K_\lambda, E)_p^{\pm} \longrightarrow \mathbf{Z}/p\mathbf{Z}.$$

In particular, if $d_\lambda \neq 0$ lies in $H^1(K_\lambda, E)_p^{\pm}$ and $s_\lambda \in (E(K_\lambda)/pE(K_\lambda))^{\pm}$ satisfies $\langle s_\lambda, d_\lambda\rangle = 0$, then $s_\lambda \equiv 0 \pmod{pE(K_\lambda)}$.

Proof (1) We have isomorphisms of $\mathrm{Gal}(K_\lambda/\mathbf{Q}_\ell)$-modules:

$$E(K_\lambda)/pE(K_\lambda) \xrightarrow{\sim} E(K_\lambda)_p \text{ and } H^1(K_\lambda, E)_p \xrightarrow{\sim} \mathrm{Hom}(\mu_p(K_\lambda), E(K_\lambda)_p).$$

Since $\ell + 1 \equiv 0 \pmod{p}$, $\mu_p(K_\lambda) = \mu_p(K_\lambda)^-$. Hence $E(K_\lambda)_p^\pm \simeq (E(K_\lambda)/pE(K_\lambda))_p^\pm \simeq H^1(K_\lambda, E)_p^\mp$, and all eigenspaces have dimension 1.

(2) It suffices to check that the $+$ and $-$ eigenspaces for τ are orthogonal under $\langle , \rangle$. But the Tate pairing satisfies $\langle c_1^\tau, c_2^\tau \rangle = \langle c_1, c_2 \rangle$, as τ acts trivially on $H^2(K_\lambda, \mu_p) = \mathbb{Z}/p\mathbb{Z}$. Since p is odd, the result follows. (Alternately, one can use the formula for the Weil pairing: $\{e_1^\tau, e_2^\tau\} = \{e_1, e_2\}^\tau = -\{e_1, e_2\}$ and Kolyvagin's formula (7.6).)

Actually, we will use the following version of Proposition 8.1, which uses the full power of global class field theory.

Proposition 8.2 Assume that the class $d \in H^1(K, E)_p^\pm$ is locally trivial for all places $v \neq \lambda$ of K, but that $d_\lambda \neq 0$ in $H^1(K_\lambda, E)_p^\pm$. Then for any class s in the subgroup $\mathrm{Sel}(E/K)_p^\pm \subset H^1(K, E_p)^\pm$ we have $s_\lambda = \mathrm{Res}_{K_\lambda}(s) = 0$ in $H^1(K_\lambda, E_p)^\pm$.

Proof The restriction s_λ lies in $(E(K_\lambda)/pE(K_\lambda))^\pm$, by the definition of the global Selmer group. Hence it suffices, by Proposition 8.1, to show that $\langle s_\lambda, d_\lambda \rangle = 0$.

To do this, we lift d to a class c in $H^1(K, E_p)$, which is well-defined modulo the image of $E(K)/pE(K)$. The global pairing $\langle s, c \rangle_K$ induced by cup-product lies in $H^2(K, \mu_p) = \mathrm{Br}(K)_p$, and is completely determined by its local components $\langle s_v, c_v \rangle \in \mathrm{Br}(K_v)_p$ for all places v of K. But $\langle s_v, c_v \rangle = 0$ for all $v \neq \lambda$, as $d_v = 0$ in $H^1(K_v, E)_p$. Since the sum of local invariants is zero, by the reciprocity law of global class field theory, we must have $\langle s_\lambda, c_\lambda \rangle = \langle s_\lambda, d_\lambda \rangle = 0$ also.

Kolyvagin's idea is to use global classes d satisfying Proposition 8.2 to bound the order of $\mathrm{Sel}(E/K)_p$. The classes $d = d(n)$ are constructed using Heegner points of conductors $n \geq 1$ for K in §4–5, and their local behavior is analyzed in Proposition 6.2.

9. In this section we give a concrete description of the Selmer group $\mathrm{Sel}(E/K)_p$ in $H^1(K, E_p)$, under the hypothesis that p is odd and that the Galois group of $\mathbb{Q}(E_p)$ is isomorphic to $GL_2(\mathbb{Z}/p\mathbb{Z}) \simeq \mathrm{Aut}(E_p)$. Let $L = K(E_p)$; the hypothesis that D is prime to Np implies that the numberfields K and $\mathbb{Q}(E_p)$ are disjoint. Hence $\mathcal{G} = \mathrm{Gal}(L/K)$ is isomorphic to $GL_2(\mathbb{Z}/p\mathbb{Z})$ and contains the central subgroup $Z \simeq (\mathbb{Z}/p\mathbb{Z})^*$ of homotheties of E_p. Since Z has order $p-1$, which is prime to p, $H^n(Z, E_p) = 0$ for $n \geq 1$. Since p is odd, $Z \neq 1$ and $E_p^Z = H^0(Z, E_p) = 0$.

Proposition 9.1 We have $H^n(\mathcal{G}, E_p) = 0$ for all $n \geq 0$. The restriction of classes gives an isomorphism of $\mathrm{Gal}(K/\mathbf{Q})$-modules:

$$\mathrm{Res}: H^1(K, E_p) \xrightarrow{\sim} H^1(L, E_p)^{\mathcal{G}} = \mathrm{Hom}_{\mathcal{G}}(\mathrm{Gal}(\overline{\mathbf{Q}}/L), E_p(L)).$$

Proof The spectral sequence $H^m(\mathcal{G}/Z, H^n(Z, E_p)) \Longrightarrow H^{m+n}(\mathcal{G}, E_p)$, and the vanishing of $H^n(Z, E_p)$ for all $n \geq 0$, gives the vanishing of the cohomology of $\mathcal{G}$ in E_p. (This elegant proof is due to Serre.) The fact that restriction is an isomorphism follows, as its kernel is $H^1(\mathcal{G}, E_p)$ and its cokernel injects into $H^2(\mathcal{G}, E_p)$.

From Proposition 9.1 we obtain a pairing:

$$(9.2) \qquad [\ ,\ ]: H^1(K, E_p) \times \mathrm{Gal}(\overline{\mathbf{Q}}/L) \longrightarrow E_p(L),$$

which satisfies $[s^\sigma, \rho^\sigma] = [s, \rho^\sigma] = [s, \rho]^\sigma$ for all $s \in H^1(K, E_p)$, $\rho \in \mathrm{Gal}(\overline{\mathbf{Q}}/L)$, and $\sigma \in \mathcal{G} = \mathrm{Gal}(L/K)$. If $[s, \rho] = 0$ for all $\rho \in \mathrm{Gal}(\overline{\mathbf{Q}}/L)$, then $s \equiv 0$ by the injectivity of restriction.

Let $S \subset H^1(K, E_p)$ be a finite subgroup (= finite dimensional vector space over $\mathbf{Z}/p\mathbf{Z}$). Let $\mathrm{Gal}_S(\overline{\mathbf{Q}}/L)$ be the subgroup of $\rho \in \mathrm{Gal}(\overline{\mathbf{Q}}/L)$ such that $[s, \rho] = 0$ for all $s \in S$, and let L_S be the fixed field of $\mathrm{Gal}_S(\overline{\mathbf{Q}}/L)$. Then L_S is a finite normal extension of L.

Proposition 9.3 The induced pairing

$$[\ ,\ ]: S \times \mathrm{Gal}(L_S/L) \longrightarrow E_p(L)$$

is non-degenerate: it induces an isomorphism of $\mathcal{G}$-modules:

$$\mathrm{Gal}(L_S/L) \xrightarrow{\sim} \mathrm{Hom}(S, E_p(L)),$$

as well as an isomorphism of $\mathrm{Gal}(K/\mathbf{Q})$-modules:

$$S \xrightarrow{\sim} \mathrm{Hom}_{\mathcal{G}}(\mathrm{Gal}(L_S/L), E_p(L)).$$

Proof From the definition of L_S, and the injectivity of restriction proved in 9.1, the pairing $[\,,]: S \times \mathrm{Gal}(L_S/L) \to E_p(L)$ induces injections $\mathrm{Gal}(L_S/L) \hookrightarrow \mathrm{Hom}(S, E_p)$ and $S \hookrightarrow \mathrm{Hom}_{\mathcal{G}}(\mathrm{Gal}(L_S/L), E_p)$. If $r = \dim(S)$, this shows that $\mathrm{Gal}(L_S/L)$ is a $\mathcal{G}$-submodule of $\mathrm{Hom}(S, E_p) \simeq E_p^r$. Since E_p is a simple $\mathcal{G}$-module, E_p^r is semi-simple. Since any submodule of a semi-simple module is semi-simple, we have an isomorphism of $\mathcal{G}$-modules: $\mathrm{Gal}(L_S/L) \xrightarrow{\sim} E_p^s$ for $s \leq r$. Hence $\mathrm{Hom}_{\mathcal{G}}(\mathrm{Gal}(L_S/L), E_p) \simeq (\mathbf{Z}/p\mathbf{Z})^s$; since this contains $S \simeq (\mathbf{Z}/p\mathbf{Z})^r$

we must have $s \geq r$. Consequently $s = r$ and the injections induced by $[\ ,\ ]$ are both isomorphisms.

We apply Proposition 9.3 to the finite subgroup $S = \mathrm{Sel}(E/K)_p$ of $H^1(K, E_p)$. For simplicity in notation, we let $M = L_S$ and $H = \mathrm{Gal}(M/L) = \mathrm{Gal}(L_S/L)$. We assume, in preparation for the proof of Proposition 2.3, that y_K is not divisible by p in $E(K)$, and let δy_K be its non-zero image in $\mathrm{Sel}(E/K)_p$. Let I be the subgroup of H which fixes the subfield $L(\frac{1}{p}y_K) = L_{\langle \delta y_K \rangle}$ of M. Here is a field diagram.

(9.4)

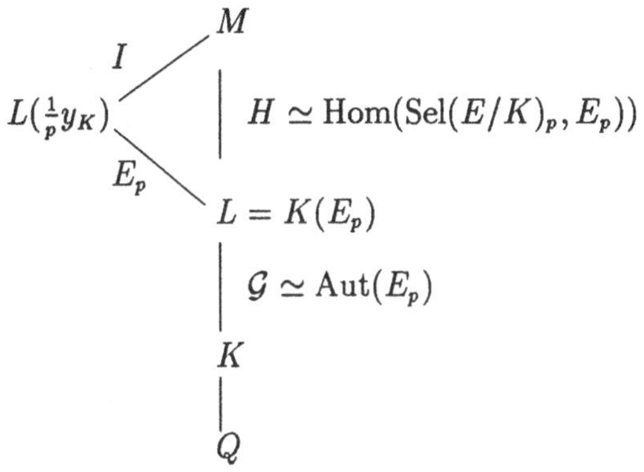

Let τ be a fixed complex conjugation in $\mathrm{Gal}(M/\mathbf{Q})$, and let H^+ and I^+ denote the $+1$ eigenspace for τ (acting by conjugation) on H and I.

Proposition 9.5 (1) We have $H^+ = \{(\tau h)^2 : h \in H\}$, $I^+ = \{(\tau i)^2 : i \in I\}$, and $H^+/I^+ \simeq \mathbf{Z}/p\mathbf{Z}$.
(2) Let $s \in \mathrm{Sel}(E/K)_p^{\pm}$. Then the following are equivalent:
 (a) $[s, \rho] = 0$ for all $\rho \in H$
 (b) $[s, \rho] = 0$ for all $\rho \in H^+$
 (c) $[s, \rho] = 0$ for all $\rho \in H^+ - I^+$
 (d) $s = 0$.

Proof (1) Since p is odd, $H^+ = H^{\tau+1} = \{h^\tau \cdot h : h \in H\}$. But $h^\tau = \tau h \tau^{-1} = \tau h \tau$ as $\tau^2 = 1$, so $h^\tau h = (\tau h)^2$. The same works for I^+. Finally, $H^+/I^+ = (H/I)^+ = E_p^+ \simeq \mathbf{Z}/p\mathbf{Z}$.

(2) Clearly (d) $\Longleftrightarrow$ (a) $\Longrightarrow$ (b) $\Longrightarrow$ (c), so it suffices to prove that (c) $\Longrightarrow$ (b) $\Longrightarrow$ (a). Since $s : H^+ \longrightarrow E_p$ is a group homomorphism and $I^+ \neq H^+$, the fact that s vanishes on $H^+ - I^+$ implies that it vanishes on the entire group H^+. Since $s \in \mathrm{Sel}(E/K)_p^{\pm}$, it induces a $\mathcal{G}$-homomorphism $H \longrightarrow E_p$ which maps $H^+ \longrightarrow E_p^{\pm}$ and $H^- \longrightarrow E_p^{\mp}$. If s vanishes on H^+, the image

$s(H)$ is therefore contained in $E_p^{\mp}$. But $s(H)$ is a $\mathcal{G}$-submodule of the simple module E_p, so if $s(H) \neq E_p$ we must have $s(H) = 0$.

Let λ be a prime of K which does not divide Np. Then λ is unramified in M/K; we assume further that λ splits completely in L/K and let λ_M be a prime factor of λ in M. The Frobenius substitution of λ_M in $\mathrm{Gal}(M/K)$ lies in the subgroup H, and its $\mathcal{G}$-orbit – which we denote by $\mathrm{Frob}(\lambda)$ – depends only on the place λ of K. We write $[s, \mathrm{Frob}(\lambda)] = 0$ iff $[s, \rho] = 0$ for all $\rho \in \mathrm{Frob}(\lambda)$.

Proposition 9.6 For $s \in \mathrm{Sel}(E/K)_p \subset H^1(K, E_p)$ the following are equivalent:

(a) $[s, \rho] = 0$, where ρ is the Frobenius substitution associated to the factor λ_M of λ in $\mathrm{Gal}(M/L) = H$.
(b) $[s, \mathrm{Frob}(\lambda)] = 0$.
(c) $s_\lambda \equiv 0 \quad \text{in } H^1(K_\lambda, E_p)$.

Proof Clearly (a) and (b) are equivalent, as for all $\sigma \in \mathcal{G}$ we have $[s, \rho^\sigma] = [s, \rho]^\sigma$. To prove the equivalence of (a) and (c) we assume $s_\lambda \equiv P_\lambda$ in $E(K_\lambda)/pE(K_\lambda) \hookrightarrow H^1(K_\lambda, E_p)$. Then $\frac{1}{p}P_\lambda$ is rational over M_{λ_M}, and $[s, \rho] = (\frac{1}{p}P_\lambda)^{\rho-1}$ in $E(M_{\lambda_M})_p \simeq E(M)_p$. Hence $[s, \rho] = 0$ if and only if $P_\lambda \in pE(K_\lambda)$.

10. We now give the proof of Proposition 2.3, treating the eigenspaces of $\mathrm{Sel}(E/K)_p$ in turn. Recall that the Heegner point $y_K = P_1$ lies in the ϵ-eigenspace for complex conjugation on $E(K)/pE(K)$ (where ϵ is the eigenvalue of the Fricke involution on the eigenform f associated to E). Hence δy_K lies in the ϵ-eigenspace of $\mathrm{Sel}(E/K)_p$.

Claim 10.1 $\mathrm{Sel}(E/K)_p^{-\epsilon} = 0$.

Proof Assume that $s \in \mathrm{Sel}(E/K)_p^{-\epsilon}$. To show $s = 0$ it suffices, by Proposition 9.5, to show that $[s, \rho] = 0$ for all $\rho \in H^+ - I^+$. Such elements have the form $\rho = (\tau h)^2$, for some $h \in H$.

Let ℓ be a rational prime which is unramified in the extension $M/\mathbb{Q}$, and has a factor λ_M whose Frobenius substitution is equal to τh in $\mathrm{Gal}(M/\mathbb{Q})$. Such primes exist, and have positive density, by Čebotarev's density theorem. Then $(\ell) = \lambda$ is inert in K and λ splits completely in L. The Frobenius substitution of F_{λ_M}/F_λ is equal to $(\tau h)^2$, so to prove that $[s, \rho] = 0$ it suffices, by Proposition 9.6, to show that $s_\lambda \equiv 0$ in $H^1(K_\lambda, E_p)$.

Let $c(\ell)$ be the global cohomology class in $H^1(K, E_p)$ constructed in §4, and let $d(\ell)$ be its image in $H^1(K, E)_p$. By Proposition 5.4, both classes lie in the $-\epsilon$-eigenspace for complex conjugation and, by Proposition 6.2, $d(\ell)$ is locally trivial except at λ. We claim that $d(\ell)_\lambda \neq 0$ in $H^1(K_\lambda, E)_p$. Indeed, by Proposition 6.2, $d(\ell)_\lambda$ is trivial if and only if $y_K = P_1 \in p\ E(K_\lambda)$, or equivalently, if the prime λ splits completely in the extension $L(\frac{1}{p}y_K)$. Since $\mathrm{Frob}(\lambda) = \rho$ is not in $I^+ = I \cap H^+$ by hypothesis, this splitting does not occur.

We therefore may apply Proposition 8.2, with $d = d(\ell)$, to conclude that $s_\lambda \equiv 0$. Since this argument works to show $[s, \rho] = 0$ for any $\rho \in H^+ - I^+$ (choosing ℓ correctly) we have shown that $s = 0$.

Proposition 10.2 Assume that y_K is not divisible by p in $E(K)$. Let ℓ be a rational prime which is unramified in $M/\mathbf{Q}$ and has a factor λ_M whose Frobenius substitution is equal to τh in $\mathrm{Gal}(M/\mathbf{Q})$, with $h \in H$. Then $(\ell) = \lambda$ is inert in K and λ splits completely in $L = K(E_p)$. The following are all equivalent:

(1) $c(\ell) \equiv 0$ in $H^1(K, E_p)$
(2) $c(\ell) \in \mathrm{Sel}(E/K)_p \subset H^1(K, E_p)$
(3) P_ℓ is divisible by p in $E(K_\ell)$
(4) $d(\ell) \equiv 0$ in $H^1(K, E_p)$
(5) $d(\ell)_\lambda \equiv 0$ in $H^1(K_\lambda, E_p)$
(6) $P_1 = y_K$ is locally divisible by p in $E(K_\lambda)$
(7) $h^{1+\tau}$ lies in the subgroup $I^+ = H^+ \cap I$ of H^+.

Proof We have (1) $\iff$ (2) as $\mathrm{Sel}(E/K)_p^{-\epsilon} = 0$ by 10.1. But $c(\ell) \equiv 0$ if and only if $P_\ell \in p\ E(K_\ell)$, so (1) $\iff$ (3).

Since $(E(K)/pE(K))^{-\epsilon} = 0$ by 10.1, $c(\ell) \equiv 0$ is equivalent to $d(\ell) \equiv 0$. Since $d(\ell)$ is locally trivial except perhaps at λ, and $\text{Ш}(E/K)_p^{-\epsilon} = 0$ by 10.1, we have $d(\ell) \equiv 0$ if and only if $d(\ell)_\lambda \equiv 0$. Conditions (6) and (7) are equivalent to $d(\ell)_\lambda \equiv 0$, by Proposition 6.2.

Claim 10.3 $\mathrm{Sel}(E/K)_p^\epsilon \simeq \mathbf{Z}/p\mathbf{Z} \cdot \delta y_K$.

Proof Let $s \in \mathrm{Sel}(E/K)_p^\epsilon$. To show s is a multiple of δy_K it suffices to prove that $[s, \rho] = 0$ for all $\rho \in I$. For then $s \in \mathrm{Hom}_{\mathcal{G}}(H/I, E_p) \simeq \mathbf{Z}/p\mathbf{Z} \cdot \delta y_K$. By the argument of Proposition 9.5, it suffices to show $[s, \rho] = 0$ for all $\rho \in I^+$. These elements all have the form $\rho = (\tau i)^2$, for $i \in I$.

Let ℓ' be a prime such that $c(\ell')$ is non-trivial in $H^1(K, E_p)$; by Proposition 10.2 we may obtain ℓ' by insisting that its Frobenius substitution is conjugate to τh in $\mathrm{Gal}(M/\mathbf{Q})$, where $h \in H$ and $h^{1+\tau} \notin I^+$. Then $c(\ell')$ is not in $\mathrm{Sel}(E/K)_p$, so the extension $L' = L_{(c(\ell'))}$ of $L = K(E_p)$ described in (9.3) has Galois group isomorphic to E_p and is disjoint from the extension M/L. A prime ideal $(\ell) = \lambda$ of K, which splits completely in L, splits completely in L' if and only if $P_{\ell'}$ is locally a pth power in $E(K_{\lambda_{\ell'}}) = E(K_\lambda)$, for all factors $\lambda_{\ell'}$ of λ in $K_{\ell'}$.

Let ℓ be a prime whose Frobenius substitution is conjugate to τi in $\mathrm{Gal}(M/\mathbf{Q})$, with $i \in I$ *and* whose Frobenius substitution is conjugate to τj in $\mathrm{Gal}(L'/\mathbf{Q})$, where $j \in \mathrm{Gal}(L'/L)$ satisfies $j^{1+\tau} \neq 1$. (Since $L' \cap M = L$, these two conditions may be satisfied simultaneously.) We claim that the class $d(\ell\ell')$ in $H^1(K,E)_p^\epsilon$ is locally trivial for all places $v \neq \lambda$, but that $d(\ell\ell')_\lambda \neq 0$. The local triviality for $v \neq \lambda, \lambda'$ follows from Proposition 6.2. Since $i \in I$, the global class $c(\ell)$ is zero by Proposition 10.2, and P_ℓ is divisible by p in $E(K_\ell)$. Hence it is locally divisible by p in the completion at a place dividing λ', and $d(\ell\ell')_{\lambda'} = 0$ by Proposition 6.2. Finally $d(\ell\ell')_\lambda$ is trivial if and only if $P_{\ell'}$ is locally divisible by p in $E(K_\lambda)$. But this implies that λ splits in L', or equivalently that $(\tau j)^2 = j^{1+\tau} = 1$. This contradicts our hypothesis on j.

We may now apply Proposition 8.2, with $d = d(\ell\ell')$, to conclude that $s_\lambda = 0$. Consequently $[s, \rho] = 0$, where $\rho = (\tau i)^2$. Since this argument works for any $\rho \in I^+$ (choosing ℓ judiciously) we have shown that $s(I^+) = s(I) = 0$.

11. When the Heegner point y_K has infinite order, but is divisible by p in $E(K)$, the cohomology classes $d(n)$ constructed by Kolyvagin in §3–4 are candidates for non-trivial elements in $Ш(E/K)_p$. Indeed, the condition that $P_1 \in pE(K)$ is equivalent to $c(1) = 0$. This implies, by Proposition 6.2, that the classes $d(\ell)$ all lie in $Ш(E/K)_p$. Similarly, if $c(\ell_1) = c(\ell_2) = 0$ then the class $d(\ell_1\ell_2)$ lies in $Ш(E/K)_p$, if $c(\ell_1\ell_2) = c(\ell_1\ell_3) = c(\ell_2\ell_3) = 0$, then the class $d(\ell_1\ell_2\ell_3)$ lies in $Ш(E/K)_p$, etc. What subgroup of $Ш(E/K)_p$ can be constructed in this manner?

A related question is the following. Assume that p does not divide the integer $c \cdot \prod_{q|N} m_q$ in Conjecture 1.2. Can one show that the class $c(n)$ is non-zero in $H^1(K, E_p)$ for *some* value of $n = \ell_1\ell_2 \cdots \ell_r$?

12. Let A be an abelian variety of dimension $d \geq 1$ over $\mathbf{Q}$ such that the algebra $\mathrm{End}_{\mathbf{Q}}(A) \otimes \mathbf{Q}$ is isomorphic to a totally real numberfield of degree d. A generalization of the conjecture of Taniyama and Weil states that A is

the quotient of the Jacobian $J_0(N)$ of some $X_0(N)$. Assume that a surjective homomorphism $\varphi : J_0(N) \to A$ exists, and define the points $y_1 = \varphi((x_1) - (\infty))$ in $A(K_1)$ and $y_K = \mathrm{Tr}_{K_1/K} y_1$ in $A(K)$ as in §1. It is easy to show that the L-function of A over K vanishes to order $\geq d$ at $s = 1$. Zagier and I proved that the order of vanishing is equal to d if and only if the Heegner point y_K has infinite order in $A(K)$ [GZ; V, 2.4]. Assuming this, Kolyvagin's method can be used to show that the finitely generated group $A(K)$ has rank d and that Ш(A/K) is finite [K2].

Another generalization is the following. Let χ be a complex character of $\mathrm{Gal}(K_1/K)$, and define the point $y_\chi = \sum_\sigma \chi^{-1}(\sigma) y_1^\sigma$ in $(E(K_1) \otimes \mathbf{C})^\chi$. Following Kolyvagin, one can show [B-D] that the hypothesis $y_\chi \neq 0$ implies that the complex vector space $(E(K_1) \otimes \mathbf{C})^\chi$ has dimension one.

13. *Acknowledgements.* I would like to thank K. Rubin, J.-P. Serre, and J. Tate for their help.

BIBLIOGRAPHY

[B-D] M. Bertolini and H. Darmon, 'Kolyvagin's descent and Mordell–Weil groups over ring class fields'. To appear

[G] B. H. Gross, 'Heegner points on $X_0(N)$', in *Modular Forms* (R. A. Rankin, Ed.) Chichester, Ellis Horwood, 1984, 87–106.

[GZ] B. H. Gross and D. Zagier, 'Heegner points and derivatives of L-series'. Invent. Math., **84** (1986), 225–320.

[K1] V. A. Kolyvagin, 'Euler systems', (1988). To appear in The Grothendieck Festschrift. Prog. in Math., Boston, Birkhauser (1990).

[K2] V. A. Kolyvagin, 'Finiteness of $E(\mathbf{Q})$ and Ш$(E/\mathbf{Q})$ for a class of Weil curves'. Izv. Akad. Nauk SSSR, **52** (1988).

[K3] V. A. Kolyvagin and D.Y. Logachev, 'Finiteness of the Shafarevich–Tate group and group of rational points for some modular abelian varieties'. Algebra and analysis (USSR), No. 5. (1989).

[Ma] B. Mazur, 'Rational isogenies of prime degree'. Invent. Math., **44** (1978), 129–62.

[M] J. S. Milne, 'Arithmetic duality theorems'. Perspectives in Mathematics, Academic Press, 1986.

[R1] K. Rubin, 'The work of Kolyvagin on the arithmetic of elliptic curves'. To appear in *Arithmetic of Complex Manifolds*, Erlangen, 1988, Springer Lecture Notes.

[R2] K. Rubin, Appendix to S. Lang, *Cyclotomic Fields* I *and* II. Springer-Verlag, 1990.

[S] J.-P. Serre, 'Propriétés galoisiennes des points d'ordre fini des courbes elliptiques'. Invent. Math., **15** (1972), 259–331 (= *Oe.*94)

[T] J. Tate, 'Duality theorems in Galois cohomology over number fields'. Proc. ICM Stockholm (1962), 288–95.

[W] L. C. Washington, 'Number fields and elliptic curves', in *Number Theory and Applications*, R.A. Mollin ed., Kluwer Academic Publishers, 1989, 245–78.

[Z] D. Zagier, 'Modular points, modular curves, modular surfaces, and modular forms'. Springer LNM 1111 (1985), 225–48.

Index theory, potential theory, and the Riemann hypothesis

SHAI HARAN

In this survey we would like to paint, in expressionistic brushstrokes, our hunch concerning the problem of the Riemann hypothesis. Langlands said it best [38]: '... I have exceeded my commission and been seduced into describing things as they may be and, as seems to me at present, are likely to be. They could be otherwise. Nonetheless, it is useful to have a conception of the whole to which one can refer during the daily, close work with technical difficulties, provided one does not become too attached to it ... I have simply fused my own observations and reflections with ideas of others and with commonly accepted tenets'.

Let us begin by recalling the well known analogies between number fields and function fields. For function fields the Riemann hypothesis was solved by Weil, over a finite field [49], and by Selberg, over the complex numbers [43]. Most attempts to date in solving the Riemann hypothesis for number fields follow Hilbert's old suggestion: find an operator, A, acting on a Hilbert space such that $\langle Ax, y\rangle + \langle x, Ay\rangle = \langle x, y\rangle$, and such that $i(\frac{1}{2} - A)$ is self adjoint, and identify its eigenvalues with the zeros of the zeta function. This approach received scrutiny [22; 25], especially after the success of Selberg's theory, where such an operator, the Laplacian, does in fact exists. Such an operator also exists in the context of Weil's theory, namely the Frobenius operator acting on ℓ-adic cohomology (or equivalently, the ℓth power torsion points of the Jacobian), but here such a realization exists only over $\mathbf{Q}_\ell$, $\ell \neq p$, ∞, a fact which hints of the difficulties of this approach to number fields.

Let us review the 'roundabout' proof of the Riemann hypothesis for a curve C over a finite field $\mathbf{F}_p$, as elucidated in [23; 40]. Given a function $f : p^{\mathbf{Z}} \longrightarrow \mathbf{Z}$ of finite support, we associate with it its Mellin transform $\hat{f}(s) = \sum_n f(p^n) \cdot p^{ns}, s \in \mathbf{C}$, and a divisor $\hat{f}(A) = \sum_n f(p^n) \cdot A^n$ on the surface $C \times C$, where A^n are the Frobenius correspondences given by $A^n = \{(x, x^{p^n})\}$, $A^{-n} = p^{-n} \cdot (A^n)^*$, $n \geq 0$; $*$ denoting the involution $(x, y)^* = (y, x)$. On the surface $C \times C$ we have intersection theory which is given explicitly for our divisors by:

$$\text{(i)} \qquad \langle \hat{f}(A),\ \hat{g}(A)\rangle = \langle (f * g^*)^\wedge(A),\ \mathrm{Diag}\rangle$$

where $g^*(p^n) = g(p^{-n}) \cdot p^{-n}$ (so $(g^*)^\wedge(s) = \hat{g}(1-s)$), and $f * g(p^n) = \sum_m f(p^m)$ $g(p^{n-m})$ (so $(f * g^*)^\wedge(s) = \hat{f}(s) \cdot \hat{g}(s)$); i.e., knowledge of the intersection numbers reduces to those with the diagonal $\text{Diag} = A^0$.

(ii)
$$\langle \hat{f}(A),\ \text{Diag} \rangle = \hat{f}(0) + \hat{f}(1) - \sum_{\zeta(s)=0} \hat{f}(s)$$

where the sum is extended over the zeros of the zeta function $\zeta(s)$ of C.

Letting $h^0(f) = \dim_{\mathbb{F}_p} H^0(C \times C,\ \mathcal{O}(\hat{f}(A)))$ denote the dimension of the space of global sections of the line bundle $\mathcal{O}(\hat{f}(A))$, we have:

Riemann–Roch inequality

$$h^0(f) + h^0(\omega - f) \geq \frac{1}{2} \langle \hat{f}(A),\ \hat{f}(A) - \omega \rangle$$

where ω is a canonical divisor on $C \times C$ which can be thought of as the distribution on $p^{\mathbb{Z}}$ give by $\langle \omega, f \rangle = (2g_c - 2)(\hat{f}(0) + \hat{f}(1))$.

Monotoneness $\quad h^0(f) > 0 \Longrightarrow h^0(f+g) \geq h^0(g)$.

Ampleness $\quad \langle \omega, f \rangle = 0 \Longrightarrow h^0(m \cdot f)$ is bounded independently of $m \in \mathbb{Z}$.

Given the above three properties it is possible to derive the

Fundamental inequality $\quad \hat{f}(0) \cdot \hat{f}(1) \geq \frac{1}{2} \langle \hat{f}(A),\ \hat{f}(A) \rangle$ or equivalently using (i), (ii):

$$\sum_{\zeta(s)=0} \hat{f}(s) \cdot \hat{f}(1-s) \geq 0$$

which is easily shown to be equivalent to the Riemann hypothesis:

$$\zeta(s) = 0 \Longrightarrow \text{Re}\, s = \frac{1}{2}.$$

Turning back to number fields, and for simplicity let us consider only the case of the rational numbers, $\mathbb{Q}$, let $\zeta(s) = \prod_{p \leq \infty} \zeta_p(s)$, $\text{Re}\, s > 1$, denote the classical Riemann zeta function (completed at ∞); $\zeta_p(s) = (1 - p^{-s})^{-1}$ for $p < \infty$, $\zeta_\infty(s) = \pi^{-\frac{s}{2}} \Gamma(\frac{s}{2})$. The exact relation between the zeros of $\zeta(s)$ and the distribution of the primes was known to Riemann, but it was Weil who crystallized it in his 'explicit sums' [50; 51]. Given a function $f : \mathbb{R}^+ \longrightarrow \mathbb{R}$ smooth and compactly supported, we associate with t its Mellin transform $\hat{f}(s) = \int_0^\infty f(x)x^s \frac{dx}{x}$, and we have by the functional equation $\zeta(s) = \zeta(1-s)$:

$$\hat{f}(0) + \hat{f}(1) - \sum_{\zeta(s)=0} \hat{f}(s) = -\frac{1}{2\pi i} \oint \hat{f}(s) d \log \zeta(s)$$

$$= \sum_{p \le \infty} \frac{1}{2\pi i} \int_{\frac{1}{2}-i\infty}^{\frac{1}{2}+i\infty} \hat{f}(s) d \log \frac{\zeta_p(1-s)}{\zeta_p(s)} \stackrel{def}{=} \sum_{p \le \infty} W_p(f) \stackrel{def}{=} W(f).$$

Using Mellin inversion it is easy to see that [26]:

$$W_p(f) = \log p \cdot \sum_{n \ne 0} f(p^n) \cdot \min(1, p^n) \qquad \text{for } p < \infty,$$

$$W_\infty(f) = (\gamma + \log \pi) f(1) + \int_0^1 \frac{f(x) - xf(1)}{1 - x^2} dx + \int_1^\infty \frac{x^2 f(x) - f(1)}{x^2 - 1} \frac{dx}{x}.$$

And again, as pointed out by Weil [50], setting $f^*(x) = f(x^{-1}) \cdot x^{-1}$, the

Fundamental inequality $\hat{f}(0) \cdot \hat{f}(1) \ge \frac{1}{2} W(f * f^*)$ or equivalently:

$$\sum_{\zeta(s)=0} \hat{f}(s) \cdot \hat{f}(1-s) \ge 0$$

is easily shown to be equivalent to the Riemann hypothesis.

It will not be an exaggeration to say that the greatest mystery of arithmetic is the simple fact that $\mathbf{Z} \otimes \mathbf{Z} \cong \mathbf{Z}$, or equivalently, that from the point of view of algebraic geometry, spec $\mathbf{Z} \times$ spec $\mathbf{Z} \cong$ spec $\mathbf{Z}$, i.e., the surface reduces to the diagonal! Nevertheless, for functions $f, g : \mathbf{R}^+ \longrightarrow \mathbf{R}$ smooth and compactly supported, to be thought of as representing 'Frobenius divisors' on the non-existing surface, we can define their intersection number: $\langle f, g \rangle \stackrel{def}{=} W(f * g^*)$, and again, associating with such a function, f, a real number $h^0(f) \ge 0$ satisfying the above three properties will lead to the solution of the Riemann hypothesis. Ergo our main point is: a two dimensional Riemann–Roch for spec $\mathbf{Z}$ may very well exist!

Let us briefly review the proof of the classical Riemann–Roch theorem, i.e., over $\mathbf{C}$, in a somewhat bias manner [2; 7; 24; 44]. Given a compact Riemannian manifold X, and a Dirac operator $\not{D} : E_+ \longrightarrow E_-$, acting on the section of two vector bundles $E_\pm$ over X, let $\Delta_+ = \not{D}^* \not{D}$, $\Delta_- = \not{D}\not{D}^*$ denote the associated Laplacians, and consider $R_\pm^\alpha = \Delta_\pm^{-\alpha}$ as operators on $L^2(E_\pm)$. A basic fact is 'supersymmetry': $\text{tr}(R_+^\alpha) - \text{tr}(R_-^\alpha)$ is independent of $\alpha > \dim X$; in fact, $R_\pm^\alpha$ have discrete spectra and the non-zero eigenvalues of R_+^α and R_-^α are the same including multiplicities. Lettting $\alpha \longrightarrow \infty$, $R_\pm^\alpha$ converge to the projections onto the kernels of $\Delta_\pm$, hence

$$\text{tr}(R_+^\alpha) - \text{tr}(R_-^\alpha) \underset{\alpha \to \infty}{\longrightarrow} \dim \ker \not{D} - \dim \ker \not{D}^* = \text{index } \not{D}.$$

On the other hand, for $\text{Re}\,\alpha > \dim X$, $R^\alpha_\pm$ are given by continuous kernels, $R^\alpha_\pm(x,y)$ which have meromorphic continuations to all α, hence upon letting $\alpha \longrightarrow 0$ we obtain:

$$\text{tr}(R^\alpha_+) - \text{tr}(R^\alpha_-) \underset{\alpha\to 0}{\longrightarrow} \int_X Pf_{\alpha=0}(R^\alpha_+(x,x) - R^\alpha_-(x,x))dx.$$

Putting these together, and explicitly evaluating the last integrand, one obtains the Atiyah–Singer index theorem. For the special case of a Kahler manifold, and the $\bar\partial$ operator, one obtains the Hirzebruch–Riemann–Roch theorem. Replacing the bounded operators in a Hilbert space by a II_∞ factor, Atiyah [1] has proved a real valued index theorem for non-compact X, from which a host of index theorems sprung [4; 13; 33–36].

The index theorem relates global information (the limit $\alpha \to \infty$) with local information (the limit $\alpha \to 0$) and it is usually the case that the precise determination of the latter, i.e., of $Pf_{\alpha=0}(R^\alpha_+(x,x) - R^\alpha_-(x,x))$, is the deepest step which includes passage to a new geometrical space, X', the cotangent space to X. A hint on what X' might be in the arithmetical setting is gained by an inspection of the proof of the K-amenability of $SL_2(\mathbb{Q}_p)$ [31; 32] which we next (biasly) describe. Let $X_p = SL_2(\mathbb{Q}_p)/SL_2(\mathbb{Z}_p)$ denote the 'p-adic hyperbolic plane', it is a $(p+1)$-regular tree, and let $d(x,y)$ denote the natural $SL_2(\mathbb{Q}_p)$-invariant metric. Acting on functions on X_p we have the Hecke operator, $T_pf(y) = \sum_{d(x,y)=1} f(x)$, and the 'infinitesimal generator' of random walk on X_p, the Laplacian $\Delta_p = \frac{\zeta_p(2)}{\zeta_p(1)}T_p - 1$, whose associated potential operator is given by the kernel $\Delta_p^{-1} = \zeta_p(1)\cdot p^{-d(x,y)}$ [10; 11]. The function $d(x,y)$ is negative definite, hence for each $\alpha > 0$ we have a positive self-adjoint operator on $L^2(X_p)$ given by the kernel $R^\alpha_p = p^{-\alpha d(x,y)}$, and an associated Hilbert space $H^\alpha = \{f : X_p \to \mathbb{C} \,|\, (f, R^\alpha_p f)_{L^2(X_p)} < \infty\}$. Letting $\alpha \longrightarrow \infty$ we obtain $H^\infty = L^2(X_p)$. On the other hand, letting $\alpha \longrightarrow 0$, we get $H^0 = \mathbb{C} \oplus \frac{\partial}{\partial\alpha}\big|_{\alpha=0}H^\alpha$, where $\frac{\partial}{\partial\alpha}\big|_{\alpha=0}H^\alpha$ is the Hilbert space completion of $S_0(X_p) = \{f : X_p \to \mathbb{C} \,|\, \text{supp}\, f \text{ finite}, \int_{X_p} f(x)dx = 0\}$ with respect to $(f,g) = \frac{\partial}{\partial\alpha}\big|_{\alpha=0}(f, R^\alpha_p g) = -\log p \cdot \int\int_{X_p\times X_p} f(x)\overline{g(y)}d(x,y)$. This latter space is the analogue of the cotangent space to the p-adic hyperbolic plane. To make this analogy more suggestive let $X'_p = SL_2(\mathbb{Q}_p)/\Gamma_0(p)$ denote the oriented edges, $\Omega_c = \{\omega : X'_p \to \mathbb{C} \,|\, \text{supp}\,\omega \text{ finite}, \omega(x,y) = -\omega(y,x)\}$ the 'differential forms' with compact support, and $L^2(\Omega)$ the closure of Ω_c in $L^2(X'_p)$. Then 'exterior differentiation' $df(x,y) = f(x) - f(y)$ induces isomorphism $d : S_0(X_p) \xrightarrow{\sim} \Omega_c$ and an isometry $2(\log p)^{-\frac{1}{2}}d : \frac{\partial}{\partial\alpha}\big|_{\alpha=0}H^\alpha \xrightarrow{\sim} L^2(\Omega)$.

We now arrive at the potentials relevant to arithmetic. These are M. Riesz's potentials given by

$$R^\alpha_p(x) = \frac{\zeta_p(1-\alpha)}{\zeta_p(\alpha)}|x|^{\alpha-1}dx, \qquad \text{Re}\,\alpha > 0.$$

For a smooth (i.e., locally constant, if $p < \infty$) function φ on $\mathbf{Q}_p$, such that $\int_{|x|>1} |\varphi(x)| \cdot |x|^{\text{Re }\alpha-1}dx < \infty$, we can form the convolution $R_p^\alpha * \varphi$:

$$R_p^\alpha * \varphi(y) = \frac{\zeta_p(1-\alpha)}{\zeta_p(\alpha)} \int \varphi(x) \cdot |y-x|^{\alpha-1}dx, \qquad \text{Re } \alpha > 0.$$

This can be meromorphically continued to all α, picking up a δ function as we cross the line Re $\alpha = 0$, i.e.,

$$R_p^\alpha * \varphi(y) = \frac{\zeta_p(1-\alpha)}{\zeta_p(\alpha)} \int (\varphi(x) - \varphi(y)) \cdot |y-x|^{\alpha-1}dx, \qquad \text{Re } \alpha < 0$$

(for $p = \infty$, we pick up the distribution $\left(\frac{\partial^2}{\partial x^2}\right)^n$ as we cross Re $\alpha = -2n$).

Letting ψ_p denote the basic additive character of $\mathbf{Q}_p$, and $\psi_p^{(x)}(y) = \psi_p(x \cdot y)$, a straightforward computation gives $R_p^{-\alpha} * \psi_p^{(x)} = |x|^\alpha \cdot \psi_p^{(x)}$, i.e., $\psi_p^{(x)}$ is a (generalized) eigenvector for $R_p^{-\alpha}$ with eigenvalue $|x|^\alpha$, reflecting the fact that, as a distribution on $\mathbf{Q}_p$, $R_p^{-\alpha}$ is the Fourier transform of $|x|^\alpha$. Similarly, we have M. Riesz's reproduction formula $R_p^\alpha * R_p^\beta = R_p^{\alpha+\beta}$, Re$(\alpha+\beta) < 1$. Written out explicitly, $|x|^\alpha = R_p^{-\alpha} * \psi_p^{(x)}(0)$, becomes

$$|x|^\alpha = \int (1 - \psi_p(xy)) \left[-\frac{\zeta_p(1+\alpha)}{\zeta_p(-\alpha)} \frac{dy}{|y|^{1+\alpha}} \right].$$

Note that the measure inside the brackets is positive for $\alpha \in (0, \alpha_p)$, where $\alpha_p = [\overline{\mathbf{Q}}_p : \mathbf{Q}_p]$, i.e., $\alpha_p = \infty$ for $p < \infty$, and $\alpha_\infty = 2$ (but for more general number fields having a complex place v, $\alpha_v = 1$). From the Levi–Khinchin theorem [6] we deduce that $|x|^\alpha$ is a negative definite function on $\mathbf{Q}_p$ for $\alpha \in (0, \alpha_p)$, hence we have a 'heat conduction' semi-group [6; 27]

$$\mu_{p,t}^\alpha(y) = \int_{\mathbf{Q}_p} e^{-t|x|^\alpha} \psi_p(xy)dx = \sum_{n\geq 1} \frac{(-t)^n}{n!} \frac{\zeta_p(1+n\alpha)}{\zeta_p(-n\alpha)} |y|^{-(1+n\alpha)}dy$$

$\mu_{p,t}^\alpha$ has infinitesimal generator $R_p^{-\alpha}$, and for $\alpha \in (0,1)$ it is transient with potential kernel R_p^α. The potential theory of R_p^α can be developed purely analytically [27], without any recourse to probability, and in many respects is in fact easier than the classical theory; e.g., Harnack's inequality is a triviality in a non-archimedean situation. Here, on the other hand, for the sake of added intuition, we recall briefly the path space formulation of the solution to the Dirichlet problem [29; 42]. Let Λ_x denote the space of 'paths' $X : [0, \infty) \longrightarrow \mathbf{Q}_p$, X right continuous, having a left limit at each point, and $X(0) = x$. We think about such a path as describing the history of a particle jumping around in $\mathbf{Q}_p$. By Kolmogorof's theorem there exists a unique probability measure P_x on Λ_x such that $P_x[X(t) \in dy] = \mu_{p,t}^\alpha(dy - x)$. We now have the probabilistic interpretation:

$$\mu_{p,t}^\alpha * \varphi(x) = E_x[\varphi(X(t))], \qquad R_p^\alpha * \varphi(x) = E_x\left[\int_0^\infty \varphi(X(t))dt\right],$$

where E_x denotes the expectation with respect to P_x. For a compact subset $K \subseteq \mathbb{Q}_p$ and a continuous function $\varphi : K \longrightarrow \mathbb{C}$, there exists a unique solution h_φ to the Dirichlet problem with 'boundary' condition $\varphi : h_\varphi(x) = \varphi(x)$ for $x \in K$, and h_φ is α-harmonic outside K, i.e., for $x \notin K$, h_φ is smooth near x and $R_p^{-\alpha} * h_\varphi(x) = 0$, or equivalently, for $p < \infty$:

$$h_\varphi(x) = (1 - p^{-\alpha}) \int_{|y|\geq 1} h_\varphi(x + p^N y)|y|^{-\alpha} d^* y, \qquad \text{for all } N \gg 0.$$

Probabilistically, h_φ is given by $h_\varphi(x) = E_x[\varphi(X(t_K));\ t_K < \infty]$, where $t_K(X) = \inf\{t > 0,\ X(t) \in K\}$ is the 'hitting time' of K; e.g., $R_p^{-\alpha} * \mathbb{1} = 0$, and $\mathbb{1}$ (= the constant function 1) is the only α-harmonic function throughout $\mathbb{Q}_p$, bounded at infinity.

Analytically, the whole information is encoded by the associated Dirichlet form [17] which is given for $\alpha \in (0, \alpha_p)$ by

$$\mathcal{E}_p^\alpha(\varphi_1, \varphi_2) = -\frac{1}{2}\frac{\zeta_p(1+\alpha)}{\zeta_p(-\alpha)} \int\int (\varphi_1(x) - \varphi_1(y)) \cdot (\varphi_2(x) - \varphi_2(y)) \cdot |x - y|^{-\alpha-1} dx dy$$

or dually, by the Hilbert space of distribution of finite energy [27], the completion of $\mathcal{S}(\mathbb{Q}_p) = \{$smooth, fast decreasing functions on $\mathbb{Q}_p\}$ with respect to $(\varphi_1, \varphi_2)_{H^\alpha} = (\varphi_1, R_p^\alpha * \varphi_2)_{L^2} = \frac{\zeta_p(1-\alpha)}{\zeta_p(\alpha)} \int\int \varphi_1(x)\varphi_2(y)|x-y|^{\alpha-1} dx dy$.

Our basic belief that an analogue of the index theorem exists for number fields, stems from a simple formula relating the intersection numbers $W(f) = \hat{f}(0) + \hat{f}(1) - \sum_{\zeta(s)=0} \hat{f}(s)$ with Riesz potentials. Namely, suppose we start with a Schwartz function $\tilde{f}$ on the ideles $\mathbb{A}^*$, such that f is the projection of $\tilde{f}$ onto $\mathbb{Q}^* \backslash \mathbb{A}^* / \Pi_p \mathbb{Z}_p^* \cong \mathbb{R}^+$. We have [26]:

$$W(f) = \sum_{q \in \mathbb{Q}^*} \frac{\partial}{\partial \alpha}\Big|_{\alpha=0} R^\alpha_{\mathbb{A}^*} \tilde{f}(q)$$

where $R^\alpha_{\mathbb{A}^*} = \otimes_p \frac{1}{c_p(\alpha)} R_p^\alpha$, with the renormalization constants $c_p(\alpha) = R_p^\alpha \phi_p^*(1) = 1 + \frac{(p^\alpha - 1)(1 - p^{-\alpha})}{p - p^\alpha}$ for $p < \infty$, $c_\infty(\alpha) = 1$.

Note that $c_p(0) = 1$, $\frac{\partial}{\partial\alpha}\big|_{\alpha=0} c_p(\alpha) = 0$, so that the $c_p(\alpha)$'s do not really affect the above formula and are plugged in only for the sake of convergence. We can also use $\tilde{c}_p(\alpha) = \frac{\zeta_p(1+\alpha)\zeta_p(1-\alpha)}{\zeta_p(1)^2}$, noting that $\Pi_p \frac{\tilde{c}_p(\alpha)}{c_p(\alpha)}$ converges for Re $\alpha > -\frac{1}{2}$.

We shall analyze next the limit case $\alpha \longrightarrow \alpha_p$ where one obtains the 'normal law', and the prospect of supersymmetry in arithmetic. For $p = \infty$, as $\alpha \to \alpha_p = 2$, $\mu^\alpha_{p,t}$ vaguely converges to the classical normal law. On the other hand, for $p < \infty$, as $\alpha \to \alpha_p = \infty$, $\mu^\alpha_{p,t}$ vaguely converges to a $\mathbb{Z}_p$-invariant probability measure $\mu^\infty_{p,t}$, i.e., the 'p-adic normal law' degenerates to a semi-group of probability measures on $\mathbf{Q}_p/\mathbf{Z}_p$.

We set

$$\phi_p = \mu^{\alpha_p}_{p,1/\alpha_p} = \begin{cases} e^{-\pi x^2} & p = \infty \\ \text{characteristic function of } \mathbf{Z}_p, & p < \infty; \end{cases}$$

$\phi_{\mathbb{A}} = \otimes_p \phi_p$ the 'normal law' on the adeles $\mathbb{A}$; and for a 'divisor on spec $\mathbf{Z}$' $a \in \mathbb{A}^*/\Pi_p \mathbf{Z}_p^*$, $\phi^{(a)}_{\mathbb{A}}(x) = |a|^{-1}\phi_{\mathbb{A}}(a^{-1}x)$ the 'normal law with background metric perturbed' by a. Letting $\text{pr} : \mathbb{A} \longrightarrow \mathbb{A}/\mathbf{Q}$ denote the natural projection, $\text{pr}_* \phi^{(a)}_{\mathbb{A}}$ is a probability measure on $\mathbb{A}/\mathbf{Q}$, acting on $L^2(\mathbb{A}/\mathbf{Q})$ via convolution, and we have Tate's (one dimensional) Riemann–Roch for spec $\mathbf{Z}$, [48]:

$$h^0(a) - h^0(a^{-1}) = \deg a,$$

where $h^0(a) = \log \operatorname{tr}\,(\text{pr}_* \phi^{(a)}_{\mathbb{A}})$, $\deg a = -\log |a|$, which is precisely the functional equation of the classical (log) theta function disguised in a fancy language. The Riemann zeta function is nothing but the Mellin transform of $\operatorname{tr}(\text{pr}_* \phi^{(a)}_{\mathbb{A}}) - 1$, and the above Riemann–Roch is equivalent to the functional equation $\zeta(s) = \zeta(1-s)$. Note that the cancellation of eigenvalues is not additive but multiplicative, and we have 'log tr', rather than 'tr' as in the index theorem. That $h^0(a) > 0$, or equivalently $\operatorname{tr}(\text{pr}_* \phi^{(a)}_{\mathbb{A}}) > 1$, follows from the fact that $\text{pr}_* \phi^{(a)}_{\mathbb{A}}$ is positive and always has 1 as an eigenvalue since it is a probability measure: $\text{pr}_* \phi^{(a)}_{\mathbb{A}} * \mathbb{1} = \mathbb{1}$.

We note on passing that R^α_p has a simple pole at $\alpha = 1$, with residue Haar measure, and $Pf_{\alpha=1}\ R^\alpha_p$ is the logarithmic kernel which enters in the (two dimensional) Arakalov–Faltings Riemann–Roch for $\mathbf{P}^1$ over spec $\mathbf{Z}$ [12]. Alternatively, $-\log \Pi_p \rho_p(q_1, q_2)$ is the intersection number of $q_1, q_2 \in \mathbf{P}^1(\mathbf{Q})$, considered as horizontal divisors on $\mathbf{P}^1$ over spec $\mathbf{Z}$, where

$$\rho_p : \mathbf{P}^1(\mathbf{Q}_p) \times \mathbf{P}^1(\mathbf{Q}_p) \longrightarrow [0,1], \quad \rho_p(x_1 : x_2, y_1 : y_2) = \frac{|x_1 y_2 - x_2 y_1|_p}{|(x_1, x_2)|_p \cdot |(y_1, y_2)|_p}$$

is the natural metric on $\mathbf{P}^1(\mathbf{Q}_p)$, and where the two dimensional absolute value is the 'L^{α_p}' one: $|(x_1, x_2)|_p = \sup(|x_1|_p, |x_2|_p)$ for $p < \infty$,

$$|(x_1, x_2)|_\infty = (|x_1|^2_\infty + |x_2|^2_\infty)^{\frac{1}{2}}.$$

We next survey the 'compactification' (or rather the 'quantization') of the Riesz potentials and their connection with the representation theory of SL_2. Let $\mathcal{S}^\alpha(\mathbf{Q}_p)$ denote the space of smooth function $\varphi : \mathbf{Q}_p \longrightarrow \mathbf{C}$, such that $\varphi(x^{-1}) \cdot |x|^{-(1+\alpha)}$ is also smooth (i.e., for $p < \infty : \varphi(x) = \text{cons} \cdot |x|^{-(1+\alpha)}$ for $|x| \gg 1$), and set

$$\phi^\alpha_p(x) = \zeta_p(1+\alpha) \cdot |(1,x)|^{-(1+\alpha)} \in \mathcal{S}^\alpha(\mathbf{Q}_p).$$

An easy verification gives

$$R_p^\alpha : \mathcal{S}^\alpha(\mathbf{Q}_p) \xrightarrow{\sim} \mathcal{S}^{-\alpha}(\mathbf{Q}_p), \quad R_p^\alpha * \phi_p^\alpha = \phi_p^{-\alpha}.$$

We can identify $\mathcal{S}^\alpha(\mathbf{Q}_p)$ with the smooth functions $\varphi : \mathbf{Q}_p^{\oplus 2} \setminus \{(0,0)\} \longrightarrow \mathbf{C}$ such that $\varphi(ax_1, ax_2) = |a|^{-(1+\alpha)}\varphi(x_1,x_2)$ (via $\varphi(x) = \varphi(1,x)$; $\varphi(x_1,x_2) = \varphi(\frac{x_2}{x_1}) \cdot |x_1|^{-(1+\alpha)}$), which in turn can be identified with the smooth sections of a line bundle $\mathcal{O}_p(\alpha)$ over $\mathbf{P}^1(\mathbf{Q}_p)$. The bundle $\mathcal{O}_p(\alpha)$ can be trivialized by means of the never vanishing section corresponding to ϕ_p^α, and we obtain:

$$\tilde{R}_p^\alpha \stackrel{def}{=} (\phi_p^{-\alpha})^{-1} \circ R_p^\alpha \circ \phi_p^\alpha : \mathcal{S}(\mathbf{P}^1(\mathbf{Q}_p)) \xrightarrow{\sim} \mathcal{S}(\mathbf{P}^1(\mathbf{Q}_p)), \quad \tilde{R}_p^\alpha \mathbb{1} = \mathbb{1}$$

where $\mathcal{S}(\mathbf{P}^1(\mathbf{Q}_p))$ denotes the smooth functions on $\mathbf{P}^1(\mathbf{Q}_p)$, e.g., for $p < \infty$: $\mathcal{S}(\mathbf{P}^1(\mathbf{Q}_p)) = \mathcal{S}(\mathbf{Q}_p) \oplus \mathbf{C}\cdot\mathbb{1}$. Let $SL_2(\mathbf{Z}_p) = \{g \in SL_2(\mathbf{Q}_p) \,\big|\, |g(x_1,x_2)| = |(x_1,x_2)|\}$ denote the maximal compact subgroup of $SL_2(\mathbf{Q}_p)$, $p \le \infty$; and $dx_1 : x_2$ the unique $SL_2(\mathbf{Z}_p)$-invariant probability measure on $\mathbf{P}^1(\mathbf{Q}_p)$. Then we have,

$$\tilde{R}_p^\alpha \varphi(y_1 : y_2) = \frac{1}{r_p(\alpha)} \int_{\mathbf{P}^1(\mathbf{Q}_p)} \varphi(x_1 : x_2)\rho_p(y_1 : y_2, x_1 : x_2)^{\alpha - 1} dx_1 : x_2, \quad \mathrm{Re}\,\alpha > 0$$

where $r_p(\alpha) = \frac{\zeta_p(2)}{\zeta_p(1)} \cdot \frac{\zeta_p(\alpha)}{\zeta_p(1+\alpha)} = \int \rho_p(x_1 : x_2, y_1 : y_2)^{\alpha-1} dx_1 : x_2$ for any $y_1 : y_2$.

Note that while R_p^α was diagonalizable with respect to Fourier transform carrying $\mathcal{S}(\mathbf{Q}_p)$ to itself, $\tilde{R}_p^\alpha$ is diagonalizable with respect to Fourier transform carrying $\mathcal{S}(\mathbf{P}^1(\mathbf{Q}_p))$ into $\mathcal{S}(\mathbf{P}^1(\mathbf{Q}_p)^\vee)$, where $\mathbf{P}^1(\mathbf{Q}_p)^\vee$ is the (discrete) group of characters of $\mathbf{P}^1(\mathbf{Q}_p)$; here $\mathbf{P}^1(\mathbf{Q}_p)$ is considered as a group via $\mathbf{P}^1(\mathbf{Q}_p) \cong \mathbf{Z}_p[\sqrt{\varepsilon_p}]^*/\mathbf{Z}_p^*$, $\varepsilon_p \in \mathbf{Z}_p^* \setminus (\mathbf{Z}_p^*)^2$ [19]. Indeed, for $p < \infty$, and a character $\chi \ne \mathbb{1}$, primitive of conductor p^N, $\tilde{R}_p^\alpha \chi = \frac{\zeta_p(1+\alpha)}{\zeta_p(1-\alpha)} p^{-N\alpha}\chi$; for $p = \infty$, and the character $\chi_n(z) = z^{2n}$, $\tilde{R}_\infty^\alpha \chi_n = \frac{\zeta_\infty(1+\alpha)}{\zeta_\infty(1-\alpha)} \cdot \frac{\zeta_\infty(1-\alpha+2n)}{\zeta_\infty(1+\alpha+2n)} \cdot \chi_n$. Thus, while R_p^α had spectrum $p^{\alpha\mathbf{Z}}$ ($\mathbf{R}^+$, for $p = \infty$) with infinite multiplicity, $\tilde{R}_p^\alpha$ has discrete spectrum; while R_p^α was unbounded, $\|\tilde{R}_p^\alpha\|_{L^2(\mathbf{P}^1)} = 1$ for all $\alpha \ge 0$, and moreover $\tilde{R}_p^\alpha$ is Hilbert–Schmidt for $\mathrm{Re}\,\alpha > \frac{1}{2}$ (and trace class for $\mathrm{Re}\,\alpha > 1$). Note also that we lose the semi-group property $R_p^\alpha * R_p^\beta = R_p^{\alpha+\beta}$, and we are left only with $\tilde{R}_p^\alpha * \tilde{R}_p^{-\alpha} = \tilde{R}_p^0 = id$; we lose the positivity $R_p^\alpha \ge 0$, and we are left only with $\tilde{R}_p^\alpha \ge 0$ for $\alpha \in (-1,1)$, corresponding to the complementary series representation of SL_2; we lose the markovian nature of $R_p^{-\alpha}$, $\alpha \in (0, \alpha_p)$: $\tilde{R}_p^{-\alpha}$ doesn't generate a markovian semi-group even for $\alpha \in (0,1)$.

Shifting attention to imaginary values of α, note that $\tilde{R}_{\mathbf{A}}^{it} = \otimes_p \tilde{R}_p^{it}$ is the well known intertwining operator for the unramified principal series representation of $SL_2(\mathbf{A})$ [19; 21; 53]. Namely, the two dimensional symplectic Fourier transform (the 'isotropic symbol' [28])

$$\mathcal{F}\varphi(y_1,y_2) = \int\!\!\int_{\mathbf{A}\oplus\mathbf{A}} \varphi(x_1,x_2)\psi(x_1y_2 - x_2y_1)dx_1dx_2$$

descends to a unitary operator $\mathcal{F}$ on the space $\Omega = SL_2(\mathbb{A})/\mathbb{Q}^*\Pi_p\mathbb{Z}_p^* \ltimes \mathbb{A}$, which is a fibration of $\mathbb{P}^1(\mathbb{A}) = SL_2(\mathbb{A})/\mathbb{A}^* \ltimes \mathbb{A}$, with fiber $\mathbb{A}^*/\mathbb{Q}^*\Pi_p\mathbb{Z}_p^* \cong \mathbb{R}^+$. Thus, via Mellin transform, $L^2(\Omega) = \int_0^\infty H^{it} \oplus H^{-it}$, where H^{it} denotes $L^2(\mathbb{P}^1(\mathbb{A}))$ with the irreducible unitary $SL_2(\mathbb{A})$-action

$$\pi^{it}(g)\varphi(x_1 : x_2) = \varphi(g^{-1}(x_1 : x_2)) \cdot \left[\frac{|(x_1, x_2)|}{|g^{-1}(x_1, x_2)|}\right]^{1-it},$$

and $\mathcal{F}$ decomposes as $\begin{pmatrix} 0 & \tilde{R}_{\mathbb{A}}^{it} \\ \tilde{R}_{\mathbb{A}}^{-it} & 0 \end{pmatrix}$.

We end this survey by transforming our basic formula relating the explicit sums to Riesz potentials into a formula involving a trace. A straightforward computation gives for Re $\alpha > 0$:

$$\frac{1}{\tilde{c}_p(\alpha)} R_p^\alpha(\tilde{f}_p)(q) = \frac{\zeta_p(2\alpha)}{\zeta_p(\alpha)^2} \frac{\zeta_p(1)^2}{\zeta_p(1+\alpha)^2} \operatorname{tr}(\tilde{R}_p^\alpha \pi_p^\alpha(\tilde{f}_p)\pi_p^\alpha(q))$$

where

$$\pi_p^\alpha(a)\varphi(x_1 : x_2) = \varphi(a^{-1}x_1 : x_2) \left[\frac{|(x_1, x_2)|}{|(a^{-1}x_1, x_2)|}\right]^{1+\alpha},$$

$$\pi_p^\alpha(\tilde{f}_p) = \int_{\mathbb{Q}_p^*} \tilde{f}_p(a)\pi_p^\alpha(a)d^*a.$$

Using the fact that $\frac{\zeta(2\alpha)}{\zeta(\alpha)^2} = -\frac{\alpha}{2}(1 + O(\alpha^2))$ as $\alpha \to 0$, we obtain:

$$W(f * f^*) = -\frac{1}{2} \sum_{q \in \mathbb{Q}^*} Pf_{\alpha=0} \prod_p \frac{\zeta_p(1)^2}{\zeta_p(1+\alpha)^2} \operatorname{tr}(\tilde{R}_p^\alpha \pi_p^\alpha(\tilde{f}_p^*)\pi_p^\alpha(\tilde{f}_p^{(q)})).$$

Setting $S_p^\alpha \stackrel{def}{=} \frac{\zeta_p(1-\alpha)}{\zeta_p(1+\alpha)} \tilde{R}_p^\alpha + \left(1 - \frac{\zeta_p(1-\alpha)}{\zeta_p(1+\alpha)}\right) \tilde{R}_p^1$, $\tilde{R}_p^1$ being the projection onto the space of constant functions, we have for $p < \infty$: $S_p^\alpha \mathbb{1} = \mathbb{1}$, $S_p^\alpha \chi = p^{-N\alpha} \cdot \chi$ for a character χ primitive of conductor p^N, so that S_p^α forms a bounded positive semi-group for $\alpha > 0$. Again a straightforward computation gives for Re $\alpha > 0$:

$$\frac{\zeta_p(2\alpha)}{\zeta_p(\alpha)^2} \operatorname{tr}(S_p^\alpha \pi_p^\alpha(\phi_p^*)) = 1$$

$$\frac{\zeta_p(2\alpha)}{\zeta_p(\alpha)^2} \operatorname{tr}(S_p^\alpha \pi_p^\alpha(\tilde{f}_p^* * \tilde{f}_p^{(q)})) = R_p^\alpha(\tilde{f}_p^* * \tilde{f}_p^{(q)})(1) + O(\alpha^2).$$

Hence we obtain:

$$W(f * f^*) = -\frac{1}{2} \sum_{q \in \mathbb{Q}^*} Pf_{\alpha=0} \operatorname{tr}(\pi^\alpha(\tilde{f}^*) S_{\mathbb{A}}^\alpha \pi^\alpha(\tilde{f}^{(q)}))$$

where $S^\alpha_{\mathbb{A}} = \otimes_p S^\alpha_p$, $\pi^\alpha(f) = \otimes_p \pi^\alpha_p(\tilde{f}_p)$, as operators on $L^2(\mathbf{P}^1(\mathbb{A})) = \otimes_p L^2(\mathbf{P}^1(\mathbb{Q}_p))$ (Hilbert space $\otimes_p$ w.r.t. $1\!\!1$). Note that for Re $\alpha > 1$, S^α_p is given by the kernel

$$S^\alpha_p(y,x) = 1\!\!1 - \frac{\zeta_p(1-\alpha)}{\zeta_p(1+\alpha)} \left(1\!\!1 - \frac{\zeta_p(1)}{\zeta_p(2)} \frac{\zeta_p(1+\alpha)}{\zeta_p(\alpha)} \rho_p(y,x)^{\alpha-1}\right)$$

and that for Re $\alpha > 2$, $S^\alpha_{\mathbb{A}}$ is given by the kernel $\Pi_p S_p(y_p, x_p)$; note also the 'smoothing effect': $S^\alpha_p \pi^\alpha_p(\tilde{f}_p)$ is given by a continuous kernel for Re $\alpha > 0$. It is easy to check that, for $\alpha \in (0, \alpha_p)$, $S^{-\alpha}_p$ generates a Dirichlet form, and we obtain a Hunt process on $\mathbf{P}^1(\mathbb{Q}_p)$ [17], and the associated 'Schrödinger semi-group' [45] on the 'weighted' L^2 space $L^2\left(\mathbf{P}^1(\mathbb{Q}_p), dx_1 : x_2\right) \cong L^2\left(\mathbb{Q}_p, \frac{\zeta_p(2)}{\zeta_p(1)} \frac{dx}{|(1,x)|^2}\right)$. Having a process on $\mathbf{P}^1(\mathbb{Q}_p)$ for each p, we obtain a process on $\mathbf{P}^1(\mathbb{A}) = \Pi_p \mathbf{P}^1(\mathbb{Q}_p)$, whose infinitesimal generator is $\sum_p S^\alpha_p$, but for our purpose it is more interesting to consider the process generated by $S^{-\alpha}_{\mathbb{A}}$.

Finally, we remark that the operators we seek are only roughly approximated by the $\tilde{R}^\alpha_{\mathbb{A}}$ or $S^\alpha_{\mathbb{A}}$. It is possible to similarly construct operators on the adeles $\mathbb{A}$ (rather than $\mathbf{P}^1(\mathbb{A})$) with trace related to the explicit sums as above. What is missing in order to complete our scheme for attacking the Riemann hypothesis is precisely the analogue of supersymmetry in our context.

REFERENCES

1. M. F. Atiyah, 'Elliptic operators, discrete groups, and von Neumann algebras', Asterisque, **32** (1976), 43–72.

2. M. F. Atiyah, R. Bott and V. K. Patodi, 'On the heat equation and the index theorem', Invent. Math., **19** (1973), 279–330; also errata, idid. **28**, 277–80.

3. N. Aronszajn and K. T. Smith, 'Theory of Bessel potentials. Part I', Ann. Inst. Fourier, **11** (1961), 385–475.

4. P. Baum and R. Douglas, 'K-homology and index theory', Proceedings of A.M.S., **38** (1980), 117–73.

5. Yu. M. Berezanskii, *Selfadjoint Operators in Spaces of Functions of Infinitely Many Variables*, A.M.S Translations of Math. Monographs, 63 (1986).

6. C. Berg and G. Forst, *Potential Theory on Locally Compact Abelian Groups*, Springer-Verlag (1975).

7. J. M. Bismut, 'The Atiyah–Singer theorems for classical elliptic operators: a probabilistic approach', J. Func. Anal., **57** (1984), 56–99.

8. J. Bliedtner and W. Hansen, *Potential Theory*, Springer-Verlag (1986).

9. D. Cantor, 'On an extension of the definition of transfinite diameter and some applications', J. Reine. Angew. Math., **316** (1980), 160–207.

10. P. Cartier, 'Functions harmoniques sur un arbre', Symposia Mathematics, **9** (1972), 203–70.

11. P. Cartier, 'Géométrie et analyse sur les arbres', Sem. Bourbaki, 1971/72, Exposé 407.

12. T. Chinburg, 'Intersection theory and capacity theory on arithmetic surfaces', Proc. Canadian Math. Soc. Summer Seminar in Number Theory, 7, A.M.S. (1986).

13. A. Connes, 'Non commutative differential geometry', Publ. Math. de l'IHES, **62** (1985), 41–144.

14. P. Deligne, 'Le déterminat de la cohomologie', Contemporary Math., **67** (1987), 93–177.

15. W. Feller, 'On a generalization of M. Riesz' potentials and the semi-groups generated by them', Proc. R. Physiogr. Soc. Lund, **21** (1952), 73–81.

16. W. Feller, *An Introduction to Probability Theory and its Applications*, Vol. 2, John Wiley & Sons (1970).

17. M. Fukushima, *Dirichlet Forms and Markov Processes*, North-Holland Publ. (1980).

18. S. Gelbart and I. I. Piatetskii-Shapiro, 'Distinguished representations and modular forms of half-integral weight', Invent. Math., **59** (1980) 145–88.

19. I. M. Gel'fand, M. I. Graev and I. I. Piatetskii-Shapiro, *Representation Theory and Automorphic Functions*, Saunders Com. (1969).

20. I. M. Gel'fand, M. I. Graev and N. Ya. Vilenkin, *Generalized Functions* vol. 5, Academic Press (1966).

21. R. Godement, 'The decomposition of $L^2(G/\Gamma)$ for $\Gamma = SL(2, \mathbb{Z})$', Proc. Symp. Pure Math. IX (1966), 211–24.

22. D. Goldfeld, 'Explicit formulae as trace formulae', in *The Selberg Trace Formula and Related Topics*, D.A. Hajhal *et al.* eds., A.M.S. (1989).

23. A. Grothendieck, 'Sur une note de Mattuck–Tate', J. Reine Angew. Math., **200** (1958), 208–15.

24. E. Getzler, 'A short proof of the local Atiyah–Singer index theorem', Topology, **25** (1986), 111–7.

25. D. Hajhal, 'The Selberg trace formula and the Riemann zeta function', Duke Math. J., **43** (1976), 441–82.

26. S. Haran, 'Riesz potentials and explicit sums in arithmetic', to appear in Invent. Math.

27. S. Haran, 'Analytic potential theory over the p-adics', preprint.

28. R. Howe, 'On the role of the Heisenberg group in harmonic analysis', Bulletin of A.M.S, **3** (1980), no.2, 821–43.

29. G. A. Hunt, 'Markoff processes and potentials, I, II, and III', Illinois J. Math., **1** (1957), 44–93; **1** (1957), 316–69; **2** (1958), 151–213.

30. H. Jacquet and R. P. Langlands, *Automorphic forms on $GL(2)$*, Lecture Notes in Math., Springer, 114 (1970).

31. P. Julg and A. Valette, 'K-amenability for $SL_2(\mathbb{Q}_p)$ and the action on the associated tree', J. Funct. Anal., **58** (1984), 194–215.

32. P. Julg and A. Valette, 'Twisted coboundary operator on a tree and the Selberg principle', J. Operator Theory, **16** (1986), 285–304.

33. G. G. Kasparov, 'Operator K-theory and its applications: elliptic operators, group representations, higher signatures, C^*-extensions', Proc. Int. Cong. Math. Warszawa (1983), 987–1000.

34. G. G. Kasparov, 'An index for invariant elliptic operators, K-theory, and representations of Lie groups', Soviet Math. Dokl., **27** (1983) No.1, 105–9.

35. G. G. Kasparov, 'Lorentz groups: K-theory of unitary representations and crossed products', Soviet Math. Dokl., **29** (1984) No.2, 256–60.

36. G. G. Kasparov, 'The operator K-functor and extensions of C^*-algebras', Math. USSR Izvestija, **16** (1981), No.3, 513–72.

37. N. S. Landkof, *Foundations of Modern Potential Theory*, Springer-Verlag (1972).

38. R. P. Langlands, 'Automorphic representations, Shimura varieties, and motives', Ein Märchen, Proc. Symp. Pure Math. XXIII (1979), 205–46.

39. Yu. I. Manin, *New Dimensions in Geometry*, Proc. Arbeitstagung Bonn, Lecture Notes in Math., Springer, 1111 (1984).

40. A. Mattuck and J. Tate, 'On the inequality of Castelnuovo–Severi', Hamb. Abh., **22** (1958), 295–9.

41. S. J. Patterson, *Introduction to the Theory of the Riemann Zeta Functions*, Cambridge Univ. Press (1988).

42. S. C. Port and C. J. Stone, 'Infinitely divisible processes and their potential theory, I and II, Ann. Inst. Fourier, **21** (1971), no.2, 157–275; no.4, 179–265.

43. A. Selberg, 'Harmonic analysis and discontinuous groups in weakly symmetric Riemannian spaces', J. Indian Math. Soc., **20** (1956), 47–87.

44. R. T. Seeley, 'Complex powers of an elliptic operator', Proc. Symp. Pure Math. X (1967), 288–307.

45. B. Simon, 'Schrödinger semigroups', A.M.S. Bulletin, **7** (1982), No.3, 447–526.

46. B. Simon, *Functional Integration and Quantum Physics*, Academic Press (1979).

47. M. H. Taibleson, *Fourier Analysis on Local Fields*, Princeton Univ. Press (1975).

48. J. Tate, 'Fourier analysis in number fields and Hecke's zeta-functions', thesis reproduced in J.W.S. Cassels and A. Fröhlich *Algebraic Number Theory*, Thompson Book Co. (1967).

49. A. Weil, *Sur les Courbes Algébriques et les Variétés qui s'en Déduisent*, Hermann (1948).

50. A. Weil, 'Sur les formules explicites de la théorie des nombres premiers', Proc. R. Physiogr. Soc. Lund, **21** (1952), 252–65.

51. A. Weil, 'Sur les formules explicites de la théorie des nombres', Izv. Mat. Nauk., **36** (1972) 3–18.

52. A. Weil, 'Function zeta et distributions', Sem. Bourbaki 312 (1966).

53. D. Zagier, 'Eisenstein series and the Riemann zeta function', in *Automorphic forms, Representation theory and Arithmetic*, Bombay Colloquium 1979, Springer (1981).

Katz p-adic L-functions, congruence modules and deformation of Galois representations

H. HIDA* AND J. TILOUINE

0. Although the two-variable main conjecture for imaginary quadratic fields has been successfully proven by Rubin [R] using brilliant ideas found by Thaine and Kolyvagin, we still have some interest in studying the new proof of a special case of the conjecture, i.e., the anticyclotomic case given by Mazur and the second named author of the present article ([M-T], [T1]). Its interest lies firstly in surprizing amenability of the method to the case of CM fields in place of imaginary quadratic fields and secondly in its possible relevance for non-abelian cases. In this short note, we begin with a short summary of the result in [M-T] and [T1] concerning the Iwasawa theory for imaginary quadratic fields, and after that, we shall give a very brief sketch of how one can generalize every step of the proof to the general CM-case. At the end, coming back to the original imaginary quadratic case, we remove some restriction of one of the main result in [M-T]. The idea for this slight amelioration to [M-T] is to consider deformations of Galois representations not only over finite fields but over any finite extension of $\mathbf{Q}_p$. Throughout the paper, we assume that $p > 2$.

1. Let M be an imaginary quadratic field and p be an odd prime which splits in M; i.e., $p = \overline{\mathfrak{p}}\mathfrak{p}(\mathfrak{p} \neq \overline{\mathfrak{p}})$. We always fix the algebraic closures $\overline{\mathbf{Q}}$ and $\overline{\mathbf{Q}}_p$ and embeddings of $\overline{\mathbf{Q}}$ into $\mathbf{C}$ and $\overline{\mathbf{Q}}_p$. Any algebraic number field will be considered to be inside $\overline{\mathbf{Q}}$. Suppose the factor $\mathfrak{p}$ of p is compatible with this embedding M into $\overline{\mathbf{Q}}_p$. The scheme of the new proof of the main conjecture for the anti-cyclotomic $\mathbf{Z}_p$-tower of M consists in proving two divisibility theorems between the following three power series:

$$(1.1) \qquad L^-|H|Iw^-,$$

where

* The first named author is supported in part by an NSF grant

(i) L^- is the Katz–Yager p-adic L-function (which interpolates p-adically Hurwitz–Damerell numbers) projected to one branch of the anti-cyclotomic line of the imaginary quadratic field M;

(ii) H is the characteristic power series of the congruence module attached to M (and the branch in (i)) constructed via the theory of Hecke algebras for $GL(2)_{/\mathbf{Q}}$;

(iii) Iw^- is the characteristic power series (of the branch in (i)) of the maximal $\mathfrak{p}$-ramified extension of the anti-cyclotomic $\mathbf{Z}_p^\times$-tower over M.

Once these divisibilities are assumed, the proof is fairly easy: Under a suitable branch condition, we know from the analytic class number formula that the λ and μ-invariants of G and Iw^- are the same and hence

$$Iw^- = L^- \quad \text{up to a unit power series} \tag{1.2}$$

as the anticyclotomic main conjecture predicts.

Strictly speaking, the equality (1.2) is proven in [M-T] and [T] under the assumption that the class number of M is equal to 1. In fact, if the class number h of M is divisible by p, we need to modify (1.1) as

$$h \cdot L^- | H | h \cdot Iw^- \quad \text{for the class number } h \text{ of } M. \tag{1.3}$$

In [M-T], the second divisibility assertion: $H \mid Iw^-$ is proven under the milder assumption that h is prime to p but there is another assumption that the branch character ψ of L^- must be non-trivial on the inertia group $I_\mathfrak{p}$ at $\mathfrak{p}$. We will prove the divisibility (1.3) outside the trivial zero of L^- (if any) without hypothesis in Appendix.

2. In this section, we deal with the generalization of the first divisibility result: $L^- \mid H$ in the general CM case. The second divisibility: $H \mid Iw^-$ will be dealt with in the following paragraphs. To state the result precisely, we fix a prime p and write the fixed embeddings as $\iota_p : \overline{\mathbf{Q}} \to \overline{\mathbf{Q}}_p$ and $\iota_\infty : \overline{\mathbf{Q}} \to \mathbf{C}$. We consider $\overline{\mathbf{Q}}$ as a subfield of $\overline{\mathbf{Q}}_p$ and $\mathbf{C}$ by these embeddings. Let F be a totally real number field with class number $h(F)$ and M/F be a totally imaginary quadratic extension whose class number is denoted by $h(M)$. Let c be the complex conjugation which induces the unique non-trivial automorphism of M over F. We assume the following ordinarity condition:

Ordinarity hypothesis All prime factor $\mathfrak{p}$ of p in F splits in M.

Thus we can write the set of prime factors of p in M as a disjoint union $S \cup S^c$ of two subsets of prime ideals so that $\mathfrak{P} \in S$ if and only if $\mathfrak{P}^c \in S^c$. If a is the number of prime ideals in F over p, there are 2^a choices of such subset S. Such an S will be called a p-adic CM-type. Considering S as a set of p-adic places of M, let Σ be the set of embeddings of M into $\overline{\mathbf{Q}}$ which give rise to places in S after combining with ι_p. Then $\Sigma \cup \Sigma \circ c$ is the total set of embeddings of M into $\overline{\mathbf{Q}}$ and hence gives a complex CM-type of M. Hereafter we fix a p-adic CM-type S and compatible complex CM-type Σ. Let $\mathbf{G}$ be the Galois group of the maximal p-ramified abelian extension M_∞ of M. Then we fix a decomposition $\mathbf{G} = \mathbf{G}_{\mathrm{tor}} \times \mathbf{W}$ for a finite group $\mathbf{G}_{\mathrm{tor}}$ and a $\mathbf{Z}_p$-free module $\mathbf{W}$. Let $K/\mathbf{Q}_p$ be a p-adically complete extension in the p-adic completion Ω of $\overline{\mathbf{Q}}_p$ containing all the images $\sigma(M)$ for $\sigma \in \Sigma$ and $\mathfrak{O} = \mathfrak{O}_K$ be the p-adic integer ring of K. We now consider the continuous group algebras $\Lambda = \mathfrak{O}[[\mathbf{W}]]$ and $\mathfrak{O}[[\mathbf{G}]] = \Lambda[\mathbf{G}_{\mathrm{tor}}]$. By choosing a basis of $\mathbf{W}$, we have $\mathbf{W} \cong \mathbf{Z}_p^r$ and $\Lambda \cong \mathfrak{O}[[X_1, \ldots, X_r]]$. Here $r = [F : \mathbf{Q}] + 1 + \delta$, where δ is the defect of the Leopoldt conjecture for F; i.e., $\delta \geq 0$ and $\delta = 0$ if and only if the Leopoldt conjecture holds for F and p. Fix a character $\lambda : \mathbf{G}_{\mathrm{tor}} \to \mathfrak{O}^\times$ and define the projection $\lambda_* : \mathfrak{O}[[\mathbf{G}]] \to \mathfrak{O}[[\mathbf{W}]] = \Lambda$ by $\lambda_*(g, w) = \lambda(g)[w]$ for the group element $[w]$ in Λ for $w \in \mathbf{W}$ and $g \in \mathbf{G}_{\mathrm{tor}}$. We consider two anti-cyclotomic characters of $\mathbf{G}$ given by $\lambda_- = \lambda(\lambda^c)^{-1}$ and $\alpha = \lambda_*(\lambda_*^c)^{-1}$, where $\lambda^c(\sigma) = \lambda(c\sigma c^{-1})$ and $\lambda_*^c(\sigma) = \lambda_*(c\sigma c^{-1})$. Let $M^-(\lambda_-)$ be the subfield of M_∞ fixed by $\mathrm{Ker}(\alpha)$. Let $\mathbf{M}_S(\lambda_-)/M^-(\lambda_-)$ be the maximal p-abelian extension unramified outside S. Naturally $X_S = \mathrm{Gal}(\mathbf{M}_S(\lambda_-)/M^-(\lambda_-))$ is a continuous module over $\mathbf{Z}_p[[\mathbf{H}]]$ of $\mathbf{H} = \mathrm{Im}(\alpha)$. We consider the λ_--branch of X_S defined by

$$X_S(\lambda_-) = X_S \otimes_{\mathbf{Z}_p[\mathbf{G}_{\mathrm{tor}}]} \mathfrak{O}(\lambda^-)$$

where $\mathfrak{O}(\lambda_-)$ is the $\mathfrak{O}$-free module of rank one on which $\mathbf{G}_{\mathrm{tor}}$ acts via λ_-. Once we are given a p-adic CM-type S, we have the following 3 objects as in the imaginary quadratic case:

(i) The λ_--branch of the projection L^- of the Katz p-adic L-function $L \in \mathfrak{O}_\Omega[[\mathbf{G}]]$ to the anti-cyclotomic tower $M^-(\lambda_-)$;

(ii) The congruence power series $H \in \Lambda$ attached to the λ-branch of the nearly ordinary Hecke algebra of CM-type S;

(iii) The characteristic power series Iw^- of $X_S(\lambda_-)$ in Λ.

Note that $\mathrm{Ker}(\alpha)$ contains $\mathbf{G}_+ = \{x \in \mathbf{G} \mid cxc^{-1} = x\}$ and we can realize the quotient $\mathbf{G}/\mathbf{G}_+$ inside $\mathbf{G}$ by the subgroup of commutators $[x, c] = xcx^{-1}c^{-1}$.

Especially the maximal torsion-free quotient $\mathbf{W}^-$ of $\mathbf{H}$ can be thought of a direct factor of $\mathbf{W}$ via this map. For a technical reason (namely, H resides in Λ), we regard L^- and Iw^- as elements in Λ via this inclusion although they belong to $\Lambda_- = \mathfrak{O}[[\mathbf{W}^-]]$. Moreover, to have a non-zero Iw^-, we need to suppose a weak version of the Leopoldt conjecture (depending on S) for the anti-cyclotomic tower. This weak form of Leopoldt's conjecture holds true if the CM field M is abelian over $\mathbf{Q}$. On the other hand, one can prove unconditionally (i.e. without supposing the weak Leopoldt conjecture) the non-vanishing of the characteristic power series Iw of the maximal S-ramified abelian extension over the full $\mathbf{Z}_p^r$-tower of M. Before giving the precise definition of L^- and H, we state the first theorem:

Theorem 2.1 L^- divides H in $\mathfrak{O}_\Omega[[\mathbf{W}]] \otimes_{\mathbf{Z}} \mathbf{Q}$. Moreover if the μ-invariant of every branch of the Katz p-adic L-function of M vanishes, then we have the strong divisibility:

$$(h(M)/h(F))L^- \mid H \text{ in } \mathfrak{O}_\Omega[[\mathbf{W}]].$$

The following conjecture is obviously motivated by (1.1):

Conjecture 2.2 $H = (h(M)/h(F))L^-$ up to a unit in $\mathfrak{O}_\Omega[[\mathbf{W}]]$ if $p > 2$, where $h(M)$ (resp. $h(F)$) is the class number of M (resp. F).

This conjecture is known to be true if $F = \mathbf{Q}$, $p \geq 5$ and the class number $h(M)$ of M is prime to p under a certain branch condition.

First, let us explain the definition of L^-. Although we will not make the identification with the power series ring due to the lack of canonical coordinates of $\mathbf{W}$, we may regard any element of Λ as a p-adic analytic function of several variables. There are two different ways of viewing $\Phi \in \Lambda$ as an analytic object: For $G = \mathbf{G}$ or $\mathbf{W}$, let $\mathfrak{X}(G)$ be the set of all continuous characters of G with values in $\overline{\mathbf{Q}}_p$. If one fixes a $\mathbf{Z}_p$-basis (w_i) of $\mathbf{W}$, then each character $P \in \mathfrak{X}(\mathbf{W})$ is determined by its value $(P(w_i)) \in D^r$, where $D = \{x \in \overline{\mathbf{Q}}_p \mid |x-1|_p < 1\}$. Thus $\mathfrak{X}(\mathbf{W}) \cong D^r$. Each character $P : G \to \overline{\mathbf{Q}}_p^\times$ induces an $\mathfrak{O}$-algebra homomorphism $P : \mathfrak{O}[[\mathbf{G}]] \to \overline{\mathbf{Q}}_p$ such that $P|_G$ is the original character of G. In this way, we get an isomorphism:

$$\mathfrak{X}(G) \cong \mathrm{Spec}(\mathfrak{O}[[\mathbf{G}]])(\overline{\mathbf{Q}}_p) = \mathrm{Hom}_{\mathfrak{O}\text{-alg}}(\mathfrak{O}[[\mathbf{G}]], \overline{\mathbf{Q}}_p).$$

Then

(A1) Φ is an analytic function on $\mathfrak{X}(G)$ whose value at P is $P(\Phi) \in \overline{\mathbf{Q}}_p$.

On the other hand, we can view Λ as a space of measures on G in the sense of Mazur so that

$$\int_G P(g)\, d\Phi(g) = P(\Phi) = \Phi(P). \tag{A2}$$

By class field theory, we can identify, via the Artin symbol, the group $\mathbf{G}$ with the quotient of the idele group $M_A^\times$. For a given A_0-type Hecke character $\varphi : M_A^\times/M^\times \to \mathbf{C}^\times$ of p-power conductor whose infinity type is given by

$$\varphi(x_\infty) = x_\infty^{-\xi} = \Pi_{\sigma\in\Sigma\cup\rho\Sigma}(x_\infty^\sigma)^{-\xi_\sigma} \text{ for } \xi = (\xi_\sigma)_{\sigma\in\Sigma\cup\rho\Sigma} \in \mathbf{Z}^{\Sigma\cup\rho\Sigma},$$

as shown by A. Weil in 1955, φ has values in $\overline{\mathbf{Q}}$ on finite ideles and we have a unique p-adic avatar $\hat{\varphi} : \mathbf{G} \to \overline{\mathbf{Q}}_p^\times$ which satisfies $\hat{\varphi}(x) = \varphi(x)$ if $x_p = x_\infty = 1$, and if $x_p \in M_p^\times$ is close enough to 1, then

$$\hat{\varphi}(x_p) = x_p^{-\xi} = \Pi_{\sigma\in\Sigma\cup\rho\Sigma}(x_p^\sigma)^{-\xi_\sigma}.$$

In 1978, Katz showed in [K] the existence of a unique p-adic L-function given by an element L of $\mathfrak{O}_\Omega[[\mathbf{G}]]$ such that

$$\frac{L(\hat{\varphi})}{\text{suitable } p\text{-adic period}} = c(\varphi)\frac{L(0,\varphi)}{\text{suitable complex period}}$$

whenever φ is critical at 0 (i.e. if either $\xi_{\sigma\rho} \geq \xi_\sigma + \xi_{\sigma\rho} + 1 \geq 0$ or $\xi_\sigma + 1 \leq \xi_\sigma + \xi_{\sigma\rho} + 1 \leq 0$ for all $\sigma \in \Sigma$). Here, $c(\varphi)$ is a simple constant including a modifying Euler p-factor, local Gauss sum, Γ-factor and a power of π. See [K, (5.3.0), (5.7.8-9)] for details. To define L^-, we first project Katz's L to Λ. Namely, we fix once and for all a finite order character $\lambda : \mathbf{G}_{\text{tor}} \to \mathfrak{O}^\times$. Then we have a continuous character $\lambda_* : \mathbf{G} = \mathbf{G}_{\text{tor}} \times \mathbf{W} \to \mathfrak{O}[[\mathbf{W}]]$ given by

$$\lambda_*(g,w) = \lambda(g)w \in \mathfrak{O}[[\mathbf{W}]],$$

where we consider $\lambda(g)$ for $g \in \mathbf{G}_{\text{tor}}$ as a scalar in $\mathfrak{O}$ but w as a group element in $\mathbf{W}$. This character induces the projection to Λ

$$\lambda_* : \mathfrak{O}[[\mathbf{G}]] = \Lambda[\mathbf{G}_{\text{tor}}] \to \Lambda.$$

Then for any point $P \in \mathfrak{X}(\mathbf{W})$, $\lambda_P = P \circ \lambda_* : \mathbf{G} \to \overline{\mathbf{Q}}_p$ is a p-adic character of $\mathbf{G}$. When λ_P is the avatar of an A_0-type Hecke character, we say that P is arithmetic (this notion of arithmeticity is independent of the choice of λ). Let c denote the complex conjugation in $\mathrm{Gal}(\overline{\mathbf{Q}}/F)$ and write $\lambda^c(x) = \lambda(cxc^{-1})$. We then consider the anti-cyclotomic character α attached to λ_* given by

$$\alpha(x) = \lambda_*^{-1}\lambda_*^c(x) = \lambda_*(cxc^{-1}x^{-1})$$

and the corresponding Λ-algebra homomorphism

$$\alpha_* : \mathfrak{O}[[\mathbf{G}]] \to \mathfrak{O}[[\mathbf{W}]].$$

This α_* actually has values in the anti-cyclotomic part $\mathfrak{O}[[\mathbf{W}^-]]$, where

$$\mathbf{W}^- = \{w \in \mathbf{W} \mid w^c = cwc^{-1} = w^{-1}\}.$$

Then we define

$$L^- = \alpha_*(L) \in \mathfrak{O}_\Omega[[\mathbf{W}^-]].$$

Although the divisibility of Theorem 2.1 is stated as taking place in the bigger ring $\mathfrak{O}[[\mathbf{W}]] \supset \mathfrak{O}[[\mathbf{W}^-]]$, actually the congruence power series H itself also falls in the subring $\mathfrak{O}[[\mathbf{W}^-]]$. However we will know this fact *after* proving the second divisibility: $H \mid (h(M)/h(F))Iw^-$ and we do not know this fact *a priori.* Thus we continue to formulate our result using $\mathfrak{O}_\Omega[[\mathbf{W}]]$ as the base ring. This power series L^- satisfies the following interpolation property:

$$\frac{L^-(P)}{p\text{-adic period}} = c(\lambda_P^c\lambda_P^{-1})\frac{L(0,\lambda_P^c\lambda_P^{-1})}{\text{complex period}}$$

whenever P is arithmetic and $\lambda_P^c\lambda_P^{-1}$ is critical at P.

We now define the p-adic Hecke algebra and the congruence power series and then give a sketch of the proof of the theorem. To define Hecke algebra, we explain first a few things about Hilbert modular forms. Let I be the set of all field embeddings of F into $\overline{\mathbf{Q}}$. The weight $k = (k_\sigma)_{\sigma\in I}$ of a modular form will be an element of the free $\mathbf{Z}$-module $\mathbf{Z}[I]$ generated by elements of I. Actually, our holomorphic modular forms have double digit weight $(k,v) \in \mathbf{Z}[I]^2$ associated to the following automorphic factor:

$$J_{k,v}\left(\begin{pmatrix} a & b \\ c & d\end{pmatrix}, z\right) = \Pi_{\sigma\in I}\{\det(\gamma_\sigma)^{v_\sigma-1}(c^\sigma z_\sigma + d^\sigma)^{k_\sigma}\},$$

where $\gamma = \begin{pmatrix} a & b \\ c & d\end{pmatrix} \in GL_2(F_\infty)$ ($F_\infty = F\otimes_{\mathbf{Q}}\mathbf{R} = \mathbf{R}^I$) with totally positive determinant and $z = (z_\sigma)_{\sigma\in I} \in \mathfrak{H}^I$ is a variable on the product of copies of upper half complex planes $\mathfrak{H}^I$ indexed by I. For each open compact subgroup V of the finite part of the adele group $GL_2(F_A)$, let $S_{k,v}(V)$ be the space of holomorphic cusp forms f of weight (k,v) defined in [H1, §2]. Namely f is a function on $GL_2(F_A)$ satisfying

$$f(\alpha x u) = f(x)J_{k,v}(u_\infty, z_0)^{-1} \quad\text{for}\quad \alpha\in GL_2(F) \quad\text{and}\quad u \in V\times C,$$

where C is the stabilizer of $z_0 = (\sqrt{-1},\dots,\sqrt{-1}) \in \mathfrak{H}^I$ in $GL_2(F_\infty)$, which is isomorphic to the product of the center ($\cong (\mathbf{R}^\times)^I$) of $GL_2(F_\infty)$ and $SO_2(\mathbf{R})^I$.

We can associate to f and each finite idele $t \in GL_2(F_{A_f})$, a function f_x on $\mathfrak{H}^I$ by

$$f_t(z) = f(t\begin{pmatrix} y & x \\ 0 & 1 \end{pmatrix})J_{k,v}(\begin{pmatrix} y & x \\ 0 & 1 \end{pmatrix}, z_0).$$

It is easy to check that f_t satisfies the automorphic condition:

$$f_t(\gamma(z)) = f_t(z)J_{k,v}(\gamma, z) \quad \text{for} \quad \gamma \in \Gamma_t = t^{-1}VtGL_2^+(F_\infty) \cap GL_2(F),$$

where $GL_2^+(F_\infty)$ is the connected component of $GL_2(F_\infty)$ with identity. Similarly we write $F_{\infty+}^\times$ for the connected component with identity of $F_\infty^\times$. Then we suppose for $f \in S_{k,v}(V)$ that, for all $t \in GL_2(F_{A_f})$,

(i) f_t is holomorphic on $\mathfrak{H}^I$ (holomorphy),
(ii) $f_t(z)$ has the following Fourier expansion:

$$\Sigma_{\xi \in F}\ c(\xi, f_t)\exp(2\pi i \mathrm{Tr}(\xi z))$$

with $c(\xi, f_t) = 0$ unless $\xi^\sigma > 0$ for all $\sigma \in I$ (cuspidality).

Let D be the relative discriminant of M/F and let $\mathfrak{r}$ and $\mathfrak{R}$ be the integer ring of F and M, respectively. As the open compact subgroup V, we take the group V_α given by

$$\{\begin{pmatrix} a & b \\ c & d \end{pmatrix} \in GL_2(\hat{\mathfrak{r}}) \mid c \in Dp^\alpha\hat{\mathfrak{r}}, \quad a \equiv 1 \bmod p^\alpha\hat{\mathfrak{r}}, \quad d \equiv 1 \bmod Dp^\alpha\hat{\mathfrak{r}}\},$$

where $\mathfrak{r}$ is the integer ring of F and $\hat{\mathfrak{r}} = \varprojlim_N \mathfrak{r}/N\mathfrak{r}$ is the product of $\mathfrak{l}$-adic completion of $\mathfrak{r}$ over all primes $\mathfrak{l}$. Let $\chi : F_A^\times/F^\times \rightarrow \mathbf{Z}_p^\times$ be the cyclotomic character. If $c(\xi, f_t) \in \overline{\mathbf{Q}}$ for all $t \in GL_2(F_{A_f})$, we can associate to each f as above the following p-adic q-expansion (cf. [H4, §1]):

$$f(y) = \Sigma_{0 \ll \xi \in F}\ a(\xi y d, f)q^\xi \quad \text{with} \quad a(\xi y d, f) \in \overline{\mathbf{Q}}_p,$$

where d is any differental idele of F (i.e., its ideal is the different of $F/\mathbf{Q}$) and $y \mapsto a(y, f)$ is a function on finite ideles, vanishing outside integral ideles, given by

$$a(y, f) = c(\xi, f_t)y_p^{-v}\xi^v\chi(\det(t)) \quad \text{for} \quad t = \begin{pmatrix} a & 0 \\ 0 & 1 \end{pmatrix}$$
$$\text{with} \quad y \in \xi a d(V_\alpha \cap F_{A_f}^\times)F_{\infty+}^\times.$$

Out of this q-expansion, we can recover the Fourier expansion of f:

$$f\left(\begin{pmatrix} y & x \\ 0 & 1 \end{pmatrix}\right) = |y|_A\{\Sigma_{0 \ll \xi \in F}\ a(\xi y d, f)(\xi d y)_p^v(\xi y_\infty)^{-v}\mathbf{e}_F(i\xi y_\infty)\mathbf{e}_F(\xi x)\}.$$

Here note that $a(\xi yd, f)(\xi dy)_p^v$ is an algebraic number which is considered to be a complex number via the fixed embedding of $\overline{\mathbf{Q}}$ into $\mathbf{C}$. When $a(y,f)$ is algebraic for all y with $y_p = 1$, f is called algebraic (this is equivalent to asking that $c(\xi, f_t)$ are algebraic for all t). We consider the union $S(\overline{\mathbf{Q}})$ of all algebraic forms of all weight (k,v) inside the space of formal q-expansions. Then putting a p-adic uniform norm

$$|f|_p = \mathrm{Sup}_y |a(y,f)|_p$$

on $S(\overline{\mathbf{Q}})$, we define the space S of p-adic modular forms by the completion of $S(\overline{\mathbf{Q}})$ under the norm $|\ \ |_p$.

Now we define the Hecke operators. For each $x \in F_A^\times$ with $x_\infty = 1$, we can define the Hecke operator $T(x) = T_\alpha(x)$ acting on $S_{k,v}(V_\alpha)$ as follows: First take the double coset $V_\alpha \begin{pmatrix} x & 0 \\ 0 & 1 \end{pmatrix} V_\alpha$ and decompose it into a disjoint union of finite right cosets $\cup_i x_i V_\alpha$. Then we define $T_\alpha(x)$ by

$$f \mid T_\alpha(x)(g) = \Sigma_i f(gx_i).$$

Since we have taken the average of right translation of f on a double coset, we can check easily that $T_\alpha(x)$ is a linear operator acting on $S_{k,v}(V_\alpha)$. Especially the action of $T(u)$ for $u \in \mathfrak{r}_p^\times$ factors through $(\mathfrak{r}/p^\alpha\mathfrak{r})^\times$. Similarly, the center $F_A^\times$ acts on $S_{k,v}(V_\alpha)$ so that $f \mid z(g) = f(gz)$. This action factors through $Z = F_A^\times / \overline{F^\times U(D)^{(p)} F_\infty^\times}$ for

$$U(D)^{(p)} = \{u \in \hat{\mathfrak{r}}^\times \mid u \equiv 1 \bmod D\hat{\mathfrak{r}} \text{ and } u_p = 1\}.$$

Thus $S_{k,v}(V_\alpha)$ has an action of the group $G = Z \times \mathfrak{r}_p^\times$ and Hecke operators $T(x)$. The group G is a profinite group and we can decompose

$$G = G_{\mathrm{tor}} \times W$$

so that $W \cong \mathbf{Z}_p^{[F:\mathbf{Q}]+1+\delta}$ and G_{tor} is a finite group. Since $M_A \supset F_A$, we have a natural homomorphism of Z into $\mathbf{G}$. On the other hand, by our choice of p-adic CM-type, we can identify $\mathfrak{r}_p = \mathfrak{r} \otimes_{\mathbf{Z}} \mathbf{Z}_p$ with $\mathfrak{R}_S = \Pi_{\mathfrak{P} \in S} \mathfrak{R}_{\mathfrak{P}}$. This identification gives an injection of $\mathfrak{r}_p^\times$ into $M_A^\times$ and yields a homomorphism of $\mathfrak{r}_p^\times$ into $\mathbf{G}$. Thus we have natural morphisms:

$$\iota : G = Z \times \mathfrak{r}_p^\times \to \mathbf{G} \quad \text{and} \quad \iota_* : \mathfrak{O}[[G]] \to \mathfrak{O}[[\mathbf{G}]].$$

We can easily check that ι takes W into a subgroup of finite index of $\mathbf{W}$ and ι_* is an $\mathfrak{O}[[\mathbf{W}]]$-algebra homomorphism.

We take the Galois closure Φ of F in $\overline{\mathbf{Q}}$ and let $\mathfrak{W}$ be the valuation ring of Φ corresponding to the embedding Φ into $\overline{\mathbf{Q}}_p$. We pick an element $\varpi_\mathfrak{p}$ for each prime factor $\mathfrak{p}$ of p in F such that $\varpi_\mathfrak{p}\mathfrak{r} = \mathfrak{p}\mathfrak{a}$ for an ideal $\mathfrak{a}$ prime to p. We consider $\varpi_\mathfrak{p}$ as a prime element in $F_\mathfrak{p}$. Then the p-adic Hecke algebra $\mathbf{h}_{k,v}(Dp^\alpha; \mathfrak{W})$ with coefficients in $\mathfrak{W}$ is by definition the $\mathfrak{W}$-subalgebra of $\mathrm{End}_\mathfrak{W}(S_{k,v}(V_\alpha))$ generated by

(a) Hecke operators $T(x)$ for all $x \in \hat{\mathfrak{r}} \cap F_{A_f}^\times$,
(b) the Hecke operator $\varpi_\mathfrak{p}^{-v}T(\varpi_\mathfrak{p})(\varpi_\mathfrak{p} \in \mathfrak{r}_\mathfrak{p})$ for all $\mathfrak{p} \mid p$,
(c) the action of the group $G = Z \times \mathfrak{r}_p^\times$.

It is well known that $\mathbf{h}_{k,v}(Dp^\alpha; \mathfrak{W})$ is free of finite rank over $\mathfrak{W}$ (cf [H1, Th.3.1]). Especially $T(\varpi_\mathfrak{p})$ is divisible by $\varpi_\mathfrak{p}^v = \Pi_{\sigma \in I}\varpi_\mathfrak{p}^{\sigma v_\sigma}$. For each extension K of $\mathbf{Q}_p$ containing Φ, let $\mathfrak{O}$ be the p-adic integer ring of K. Then the p-adic Hecke algebra of level Dp^α is defined by

$$\mathbf{h}_{k,v}(Dp^\alpha; \mathfrak{O}) = \mathbf{h}_{k,v}(Dp^\alpha; \mathfrak{W}) \otimes_\mathfrak{W} \mathfrak{O}.$$

By definition, the restriction of $T_\beta(x)$ acting on $S_{k,v}(V_\beta)$ to $S_{k,v}(V_\alpha)$ for $\beta > \alpha > 0$ coincides $T_\alpha(x)$. Thus the restriction induces a surjective $\mathfrak{O}$-algebra homomorphism:

$$\mathbf{h}_{k,v}(Dp^\beta; \mathfrak{O}) \to \mathbf{h}_{k,v}(Dp^\alpha; \mathfrak{O})$$

which takes $T_\beta(x)$ to $T_\alpha(x)$. Thus we can take the projective limit

$$\mathbf{h}_{k,v}(Dp^\infty; \mathfrak{O}) = \varprojlim_\alpha \mathbf{h}_{k,v}(Dp^\alpha; \mathfrak{O}),$$

which is naturally an algebra over the continuous group algebra $\mathfrak{O}[[G]]$. For each α, we can decompose

$$\mathbf{h}_{k,v}(Dp^\alpha; \mathfrak{O}) = \mathbf{h}_{k,v}^{n.ord}(Dp^\alpha; \mathfrak{O}) \times \mathbf{h}_{k,v}^{s}(Dp^\alpha; \mathfrak{O})$$

so that $p^{-v}T(p)$ is a unit in $\mathbf{h}_{k,v}^{n.ord}(Dp^\alpha; \mathfrak{O})$ and is topologically nilpotent in $\mathbf{h}_{k,v}^{s}(Dp^\alpha; \mathfrak{O})$. Then basic known facts are (see [H2]):

(H1) The pair $(\mathbf{h}_{k,v}(Dp^\infty; \mathfrak{O}), x_p^{-v}T(x))$ is independent of (k, v) if $k \geq 2t$, where $t = \Sigma_{\sigma \in I}\sigma$.

In fact, the union $S_{k,v}(V_\infty) = \cup_\alpha S_{k,v}(V_\alpha; \overline{\mathbf{Q}})$ of all algebraic modular forms of weight (k, v) is dense in S and thus the algebra $\mathbf{h}_{k,v}(Dp^\infty; \mathfrak{O})$ can be considered as a subalgebra of $\mathrm{End}(S)$ topologically generated by $x_p^{-v}T(x)$ and it is independent of (k, v). Now we can remove the suffix (k, v) from notation of the Hecke algebra and we write $\mathbf{h}(D; \mathfrak{O})$ (resp. $\mathbf{h} = \mathbf{h}^{n.ord}(D; \mathfrak{O})$)

for $\mathbf{h}_{k,v}(Dp^\infty;\mathfrak{O})$ (resp. $\mathbf{h}^{n.\mathrm{ord}}_{k,v}(Dp^\infty;\mathfrak{O})$). In other words, there is a universal Hecke operator $\mathbf{T}(x) \in \mathbf{h}(D;\mathfrak{O})$ which is sent to $x_p^{-v}T(x)$ under the isomorphism: $\mathbf{h}(D;\mathfrak{O}) \cong \mathbf{h}_{k,v}(Dp^\infty;\mathfrak{O})$.

(H2) $\mathbf{h}$ is of finite type and torsion-free as $\mathfrak{O}[[W]]$-module.

(H3) There exists an $\mathfrak{O}[[W]]$-algebra homomorphism $\theta^* : \mathbf{h} \to \mathfrak{O}[[\mathbf{G}]]$ such that for primes outside Dp

$$\theta^*(T(\mathfrak{q})) = \begin{cases} [\mathfrak{Q}] + [\mathfrak{Q}^c] & \text{if } \mathfrak{q} = \mathfrak{Q}\mathfrak{Q}^c, \\ 0 & \text{if } \mathfrak{q} \text{ remains prime in } M \end{cases}$$

where $[\mathfrak{Q}]$ is the image of the prime ideal $\mathfrak{Q}$ under the Artin symbol.

This statement is just an interpretation of the existence of theta series $\theta(\varphi)$ for each A_0-type Hecke character φ of $\mathbf{G}$ characterized by

$$\theta(\varphi) \mid T(\mathfrak{q}) = \begin{cases} (\varphi(\mathfrak{Q}) + \varphi(\mathfrak{Q}^c))\theta(\varphi) & \text{if } \mathfrak{q} = \mathfrak{Q}\mathfrak{Q}^c, \\ 0 & \text{if } \mathfrak{q} \text{ remains prime in } M. \end{cases}$$

By (H3), we may consider the composite $\lambda_* \circ \theta^* : \mathbf{h} \to \Lambda$.

(H4) After tensoring the quotient field $\mathbf{L}$ of Λ over $\Lambda_0 = \mathfrak{O}[[\mathbf{W}]]$, we have a Λ_0-algebra decomposition

$$\mathbf{h} \otimes_{\Lambda_0} \mathbf{L} \cong \mathbf{L} \oplus \mathbf{B} \quad \text{for a complementary summand } \mathbf{B},$$

where the projection to the first factor is given by $\lambda_* \circ \theta^*$.

Then the congruence module of λ_* is defined by

(H5) $\mathfrak{C}(\lambda_*;\Lambda) = \Lambda/(\mathbf{h} \otimes_{\Lambda_0} \Lambda \cap \mathbf{L})$.

The congruence power series H is then defined by the characteristic power series of $\mathfrak{C}(\lambda_*;\Lambda)$. By definition, the principal ideal $H\Lambda$ is the reflexive closure of the ideal $\mathbf{h} \otimes_{\Lambda_0} \Lambda \cap \mathbf{L}$ in Λ.

3. We now give a sketch of the proof of Theorem 2.1. The idea of the proof is the comparison of two p-adic interpolations of Hecke L-functions of M. One is Katz's way and the other is the p-adic L-function attached to the Rankin product L-function of $\theta(\lambda_P)$ and $\theta(\mu_Q)$. Here μ is another character of $\mathbf{G}_{\mathrm{tor}}$ and we extend it to a character $\mu_* : \mathbf{G} \to \Lambda^\times$ similarly to λ_*. In fact, we can show by the method of p-adic Rankin convolution ([H4, Theorem I]) that there exists a power series Δ in $\mathfrak{O}[[\mathbf{W} \times \mathbf{W}]]$ such that

$$\frac{\Delta(P,Q)}{H(P)} = c(P,Q)\frac{D(1 + \frac{m(Q)-m(P)}{2}, \theta(\lambda_P), \theta(\mu_Q \circ c))}{(\theta(\lambda_P), \theta(\lambda_P))}$$

whenever both P and Q are arithmetic and both $\lambda_P^{-1}\mu_Q$ and $\lambda_P^{-1}(\mu_Q^c)$ are critical. Here $D(s,\theta(\lambda_P),\theta(\mu_Q)^c)$ is the Rankin product of $\theta(\lambda_P)$ and $\theta(\mu_Q)^c$, i.e., the standard L-function for $GL(2)\times GL(2)$ attached to the tensor product of automorphic representations spanned by $\theta(\lambda_P)$ and $\theta(\mu_Q)^c$; $(\theta(\lambda_P),\theta(\lambda_P))$ is the self Petersson inner product of $\theta(\lambda_P)$ and $c(P,Q)$ is a simple constant including the modifying Euler p-factor, Gauss sums, Γ-factors and a power of π. The integer $m(P)$ is given as follows: Write the infinity type of λ_P as ξ and $m(P)=\xi_\sigma+\xi_{\sigma\rho}$ for $\sigma\in\Sigma$ (this value is independent of σ). Similarly $m(Q)$ is defined for μ_Q. Now looking at the Euler product of D and the functional equation of Hecke L-functions, we see

$$D(1+\frac{m(Q)-m(P)}{2},\theta(\lambda_P),\theta(\mu_Q^c))\approx L(0,\lambda_P^{-1}\mu_Q)L(0,\lambda_P^{-1}\mu_Q^c).$$

It is also well known that, with a simple constant $c(P)$ similar to $c(P,Q)$,

$$(\theta(\lambda_P),\theta(\lambda_P))=c(P)(2^{1-[F:\mathbf{Q}]}h(M)/h(F))L(0,\lambda_P^c\lambda_P^{-1}).$$

Modifying the Katz measure L in $\mathfrak{O}[[\mathbf{G}]]$, we can find two power series L' and L'' in $\mathfrak{O}[[\mathbf{W}\times\mathbf{W}]]$ interpolating $L(0,\lambda_P^{-1}\mu_Q)$ and $L(0,\lambda_P^{-1}\mu_Q^c)$, respectively. Then out of the above formulas, we get the following identity:

$$\frac{\Delta}{H}=U\frac{L'L''}{2^{1-[F:\mathbf{Q}]}h(M)/h(F))L^-}\quad\text{in }\mathfrak{O}[[\mathbf{W}\times\mathbf{W}]],$$

where U is a unit in $\mathfrak{O}[[\mathbf{W}\times\mathbf{W}]]$. Thus if L' and L'' are prime to L^- in $\mathfrak{O}[[\mathbf{W}\times\mathbf{W}]]\otimes_{\mathbf{Z}}\mathbf{Q}$, we get the desired divisibility. Almost immediately from the construction of L' and L'', we know that for any character $P\in\mathfrak{X}(\mathbf{W})$ the half specialized power series $L'_P(X)=L'(P,X)$ and $L''_P(X)$ in $\mathfrak{O}[[\mathbf{W}]]$ have their μ-invariants *independent* of P, equal to the μ-invariant of the Katz measure along the irreducible component of $\lambda^{-1}\mu$ and $\lambda^{-1}\mu^c$. If (a characteristic 0) prime factor $\mathbf{P}(Y)$ (in $\mathfrak{O}_\Omega[[\mathbf{W}]]$) of $L^-(Y)$ divides L', then by letting P approach to a zero of $\mathbf{P}$, we observe that the μ-invariant of L'_P goes to infinity, which contradicts the constancy of the μ-invariant of L'_P. Thus L^- is prime to $L'L''$ in $\mathfrak{O}[[\mathbf{W}\times\mathbf{W}]]\otimes_{\mathbf{Z}}\mathbf{Q}$, which shows the desired assertion. Especially if the μ-invariant of the Katz measure vanishes, then we know the strong divisibility as in the theorem.

4. Now we explain briefly how one can show the other divisibility: $H\mid Iw^-$ by using Mazur's theory of deformation of Galois representations. We keep the notations and assumptions introduced above. In particular, we assume the ordinarity hypothesis and fix a p-adic CM-type S. To the pair (S,λ), where λ is a given character of $\mathbf{G}_{\text{tor}}$, we attached a congruence module, with characteristic power series H. On the other hand, let M_∞ be the maximal abelian

extension of M unramified outside p of M; so, we have $\mathbf{G} = \mathrm{Gal}(M_\infty/M)$. We have defined a character $\lambda_* : \mathbf{G} \to \Lambda^\times$ for $\Lambda = \mathfrak{O}[[\mathbf{W}]]$ for a fixed finite order character $\lambda : \mathbf{G}_{\mathrm{tor}} \to \mathfrak{O}^\times$. In fact, on $\mathbf{G}_{\mathrm{tor}}$, λ_* coincides with λ and on $\mathbf{W}$, it is the tautological inclusion of $\mathbf{W}$ into Λ. Define the '$-$' part of λ_*, which we write as $\alpha = \alpha_\lambda$, by

$$\alpha = \lambda_*(\lambda_*^c)^{-1} \quad \text{for } \lambda_*^c(\sigma) = \lambda_*(c\sigma c^{-1}).$$

Let $M^- = M^-(\lambda_-)$ be the fixed part of M_∞ by $\mathrm{Ker}(\alpha)$, which contains $\mathbf{G}^+ = \{\sigma \in \mathbf{G} \mid c(\sigma) = \sigma\}$. We write $\mathbf{H} = \mathrm{Gal}(M^-(\lambda_-)/M) \cong \mathrm{Im}(\alpha)$. Let $\mathbf{M}_S(\lambda_-)$ be the maximal p-abelian extension of M^- unramified outside S. One can prove, under a weak Leopoldt type assumption for the extension $M^-(\lambda_-)/M$ (for details of this assumption, see our forthcoming paper), that $X_S = \mathrm{Gal}(\mathbf{M}_S(\lambda_-)/M^-)$ is torsion over $\mathbb{Z}_p[[\mathbf{H}]]$. The character $\lambda^- = \lambda/\lambda^c$: $\mathbf{G}_{\mathrm{tor}} \to \mathfrak{O}^\times$ factors through the torsion part $\mathbf{H}_{\mathrm{tor}}$ of $\mathbf{H}$ and the characteristic power series of the λ^--part $X_S(\lambda^-) = X_S \otimes_{\mathbb{Z}_p[\mathbf{H}_{\mathrm{tor}}]} \mathfrak{O}(\lambda^-)$ of X_S is nothing but Iw^-. Then the precise result, we can obtain at this date is as follows:

Theorem 4.1 (i) If $[F:\mathbb{Q}] > 1$, H divides $(h(M)/h(F))Iw^-$ in $\mathfrak{O}[[\mathbf{W}]]$.
(ii) If M is imaginary quadratic, then H divides $h(M)Iw^-$ in $\mathfrak{O}[[\mathbf{W}]]$ unless $\lambda^- = \lambda/\lambda^c$ restricted to the decomposition group D of $\mathfrak{P}$ in $\mathbf{G}$ is congruent to 1 modulo the maximal ideal $\pi\mathfrak{O}$ of $\mathfrak{O}$. In this exceptional case, we need to exclude the trivial zero, i.e., the divisibility holds in $\mathfrak{O}[[\mathbf{W}]][\frac{1}{P_\lambda}]$, where P_λ is a generator of the unique height one prime ideal corresponding to the character $\tau_\lambda : \mathbf{H} \to \mathfrak{O}^\times$ such that $\tau_\lambda(D) = 1$ and $\tau_\lambda \mid \mathbf{H}_{\mathrm{tor}} = \lambda^-$.

Comments (a) By a base change argument in Iwasawa theory, one can probably include the 'trivial zero' P_λ. Nevertheless, the argument possibly needs a sort of multiplicity one result for 'trivial zeros' of the Katz–Yager p-adic L-function which needs to be verified.

(b) The reason why things become easier when $[F:\mathbb{Q}] > 1$ is contained in the following easy lemma. To state the lemma, let us recall the character $\lambda_* : G \to \mathfrak{O}[[\mathbf{W}]]$ given by

$$\lambda_*(g, w) = \lambda(g)w \in \mathfrak{O}[[\mathbf{W}]] \quad \text{for } g \in \mathbf{G}_{\mathrm{tor}} \text{ and } w \in \mathbf{W} \text{ in §2.}$$

Lemma 4.2 (i) If $[F:\mathbb{Q}] > 1$, the ideal $\mathfrak{I}$ generated by the values $\lambda_*(\sigma) - \lambda_*(c(\sigma))$, σ running over the decomposition group $D_\mathfrak{p}$ at $\mathfrak{p}$ in $\mathbf{G}$ is of height greater than 1, i.e., is not contained in any prime of height one in $\Lambda = \mathfrak{O}[[\mathbf{W}]]$.
(ii) If $F = \mathbb{Q}$, this ideal is contained in $P_\lambda\Lambda$.

The outline of the proof of Theorem 4.1 runs as follows. Let us fix a prime P of height one such that the restrictions of λ and λ^c to $D_{\mathfrak{p}}$ are not congruent modulo P for all prime $\mathfrak{p}$ in F over p (by Lemma 4.2, this gives no restriction when $[F:\mathbb{Q}] \geq 2$). We consider the complete discrete valuation ring Λ_P with residue field $k(P)$ and look at the residual representation

$$\overline{\rho}_0 : \mathrm{Gal}(\overline{\mathbb{Q}}/F) \to GL_2(k(P))$$

given as the reduction modulo P of the induced representation ρ_0 of $\lambda_* : \mathrm{Gal}(\overline{\mathbb{Q}}/M) \to \Lambda^\times$. By the choice of P, $\lambda \not\equiv \lambda^c \bmod P$, hence $\overline{\rho}_0$ is irreducible. The main step in the proof of Theorem 4.1 is to relate $X_S(\lambda^-) \otimes_\Lambda \Lambda_P$ to a module of Kähler differentials attached to some deformation problem of $\overline{\rho}_0$ over Λ_P. Since $k(P)$ is not a finite field, the study of this deformation problem, though very similar to the one made by B. Mazur in [M], is slightly trickier. To define this problem, we need to introduce some notations. First, let N be the ray class field of M of conductor p (one has of course N inside M_∞) and $N^{(p)}$ be the maximal p-extension of N unramified outside p. It is clear that $\overline{\rho}_0$ restricted to $\mathrm{Gal}(\overline{\mathbb{Q}}/N)$ factors through $\Pi_N = \mathrm{Gal}(N^{(p)}/N)$, so that, for the deformation problem of $\overline{\rho}_0$, we can restrict ourselves to representations ρ of $\Pi = \mathrm{Gal}(N^{(p)}/F)$. The great advantage of such a limitation in the choice of ρ's is that Π is topologically of finite type (Π_N is a pro-p-group and its Frattini quotient is finite by Kummer theory over N). Now, let Art be the category of local artinian Λ_P-algebras with residue field $k(P)$, Sets the category of sets and $\mathfrak{F}$ the covariant functor

$$\mathfrak{F} : \mathsf{Art} \to \mathsf{Sets}$$

given for $A \in Ob(\mathsf{Art})$ with maximal ideal $\mathfrak{m}_A$ by

$$\mathfrak{F}(A) = \{\rho : \Pi \to GL_2(A) \mid \rho \text{ is finitely continuous and } \rho \bmod \mathfrak{m}_A = \overline{\rho}_0\}/\approx .$$

Here (i) '$\approx$' denotes the strict equivalence of representations, that is, conjugation by a matrix in $GL_2(A)$ congruent to 1 modulo $\mathfrak{m}_A$.

(ii) The phrase 'finitely continuous' means that there exists a Λ-submodule L in A^2 of finite type stable by ρ generating A^2 over Λ_P. The reason for this definition instead of usual P-adic continuity is that Λ_P is not locally compact for the P-adic topology, but Π is even compact. Hence a P-adically continuous representation should have a very small image, and in some sense, we look for representations with open image (over Λ). Note that a finitely continuous representation induces a continuous representation: $\Pi \to GL(L)$, L being endowed with the usual $\mathfrak{m}$-adic topology for the maximal ideal $\mathfrak{m}$ of Λ.

This notion of 'finite continuity' does not depend on the choice of the lattice L by the Artin–Rees lemma. We can extend this notion of finite continuity to any map u of Π to an A-module V requiring that u having values in a Λ-finite submodule L in V and the induced map $u : \Pi \to L$ is continuous under the $\mathfrak{m}$-adic topology on L. This generalized notion will be used later to define finitely continuous cohomology.

By using the fact that $\overline{\rho}_0$ is induced from a finitely continuous character: $\Pi \to k(P)^\times$, it is not so difficult to check by Schlessinger's criterion the following fact:

Theorem 4.3 The functor $\mathfrak{F}$ is pro-representable; that is, there exists a unique universal couple (R', ρ') where R' is a local noetherian complete Λ_P-algebra with residue field $k(P)$ and $\rho' \in \mathfrak{F}(R') = \varprojlim_\alpha \mathfrak{F}(R'/\mathfrak{m}_{R'}^\alpha)$.

Comments a) The 'continuity' property ρ' enjoys should be called 'profinite continuity', meaning that for any artinian quotient $\varphi : R' \to A$ of R', $\varphi \circ \rho'$ is finitely continuous. There is also an obvious notion of profinite continuity of maps from Π to any Λ_P-module.

b) It is natural to ask for the pro-representability of this problem starting from an *arbitrary* irreducible finitely continuous representation $\overline{\rho}_0$. The answer is not known in general because of the lack of a cohomology theory adapted to finite continuous representations and subgroups of $GL_2(k(P))$. Such theory is available when $k(P)$ is a p-adic field, due to Lazard [L], and allows us to give a positive answer in this case. See Appendix below.

In fact, the universal ring we need is smaller than R'. It will pro-represent a subfunctor $\mathfrak{F}_S$ of $\mathfrak{F}$ requiring local conditions at primes of F above p (these conditions involve the choice we made of a p-adic CM-type S). We call this problem the *S-nearly ordinary* deformation problem of $\overline{\rho}_0$. For $\mathfrak{P}$ in S, recalling $\mathfrak{p} = \mathfrak{P} \cap F$, we choose $D_\mathfrak{p}$ so that $D_\mathfrak{p}$ is the decomposition group of $\mathfrak{P}$ in $\Pi_M = \mathrm{Gal}(N^{(p)}/M)$. A strict equivalence class $[\rho]$ in $\mathfrak{F}(A)$ belongs to $\mathfrak{F}_S(A)$ if and only if for any representative ρ, the following conditions are satisfied:

(4.1a) For each prime $\mathfrak{p}$ above p in F, there exists a finitely continuous character $\delta_\mathfrak{p} : D_\mathfrak{p} \to A^\times$ such that ρ restricted to $D_\mathfrak{p}$ is equivalent (but not necessarily strictly) to $\begin{pmatrix} * & * \\ 0 & \delta_\mathfrak{p} \end{pmatrix}$;

(4.1b) $\delta_\mathfrak{p}$ is congruent modulo $\mathfrak{m}_A$ to the restriction of λ_*^c to $D_\mathfrak{p}$ and $\delta_\mathfrak{p}$ restricted to the inertia subgroup $I_\mathfrak{p}$ of $D_\mathfrak{p}$ coincides with the restriction to $I_\mathfrak{p}$ of λ_*^c;

(4.1c) $\det(\rho) = \det(\rho_0)$ (considered as having values in A via the structural morphism: $\Lambda \to \Lambda_P \to A$).

One can deduce from Theorem 4.3 that $\mathfrak{F}_S$ is pro-representable. We denote by (R_S, ρ_S) the corresponding universal couple. Let us define a Λ_P-module $\mathfrak{W}_P$ by

$$\mathfrak{W}_P = \cup_{m=1}^{\infty} P^{-m}\Lambda_P/\Lambda_P = \mathbf{L}/\Lambda_P,$$

where $\mathbf{L}$ is the quotient field of Λ. Then $\mathfrak{W}_P$ is the injective envelope of $k(P)$. We consider the algebra $R_S[\mathfrak{W}_P] = R_S \oplus \mathfrak{W}_P$ with $\mathfrak{W}_P^2 = 0$. One can consider, by abusing the notation, $\mathfrak{F}_S(R_S[\mathfrak{W}_P])$. Namely $\mathfrak{F}_S(R_S[\mathfrak{W}_P])$ is a set of profinitely continuous deformations of $\overline{\rho}$ satisfying the above conditions (i), (ii) and (iii). Since Π is topologically finitely generated, by the profinite continuity, ρ has image in a noetherian subring $R_m = R_S[P^{-m}\Lambda_P/\Lambda_P]$ for sufficiently large m. Thus we have a local Λ-algebra homomorphism φ_ρ : $R_S \to R_S[\mathfrak{W}_P]$ such that $\rho \approx \varphi_\rho \circ \rho_S$. Now we consider the subset

$$\mathfrak{F}_0(R_S[\mathfrak{W}_P]) = \{\rho \in \mathfrak{F}_S(R_S[\mathfrak{W}_P]) \mid \rho \bmod \mathfrak{W}_P = \rho_S\}.$$

We also define $\mathsf{Sect}_\Lambda(R_S[\mathfrak{W}_P]/R_S)$ to be the set of continuous sections (under the $\mathfrak{m}_{R_S}$-adic topology) $\varphi : R_S \to R_S[\mathfrak{W}_P]$ as R_S-algebras whose projection to $\mathfrak{W}_P$ is contained in $P^{-m}\Lambda_P/\Lambda_P$ for m sufficiently large. We put

$$sl_2(\mathfrak{W}_P) = \{x \in M_2(\mathfrak{W}_P) \mid \mathrm{Tr}(x) = 0\},$$

which is a module over Π under the action: $\sigma x = \rho_S(x)x\rho_S(x)^{-1}$. We consider the cohomology group $H^1(\Pi, sl_2(\mathfrak{W}_P))$, which is the quotient of the module of profinitely continuous 1-cocycles on Π having values in $sl_2(P^{-m}\Lambda_P/\Lambda_P)$ for sufficiently large m modulo usual coboundaries. In fact, for each $\rho \in \mathfrak{F}_S(R_S[\mathfrak{W}_P])$, $\varphi_\rho : R_S \to R_S[\mathfrak{W}_P]$ is a Λ-algebra homomorphism. If $\rho \in \mathfrak{F}_0(R_S[\mathfrak{W}_P])$, then by the fact that $\rho \bmod \mathfrak{W}_P = \rho_S$, $\pi \circ \varphi = \mathrm{id}_{R_S}$. Thus we have a morphism: $\mathfrak{F}_0(R_S[\mathfrak{W}_P]) \to \mathsf{Sect}_\Lambda(R_S[\mathfrak{W}_P]/R_S)$. This morphism is of course a surjective isomorphism because for $\varphi \in \mathsf{Sect}_\Lambda(R_S[\mathfrak{W}_P]/R_S)$, $\varphi \circ \rho_S$ is an element of $\mathfrak{F}_0(R_S[\mathfrak{W}_P])$. Therefore we know that

(4.2) $\quad \mathfrak{F}_0(R_S[\mathfrak{W}_P]) \cong \mathsf{Sect}_\Lambda(R_S[\mathfrak{W}_P]/R_S)$.

For each $\mathfrak{p} \in \Sigma$, we can find $\alpha_\mathfrak{p} \in GL_2(R_S)$ such that

$$\alpha_\mathfrak{p}\rho_S(\sigma)\alpha_\mathfrak{p}^{-1} = \begin{pmatrix} * & * \\ 0 & \delta_\mathfrak{p}^S(\sigma) \end{pmatrix} \quad \text{for all } \sigma \in D_\mathfrak{p}$$
$$\text{and } \delta_\mathfrak{p}^S \equiv \lambda^c \bmod \mathfrak{m}_{R_S}.$$

We fix such a $\alpha_\mathfrak{p}$ for each $\mathfrak{p}$. Then we define the ordinary cohomology subgroup $H^1_{\mathrm{ord}}(\Pi, sl_2(\mathfrak{W}_P))$ by the subgroup of cohomology classes of cocycle u satisfying, for every $\mathfrak{p}$ dividing p in F,

$$\{\begin{pmatrix} * & * \\ 0 & * \end{pmatrix}\} \supset \alpha_\mathfrak{p} u(D_\mathfrak{p})\alpha_\mathfrak{p}^{-1} \quad \text{and} \quad \{\begin{pmatrix} 0 & * \\ 0 & 0 \end{pmatrix}\} \supset \alpha_\mathfrak{p} u(I_\mathfrak{p})\alpha_\mathfrak{p}^{-1}.$$

Theorem 4.4 We have a canonical isomorphism:

$$\mathrm{Hom}_{R_S}(\Omega_{R_S/\Lambda_P}, \mathfrak{W}_P) \cong H^1_{\mathrm{ord}}(\Pi, sl_2(\mathfrak{W}_P)),$$

where the Kähler differential module Ω_{R_S/Λ_P} is defined to be the module of continuous differentials, i.e., $\Omega_{R_S/\Lambda_P} = I/I^2$ for the kernel I of the multiplication map of the completed tensor product $R_S \hat{\otimes}_{\Lambda_P} R_S$ (under the adic topology of the maximal ideal of $R_S \otimes_{\Lambda_P} R_S$) to R_S.

Proof For each $\rho \in \mathfrak{F}_0(R_S[\mathfrak{W}_P])$, we define $u : \Pi \to M_2(\mathfrak{W}_P)$ by

$$\rho(\sigma) = (1 \oplus u(\sigma))\rho_S(\sigma) \quad \text{for } \sigma \in \Pi.$$

Since ρ_S and ρ are both profinitely continuous, u has values in $P^{-m}\Lambda_P/\Lambda_p$ for sufficiently large m and is profinitely continuous. Then by (6.1c), we know that $\det(\rho_S) = \det(\rho)$. This shows that u has values in $sl_2(\mathfrak{W}_P)$. Similarly by (6.1b), we know that $\{\begin{pmatrix} 0 & * \\ 0 & 0 \end{pmatrix}\} \supset \alpha_\mathfrak{p} u(I_\mathfrak{p})\alpha_\mathfrak{p}^{-1}$. By the multiplicativity: $\rho(\sigma)\rho(\tau) = \rho(\sigma\tau)$, we see easily that u is a cocycle and u is a coboundary if and only if $\rho \approx \rho_S$. Thus the map $\mathfrak{F}_0(R_S[\mathfrak{W}_P]) \to H^1_{\mathrm{ord}}(\Pi, sl_2(\mathfrak{W}_P))$ is injective. Surjectivity follows from the fact that we can recover a profinitely continuous representation out of a profinitely continuous cocycle by the above formula. Namely we know that

$$\mathfrak{F}_0(R_S[\mathfrak{W}_P]) \cong H^1_{\mathrm{ord}}(\Pi, sl_2(\mathfrak{W}_P)).$$

If we have a section $\varphi \in \mathrm{Sect}_\Lambda(R_S[\mathfrak{W}_P]/R_S)$, we can write $\varphi(r) = r \oplus d_\varphi(r)$. Then $d_\varphi \in \mathrm{Der}_\Lambda(R_S, \mathfrak{W}_P) = \mathrm{Hom}_{R_S}(\Omega_{R_S/\Lambda}, \mathfrak{W}_P)$. It is easy that from any derivation $d : R_S \to \mathfrak{W}_P$, we can reconstruct a section by the above formula. Thus we know that

$$\mathrm{Hom}_{R_S}(\Omega_{R_S/\Lambda}, \mathfrak{W}_P) \cong \mathrm{Sect}_\Lambda(R_S[\mathfrak{W}_P]/R_S)$$

which conclude the proof by (4.2).

We have an injection

$$\mathrm{res} : H^1(\Pi, sl_2(\mathfrak{W}_P)) \to H^1(\Pi_M, sl_2(\mathfrak{W}_P))^{\mathrm{Gal}(M/F)}.$$

Note that as Π_M-module, $sl_2(\mathfrak{W}_P) \cong \mathfrak{W}_P(\alpha) \oplus \mathfrak{W}_P(\alpha^{-1}) \oplus \mathfrak{W}_P$, where $\alpha = \lambda_*(\lambda_*^c)^{-1}$ and $\mathfrak{W}_P(\alpha) \cong \mathfrak{W}_P$ as Λ-module but Π acts via the one-dimensional abelian character α. The action of c interchanges $\mathfrak{W}_P(\alpha)$ and $\mathfrak{W}_P(\alpha^{-1})$ and acts by -1 on $\mathfrak{W}_P$. Thus we see

$$H^1(\Pi_M, sl_2(\mathfrak{W}_P))^{\mathrm{Gal}(M/F)} = H^1(\Pi_M, \mathfrak{W}_P(\alpha)) \oplus \mathrm{Hom}_{\mathrm{conti}}(\mathbf{G}/(1+c)\mathbf{G}, \mathfrak{W}_P).$$

Recall that $M^-(\lambda_-)/M$ is the extension corresponding to $\mathrm{Ker}(\alpha)$. The inclusion of $H^1(\Pi_M, \mathfrak{W}_P(\alpha))$ into $H^1(\Pi_M, sl_2(\mathfrak{W}_P))^{\mathrm{Gal}(M/F)}$ is given in terms of cocycle by the cocycle U such that $U(\sigma) = \begin{pmatrix} 0 & u(\sigma) \\ u({}^c\sigma) & 0 \end{pmatrix}$ for ${}^c\sigma = c\sigma c^{-1}$. From this, it follows, by the ordinarity condition,

$$\mathrm{res}_{\Pi_{M^-}}(u)(cI_{\mathfrak{P}}c^{-1}) = \mathrm{res}_{\Pi_{M^-}}(u)(I_{\mathfrak{P}^c}) = 0 \text{ for all } \mathfrak{P} \in S,$$

where $M^- = M^-(\lambda_-)$. Namely $\mathrm{res}_{\Pi_{M^-}}(u)$ is unramified outside S. Thus we have a natural map:

$$\mathrm{res} : H^1_{\mathrm{ord}}(\Pi_M, \mathfrak{W}_P(\alpha)) \to \mathrm{Hom}_H(X_S, \mathfrak{W}_P(\alpha)) = \mathrm{Hom}_{\Lambda_-}(X_S(\lambda^-), \mathfrak{W}_P).$$

Comments We omitted α from the module of extreme right, because the Λ_--module structure on $\mathfrak{W}_P$ given by α coincides with the natural structure given by the inclusion $\Lambda_- = \mathfrak{O}[[\mathbf{W}^-]]$ into Λ through the Λ-module structure of $\mathfrak{W}_P$. Moreover we can write the extreme right as

$$\mathrm{Hom}_{\Lambda_-}(X_S(\lambda^-), \mathfrak{W}_P) = \mathrm{Hom}_{\Lambda}(X_S(\lambda^-) \otimes_{\Lambda_-} \Lambda, \mathfrak{W}_P).$$

Thus the variable coming from the '+' part $\mathbf{W}^+$ in Λ is just a 'fake' and the divisibility we will obtain is in fact the divisibility in Λ_- although we have variables coming from $\mathbf{W}^+$ inside the Hecke algebra. This is natural because $X_S(\lambda^-)$ is a Λ_--module. The use of '+'-variables is inevitable because we do not know *a priori* that the congruence power series belongs to the '−' part. In the appendix, we prove that when $F = \mathbf{Q}$, the congruence power series belongs to the ordinary part of the Hecke algebra, which can be regarded as the '−' part in our situation.

It is not difficult to show that the above map: res. is injective; namely,

Corollary 4.5 $\mathrm{Hom}_{\Lambda}(X_S(\lambda^-) \otimes_{\Lambda_-} \Lambda, \mathfrak{W}_P) \supset H^1_{\mathrm{ord}}(\Pi_M, \mathfrak{W}_P(\alpha))$.

Since the inclusion of $\mathrm{Hom}_{\mathrm{conti}}(\mathbf{G}/(1+c)\mathbf{G}, \mathfrak{W}_P)$ into $H^1(\Pi_M, sl_2(I_P))^{\mathrm{Gal}(M/F)}$ is given in terms of cocycle by

$$\mathrm{Hom}_{\mathrm{conti}}(\mathbf{G}/(1+c)\mathbf{G}, \mathfrak{W}_P) \ni u \mapsto U(\sigma) = \begin{pmatrix} u(\sigma) & 0 \\ 0 & -u(\sigma) \end{pmatrix},$$

we know that if U is ordinary, then u is unramified everywhere. Let Cl^- be the '−' quotient of the ideal class group of K. We thus know that

Theorem 4.6 $H^1_{\mathrm{ord}}(\Pi, sl_2(\mathfrak{W}_P)) \cong \mathrm{Hom}_{\Lambda_P}(\Omega_{R_S/\Lambda_P} \otimes_{R_S} \Lambda_P, \mathfrak{W}_P)$ injects naturally into

$$\mathrm{Hom}_\Lambda(X_S(\lambda^-) \otimes_{\Lambda_-} \Lambda, \mathfrak{W}_P) \oplus \mathrm{Hom}(Cl^-, \mathfrak{W}_P)$$

as Λ-module.

To relate $X_S(\lambda^-)$ to the congruence power series, we recall the morphism $\lambda_* \circ \theta^* : \mathbf{h} \to \Lambda$ seen in §2, H3. Let R_0 be the local ring of $\mathbf{h}$ through which the above morphism factors. To make R_0 a Λ-algebra, we consider $R = R_0 \otimes_{\Lambda_0} \Lambda$, which is still a complete local ring. Consider the module of differentials $\mathfrak{C}_1 = \Omega_{R/\Lambda} \otimes_R \Lambda$ introduced in [H1, p. 319], where the tensor product is taken via

$$R \to \Lambda \otimes_{\Lambda_0} \Lambda \to \Lambda,$$

which is $\lambda_* \circ \theta^*$ composed with the multiplication on Λ. Let R_P be the completion of the localization of R at P. In [H3, Th.I], an S-nearly ordinary deformation $\rho^{\mathrm{mod}} : \Pi \to GL_2(R_P)$ of $(k(P), \overline{\rho}_0)$ has been constructed. Especially R_P is generated over Λ_P by $\mathrm{Tr}(\rho^{\mathrm{mod}})$, and hence, the natural map $\varphi : R_S \to R_P$ which induces the equality $[\varphi \circ \rho_S] = [\rho^{\mathrm{mod}}]$ is surjective. Then φ induces another surjection

$$\varphi_* : \Omega_{R_S/\Lambda_P} \otimes_{R_S} \Lambda_P \to \mathfrak{C}_1 \otimes_\Lambda \Lambda_P.$$

This combined with Theorem 4.6 yields

Theorem 4.7 We have a surjective homomorphism of Λ-modules:

$$(X_S(\lambda^-) \otimes_{\Lambda_-} \Lambda') \oplus (Cl^- \otimes_{\mathbf{Z}} \Lambda') \to \mathfrak{C}_1,$$

where Λ' is either Λ or $\Lambda[\frac{1}{P_\lambda}]$ in Lemme 4.2 according as $F \neq \mathbf{Q}$ or $F = \mathbf{Q}$ and λ_- mod $\pi\mathfrak{O}$ is trivial on $D_\mathfrak{p}$.

As explained in [T2], there is a divisibility theorem proven by M. Raynaud:

Theorem 4.8 H divides the characteristic power series of $\mathfrak{C}_1$ in Λ.

Then Theorems 4.7 and 4.8 prove Theorem 4.1.

Although we have concentrated to the anti-cyclotomic tower, there is a (hypothetical) way to include the case of the cyclotomic tower. To show the dependence on F, we add subscript F to each notation, for example L_F^- for L^- over F. Supposing the strong divisibility in $\Lambda : L_{F_n}^- \mid Iw_{F_n}^-$ for the nth layer F_n of the cyclotomic $\mathbf{Z}_p$-extension of F for all n, we hope that we could eventually get the full divisibility: $L \mid Iw$ over F? But for the moment, this is still far away.

APPENDIX

Let $F/\mathbf{Q}$ be a finite extension and fix an arbitrary finite Galois extension N/F. Let $N^{(p)}/N$ be the maximal p-profinite extension of N unramified outside p and ∞. Put $\Pi = \mathrm{Gal}(N^{(p)}/F)$. In this appendix, we shall prove the existence of the universal deformation for any (continuous) *absolutely irreducible* Galois representation $\overline{\rho} : \Pi \to GL_n(K)$ for a finite extension $K/\mathbf{Q}_p$ and then we prove the divisibility in Λ' (as in Theorem 4.7) of $h(M)Iw^-$ by H when M is an imaginary quadratic field. Let Λ be a noetherian local ring with residue field K and suppose that Λ is complete under the $\mathfrak{m}$-adic topology for the maximal ideal $\mathfrak{m}$ of Λ. We consider the category Art_Λ of artinian local Λ-algebras with residue field K. For any object A in Art_Λ, the p-adic topology on A gives a locally compact topology on $GL_n(A)$. We consider the covariant functor

$$\mathfrak{F} : \mathsf{Art}_\Lambda \to \mathsf{Sets}$$

which associates to each object A in Art_Λ a set of strict equivalence classes of continuous representations $\rho : \Pi \to GL_n(A)$ such that $\rho \bmod \mathfrak{m}_A = \overline{\rho}$. Then we have

Theorem A.1 $\mathfrak{F}$ is pro-representable on Art_Λ.

Proof We verify the Schlessinger's criterion H_i ($i = 1, 2, \ldots, 4$) for pro-representability ([Sch]). The conditions H_1, H_2 and H_4 can be checked in exactly the same manner as in [M, 1.2]. We verify the finiteness of tangential dimension; i.e.,

H3: $\dim_K \mathfrak{F}(K[\varepsilon])$ is finite, where $K[\varepsilon] = K \oplus K\varepsilon$ with $\varepsilon^2 = 0$.

If $\rho \in \mathfrak{F}(K[\varepsilon])$, then we define a map $u = u_\rho : \Pi \to M_n(K)$ by $\rho(\sigma) = (1 \oplus u(\sigma)\varepsilon)\overline{\rho}(\sigma)$. Since ρ is continuous, u is a continuous 1-cocycle with values in the Π-module $M_n(K)$, where Π acts on $M_n(K)$ by $\sigma x = \overline{\rho}(\sigma)x\overline{\rho}(\sigma)^{-1}$. On the other hand, if we have a continuous 1-cocycle u as above, we construct a representation ρ by $\rho(\sigma) = (1 \oplus u(\sigma)\varepsilon)\overline{\rho}(\sigma)$. As a map to $M_n(K)$, ρ is continuous. Then ρ is finitely continuous as a representation. Thus the map $\mathfrak{F}(K[\varepsilon]) \to H^1_c(\Pi, M_n(K))$ is surjective. Here 'H_c' indicates the continuous cohomology. We see easily that $u(\sigma) = (\sigma - 1)m$ if and only if $(1 \oplus m)^{-1}\overline{\rho}(1 \oplus m) = \rho$ (i.e., ρ is strictly equivalent to $\overline{\rho}$, which is the 'zero' element in $\mathfrak{F}(K[\varepsilon])$). Thus we have

$$\mathfrak{F}(K[\varepsilon]) \cong H^1_c(\Pi, M_n(K))$$

and

$$H^1_c(\Pi, M_n(K)) \cong H^1_c(\Pi, sl_n(K)) \oplus \mathrm{Hom}_c(\Pi, K). \tag{A.1}$$

By class field theory, $\mathrm{Hom}_c(\Pi, K)$ is finite dimensional. We now claim

$$\dim_K H^1_c(\Pi, sl_n(K)) < +\infty. \tag{A.2}$$

Let us prove this. Let F_∞ be the subfield of $N^{(p)}$ fixed by $\mathrm{Ker}(\overline{\rho})$. Since cohomology groups of a finite group with coefficients in finite dimensional vector space over K are finite dimensional, we may replace Π by any normal subgroup of finite index because of the inflation-restriction sequence. First we may assume that $H = \mathrm{Im}(\overline{\rho})$ is a pro-p-group without torsion and that F_∞/N is unramified outside p and ∞. Then applying a theorem of Lazard [L, III.3.4.4.4], we know that H has a subgroup of finite index which is pro-p-analytic. Hence we may even assume that H itself is pro-p-analytic. By inflation-restriction sequence, the sequence:

$$0 \to H^1_c(H, sl_n(K)) \to H^1_c(\Pi, sl_n(K)) \to \mathrm{Hom}_H(\mathrm{Ker}(\overline{\rho}), sl_n(K)) \tag{A.3}$$

is exact. Let $\mathbf{M}_\infty/F_\infty$ be the maximal p-abelian extension unramified outside p and ∞ and X be the Galois group $\mathrm{Gal}(\mathbf{M}_\infty/F_\infty)$. Let $\mathbf{A} = \mathbf{Z}_p[[H]]$. Since H is pro-p-analytic and is contained in the maximal compact subgroup of $GL_n(K)$, we know that X is a $\mathbf{A}$-module of finite type by [Ha, §3]. The maximal topological abelian quotient $\mathrm{Ker}(\overline{\rho})^{ab}$ is a quotient of X and hence of finite type over $\mathbf{A}$. This proves that

$$\dim_K \mathrm{Hom}_H(\mathrm{Ker}(\overline{\rho}), sl_n(K)) < +\infty. \tag{A.4}$$

Thus we need to show the finite dimensionality of $H^1_c(H, sl_n(K))$. Let $\mathfrak{H}$ be the Lie algebra of $G \cap H$. Then again by a result of Lazard [L, V.2.4.10], we see

$$H^1_c(H, sl_n(K)) \cong H^0(H, H^1(\mathfrak{H}, sl_n(K))),$$

which is finite dimensional.

Let $\mathbf{h}_0 = \mathbf{h}_0^{\mathrm{ord}}(D; \mathfrak{O})$ be the ordinary Hecke algebra defined in [H1, Th.3.3] for any positive integer D prime to p. In this case G in §2 is just $Z \times \mathbf{Z}_p^\times$ for $Z = ((\mathbf{Z}/D\mathbf{Z})^\times \times \mathbf{Z}_p^\times)/\{\pm 1\}$. Then we have

Theorem A.2 Suppose that $p \geq 5$ and $F = \mathbf{Q}$. Let $\chi : \mathbf{A}^\times \to \mathbf{Z}_p^\times$ be the cyclotomic character such that $\chi(\varpi_l) = l$ for the prime element ϖ_l in $\mathbf{Q}_l$ ($l \neq p$). Then we have an $\mathfrak{O}[[G]]$-algebra isomorphism:

$$\mathbf{h} \cong \mathbf{h}_0 \hat{\otimes}_{\mathfrak{O}} \mathfrak{O}[[\mathbf{Z}_p^\times]],$$

which is given by $\mathbf{T}(x) \mapsto T(x) \otimes [\chi(x)]$ for all $x \in \hat{\mathbf{Z}} \cap A_f^\times$. Here $\mathbf{h}_0 \hat{\otimes}_{\mathfrak{O}} \mathfrak{O}[[\mathbf{Z}_p^\times]]$ is the profinite completion of $\mathbf{h}_0 \otimes_{\mathfrak{O}} \mathfrak{O}[[\mathbf{Z}_p^\times]]$, i.e., $\mathfrak{m}$-adic completion for the maximal ideal $\mathfrak{m}$ of Λ_0.

Proof Let $S(\mathfrak{O}) = \{f \in S \mid a(y,f) \in O\}$ and $\mathbf{S} = eS(\mathfrak{O})$ for the idempotent e of $\mathbf{h}$ in $\mathbf{h}(D;\mathfrak{O})$. Let $\mathbf{S}_0$ be the ordinary subspace of $\mathbf{S}$ which is denoted by $S_0^{\mathrm{ord}}(D;\mathfrak{O})$ in [H1, p. 336]. Then it is known that the pairing given by

$$\langle h, f\rangle = a(1, f \mid h) \text{ on } \mathbf{h} \times \mathbf{S} \text{ and } \mathbf{h}_0 \times \mathbf{S}_0$$

is perfect in the sense that $\mathrm{Hom}_{\mathfrak{O}}(\mathbf{h},\mathfrak{O}) \cong \mathbf{S}$ and vice versa [H4, Th.3.1]. For any character $\psi : \mathbf{Z}_p^\times \to \overline{\mathbf{Q}}_p$ and $f \in \mathbf{S}$, $f \otimes \psi(y)$ given by $a(y, f \otimes \psi) = \psi(\chi(y))a(y,f)$ is again an element in S with $f \otimes \psi \mid e = f \otimes \psi$ (cf. [H4, §7.VI]). This shows that we have a natural $\mathfrak{O}[[G]]$-linear map $m : \mathbf{S}_0 \hat{\otimes}_{\mathfrak{O}} \mathfrak{C}(\mathbf{Z}_p^\times;\mathfrak{O}) \to \mathbf{S}$ given by

$$a(y, m(f \otimes \phi)) = \phi(\chi(y))a(y,f),$$

where $\mathfrak{C}(\mathbf{Z}_p^\times;\mathfrak{O})$ is the Banach $\mathfrak{O}$-module of all continuous functions on $\mathbf{Z}_p^\times$ into $\mathfrak{O}$ and $\mathbf{S}_0 \hat{\otimes}_{\mathfrak{O}} \mathfrak{C}(\mathbf{Z}_p^\times;\mathfrak{O})$ is the p-adic completion of $\mathbf{S}_0 \otimes_{\mathfrak{O}} \mathfrak{C}(\mathbf{Z}_p^\times;\mathfrak{O})$. Note that $\mathrm{Hom}_{\mathfrak{O}}(\mathbf{S}_0 \hat{\otimes}_{\mathfrak{O}} \mathfrak{C}(\mathbf{Z}_p^\times;\mathfrak{O}),\mathfrak{O}) \cong \mathbf{h}_0 \hat{\otimes}_{\mathfrak{O}} \mathfrak{O}[[\mathbf{Z}_p^\times]]$. It is easy to verify that the dual map $m^* : \mathbf{h} \to \mathbf{h}_0 \hat{\otimes}_{\mathfrak{O}} \mathfrak{O}[[\mathbf{Z}_p^\times]]$ is in fact an $\mathfrak{O}[[G]]$-algebra homomorphism. Since the projection map $\mathbf{h} \to \mathbf{h}_0$ is surjective by definition and since any $[z^{-1}] \in \mathfrak{O}[[\mathbf{Z}_p^\times]]$ for $z \in \mathbf{Z}_p^\times$ is the image of $\mathbf{T}(z)$, m^* is surjective. Note that $\mathbf{h}_0 \hat{\otimes}_{\mathfrak{O}} \mathfrak{O}[[\mathbf{Z}_p^\times]]$ is free of finite rank over Λ_0 by [H7, Th.3.1]. Since $\mathbf{h}$ is torsion-free over Λ_0 and its generic rank is equal to that of $\mathbf{h}_0 \hat{\otimes}_{\mathfrak{O}} \mathfrak{O}[[\mathbf{Z}_p^\times]]$, we conclude that m^* is an isomorphism.

Corollary A.3 The congruence power series H can be chosen inside Λ_-.

By this corollary, when $F = \mathbf{Q}$, it is sufficient to consider only ordinary Hecke algebras instead of nearly ordinary Hecke algebras and only ordinary deformations instead of nearly ordinary deformations. To make this fact more precise, let $M/\mathbf{Q}$ be an imaginary quadratic field of discriminant D satisfying the ordinarity hypothesis: $p = \mathfrak{P}\mathfrak{P}^c$. We also assume that $p \geq 5$. Let L (resp. L^*) be the maximal abelian extension of M unramified outside $\mathfrak{P}$ (resp. $\mathfrak{P}^c$). Let $G_{cw} = \mathrm{Gal}(L/M)$ and $G^*_{cw} = \mathrm{Gal}(L^*/M)$ and W_{cw} (resp. W^*_{cw}) be the maximal torsion-free quotient of G_{cw} (resp. G^*_{cw}). Then the restriction map gives an isomorphism $\mathbf{W} \cong W_{cw} \times W^*_{cw}$. Thus $\alpha : W_{cw} \ni w \mapsto wcw^{-1}c^{-1} \in \mathbf{W}_-$ gives an isomorphism. Similarly, without losing generality, we may assume that $\lambda : \mathbf{G}_{\mathrm{tor}} \to \mathfrak{O}^\times$ factors through G_{cw}. We decompose $G_{cw} = \Delta \times W_{cw}$. Let $\Lambda_- = \mathfrak{O}[[W_{cw}]]$ identifying $\mathbf{W}_-$ with W_{cw}. We consider the character $\lambda_* : G_{cw} \to \Lambda_-$ such that $\lambda_*(\delta, w) = \lambda(\delta)[w]$ for $\delta \in \Delta$ and $w \in W$. It is known that the μ-invariant of Iw^- and L^- are both trivial [G]. Thus we only worry about height one primes P (in Λ_-) of residual characteristic 0. We take $N/\mathbf{Q}$ to be the ray class field of M modulo p and consider the Galois group Π as in Theorem A.1. Let K be the quotient field of Λ_-/P. Then $K/\mathbb{Q}_p$ is a finite extension and we consider the Galois representation:

$$\rho_0 = \mathrm{Ind}_{\Pi_K}^{\Pi}(\lambda_*) : \Pi \to GL_2(\Lambda_-), \quad \text{and}$$
$$\rho_P = \mathrm{Ind}_{\Pi_K}^{\Pi}(\lambda_* \bmod P) : \Pi \to GL_2(K).$$

Suppose that $P \neq P_\lambda$ as in Lemma 4.2. Then ρ_P is absolutely irreducible. Let Λ be the P-adic completion of the localization of Λ_- at P. Let Art be the category of artinian local Λ-algebras with residue field K. Any object A in Art is a locally compact ring with respect to p-adic topology and thus we do not worry about 'finite continuity' etc. Let (R', ρ') be the universal couple representing the functor $\mathfrak{F}$: Art $\to$ Sets defined for $\overline{\rho} = \rho_P$. We consider the subfunctor of $\mathfrak{F}$

$$\mathfrak{F}^{\mathrm{ord}} : \mathsf{Art} \to \mathsf{Sets}$$

which associates to $A \in Ob(\mathsf{Art})$ the set of strict equivalence class of representations $\rho : \Pi \to GL_2(A)$ such that

(i) $\rho \bmod \mathfrak{m}_A = \rho_P$,

(ii) There exists a continuous character $\delta : D_{\mathfrak{P}} \to A^\times$ such that ρ restricted to $D_{\mathfrak{P}}$ is equivalent (but not necessarily strictly) to $\begin{pmatrix} * & * \\ 0 & \delta \end{pmatrix}$;

(iii) δ is congruent modulo $\mathfrak{m}_A$ to the restriction of λ_*^c to $D_{\mathfrak{P}}$ and δ restricted to the inertia subgroup I_p of $D_\mathfrak{p}$ coincides with the restriction to I_p of λ_*^c (i.e., δ is **unramified** at $\mathfrak{P}$)

(iv) $\det(\rho) = \det(\rho_0)$.

We say that an ideal $\mathfrak{a}$ of R' is ordinary if ρ' mod $\mathfrak{a}$ satisfies (i), (ii), (iii) and (iv). Then it is an easy exercise to verify that if $\mathfrak{a}$ and $\mathfrak{b}$ are ordinary, then $\mathfrak{a} + \mathfrak{b}$ and $\mathfrak{a} \cap \mathfrak{b}$ are ordinary. Namely for $\mathfrak{I} = \bigcap_{\mathfrak{a}:\mathrm{ordinary}} \mathfrak{a}$, $R^{\mathrm{ord}} = R'/\mathfrak{I}$ and $\rho^{\mathrm{ord}} = \rho'$ mod $\mathfrak{I}$ represents $\mathfrak{F}^{\mathrm{ord}}$. Then the same argument as in §4 prove that $H \mid h(M)Iw^-$. From Theorem 2.1 and the vanishing of the μ-invariant [G], we conclude

Theorem A.4 Suppose $p \geq 5$ and that M is an imaginary quadratic field. Let $\Lambda' = \Lambda_-[\frac{1}{P_\lambda}]$ if λ_- mod $\pi\mathfrak{O}$ is trivial on $D_{\mathfrak{P}}$ and otherwise we put $\Lambda' = \Lambda_-$. Then we have

$$h(M)L^- \mid H \text{ in } \Lambda_- \quad \text{and} \quad H \mid h(M)Iw^- \text{ in } \Lambda'.$$

Although we confined ourselves to characters λ of p-power conductor, similar result holds for any character whose conductor is prime to its complex conjugate. We hope to prove the divisibility even at the 'trivial-zero' P_λ in our subsequent paper.

REFERENCES

[G] R. Gillard, 'Fonctions L p-adiques des corps quadratiques imaginaires et de leurs extensions abéliennes', *J. reine angew. Math.* **358** (1985), 76–91.

[Ha] M. Harris, 'p-adic representations arising from descent on abelian varieties', *Compositio Math.* **39** (1979), 177–245.

[H1] H. Hida, 'On p-adic Hecke algebras for GL_2 over totally real fields', *Ann. of Math.* **128** (1988), 295-384.

[H2] H. Hida, 'On nearly ordinary Hecke algebras for $GL(2)$ over totally real fields', *Adv. Studies in Pure Math.* **17** (1989), 139–169.

[H3] H. Hida, 'Nearly ordinary Hecke algebras and Galois representations of several variables', *Proc. JAMI inaugural Conference, Supplement of Amer. J. Math.* (1989), 115–134.

[H4] H. Hida, 'On p-adic L-functions of $GL(2) \times GL(2)$ over totally real fields', preprint.

[H5] H. Hida, 'Galois representations into $GL_2(\mathbf{Z}_p[[X]])$ attached to ordinary cusp forms', *Invent. Math.* **85** (1986), 545–613.

[H6] H. Hida, 'A p-adic measure attached to the zeta functions associated with two elliptic modular forms II', *Ann. Inst. Fourier* **38**, No.1 (1988), 1–83.

[H7] H. Hida, 'Iwasawa modules attached to congruences of cusp forms', *Ann. Scient. Ec. Norm. Sup.* 4e-serie, t.19, 1986, 231–273.

[K] N. M. Katz, 'p-adic L-functions for CM fields', *Invent. Math.* **49** (1978), 199–297.

[L] M. Lazard, 'Groupes analytiques p-adiques', *Publ. Math. I.H.E.S.* No.26, 1965.

[M] B. Mazur, 'Deforming Galois representations', in *Galois Groups over* $\mathbf{Q}$, *Proc. Workshop at MSRI*, 1987, pp.385–437.

[M-T] B. Mazur and J. Tilouine, 'Représentations galoisiennes, différentielles de Kähler et "conjecture principales" ', *Publ. I.H.E.S.*

[R] K. Rubin, 'The one-variable main conjecture for elliptic curves with complex multiplication', preprint.

[Sch] M. Schlessinger, 'Functors on Artin rings', *Trans. A.M.S.* **130** (1968), 208–22.

[T1] J. Tilouine, 'Sur la conjecture principale anticyclotomique', *Duke Math. J.*

[T2] J. Tilouine, 'Théorie d'Iwasawa classique et de l'algèbre de Hecke ordinaire', *Compositio Math.* **65** (1988), 265–320.

Kolyvagin's work on Shafarevich–Tate groups

WILLIAM G. MCCALLUM

1 INTRODUCTION

Let E be an elliptic modular curve defined over $\mathbf{Q}$ of conductor N, with a fixed modular parametrization $\phi : X_0(N) \to E$ mapping the cusp ∞ on $X_0(N)$ to the origin of the group law on E. Let $K = \mathbf{Q}(\sqrt{-D})$ be a quadratic imaginary field in which all the prime factors of N are split. Let $y_K \in E(K)$ be the Heegner point associated with the maximal order in K, and let $L(E/K, s)$ be the complex L-function of E/K. In [2] Gross and Zagier proved that y_K has infinite order if and only if $L'(E/K, 1) \neq 0$, and gave a formula for the value of the derivative in terms of the height of y_K. This formula and the conjecture of Birch and Swinnerton-Dyer yield the following conjectural formula for the order of the Shafarevich–Tate group of E over K.

Conjecture Suppose that y_K has infinite order. Then $Ш(E/K)$ is finite of order

$$|Ш(E/K)| = \left(\frac{[E(K) : \mathbf{Z}y_K]}{c \cdot \prod_{q|N} m_q} \right)^2 .$$

Here m_q is the number of connected components of the special fiber of the Néron model of E at q, and c is the Manin constant of the modular parametrization, i.e., if ω is a Néron differential on E then c is the unique positive integer such that $\phi^*(\omega/c)$ is the differential associated with a normalized newform on $X_0(N)$.

Let $R = \mathrm{End}(E)$, and let F be the fraction field of R. In [3] Kolyvagin proved:

Theorem **(Kolyvagin)** Suppose y_K has infinite order. Then $E(K)$ has rank 1 and $Ш(E/K)$ is finite. Further, if p is an odd prime which is unramified in F and such that $\mathrm{Gal}(F(E_p)/F) = \mathrm{Aut}_R(E_p)$, then

$$\mathrm{ord}_p|Ш(E/K)| \leq 2\,\mathrm{ord}_p[E(K) : \mathbf{Z}y_K].$$

Gross's paper [1] provides an excellent introduction to the proof of this theorem. The purpose of this paper is to given an account of more recent work of Kolyvagin in which he determines the exact group structure of the p-part of $\text{Ш}(E/K)$ in terms of his derived Heegner points P_n. (These will be defined later.) The precise result is stated in Theorem 5.4; the following is a simple consequence of it.

Theorem **(Kolyvagin)** Suppose y_K has infinite order, and let p be an odd prime which is unramified in F and such that $\text{Gal}(F(E_p)/F) = \text{Aut}_R(E_p)$. Suppose one of Kolyvagin's points P_n satisfies $P_n \notin pE(K(P_n))$. Then

$$\text{ord}_p|\text{Ш}(E/K)| = 2\,\text{ord}_p[E(K) : \mathbf{Z}y_K].$$

In Section 2 we recall some results we need from the theory of duality of elliptic curves; in Section 3 we give an application of the Čebotarev density theorem; in Section 4 we recall the definition of Kolyvagin's cohomology classes; and in Section 5 we prove the main theorems.

Notation If m is a positive integer and G is an abelian group object, we denote the kernel of multiplication by m on G by G_m. If G is a finite group we denote by G^* the group of characters $G \to \mathbf{Q}/\mathbf{Z}$. If L/K is a galois extension of number fields, and if λ is a prime of K, we denote by $\text{Frob}(\lambda)$ the conjugacy class of Frobenius substitutions associated with λ.

Acknowledgments I would like to thank B. Gross and K. Rubin for useful conversations, and M. Bertolini and H. Darmon for pointing out an error in an earlier version of this paper. Some of this work, including Theorem 5.8, was obtained independently by the author. H. Darmon also independently discovered a related theorem.

2 GLOBAL DUALITY

In this section we consider an elliptic curve E over an arbitrary number field K. If v is a valuation of K we denote the completion by K_v. If λ is a prime ideal of K we denote the associated valuation by v_λ, and write K_λ for K_{v_λ}. Recall that for a positive integer m, the cup product

$$H^1(K_v, E_m) \cup H^1(K_v, E_m) \to H^2(K_v, \mathbf{G_m}) \overset{\text{inv}_v}{\tilde{\to}} \mathbf{Q}/\mathbf{Z},$$

induced by the Weil pairing, is a non-degenerate pairing of finite groups, and if K is galois, this pairing is $\text{Gal}(K/\mathbf{Q})$-equivariant (see [6], Chapter I, Remark 3.5). It is related to the Tate pairing

$$\langle\, ,\, \rangle_v : H^1(K_v, E) \times E(K_v) \to \mathbf{Q}/\mathbf{Z}$$

by the formula

$$\langle i_*(c), x\rangle_v = c \cup \delta(x),$$

where i is the inclusion $E_m \hookrightarrow E$ and δ is the coboundary for the Kummer sequence

$$0 \to E_m \xrightarrow{i} E \xrightarrow{m} E \to 0.$$

Proposition 2.1 Let K be a number field, and let $m > 1$ be an integer. Let w be a valuation of K such that $H^1(K_w, E_m) \neq \{0\}$, and let S be a finite set of valuations of K not containing w. For each $v \in S$, let $H_v \subset H^1(K_v, E_m)$ be a subgroup satisfying $|H_v| = (1/2)|H^1(K_v, E_m)|$. Then there exists $c \in H^1(K, E_m)$ satisfying

1. $c \neq 0$,
2. $c_v \in \delta(E(K_v))$ for all $v \notin S \cup \{w\}$, and
3. $c_v \in H_v$ for all $v \in S$.

Proof Enlarging S if necessary, and choosing $H_v = \delta(E(K_v))$ for the added valuations, we may suppose that $S \cup \{w\}$ contains the infinite primes, the primes of bad reduction of E, and the primes dividing m. Let $T = S \cup \{w\}$. It follows from Tate global duality ([6], Chapter I, Theorem 4.10) that there is a self dual exact sequence

$$H^1(K_T/K, E_m) \to \bigoplus_{v\in T} H^1(K_v, E_m) \to H^1(K_T/K, E_m)^*,$$

where K_T is the maximal extension of K unramified outside T. Hence the image of $H^1(K_T/K, E_m)$ is a maximal isotropic subgroup of

$$\bigoplus_{v\in T} H^1(K_v, E_m).$$

Since $H^1(K_w, E_m) \neq 0$ such a subgroup is strictly of larger order than

$$\bigoplus_{v\in S} \frac{H^1(K_v, E_m)}{H_v}.$$

Thus we may choose a $c \in H^1(K_T/K, E_m)$ satisfying (1) and (3). Further, c satisfies (2) because $H^1(K_v^{\text{unr}}/K_v, E_m) = \delta(E(K_v))$ if $v(m) = 0$ and E has good reduction at v. □

Proposition 2.2 Let c and c' be two elements of $H^1(K, E_p)$. Then

$$\sum_v \mathrm{inv}_v(c_v \cup c'_v) = 0.$$

Proof The sum of the invariants of a global class is zero. □

The group of classes $c \in H^1(K, E_m)$ satisfying

$$c_v \in \delta(E(K_v)) \quad \text{for all } v$$

is called the m-Selmer group of E over K, denoted $S_m(E/K)$. It fits into an exact sequence

$$0 \to E(K)/mE(K) \to S_m(E/K) \to Ш(E/K)_m \to 0,$$

where $Ш(E/K)$, the Shafarevich–Tate group of E over K, is defined to be the kernel of

$$H^1(K, E) \longrightarrow \sum_v H^1(K_v, E).$$

There is a skew-symmetric pairing on $Ш(E/K)$, the Cassels pairing, which is non-degenerate if $Ш(E/K)$ is finite. It is defined as follows. Suppose $d \in Ш(E/K)_m$, $d' \in Ш(E/K)_{m'}$. Choose $c' \in S_{m'}(E/K)$ so that $d' = i_*(c')$, and choose *local points* for each valuation v of K

$$y_v \in E(K_v), \quad \delta(y_v) = c'_v.$$

To pair d and d' we need

$$d_1 \in H^1(K, E)_{mm'}, \quad m'd_1 = d.$$

Note that since $d \in Ш(E/K)$, $d_{1,v} \in H^1(K_v, E)_{m'}$ for each v. Then the Cassels pairing of d and d' is

$$\langle d, d' \rangle = \sum_v \langle d_{1,v}, y_v \rangle_v. \tag{1}$$

It is not known in general that d_1 exists, and Tate has a rather clever trick for dealing with this (see [6], Chapter I, Proposition 6.9), but in our case it always exists.

3 AN APPLICATION OF THE ČEBOTAREV DENSITY THEOREM

Suppose now that E is defined over $\mathbf{Q}$. For the rest of the paper we will assume that E does not have complex multiplication, in order to simplify the exposition. The general case is not significantly more difficult. Let K be an imaginary quadratic field, and let p be an odd prime such that $\mathrm{Gal}(\mathbf{Q}(E_p)/\mathbf{Q}) = Gl_2(\mathbf{Z}/p\mathbf{Z})$. Let $L = K(E_{p^M})$. Then for $M > 0$, the restriction map

$$H^1(K, E_{p^M}) \rightarrow H^1(L, E_{p^M}) = \mathrm{Hom}(\mathrm{Gal}(\overline{\mathbf{Q}}/L), E_{p^M})$$

is an injection of $\mathrm{Gal}(L/\mathbf{Q})$-modules (see [1], Proposition 9.1). Hence, if C is any finite subgroup of $H^1(K, E_{p^M})$ there is a finite Galois extension L_C of L and an isomorphism of $\mathrm{Gal}(L/\mathbf{Q})$-modules

$$\begin{aligned} \mathrm{Gal}(L_C/L) &\simeq \mathrm{Hom}(C, E_{p^M}), \\ \sigma &\mapsto \phi_\sigma. \end{aligned} \tag{2}$$

Further,

$$c_\lambda = 0 \iff \phi_\sigma(c) = 0 \quad \text{for all} \quad \sigma \in G_{\lambda_L}, \tag{3}$$

where λ_L is a prime of L above λ, and G_{λ_L} is the decomposition group of a prime of L_C above λ_L.

Let $\tau \in \mathrm{Frob}(\infty) \subset \mathrm{Gal}(L/\mathbf{Q})$. Since τ acts as -1 on p^∞-th roots of unity and preserves the Weil pairing on E_{p^M}, it has both a plus and a minus eigenspace on E_{p^M}. Hence we may choose an isomorphism of $\langle\tau\rangle$-modules

$$p^{-M}\mathbf{Z}/\mathbf{Z} \oplus (p^{-M}\mathbf{Z}/\mathbf{Z})\tau \simeq E_{p^M}.$$

Using this, we may and do identify

$$\mathrm{Hom}(H^1(K, E_{p^M}), E_{p^M})^{\langle\tau\rangle}$$

with

$$\mathrm{Hom}(H^1(K, E_{p^M}), \mathbf{Q}/\mathbf{Z}) = H^1(K, E_{p^M})^*.$$

Proposition 3.1 Let $M > 1$ be an integer. Let C be a finite subgroup of $H^1(K, E_{p^M})$, and let $\phi \in C^* = \mathrm{Hom}(C, E_{p^M})^{\langle\tau\rangle}$. There exist infinitely many primes l satisfying the following.

1. $\mathrm{Frob}(l) = \mathrm{Frob}(\infty)$ in $\mathrm{Gal}(\mathbf{Q}(E_{p^M})/\mathbf{Q})$.
2. $\phi = \phi_{\mathrm{Frob}(\lambda')}$ for some prime λ' of $\mathbf{Q}(E_{p^M})$ lying above λ.

Proof By (2), there is some $\sigma \in \mathrm{Gal}(L_C/L)$ such that

$$\phi = \phi_\sigma.$$

Further, since $\phi_\sigma^\tau = \phi_\sigma$, and the order of $\mathrm{Gal}(L_C/L)$ is odd, $\sigma = \rho^\tau \cdot \rho$ for some $\rho \in \mathrm{Gal}(L_C/L)$. By the Čebotarev density theorem, there are infinitely many primes l such that $\mathrm{Frob}(l)$ contains $\tau\rho$. Since the restriction of $\tau\rho$ to L is τ, condition 1 is clearly satisfied. In particular, l has residue class degree 2 in $L/\mathbf{Q}$, so, for a suitable choice of λ',

$$\mathrm{Frob}(\lambda') = (\tau\rho)^2 = \rho^\tau \cdot \rho = \sigma.$$

This concludes the proof. □

We say a set of non-zero classes $c_1, \ldots, c_r \in H^1(K, E_{p^M})$ is *independent* if any relation

$$a_1 c_1 + \cdots + a_r c_r = 0, \quad a_i \in \mathbf{Z},$$

implies that $\mathrm{ord}\, c_i$ divides a_i, $1 \le i \le r$.

Corollary 3.2 Let $c_1, \ldots, c_r \in H^1(K, E_{p^M})$ be independent and let $p^{M_i} = \mathrm{ord}\,(c_i)$, $1 \le i \le r$. Let $N_1, \ldots, N_r$ be integers such that $0 \le N_i \le M_i$, $1 \le i \le r$. Then there are infinitely many primes l satisfying the following.

1. $\mathrm{Frob}(l) = \mathrm{Frob}(\infty)$ in $\mathrm{Gal}(\mathbf{Q}(E_{p^M})/\mathbf{Q})$.
2. For the prime λ of K lying above l,

$$\mathrm{ord}\, c_{i,\lambda} = p^{N_i}, \quad 1 \le i \le r.$$

Proof Let $C = \langle c_1, \ldots, c_r \rangle$, and choose $\phi \in C^*$ such that $\mathrm{ord}\, \phi(c_i) = p^{N_i}$, $1 \le i \le r$. Choose l as in Proposition 3.1, and different from the finitely many primes where the classes c_i ramify. Then the decomposition group of λ_L is generated by $\mathrm{Frob}(\lambda_L)$, so by (3), $p^M c_{i,\lambda} = 0 \iff \phi(p^M c_i) = 0$. □

4 KOLYVAGIN'S CLASSES

We briefly review the definition of these. Since the proofs of [1] generalize easily to our situation, we won't repeat most of them.

For a positive integer n let $\mathcal{O}_n$ be the order of conductor n in K. Choose an ideal $\mathcal{N}$ in $\mathcal{O}_n$ of norm N. The isogeny of complex tori $\mathbf{C}/\mathcal{O}_n \to \mathbf{C}/\mathcal{N}^{-1}$ gives a point x_n on $X_0(N)$, defined over K_n, the ray class field over K of conductor n. The Heegner point referred to in the introduction is $y_K = \mathrm{Tr}_{K_1/K}\phi(x_1)$.

Let p be an odd prime such that $\mathrm{Gal}(\mathbf{Q}(E_p)/\mathbf{Q}) = Gl_2(\mathbf{Z}/p\mathbf{Z})$. Let r and M be integers, $r \geq 0$, $M \geq 1$, and consider the set $S_r(M)$ of positive, square-free integers with exactly r prime factors l, each of which satisfies the following conditions:

$$l \quad \text{does not divide} \quad N \cdot D \cdot p$$
$$l \quad \text{is inert in} \quad K$$
$$a_l \equiv l + 1 \equiv 0 \pmod{p^M},$$

where a_l is the trace of $\mathrm{Frob}(l)$ on E_p. The last two conditions are equivalent to

$$\mathrm{Frob}(l) = \mathrm{Frob}(\infty) \quad \text{on} \quad \mathbf{Q}(E_{p^M}).$$

In particular, λ, the prime of K above l, splits completely in $K(E_{p^M})$. Let

$$S(M) = \bigcup_{r \geq 0} S_r(M).$$

Let $n \in S(M)$, and let $y_n = \phi(x_n) \in E(K_n)$. Let K_1 be the Hilbert class field of K, and let $G_n = \mathrm{Gal}(K_n/K_1)$. Then $G_n \simeq \prod G_l$, where $G_l = \mathrm{Gal}(K_n/K_{n/l})$ is cyclic of order $l+1$, with generator σ_l, say.

Define $D_n \in \mathbf{Z}[G_n]$ by $D_n = \prod D_l$, where

$$D_l = \sum_{i=1}^{l} i \cdot \sigma_l^i.$$

Let $\mathcal{G}_n = \mathrm{Gal}(K_n/K)$, let S be a set of coset representatives for G_n in $\mathcal{G}_n$, and define a point $P_n \in E(K_n)$,

$$P_n = \sum_{\sigma \in S} \sigma(D_n y_n).$$

Then

$$P_n \in (E(K_n)/p^M E(K_n))^{\mathcal{G}_n}. \tag{4}$$

From P_n we construct cohomology classes as follows. Consider the commutative diagram of cohomology sequences:

$$\begin{array}{ccccccc}
0 & \to & E(K)/p^M E(K) & \xrightarrow{\delta} & H^1(K, E_{p^M}) & \to & H^1(K,E)_{p^M} & \to & 0 \\
 & & \downarrow & & \wr\downarrow \text{res} & & \downarrow \text{res} & & \\
0 & \to & (E(K_n)/p^M E(K_n))^{\mathcal{G}_n} & \xrightarrow{\delta'} & H^1(K_n, E_{p^M})^{\mathcal{G}_n} & \to & H^1(K_n, E)_{p^M}^{\mathcal{G}_n} & &
\end{array}$$

The middle vertical map is an isomorphism because

$$E \text{ has no } K_n\text{-rational } p\text{-torsion} \tag{5}$$

(see [1], Lemma 4.3). The class $c_M(n)$ is defined to be the unique class in $H^1(K, E_{p^M})$ such that

$$\mathrm{res}(c_M(n)) = \delta'(P_n). \tag{6}$$

Lemma 4.1 The class $c_M(n)$ is represented by the cocycle

$$\sigma \longmapsto -\frac{(\sigma-1)P_n}{p^M} + \sigma\frac{P_n}{p^M} - \frac{P_n}{p^M}$$

where $\frac{(\sigma-1)P_n}{p^M}$ is the unique p^M-division point of $(\sigma-1)P_n$ in $E(K_n)$.

Proof The existence of the p^M-division point follows from (4), the uniqueness from (5). Uniqueness implies that

$$\sigma \mapsto -\frac{(\sigma-1)P_n}{p^M}$$

is a cocycle, hence the expression given in the statement of the lemma is a cocycle. Clearly it takes values in E_{p^M}, and the first term disappears when we restrict to K_n, hence it satisfies (6), the defining property of $c_M(n)$. □

Let $d_M(n)$ denote the image of $c_M(n)$ in $H^1(K, E)$.

Corollary 4.2 The class $d_M(n)$ is represented by the cocycle

$$\sigma \longmapsto -\frac{(\sigma-1)P_n}{p^M}.$$

Proof Regarded as a cocycle with values in E,

$$\sigma \longmapsto \sigma\frac{P_n}{p^M} - \frac{P_n}{p^M}$$

is a coboundary. □

Let λ be the unique prime of K above l and let λ_n represent a prime of K_n above λ. We denote the completion of K_n at λ_n by K_{λ_n}. Suppose that $n = l \cdot m$. The prime ideal λ is principle, generated by the number l prime to m, and hence splits completely in K_m by class field theory, and each prime factor of l in K_m ramifies totally in K_n. In particular, there is an embedding $K_m \hookrightarrow K_\lambda$, and by (4) the resulting image of P_m in $E(K_\lambda)/p^M E(K_\lambda)$ is independent of the choice of embedding.

Lemma 4.3 Let $n \in S(M)$, and let v be a valuation of K prime to n. Then $c_M(n)_v \in \delta(E(K_v))$. If in addition $v = v_\lambda$, where l is inert in K, then $c_M(n)_\lambda = \delta(P_n)$.

Proof The first statement is proved in [1], Proposition 6.2. By Lemma 4.1,

$$c_M(n) = -\frac{(\sigma-1)P_n}{p^M} + \sigma\frac{P_n}{p^M} - \frac{P_n}{p^M}.$$

If l is inert in K, then, as we saw above, λ splits in K_n, by class field theory. Thus, when we restrict this cocycle to the decomposition group at λ, the first term goes away. Thus the cocycle is $\delta(P_n)$ locally at λ. □

For primes dividing n we have the following proposition.

Proposition 4.4 (Kolyvagin, [3] Theorem 3) Let $l \in S_1(M)$. There is a homomorphism

$$\chi_l : E(K_\lambda) \to H^1(K_\lambda, E_{p^M})$$

such that

1. for all $m \in S(M)$, $(m,l) = 1$,
$$c_M(ml)_\lambda = \chi_l(P_m),$$
2. $\ker \chi_l = p^M E(K_\lambda)$ and
$$\chi_l((E(K_\lambda)/p^M E(K_\lambda))^\pm) \subset H^1(K_\lambda, E_{p^M})^\mp,$$
and
3. the composition of χ_l with $H^1(K_\lambda, E_{p^M}) \to H^1(K_\lambda, E)_{p^M}$ induces an isomorphism
$$E(K_\lambda)/p^M E(K_\lambda) \simeq H^1(K_\lambda, E)_{p^M}.$$

In particular,

$$\operatorname{ord} d_M(ml)_\lambda = \operatorname{ord} c_M(ml)_\lambda = \operatorname{ord} c_M(m)_\lambda.$$

Proof Let $n = ml$. Let $P \in E(K_\lambda)$. Let $\mathbf{F}_\lambda$ denote the residue field of λ, and let $\tilde{P}$ be the image of P in $E(\mathbf{F}_\lambda)$. Since $\mathrm{Frob}(l)^2 = 1$ on $\mathbf{F}_\lambda$,

$$(a_l - (l+1)\mathrm{Frob}(l))\tilde{P} = -\mathrm{Frob}(l)(\mathrm{Frob}(l)^2 - a_l\mathrm{Frob}(l) + 1)\tilde{P} = 0.$$

Since λ splits completely in $K(E_{p^M})$, $E_{p^M}(\overline{K}_\lambda) = E_{p^M}(K_\lambda)$. Since l is prime to p and E has good reduction at λ, the reduction map is injective on E_{p^M}. Hence there is a unique $T \in E_{p^M}$ such that

$$\frac{a_l - (l+1)\mathrm{Frob}(l)}{p^M} P \equiv T \pmod{\lambda}.$$

Define $\chi_l(P)$ to be the cocycle for $\mathrm{Gal}(K_{\lambda_l}/K_\lambda)$ which takes σ_l to T. To see that this satisfies the statement of the proposition, recall from Lemma 4.1 that $c_M(n)$ is represented by the cocycle

$$\sigma \mapsto -\frac{(\sigma-1)P_n}{p^M} + \sigma\frac{P_n}{p^M} - \frac{P_n}{p^M}.$$

Let $\overline{\lambda}$ be any prime of $\overline{K}$ above λ, and restrict this cocycle to the decomposition group of $\overline{\lambda}$. Let λ_n be the prime of K_n below $\overline{\lambda}$. Since $D_l = l(l+1)/2$ on the residue field of λ_n, and p^M divides $l+1$, $P_n \in p^M E(K_{\lambda_n})$. Hence the cocycle vanishes when restricted to K_{λ_n}, and factors through $\mathrm{Gal}(K_{\lambda_n}/K_\lambda) = \langle\sigma_l\rangle$. Furthermore, since σ_l is in the inertia group of λ, $\sigma_l(P_n/p^M) - (P_n/p^M)$ reduces to zero modulo λ_n. Hence

$$-\frac{(\sigma_l-1)P_n}{p^M} + \sigma_l\frac{P_n}{p^M} - \frac{P_n}{p^M}$$

is the unique torsion point congruent to $-((\sigma_l-1)P_n)/p^M$ modulo λ_n. But, as in [1], Proposition 6.2, we have

$$-\frac{(\sigma_l-1)P_n}{p^M} \equiv \frac{a_l-(l+1)\mathrm{Frob}(l)}{p^M}P_m \pmod{\lambda_n}.$$

This proves property (1). Property (2) follows from the fact that the $\mathrm{Gal}(K/\mathbf{Q})$-eigenspaces of $E(\mathbf{F}_\lambda)$ are cyclic of order $l+1-\mathrm{Frob}(l)a_l$, and that σ_l is in the minus eigenspace. Property (3) is clear since all the non-zero cocycles in $\mathrm{im}\,\chi_l$ are ramified, and thus $\mathrm{im}\,\chi_l \cap \delta(E(K_\lambda)) = 0$.

Finally, it follows from properties (1) and (3) that

$$\mathrm{ord}\, d_M(ml)_\lambda = \mathrm{ord}\, c_M(ml)_\lambda,$$

and that $\mathrm{ord}\, c_M(ml)$ equals the order of P_m in $E(K_\lambda)/p^M E(K_\lambda)$. By Lemma 4.3, this equals $\mathrm{ord}\, c_M(m)_\lambda$. □

Corollary 4.5 Suppose $n \in S(M)$, let l be a prime divisor of n, and let $m = n/l$. If $P_m \notin p^M E(K_\lambda)$ then $P_n \notin p^M E(K_n)$.

Proof If $P_m \notin p^M E(K_\lambda)$, then by Proposition 4.4 $c_M(n)_\lambda \neq 0$, hence $c_M(n) \neq 0$. But it is easy to see from the definition that $c_M(n) = 0$ if and only if $P_n \in p^M E(K_n)$. □

Finally, we show how to compute the Cassels pairing of Kolyvagin's classes.

Lemma 4.6 Let $M \geq M'$ be positive integers and let $n \in S(M)$. Then

$$p^{M'} d_M(n) = d_{M-M'}(n).$$

Proof Clear from the definition. □

Proposition 4.7 Let M and M' be integers ≥ 1, and let $n \in S(M+M')$, $n' \in S(M')$. Suppose that $d_M(n), d_{M'}(n') \in Ш(E/K)$. Then the Cassels pairing is

$$\langle d_M(n), d_{M'}(n') \rangle = \sum_{\substack{l|n \\ (l,n')=1}} \langle d_{M+M'}(n), P_{n'} \rangle_\lambda$$

Proof We refer to the description of the Cassels pairing given in Section 2. By Lemma 4.6, $p^{M'} d_{M+M'}(n) = d_M(n)$; hence $d_{M+M'}(n)$ plays the role of d_1. Also, $c_{M'}(n')$ plays the role of c'.

First, suppose $v \nmid n$. Then $d_{M+M'}(n)_v$ is unramified at v, since by Corollary 4.2 it splits over K_n. From [6], Chapter I, Proposition 3.8, it follows that $d_{M+M'}(n)_v$ is killed by m_v. We claim that $d_{M+M'}(n)_v = 0$; this is clear if v is a prime of good reduction, and follows from Gross's argument in [1], Proposition 6.2 otherwise. Hence there is no contribution to the Cassels pairing from v.

Now suppose that $v = \lambda$, for some $l|n$. If $l|n'$, then by Proposition 4.4, $c_{M'}(n')_\lambda = 0$, since $d_{M'}(n') \in Ш(E/K)$. If $l \nmid n'$, then by Lemma 4.3, $P_{n'}$ plays the role of y_λ. The proposition now follows from (1). □

5 STRUCTURE OF THE SHAFAREVICH–TATE GROUP

From now on we will assume that y_K has infinite order and that p is an odd prime such that $\mathrm{Gal}(\mathbf{Q}(E_p)/\mathbf{Q}) = Gl_2(\mathbf{Z}/p\mathbf{Z})$. We choose a fixed complex conjugation

$$\tau \in \mathrm{Gal}(\overline{\mathbf{Q}}/\mathbf{Q}),$$

and if M is a $\mathrm{Gal}(\overline{\mathbf{Q}}/\mathbf{Q})$-module, denote by M^+ (resp. M^-) the part of M on which τ acts by $+1$ (resp. -1). The image of the Heegner point y_K in $E(K)$/torsion is an eigenvector for τ ([1], Proposition 5.3); let $\epsilon = \pm 1$ be its eigenvalue. (It is the negative of the sign of the functional equation of $L(E/\mathbf{Q}, s)$.) By Kolyvagin's theorem [3], $Ш(E/K)$ is finite, and hence the Cassels pairing is non-degenerate. Since the pairing is skew-symmetric, the elementary divisors of $Ш(E/K)$ come in pairs. Let

$$N_1 \geq N_3 \geq N_5 \geq \cdots$$

be integers such that

$$Ш(E/K)^{-\epsilon}_{p^\infty} \simeq (\mathbf{Z}/p^{N_1}\mathbf{Z})^2 \times (\mathbf{Z}/p^{N_3}\mathbf{Z})^2 \times \cdots,$$

and let

$$N_2 \geq N_4 \geq N_6 \geq \cdots$$

be integers such that

$$Ш(E/K)^{\epsilon}_{p^\infty} \simeq (\mathbf{Z}/p^{N_2}\mathbf{Z})^2 \times (\mathbf{Z}/p^{N_4}\mathbf{Z})^2 \times \cdots.$$

We write $p^M|P_n$ if $P_n \in p^M E(K_n)$, and $p^M\|P_n$ if $P_n \in p^M E(K_n) - p^{M+1}E(K_n)$. Let

$$\mathrm{ord}_p(P_n) = \max\{M : p^M|P_n\},$$

and let

$$M_r = \min\{\mathrm{ord}_p(P_n) : n \in S_r(\mathrm{ord}_p(P_n)+1)\}.$$

Lemma 5.1 We have $M_0 = \mathrm{ord}_p[E(K) : \mathbf{Z}y_K]$ and $M_r \geq M_{r+1}$ for all $r \geq 0$.

Proof We have $P_1 = y_K$, and

$$M_0 = \mathrm{ord}_p y_K = \max\{M : y_K \in p^M E(K_1)\},$$

and

$$\mathrm{ord}_p[E(K) : \mathbf{Z}y_K] = \max\{M : y_K \in p^M E(K)\}.$$

Since $E(K_1)$ has no p-torsion, $E(K)/p^M E(K)$ injects into $E(K_1)/p^M E(K_1)$; hence these two numbers are the same.

In particular, M_0 is finite. Now suppose that M_r is finite, and let $n \in S_r(M_r+1)$ satisfy $p^{M_r}\|P_n$. By Corollary 3.2, we may choose $l \in S(M_r+1)$, prime to n, so that $c_{M_r+1}(n)_\lambda \neq 0$. Then by Lemma 4.3, $P_n \notin p^{M_r+1}E(K_\lambda)$, so by Corollary 4.5, $P_{nl} \notin p^{M_r+1}E(K_{nl})$. Hence M_{r+1} is finite and no greater than M_r. □

The goal of this section is to prove that $N_i = M_{i-1} - M_i$ for all i (Theorem 5.4). We will construct elements in $Ш(E/K)$ as follows. Suppose $M_{r-1} > M_r$ and let $n \in S_r(M_{r-1})$. Then it follows from the definition of the M_i that $p^{M_r}|P_n$ and $p^{M_{r-1}}|P_{n/l}$ for all l dividing n. It follows from Lemma 4.3 and Proposition 4.4 that $d_{M_{r-1}}(n) \in Ш(E/K)$. The order of this element is at most $p^{M_{r-1}-M_r}$. By careful choice of n we will construct such elements achieving this order and independent of each other. By (5) the natural map

$$H^1(K, E_{p^M}) \to H^1(K, E_{p^{M'}}), \quad M' \geq M$$

is an injection. We let

$$H^1(K, E_\infty) = \varinjlim H^1(K, E_{p^M}), \quad S_\infty(E/K) = \varinjlim S_{p^M}(E/K).$$

If $n \in S_r(M)$, then $c_M(n) \in H^1(K, E_{p^M})^{\epsilon_r}$, where $\epsilon_r = (-1)^r \epsilon$ ([1], Proposition 5.4).

Proposition 5.2 Let r be a positive integer, and let C be a subgroup of $S_\infty(E/K)^{\epsilon_r}$ of rank r. Let $M > M_r$. There exists $n \in S_r(M)$ such that $c_M(n)$ had order p^{M-M_r} and $\langle c_M(n) \rangle \cap C = \{0\}$.

For the proof we will need the following variant of Proposition 2.1.

Lemma 5.3 Let $l \in S_1(M)$. The Tate pairing induces a non-degenerate pairing

$$(E(K_\lambda)/p^M E(K_\lambda))^\pm \times H^1(K_\lambda, E_{p^M})^\pm \to \mathbf{Q}/\mathbf{Z}$$

which is a duality of cyclic groups of order p^M. Further, if S is a finite subset of $S_1(M)$ not containing l, then there exists $c \in H^1(K, E_{p^M})^\pm$ satisfying

1. $c \neq 0$,
2. $c_v \in \delta(E(K_v))$ for all v prime to $S \cup \{l\}$, and
3. $c_{v_\lambda} \in \operatorname{im} \chi_l$ for all $l \in S$.

Finally, $\operatorname{im} \chi_l$ is an isotropic subgroup of $H^1(K_\lambda, E_{p^M})$.

Proof The first statement is proved as in [1], Proposition 8.1. It implies that

$$|\operatorname{im}(\chi_l)^+| = \frac{1}{2}|H^1(K_\lambda, E_{p^M})^+|$$

and implies that

$$|\operatorname{im}(\chi_l)^-| = \frac{1}{2}|H^1(K_\lambda, E_{p^M})^-|.$$

Hence, in the proof of Proposition 2.1, we can add the further stipulation that $c \in H^1(K, E_{p^M})^\pm$. Finally, it follows from Proposition 4.4 that $\operatorname{im}(\chi_l)^+ \simeq \operatorname{im}(\chi_l)^- \simeq \mathbf{Z}/p^M\mathbf{Z}$. Since the cup product is skew symmetric and $\operatorname{Gal}(K/\mathbf{Q})$-equivalent, it must vanish on $\operatorname{im}(\chi_l)$. □

Proof of Proposition 5.2 By Lemma 4.6, it suffices to prove the proposition for large enough $M > M_r$. Let

$$p^M \geq \max\{\text{exponent of } C, p^{M_{r-1}}\},$$

and let $L = K(E_{p^M})$. For $n \in S_r(M)$, the class $c_M(n)$ has order p^{M-M_r} if and only if $p^{M_r} \| P_n$. By definition of M_r, there exists

$$n \in S_r(M_r + 1)$$

such that

$$p^{M_r} \| P_n.$$

Choose such an n. Let S be the set of prime factors of n, and for each $l \in S$, choose a prime factor λ_L of l in L. Let $X \subset C^*$ be the group of characters generated by

$$\phi_{\mathrm{Frob}(\lambda_L)}, \quad l \in S \cap S(M).$$

Let k be the rank of the image of X in C^*/pC^*. Suppose that $k < r$. Then there is a redundant $l_0 \in S$ such that the characters

$$\phi_{\mathrm{Frob}(\lambda_L)}, \quad l \in S \cap S(M) - \{l_0\}$$

generate X modulo pC^*. Choose $\psi \in C^*$ such that

$$\psi \notin X + pC^*;$$

and if $c_{M_r+1}(n) \in C$ choose $\psi \in C^*$ such that

$$\psi \notin X + pC^* \quad \text{and} \quad \psi \notin \langle c_{M_r+1}(n)\rangle^\perp$$

(this is possible since a finite group cannot be the union of two proper subgroups). Using Lemma 5.3, choose

$$c \in H^1(K, E_p)^{-\epsilon_r}$$

satisfying

$$c \neq 0, \tag{7}$$
$$c_v \in \delta(E(K_v)), \quad v \text{ prime to } S, \tag{8}$$
$$c_{v_\lambda} \in \mathrm{im}(\chi_l) \quad \text{for all} \quad l \in S - \{l_0\}. \tag{9}$$

Since c is in a different eigenspace,

$$C \times \langle c_{M_r+1}(n)\rangle \cap \langle c\rangle = \{0\}.$$

So we can choose

$$\phi : C \times \langle c_{M_r+1}(n) \rangle \times \langle c \rangle \to \mathbf{Q}/\mathbf{Z}$$

such that

$$\phi|_C = \psi, \tag{10}$$

$$\phi(c_{M_r+1}(n)) \neq 0, \tag{11}$$

$$\phi(c) \neq 0. \tag{12}$$

By Proposition 3.1, there exists $l' \in S_1(M)$ such that

$$\phi = \phi_{\mathrm{Frob}(\lambda'_L)}.$$

Consider

$$\sum_v c_{M_r+1}(nl')_v \cup c_v. \tag{13}$$

If $v \notin S \cup \lambda'$, then $c_{M_r+1}(nl')_v \in \delta(E(K_v))$ by Lemma 4.3 and $c_v \in \delta(E(K_v))$ by (8), so $c_{M_r+1}(nl')_v \cup c_v = 0$. If $v = v_\lambda \in S - \{\lambda_0\}$, then $c_v \in \mathrm{im}(\chi_l)$ by (9), so again $c_{M_r+1}(nl')_v \cup c_v = 0$. So the only possible non-zero terms in the sum (13) are at $v = \lambda_0, \lambda'$. Suppose $v = \lambda'$. From (11) and (3), $c_{M_r+1}(n)_{\lambda'} \neq 0$, hence by Proposition 4.4 $d_{M_r+1}(nl')_{\lambda'} \in H^1(K_{\lambda'}, E)_p^{-\epsilon_r}$ is not zero. Further, $c_{\lambda'} \in \delta(E(K_{\lambda'}))^{-\epsilon_r}$ by (8), and it is not zero by (12). Hence

$$c_{M_r+1}(nl')_{\lambda'} \cup c_{\lambda'} \neq 0.$$

Since the sum (13) is zero by Proposition 2.2, this implies

$$c_{M_r+1}(nl')_{\lambda_0} \cup c_{\lambda_0} \neq 0.$$

Hence $c_{M_r+1}(nl')_{\lambda_0} \neq 0$, and so by Proposition 4.4, $P_{nl'/l_0} \notin p^{M_r+1}E(K_{\lambda_0})$, hence $p^{M_r} \| P_{nl'/l_0}$. Replacing n by nl'/l_0, we add ψ to X and increase k to $k+1$, and hence eventually to r. But if $k = r$, then $S \subset S(M)$ and $X = C^*$, so $c_M(n)$ is defined and

$$\{c \in C : c_\lambda = 0 \quad \text{for all} \quad l \in S\} =$$
$$\{c \in C : \phi_{\mathrm{Frob}(\lambda_L)}(c) = 0 \quad \text{for all} \quad l \in S\} = \{0\}.$$

Since, by Proposition 4.4, $c_{M_{r-1}}(n)_\lambda = 0$ for all $l \in S$, we deduce

$$C \cap \langle c_{M_{r-1}}(n) \rangle = \{0\}.$$

Also, since $p^{M_r} \| P_n$, $c_M(n)$ has order p^{M-M_r} for any $M > M_r$. So the proposition is proved if $M_{r-1} > M_r$ or if $C = \{0\}$.

So suppose that $M_r = M_{r-1}$. Apply the proposition with $C = \{0\}$ to find $m \in S_{r-1}(M)$ such that $p^{M_r} \| P_m$, then use Proposition 3.1 to find $l \in S(M)$ such that $c_{M_r+1}(m)_\lambda \neq 0$ and set $n = lm$. By Proposition 4.4, $d_{M_r+1}(n)_\lambda \neq 0$, and hence $c_{M_r+1}(n) \notin S_\infty(E/K)$. Since $C \subset S_\infty(E/K)$, this implies that

$$C \cap \langle c_{M_r+1}(n) \rangle = \{0\}.$$

This proves the proposition in this case also. □

Using the Cassels pairing and a simple induction argument one can immediately deduce from Proposition 5.2 that Ш(E/K) contains a subgroup isomorphic to

$$(\mathbf{Z}/p^{M_0-M_1}\mathbf{Z})^2 \times (\mathbf{Z}/p^{M_1-M_2}\mathbf{Z})^2 \times \cdots.$$

Further, by using a slightly refined version of Kolyvagin's upper bound on the order of Ш(E/K), or by adding the hypothesis that $p \nmid P_n$ for some $n \in S(1)$, one can show that this subgroup is the full group. Thus the N_i are the $M_{i-1} - M_i$ in some order. To prove that $N_i = M_{i-1} - M_i$ requires more work. To give the basic idea, we sketch the case $i = 1$ first.

Applying Proposition 5.2 with $C = \{0\}$, we find $l \in S(M_0)$ such that $c_{M_0}(l)$ has order $p^{M_0-M_1}$ in $S_\infty(E/K)^{-\epsilon} =$ Ш$(E/K)^{-\epsilon}$. From the definition of N_1, we see

$$M_0 - M_1 \leq N_1.$$

Now let $d \in$ Ш$(E/K)^{-\epsilon}$ have order p^{N_1}. Lift d to c in the Selmer group. Using Corollary 3.2, choose a prime l such that

$$\operatorname{ord} c_{M_0+N_1}(1)_\lambda = \operatorname{ord} c_{M_0+N_1}(1), \tag{14}$$

$$\operatorname{ord} c_\lambda = \operatorname{ord} c. \tag{15}$$

(These two elements are in different eigenspaces.) Then the Cassels pairing

$$\langle d_{M_0}(l), p^M d \rangle = \langle d_{M_0+N_1-M}(l)_\lambda, y_\lambda \rangle_\lambda,$$

where $c_\lambda = \delta(y_\lambda)$. By (14) and Proposition 4.4, $d_{M_0+N_1-M}(l)_\lambda$ has order p^{N_1-M} in $H^1(K_\lambda, E)^{-\epsilon}_{p^{N_1}}$, which is cyclic of order p^{N_1}, and by (15), y_λ has order p^{N_1} in $E(K_\lambda)/p^{N_1}E(K_\lambda)^{-\epsilon}$, which is also cyclic of order p^{N_1}. Thus the pairing is non-zero for $1 \leq M \leq N_1 - 1$, and hence $d_{M_0}(l)$ has order at least p^{N_1}. Since the greatest order it can have is $M_0 - M_1$, we deduce

$$N_1 \leq M_0 - M_1,$$

and hence

$$N_1 = M_0 - M_1.$$

Now we give the theorem in general. If

$$G = G_1 \times \cdots \times G_r$$

is a product of cyclic groups, we say that a set $\{\chi_1, \ldots, \chi_r\}$ of characters of G is a triangular basis for G^* if

$$\chi_i(G_j) = 0, \quad j > i,$$

and

$$\langle \chi_1, \ldots, \chi_i \rangle = G_1^* \times \cdots \times G_i^*, \quad 1 \le i \le r.$$

Theorem 5.4 Let $p > 2$ be such that $\mathrm{Gal}(\mathbf{Q}(E_p)/\mathbf{Q}) = Gl_2(\mathbf{Z}/p\mathbf{Z})$. We have $N_i = M_{i-1} - M_i$ for $i \ge 1$.

Proof First, observe that by Lemma 5.3, if $l \in S_1(M)$, then two elements

$$y \in E(K_\lambda)/p^M E(K_\lambda), \quad d \in H^1(K_\lambda, E)_{p^M},$$

pair nontrivially if they are in the same $\mathrm{Gal}(K/\mathbf{Q})$-eigenspace and their orders multiply to more than p^M.

By definition of the N_i, there exists a maximal isotropic subgroup

$$D = D_1 \times D_2 \times \cdots$$

of $\text{Ш}(E/K)$ such that each D_i is cyclic of order p^{N_i}, $D^{-\epsilon} = D_1 \times D_3 \times \cdots$, and $D^\epsilon = D_2 \times D_4 \times \cdots$.

Let d_i be a generator of D_i and let c_i be a lifting of d_i to $S_\infty(E/K)$. For each valuation v of K and each i, choose $y_{i,v} \in E(K_v)$ such that $c_{i,v} = \delta(y_{i,v})$. By Corollary 3.2, we can choose $l_1 \in S_1(M_0 + N_1)$ so that

$$\operatorname{ord} c_{M_0+N_1}(1)_{\lambda_1} = p^{N_1}, \tag{16}$$
$$\operatorname{ord} c_{1,\lambda_1} = p^{N_1}, \tag{17}$$
$$c_{i,\lambda_1} = 0, \quad i \ge 2. \tag{18}$$

Let $n_1 = l_1$. By Proposition 4.7, for $0 \le M \le N_i - 1$,

$$\langle d_{M_0}(n_1), p^M d_i \rangle = \langle d_{M_0-M}(n_1), d_i \rangle = \langle d_{M_0-M+N_i}(n_1), y_{i,\lambda_1} \rangle_{\lambda_1}.$$

This is zero if $i \ge 2$, by (18). Let $i = 1$. By (17), y_{1,λ_1} has order p^{N_1} in $E(K_{\lambda_1})/p^{N_1}E(K_{\lambda_1})^{-\epsilon}$, and by (16) and Proposition 4.4, $d_{M_0+N_1-M}(n_1)_{\lambda_1}$ has order p^{N_1-M} in $H^1(K_{\lambda_1}, E)^{-\epsilon}$. Hence the pairing is non-trivial for $0 \le M \le N_1 - 1$. Thus we have proved that the character

$$d \mapsto \langle d_{M_0}(n_1), d\rangle$$

vanishes on $D_2 \times D_3 \cdots$, and its restriction to D_1 generates D_1^*. Hence $d_{M_0}(n_1)$ has order at least p^{N_1}. Since it has order at most $p^{M_0-M_1}$, we conclude

$$N_1 \le M_0 - M_1.$$

On the other hand, p^{N_1} is the maximum order an element of $\text{Ш}(E/K)^{-\epsilon}$ can have, and by Proposition 5.2, there is an element in $S_\infty(E/K)^{-\epsilon} = \text{Ш}(E/K)^{-\epsilon}_{p^\infty}$ of order $p^{M_0-M_1}$, which implies

$$M_0 - M_1 \le N_1.$$

Hence

$$N_1 = M_0 - M_1.$$

In particular, p^{M_1+1} does not divide P_{n_1}, and hence

$$p^{M_1} \| P_{n_1}.$$

Now suppose we have found primes $\{l_1, l_2, \ldots, l_k\} \in S_1(M)$ such that

$$c_{i,\lambda_j} = 0, \quad i > j, \quad 1 \le j \le k, \tag{19}$$

and if

$$n_j = l_1 \cdots l_j,$$

then

$$p^{M_j} \| P_{n_j}, \quad 1 \le j \le k, \tag{20}$$

and the characters

$$d \mapsto \langle d_{M_{j-1}}(n_j), d\rangle, \quad 1 \le j \le k,$$

vanish on $D_{k+1} \times \cdots$, and form a diagonal basis of $(D_1 \times \cdots \times D_k)^*$. Suppose further that we have shown that $M_{j-1} - M_j = N_j$ for $1 \le j \le k$. (We have just done all this for $k = 1$.) In particular, the order of $d_{M_{k-1}}(n_k)$ in $\text{Ш}(E/K)$ is the same as its order as a character on D. Since D is isotropic, it follows that

$$\langle d_{M_{k-1}}(n_k)\rangle \cap D = \{0\}.$$

So by Corollary 3.2, we may choose $l_{k+1} \in S_1(M_k + N_{k+1})$ satisfying

$$\operatorname{ord} c_{M_k+N_{k+1}}(n_k)_{\lambda_{k+1}} = p^{N_{k+1}}, \tag{21}$$

$$\operatorname{ord} c_{k+1,\lambda_{k+1}} = p^{N_{k+1}}, \tag{22}$$

$$c_{i,\lambda_{k+1}} = 0, \quad i > k+1. \tag{23}$$

Let $n_{k+1} = n_k l_{k+1}$. Then for $0 \le M \le N_i - 1$,

$$\langle d_{M_k}(n_{k+1}), p^M d_i \rangle =$$

$$\langle d_{M_k - M}(n_{k+1}), d_i \rangle = \sum_{j=1}^{k+1} \langle d_{M_k - M + N_i}(n_{k+1})_{\lambda_j}, y_{i,\lambda_j} \rangle_{\lambda_j}.$$

All terms but the last are zero for $i > k$ by (19), and the last term is zero for $i > k+1$ by (23). Let $i = k+1$. By (22), $y_{k+1,\lambda_{k+1}}$ has order $p^{N_{k+1}}$ in $E(K_{\lambda_{k+1}})/p^{N_1}E(K_{\lambda_{k+1}})^{\epsilon_{k+1}}$, and by (21) and Proposition 4.4, $d_{M_k + N_{k+1} - M}(n_{k+1})_{\lambda_{k+1}}$ has order $p^{N_{k+1} - M}$ in $H^1(K_{\lambda_{k+1}}, E)^{\epsilon_{k+1}}$. Hence the pairing is non-trivial for $0 \le M \le N_{k+1} - 1$.

Thus the character

$$d \mapsto \langle d_{M_k}(n_{k+1}), d \rangle$$

vanishes on $D_{k+2} \times \cdots$, and its restriction to D_{k+1} generates D_{k+1}^*, and hence extends the triangular basis to generate $(D_1 \times \ldots \times D_{k+1})^*$. Thus $d_{M_k}(n_{k+1})$ has order at least $p^{N_{k+1}}$. Since it has order at most $p^{M_k - M_{k+1}}$, we conclude

$$N_{k+1} \le M_k - M_{k+1}.$$

Let

$$C = \langle c_{M_0 + N_1}(1), c_1, \ldots, c_k, c_{M_0}(n_1), \ldots, c_{M_{k-1}}(n_k) \rangle^{\epsilon_{k+1}}.$$

Then $p^{N_{k+1}}$ is the maximum order an element $c \in S_\infty(E/K)^{\epsilon_{k+1}}$ can have if

$$\langle c \rangle \cap C = \{0\}.$$

On the other hand, by Proposition 5.2 applied to C, there is an element in $S_\infty(E/K)^{\epsilon_{k+1}}$ of order $p^{M_k - M_{k+1}}$ satisfying this condition. Hence

$$M_k - M_{k+1} \le N_{k+1},$$

and so

$$N_{k+1} = M_k - M_{k+1}.$$

By induction, this proves the theorem. □

Corollary 5.5 The numbers M_i satisfy

$$M_i - M_{i+1} \ge M_{i+2} - M_{i+3}, \quad i \ge 0,$$

and if i_0 is the first positive integer such that $M_{i_0} = M_{i_0+1} = M_{i_0+2}$, then $M_i = M_{i_0}$ for all $i \ge i_0$. We have

$$Ш(E/K)_{p^\infty} \simeq \prod_{i \ge 0} (\mathbf{Z}/p^{M_i - M_{i+1}}\mathbf{Z})^2.$$

Corollary 5.6 Let $m = \min\{M_i : i \geq 0\}$. Then

$$\operatorname{ord}_p |Ш(E/K)| = 2(M_0 - m).$$

In the course of proving Theorem 5.4, we actually proved the following more precise statement.

Proposition 5.7 If

$$D = D_1 \times D_2 \times \cdots$$

is a maximal isotropic subgroup of $Ш(E/K)$ such that D_i is cyclic of order p^{N_i}, $D^{-\epsilon} = D_1 \times D_3 \times \cdots$, and $D^{\epsilon} = D_2 \times D_4 \times \cdots$, then there exist integers $n_1|n_2|\cdots$ such that $n_i \in S_i(M_{i-1})$ and the characters

$$d \mapsto \langle d, d_{M_{i-1}}(n_i) \rangle$$

form a triangular basis of characters of D.

In particular, $Ш(E/K)$ can be generated from Kolyvagin's classes constructed from $n \in S_k$ with k less than or equal to half the rank of $Ш(E/K)$. (This fact was independently discovered by H. Darmon.) On the other hand, the following theorem shows that simply to generate $Ш(E/K)$, $k \leq 2$ will suffice.

Theorem 5.8 Let $p > 2$ be such that $\mathrm{Gal}(\mathbf{Q}(E_p)/\mathbf{Q}) = Gl_2(\mathbf{Z}/p\mathbf{Z})$. Let $M \geq 2M_0$. Then the classes $\{d_{M_0}(l) : l \in S_1(M)\}$ generate $Ш(E/K)_{p^\infty}^{-\epsilon}$ and the classes in $\{d_{M_1}(l_1 l_2) : l_1 l_2 \in S_2(M)\}$ generate $Ш(E/K)_{p^\infty}^{\epsilon}$.

Proof We will show that the dual of $Ш(E/K)$ under the Cassels pairing is generated by these classes, using the same technique as in Theorem 5.4. Since the Cassels pairing is non-degenerate, this will prove the theorem. First suppose that $d \in Ш(E/K)^{-\epsilon}$ has order exactly p^M for some $M > 0$. By Kolyvagin's upper bound [3], $M \leq M_0$. Lift d to $c \in H^1(K, E_{p^M})$. As in the proof of Theorem 5.4, choose l such that

$$\operatorname{ord} c_{M_0+M}(1)_\lambda = p^M \tag{24}$$

and

$$\operatorname{ord} c_\lambda = p^M, \tag{25}$$

and deduce that the Cassels pairing

$$\langle d_{M_0}(l), p^{M-1} d \rangle \neq 0.$$

Hence the character on $Ш(E/K)$ defined by $\{d_{M_0}(l) : l \in S(M)\}$ generates $\langle d \rangle^*$. In particular, the character group of $Ш(E/K)^{-\epsilon}$ generated by the classes

$d_{M_0}(l)$ does not vanish at d. Since d was arbitrary, this proves the first part of the theorem.

Now suppose that $d' \in \text{Ш}(E/K)^{\epsilon}$ has order exactly $p^{M'}$. Lift d' to $c' \in H^1(K, E_{p^{M'}})$. By Theorem 5.4 (in fact Proposition 5.2 suffices) there exists $d \in \text{Ш}(E/K)^{-\epsilon}$ of order exactly $p^{M_0 - M_1}$. Choose $l_1 \in S_1(M_0)$ satisfying (24) and (25) with respect to such a d and in addition

$$c'_{\lambda_1} = 0. \tag{26}$$

Then $\operatorname{ord} c_{M_0}(l_1) = p^{M_0 - M_1}$, hence $p^{M_1} \| P_{l_1}$. So we may choose l_2 satisfying

$$\operatorname{ord} c_{M_1+M'}(l_1)_{\lambda_2} = p^{M'}$$

and

$$\operatorname{ord} c'_{\lambda_2} = p^{M'}.$$

Then

$$\langle d_{M_1}(l_1 l_2), d' \rangle = \langle d_{M_1+M'}(l_1 l_2), y'_{\lambda_1} \rangle_{\lambda_1} + \langle d_{M_1+M'}(l_1 l_2), y'_{\lambda_2} \rangle_{\lambda_2}.$$

The first term is zero by (26), and the second term is non-zero by the same argument as in the proof of Theorem 5.4. Hence the classes $d_{M_1}(l_1 l_2)$ generate the dual of $\text{Ш}(E/K)^{\epsilon}$, which proves the second part of the theorem. □

Corollary 5.9 $\text{Ш}(E/K)_{p^\infty}$ is divisible in $H^1(K, E)_{p^\infty}$.

Proof Since we can choose M arbitrarily large in Theorem 5.8, this follows from Lemma 4.6. □

Corollary 5.10 Every element of $\text{Ш}(E/K)_{p^\infty}$ splits over a field ramified at at most two primes of K.

Proof By Corollary 4.2, $d_M(n)$ splits over K_n. □

REFERENCES

[1] B. H. Gross, *Kolyvagin's work on modular elliptic curves.* This volume.

[2] B. H. Gross and D. Zagier, *Heegner points and derivatives of L-series.* Invent. Math. **84**, 225–320 (1986).

[3] V. A. Kolyvagin, *Euler Systems.* To appear in a Birkhäuser volume in honor of Grothendieck.

[4] V. A. Kolyvagin, *Finiteness of* $E(\mathbf{Q})$ *and* Ш$(E/\mathbf{Q})$ *for a class of Weil curves.* Izv. Akad. Nauk SSSR **52** (1988).

[5] V. A. Kolyvagin. *On the structure of Shafarevich–Tate groups.* To appear in the proceedings of USA–USSR Symposium on Algebraic Geometry, Chicago, 1989, published in the Springer Lecture Notes series.

[6] J. S. Milne, *Arithmetic duality theorems.* Perspectives in Mathematics. Academic Press, 1986.

Arithmetic of diagonal quartic surfaces I

R. G. E. PINCH AND H. P. F. SWINNERTON-DYER

1 INTRODUCTION

This is the first of what we hope will be a number of papers on the arithmetic of diagonal quartic surfaces

$$V: \quad a_0X_0^4 + a_1X_1^4 + a_2X_2^4 + a_3X_3^4 = 0 \tag{1}$$

defined over $\mathbf{Q}$, where $a_0a_1a_2a_3 \neq 0$. In what follows we shall always assume that the a_ν are integers with no common factor, which clearly involves no loss of generality. We are interested in $K3$ surfaces more generally, as being the simplest kind of variety whose arithmetic theory is still rudimentary, but there are three advantages in confining ourselves to the narrower class (1): it is possible to write down the zeta-function explicitly, the Néron–Severi group over $\mathbf{Q}$ is frequently non-trivial, and V has a convenient form and a convenient number of parameters for numerical experimentation.

Our initial purpose in embarking on the work described in this paper was to verify the global Tate conjecture – though, as described below, there is one step still to take. The conjecture states that the rank of the Néron–Severi group of V over $\mathbf{Q}$ is equal to the order of the pole of the relevant L-function at $s = 2$. To compute the rank of the Néron–Severi group, which we do in section 2, involves a subdivision into a large number of cases; the details are lengthy and tedious, and we only give enough of them to make the methods clear and to enable a sufficiently dedicated reader to reproduce our results.

The only interesting cohomology of V is the H^2, which has rank 22. Within this there is a subgroup of rank 20, which after extension of the ground field to $\mathbf{C}$ is spanned by the classes of the curves (or even the straight lines) on V; and there is therefore a quotient group of rank 2. The De Rham cohomology even splits as a direct sum; and we have been told that the same is true of the motivic cohomology, though we do not understand why this is so. But in any case the associated L-function can be written as a product

$$L(V,s) = L'(V,s)L''(V,s)$$

where L' is associated with the subspace of rank 20 and L'' with the quotient space of rank 2. Following Weil [5], we calculate the local factors of L at a good prime (that is, a prime not dividing $2a_0a_1a_2a_3$) by counting the numbers of points defined over $\mathbf{F}_q$ on $\tilde{V}$, the reduction of V mod p; here q runs through the powers of p. If $p \equiv 1 \bmod 4$, just 20 of the characteristic roots have the form ϵp where ϵ is a root of unity; and these are the roots associated with L'. If $p \equiv 3 \bmod 4$ all 22 roots have this form, and to be rigorous it is necessary to calculate the effect of Frobenius on the set of lines on $\tilde{V}$; but it is obvious how to partition the characteristic roots so as to give L' and L'' the simplest form.

The factors at the bad primes can now be uniquely determined by the assumption that L' and L'' satisfy functional equations of standard type. It turns out that L' can be written as a product of between 9 and 20 factors, each of which is either $\zeta(\mathbf{Q}, s-1)$ or a Dirichlet L-series. If G denotes the Néron–Severi group of V over $\mathbf{C}$, this decomposition corresponds to a canonical decomposition of $G \otimes \mathbf{Q}$. It is therefore no surprise that the rank of the Néron–Severi group of V over $\mathbf{Q}$ turns out in every case to be equal to the order of the pole of $L'(V,s)$ at $s = 2$.

The function $4L''$ is equal to a Hecke L-series with Grossencharakter. We obtain its functional equation, showing in particular that it involves symmetry rather than skew-symmetry about the mid-point $s = 3/2$; this agrees with the prevailing philosophy since $3/2$ is not an integer and $L''(V,s)$ ought not to have an interesting value there – whereas skew-symmetry would force it to vanish there. We also exhibit a real number α, depending on V, such that $\alpha L''(V,2)$ is in $\mathbf{Z}$ and we give a table of values of this mysterious integer. We presume that, up to factors not worse than Tamagawa numbers, this integer is the order of a certain cohomology group; it should be possible to prove this by methods similar to those of Coates and Wiles [3] but we have not attempted to do so. It would in particular follow from such a result that $L''(V,2) \neq 0$, which is what is needed to complete the verification of the Tate conjecture for V.

We are indebted to John Coates and Barry Mazur for helpful conversation.

2 COMPUTING THE NÉRON–SEVERI GROUP

We start with a result which is undoubtedly known but for which we have no convenient reference.

Lemma 1 Let G be the Néron–Severi group of V over $\mathbf{C}$. Then G is torsion-free and has rank 20; it is spanned by the classes of the 48 lines on V, and the intersection-number matrix of a basis for G has determinant -2^6.

We sketch a proof. The result does not depend on the values of the a_ν, so it is enough to consider the special surface

$$X_0^4 + X_1^4 - X_2^4 - X_3^4 = 0. \tag{2}$$

Since this is a $K3$ surface, G is torsion-free and

$$\operatorname{rank} G \le h^{1,1} = 20.$$

Let G_0 be the subgroup of G generated by the classes of the 48 lines on (2) and let $W_0 = G_0 \otimes \mathbf{Q}$. We can decompose W_0 according to the action of $\mathrm{Gal}\,(\mathbf{Q}(\epsilon)/\mathbf{Q})$ on it, where $\epsilon = e^{\pi i/4}$; and it is straightforward to find the dimensions of the components and bases for them, provided we know enough relations between the classes of the lines. Writing π temporarily for the class of a plane section of (2) it turns out that all the relations we need are of the following two kinds:

(i) The sum of the classes of any four coplanar lines is π.

(ii) If we have two sets of four lines such that each line in one set meets each line in the other, then these eight lines lie on a quadric and the sum of their classes is equal to 2π.

It turns out that $\dim W_0 = 20$ and that we can find 20 lines for which the intersection-number matrix has determinant minus a power of 2; hence G_0 is of finite index in G and that index is a power of 2.

Now consider the plane

$$X_1 - X_2 = \lambda(X_0 - X_3), \tag{3}$$

which is the general plane through the line $X_0 = X_3$, $X_1 = X_2$ on (2). To find the residual intersection of (2) and (3) we write

$$X_0 = x + z, \quad X_1 = y + \lambda z, \quad X_2 = y - \lambda z, \quad X_3 = x - z$$

in (2); taking out the factor z we obtain

$$x^3 + \lambda y^3 + z^2(x + \lambda^3 y) = 0 \tag{4}$$

which is an elliptic curve with base-point (0,0,1). There is a natural epimorphism from G to the Mordell–Weil group of (4) over $\mathbf{C}(\lambda)$; its kernel is generated by the classes of the components of the singular fibres of the pencil (4), together with the class of the line $X_0 + X_3 = X_1 + X_2 = 0$ which is the locus of the basepoint of (4). Since this kernel is in G_0, to prove $G = G_0$ we need only prove that G and G_0 have the same image in the Mordell–Weil group; and since we already know that $[G : G_0]$ is a power of 2, it suffices to show that the image of G_0 does not admit division by 2 within the Mordell–Weil group over $\mathbf{C}(\lambda)$. This image has rank 6, because the kernel is easily shown to have rank 14.

Now write $\lambda = \mu^3$, $\omega = e^{2\pi i/3}$ and work over $\mathbf{C}(\mu)$; the equation (4) can now be written

$$(x + \mu y)(x + \omega\mu y)(x + \omega^2\mu y) = -z^2(x + \lambda^3 y) \tag{5}$$

and the 2-division points are defined over $\mathbf{C}(\mu)$. There is a well-known homomorphism from the Mordell–Weil group of (5) over $\mathbf{C}(\mu)$ to $(\mathbf{C}(\mu)^*/\mathbf{C}(\mu)^2)^3$ given in this case by

$$(x, y, z) \mapsto (-(\mu^{16} + \mu^8 + 1)(x + \mu y)(x + \lambda^3 y), \ldots, \ldots);$$

its kernel is the set of elements divisible by 2 in the Mordell–Weil group. But direct calculation shows that the image of G_0 in $(\mathbf{C}(\mu)^*/\mathbf{C}(\mu)^2)^3$ is of the form C_2^6, where C_2 denotes the cyclic group of order 2; since the image of G_0 in the Mordell–Weil group has rank 6, an element of the image is only divisible by 2 in the Mordell–Weil group over $\mathbf{C}(\mu)$ if it is already divisible by 2 in the image of G_0. Since the 2-division points are not defined over $\mathbf{C}(\lambda)$, this proves that $G_0 = G$. The last calculation also yields a base for G_0, which gives the last statement in the lemma and makes it straightforward to express the class of any line in terms of the classes of the lines in the base.

The corresponding result also holds in characteristic $p \equiv 1 \bmod 4$, though a quite different proof is needed. If the characteristic is $p \equiv 3 \bmod 4$, however, the Néron–Severi group of $\tilde{V}$ over an algebraically closed field has rank 22. An example of the extra curves that appear is the locus of

$$((1 + t)^n, \quad (1 - t)^n, \quad 2^{1/4}, \quad 2^{1/4}t^n)$$

where $p = 4n - 1$; this lies on (2).

We now return to the assumption of characteristic 0 and consider Γ, the Néron–Severi group of V over $\mathbb{Q}$, for $a_0, \ldots, a_3$ in $\mathbb{Q}^*$. With the help of the information obtained in the course of proving Lemma 1, the calculation of Γ is straightforward. We partition the 48 lines into conjugacy classes over $\mathbb{Q}$. To each conjugacy class S there corresponds an element of Γ, which is the class of the sum of the lines in S. The set of all such elements spans $\Gamma \otimes \mathbb{Q}$; and

$$\Gamma = (\Gamma \otimes \mathbb{Q}) \cap G.$$

The entire calculation takes place inside $G \otimes \mathbb{Q}$, and the details of the proof of Lemma 1 have already given us an explicit base for this vector space and expressions for the class of each line in terms of this base.

But although the calculation is straightforward in any particular case, the enumeration of all the possible cases is tedious. It depends on enumerating the possible partitions of the 48 lines into conjugacy classes over $\mathbb{Q}$, which in its turn depends on the analysis of

$$\text{Gal}\left(\mathbb{Q}(i, (-a_1/a_0)^{1/4}, (-a_3/a_2)^{1/4})/\mathbb{Q}\right) \tag{6}$$

and the two similar groups; for the large field here is the least common field of definition of those 16 lines which have the form

$$X_0 = \alpha X_1, \quad X_2 = \beta X_3 \tag{7}$$

for some α, β. The first step is to look at the possible pairs $c, \quad \mathbb{Q}(i, c^{1/4})$ with c in $\mathbb{Q}^*$; there turn out to be 5 cases according as c or $-4c$ is a fourth power in $\mathbb{Q}$, a square but not a fourth power in $\mathbb{Q}$, or none of these. The next step is to look at the possible triples

$$c_1, \quad c_2, \quad \mathbb{Q}(i, c_1^{1/4}, c_2^{1/4}). \tag{8}$$

There is a coarse classification according to the 5 cases for each of c_1 and c_2, which when account is taken of symmetry gives 15 possibilities. To refine this we need also to consider how the intersection of the two $\mathbb{Q}(i, c_\nu^{1/4})$ is embedded in each of them. This gives in all 24 possibilities for the structure of the field in (8). But in 7 of these it turns out that combinations of lines (7) contribute nothing more to the Néron–Severi group than the class of plane sections of V; and there are 3 other cases for which, though the field extensions involved are different, the contributions to the Néron–Severi group are the same. (These are merged as case VIII in Table 1.)

We have therefore 16 distinct possibilities for the part of the Néron–Severi group of V over $\mathbb{Q}$ which is generated by combinations of lines of the form

(7). The 15 cases in which we obtain more than multiples of the class of plane sections are listed in Table 1, in which we have used the allowable symmetries and the freedom to multiply any a_ν by a fourth power. In Table 1 the first column gives the case number; the second and third express a_1/a_0 and a_3/a_2 respectively in terms of supplementary parameters c_ν with values in $\mathbf{Q}^*$; the fourth gives the constraints, in terms of those expressions which are forbidden to be fourth powers; the fifth gives r, the rank of that part of the Néron–Severi group of V over $\mathbf{Q}$ which is generated by combinations of lines (7); and the sixth gives d, the order of the group (6).

Now $a_0a_1a_2a_3$ is necessarily a square in cases I to VIII, necessarily minus a square in cases IX to XIII, and necessarily neither in cases XV and XVI. Moreover every surface (1) for which $a_0a_1a_2a_3$ is a square falls under one of cases I to VIII. It is therefore advantageous to divide the surfaces (1) for which $r = 1$ into two cases: case XIV for those for which $-a_0a_1a_2a_3$ is a square and case XVII for the remainder. These cases are also included in Table 1.

The variables $X_0, \dots, X_3$ can be paired in three different ways, and any surface (1) therefore belongs to a case in Table 1 in three ways. The next step is to list the possible triples of cases, and the corresponding constraints on the a_ν. This is done in the first three columns of Table 2. The first column gives the case number; the second column gives a_1, a_2 and a_3 in terms of auxiliary parameters c_ν, where it is assumed that the surface has been normalised so that $a_0 = 1$; and the third column gives the three references back to Table 1. The constraints on the values of the c_ν are once again that certain products should not be fourth powers; for reasons of space these products have not been written down explicitly, but they can easily be read off from Table 1. Here $a_0a_1a_2a_3$ is a square in cases 1 to 23, minus a square in cases 24 to 39, and neither in cases 40 to 49.

Let Γ denote the Néron–Severi group of V over $\mathbf{Q}$. In each of the cases in Table 2, the information underlying the values of r in Table 1 enables us to write down a set of elements which span $\Gamma \otimes \mathbf{Q}$; using the information derived in the proof of Lemma 1, we can now obtain the last two columns of Table 2. These give respectively n, the rank of Γ, and Δ, the determinant of the intersection-number matrix of a base of Γ.

3 THE LOCAL L-FUNCTION

Let π be an odd prime in $\mathbf{Z}[i]$ and write $q = \text{Norm}(\pi)$; thus either $q = p^2$ with $p \equiv 3 \bmod 4$ or $q = p$ with $p \equiv 1 \bmod 4$. To any non-zero element x in

$\mathbf{F}_q$ there corresponds a unique fourth root of unity ϵ in $\mathbf{Z}[i]$ whose reduction $\bmod \pi$ is equal to $x^{(q-1)/4}$. In this way we define the multiplicative character χ on $\mathbf{F}_q^*$ by $\chi(x) = \epsilon$. Fix also a non-trivial additive character ψ on $\mathbf{F}_q$; then the Gauss sum $g(r)$ is defined to be

$$g(r) = \sum \chi^r(x)\psi(x) \qquad (r = 1, 2, 3)$$

where the sum is over the elements x of $\mathbf{F}_q^*$.

We now assume that V has a good reduction $\tilde{V}$ mod π – that is, that p does not divide $a_0a_1a_2a_3$. A special case of the main result of Weil [5] is that the roots of the L-function of $\tilde{V}$ over $\mathbf{F}_q$ are q and the 21 numbers

$$q^{-1}\overline{\chi}(a_0^{r_0} \dots a_3^{r_3})g(r_0)\dots g(r_3) \tag{9}$$

where each $r_\nu = 1, 2$ or 3 and

$$r_0 + r_1 + r_2 + r_3 \equiv 0 \bmod 4. \tag{10}$$

It is easy to see that these numbers do not depend on the choice of ψ. We must now split cases according as $p \equiv 3$ or $p \equiv 1 \bmod 4$.

Lemma 2 If $p \equiv 3 \bmod 4$ then each number (9) is equal to q.

Proof Since $(q-1)/4$ is a multiple of $(p-1)$ and each a_ν is in $\mathbf{Z}$, we have $a_\nu^{(q-1)/4} \equiv 1 \bmod p$ and hence $\chi(a_\nu) = 1$. Choose

$$\psi(x) = e^{2\pi i \mathrm{Tr}(x)/p};$$

then we must show that $g(r) = \pm p$ and $g(1) = g(3)$.

For this, let σ denote the non-trivial automorphism of $\mathbf{F}_q/\mathbf{F}_p$; then $\mathrm{Tr}(\sigma x) = \mathrm{Tr}(x)$ and $\chi(\sigma x) = \overline{\chi}(x)$, the latter because $\sigma x = x^p$. Hence if $\chi^r(x) = \pm i$, the terms from x and from σx in the sum for $g(r)$ cancel. In other words, in calculating $g(1)$ and $g(3)$ we need only take account of the x with $\chi(x) = \pm 1$; and now $g(1) = g(3)$ because the two reduced sums are term-for-term the same. Using the same idea again, for each r we have

$$\overline{g(r)} = \sum \chi^r(x)\overline{\psi}(x) = \sum \chi^r(-x)\psi(x) = g(r)$$

since $\overline{\psi}(-x) = \psi(x)$ and $\chi(-1) = 1$. Since it is well known that $|g(r)| = q^{1/2}$, this gives $g(r) = \pm p$.

Now in every product (9) the number of factors $g(1) = g(3)$ is even, and so is the number of factors $g(2)$. The lemma follows immediately.

It follows that the roots of the L-function of $\tilde{V}$ over F_p are all $\pm p$. But for any n in F_p the equations $x^4 = n$ and $y^2 = n$ have the same number of solutions in F_p; hence $\tilde{V}$ and

$$a_0 Y_0^2 + a_1 Y_1^2 + a_2 Y_2^2 + a_3 Y_3^2 = 0 \tag{11}$$

have the same number of points over F_p. But the number of points on (11) is well known to be

$$\begin{array}{ll} (p+1)^2 & \text{if} \quad a_0a_1a_2a_3 \quad \text{is a square in} \quad \mathsf{F}_p, \\ p^2+1 & \text{otherwise.} \end{array}$$

Hence the roots of the L-function of $\tilde{V}$ over F_p can be described as 11 copies of p, 10 copies of $-p$ and one copy of $\left(\frac{a_0a_1a_2a_3}{p}\right)$ where the bracket is the quadratic residue symbol.

We now turn to the case $p \equiv 1 \bmod 4$. It is well known that $|g(r)| = p^{1/2}$. In $g(2)$ the terms from x and from $-x$ are complex conjugates; so $g(2)$ is real and $(g(2))^2 = p$. Again

$$\overline{g(3)} = \sum \chi(x)\psi(-x) = \chi(-1)g(1)$$

whence $g(1)g(3) = p\chi(-1)$. Moreover the 21 quadruplets which satisfy (10) consist of $(2,2,2,2)$, $(1,1,1,1)$, $(3,3,3,3)$, 12 like $(1,2,2,3)$ and 6 like $(1,1,3,3)$; hence the roots of the L-function of $\tilde{V}$ over F_p consist of

$$\left.\begin{array}{r} p,\ p\chi^2(a_0a_1a_2a_3), \\ p^{-1}\chi^3(a_0a_1a_2a_3)(g(1))^4,\ p^{-1}\chi(a_0a_1a_2a_3)(g(3))^4, \\ \text{twelve like} \quad p\chi(-a_0a_1^2a_2^2a_3^3), \\ \text{six like} \quad p\chi(a_0a_1a_2^3a_3^3). \end{array}\right\} \tag{12}$$

Of these, it is the second pair that are in the least satisfactory form, and to improve matters we turn to Weil [6]. It is there shown that as ideals

$$(g(1))^4 = (\pi^2 p), \quad (g(3))^4 = (\overline{\pi^2 p})$$

and that as numbers $g(1)^4$ and $g(3)^4$ are Grossencharakters mod 4. By further computing their values for $\pi = 1 + 2i$ we obtain

$$g(1)^4 = \pi^2 p, \quad g(3)^4 = \bar{\pi}^2 p$$

provided $\pi \equiv 1 \bmod 2$; hence under this convention we can replace the second line of (12) by

$$\pi^2\chi^3(a_0a_1a_2a_3), \quad \bar{\pi}^2\chi(a_0a_1a_2a_3). \tag{13}$$

For future reference we also recall some facts about the biquadratic residue symbol. Assume $\pi = \lambda + i\mu \equiv 1 \bmod (2+2i)$; then

$$\chi(2) = i^{-\mu/2}, \quad \chi(-1) = \chi(2)^2.$$

For α, β in $\mathbf{Z}[i]$, coprime and with $\beta \equiv 1 \bmod (2+2i)$ we define the symbol $\left(\frac{\alpha}{\beta}\right)_4$ to be multiplicative in each variable and to satisfy

$$\left(\frac{\alpha}{\pi}\right)_4 = \chi(\alpha) \tag{14}$$

where $\pi \equiv 1 \bmod (2+2i)$. To express the law of biquadratic reciprocity, write $\alpha = \lambda + i\mu \equiv 1 \bmod (2+2i)$ and similarly for α'; then

$$\left(\frac{\alpha}{\alpha'}\right)_4 = (-1)^{\mu\mu'/4}\left(\frac{\alpha'}{\alpha}\right)_4.$$

For a modern treatment of this, see [2], pp 348–55.

4 THE GLOBAL L-FUNCTION

Up to factors corresponding to the bad primes, the L-function of V is therefore equal to

$$\prod_p{}' \prod_{\nu=1}^{22} (1 - \alpha_{p\nu} p^{-s})^{-1} \tag{15}$$

where the product is taken over all primes p not dividing $2a_0a_1a_2a_3$ and the $\alpha_{p\nu}$ are the roots calculated in the previous section. The factors at the bad primes can be defined by means of the action of Frobenius on the étale cohomology; but unfortunately it seems very difficult to compute with étale cohomology. We therefore follow another route.

Denote by $L(V, s)$ the correct L-series including the factors for the bad primes. It is a standard conjecture that $L(V, s)$ can be analytically continued to the whole s-plane as a meromorphic function, and that it satisfies a functional equation of known shape relating $L(V, s)$ and $L(V, 3-s)$. This requirement determines the missing factors uniquely. To apply it, we show that up to factors at the bad primes (15) can be written in terms of a Hecke L-series and the zeta-functions of certain algebraic number fields; if we use this statement to supply factors at the bad primes, we obtain a functional equation of the correct shape.

As a first step in this decomposition, we define W' to be the subspace of $H^2_{et}(V)$ generated by the elements of the Néron–Severi group of V over $\mathbf{C}$. Then we have an exact sequence

$$0 \to W' \to H^2_{et}(V) \to W'' \to 0$$

with $\mathrm{Gal}(\overline{\mathbf{Q}}/\mathbf{Q})$ acting on each term; and $\dim W' = 20$, $\dim W'' = 2$. The action of Frobenius on W' must be the same as on the Néron–Severi group, modulo standard conventions; so for each p we can find the 20 among the $\alpha_{p\nu}$ which belong to W' and hence the 2 which belong to W''. For $p \equiv 1 \bmod 4$ the latter have to be the two values (13) since they are the only $\alpha_{p\nu}$ which are not p times a root of unity; for $p \equiv 3 \bmod 4$ they turn out to be p and $-p$. Let $L'(V,s)$ and $L''(V,s)$ be the L-series corresponding to W' and W'' respectively; then we have

$$L(V,s) = L'(V,s)L''(V,s).$$

We now turn to these two factors separately.

5 THE FUNCTION L'

Throughout this section we adopt the convention that statements involving infinite products are to be taken modulo factors at the bad primes.

With the root p for $p \equiv 1 \bmod 4$ we associate the root p for $p \equiv 3 \bmod 4$; the resulting factor of $L'(V,s)$ is

$$\prod (1 - p^{1-s})^{-1} = \zeta(\mathbf{Q}, s-1).$$

For any non-square c in $\mathbf{Z}$ we write

$$\begin{aligned} D_1(c,s) &= \zeta(\mathbf{Q}(c^{1/2}),\, s-1)/\zeta(\mathbf{Q},\, s-1) \\ &= \prod \left(1 - \left(\frac{c}{p}\right) p^{1-s}\right)^{-1}. \end{aligned} \tag{16}$$

As a Dirichlet L-series, this has a holomorphic extension to the whole s-plane; and its functional equation can be read off from that of the zeta function. With the root $p\chi^2(a_0a_1a_2a_3)$ for $p \equiv 1 \bmod 4$ we associate the root $p\left(\frac{a_0a_1a_2a_3}{p}\right)$ for $p \equiv 3 \bmod 4$; the resulting factor of $L'(V,s)$ will be $D_1(a_0a_1a_2a_3,\, s)$ if $a_0a_1a_2a_3$ is not a square, and $\zeta(\mathbf{Q},\, s-1)$ otherwise.

Now suppose that c in $\mathbf{Z}$ is such that neither of $\pm c$ is a square, and write

$$\left.\begin{aligned} D_2(c,s) &= \frac{\zeta(\mathbf{Q}(c^{1/4}),\, s-1)}{\zeta(\mathbf{Q}(c^{1/2}),\, s-1)} = \frac{\zeta(\mathbf{Q}((-4c)^{1/4}),\, s-1)}{\zeta(\mathbf{Q}((-4c)^{1/2}),\, s-1)} \\ &= \prod_{p\equiv 1}\{(1-\chi(c)p^{1-s})(1-\overline{\chi}(c)p^{1-s})\}^{-1} \prod_{p\equiv 3}(1-p^{2-2s})^{-1}. \end{aligned}\right\} \tag{17}$$

That the middle expression in the top line is equal to the second line follows by studying the factorisation of good primes in the relevant field extensions;

for the third expression in the top line we use also $\chi(-4) = 1$. Hence the quotient of the second and third expressions on the top line is a finite product (over bad primes); but it satisfies a functional equation of a shape which is only compatible with the last statement if the finite product is trivial. As a Dirichlet L-series over $\mathbf{Q}(i)$, $D_2(c,s)$ has a holomorphic extension to the whole s-plane; and as with D_1 its functional equation can be read off from that of the zeta function.

The 12 roots in the third line of (12), and the 6 roots in the fourth line, are complex conjugate in pairs; and each pair has the form $p\chi(c)$, $p\overline{\chi}(c)$ where c is an expression in the a_ν. With each pair we associate the roots $p, -p$ for $p \equiv 3 \bmod 4$. If neither of $\pm c$ is a square, the resulting factor of $L'(V,s)$ is $D_2(c,s)$. If c is a square but not a fourth power, say $c = c_1^2$, then the resulting factor of $L'(V,s)$ is $D_1(c_1,s)\,D_1(-c_1,s)$; if c is a fourth power the resulting factor is $\zeta(\mathbf{Q},\, s-1)\,D_1(-1,s)$. If $-c$ is a square, we replace c by $-4c$ and apply the previous rules.

We have thus obtained an expression for $L'(V,s)$ as a product of between 11 and 20 classical functions, the number depending on the a_ν. In particular, the rule for determining the order of the pole at $s = 2$ is as follows: count 1 anyway, 1 if $a_0a_1a_2a_3$ is a square, and 1 for each pair of numbers like $-a_0a_1^2a_2^2a_3^3$ or $a_0a_1a_2^3a_3^3$ which have the form n^4 or $-4n^4$. If we apply this rule to the 49 cases in Table 2, we find that in each case the order of the pole of $L'(V,s)$ at $s = 2$ is equal to the rank of the Néron–Severi group of V over $\mathbf{Q}$. This reduces the Tate conjecture for V to the assertion that $L''(V,s)$ is regular and non-zero at $s = 2$. There is of course a motive associated with $L'(V,s)$, and presumably we have verified the Tate conjecture for that; but we also presume that in that case the Tate conjecture follows from the general motivic machinery.

We shall not attempt to relate the explicit formula for $L'(V,s)$ which we have just obtained to the conjectures of Beilinson, still less to those of Bloch and Kato. But it is natural to ask why $L'(V,s)$ decomposes as a product of simpler functions and what is the geometric significance of the fields which occur in the formula for $L'(V,s)$. A partial answer to the latter question comes from considering what replaces $L'(V,s)$ if $\mathbf{Q}$ is replaced by an algebraic number field k as the field of definition of V while leaving the a_ν the same. The rule is as follows. For each field E such that $\zeta(E,\, s-1)$ appears in the formula for $L'(V,s)$, write

$$E \otimes_{\mathbf{Q}} k = E_1 \oplus \ldots \oplus E_r$$

where the E_i are algebraic number fields; then in the formula for $L'(V,s)$ replace $\zeta(E, s-1)$ by $\prod \zeta(E_i, s-1)$. Hence it is the set of fields E that controls the increase in the order of the pole of $L'(V,s)$ at $s=2$ as the base field k increases – and so presumably also the rank of the Néron–Severi group of V over k.

But we can do better than this. To simplify the exposition, we change our hypotheses, so that we temporarily assume that $a_0, \ldots, a_3$ are independent indeterminates over $\mathbf{Q}$ and we consider the Néron–Severi group of V over

$$K = \mathbf{Q}(a_0, \ldots, a_3)$$

and its algebraic extensions. Let G be the Néron–Severi group of V over $\overline{K}$ and write $W = G \otimes \mathbf{Q}$. There is a unique decomposition of W as a sum of irreducible vector spaces each fixed under $\mathrm{Gal}(\overline{K}/K)$,

$$W = W_1 \oplus \ldots \oplus W_{11};$$

not surprisingly, the W_i are orthogonal with respect to intersection-number. After renumbering, we can take W_1 to be one-dimensional and generated by the Néron–Severi group of V over K – in other words, by the class of plane sections. W_2 is also one-dimensional; it contains non-zero elements of the Néron–Severi group of V over $K_1 \supset K$ if and only if $(a_0a_1a_2a_3)^{1/2}$ lies in K_1.

Each of the 9 spaces $W_3, \ldots, W_{11}$ corresponds to one of the 9 numbers c introduced in the fourth paragraph of this section. The space W_i can be written as the sum of two one-dimensional spaces each fixed under $\mathrm{Gal}(\overline{K}/K_1)$ if and only if K_1 contains at least one of $c^{1/2}$ and $(-4c)^{1/2}$; if K_1 contains both these, there are an infinity of such decompositions. W_i contains non-zero elements of the Néron–Severi group of V over K_1 if and only if K_1 contains at least one of $c^{1/4}$ and $(-4c)^{1/4}$; and W_i is spanned by elements of this Néron–Severi group if and only if K_1 contains both these expressions. These facts go a long way towards explaining the double formulae (17) for $D_2(c,s)$.

It may be useful to record the missing factors in the infinite products (16) and (17) and the functional equations for the components $\zeta(\mathbf{Q}, s-1)$, $D_1(c,s)$ and $D_2(c,s)$ over L$'(V,s)$. If $g(s)$ is any one of these three functions, its functional equation has the form

$$(\Gamma\text{-factor})\ f^{s/2} g(s) \text{ is invariant under } s \mapsto 3-s$$

where f is defined to be the *conductor* of the function $g(s)$.

For $\zeta(\mathbf{Q}, s-1)$ the Γ-factor is $\pi^{-s/2}\Gamma((s-1)/2)$ and $f=1$.

For $D_1(c,s)$ the Γ-factor is $\pi^{-s/2}\Gamma((s-1)/2)$ if $c>0$ and $\pi^{-s/2}\Gamma(s/2)$ if $c<0$. Without loss of generality we can assume that c is square-free. Then the odd part of f is the product of the odd primes dividing c; and the even part is 1, 8 or 4 according as $c \equiv 1$, 2 or 3 mod 4. There is a missing factor in the infinite product (16) if and only if $c \equiv 1 \bmod 4$; and the factor is $(1-2^{1-s})^{-1}$ if $c \equiv 1 \bmod 8$, but $(1+2^{1-s})^{-1}$ if $c \equiv 5 \bmod 8$.

For $D_2(c,s)$ the Γ-factor is $(2\pi)^{-s}\Gamma(s-1)$. Without loss of generality we can assume that c is fourth-power free. Then the odd part of f is the square of the product of the distinct odd primes which divide c; the even part of f is 2^8 if $2 \parallel c$ or $2^3 \parallel c$, 2^6 if $c \equiv 3 \bmod 4$ or $4 \bmod 16$, 2^4 if $c \equiv 5 \bmod 8$ or $12 \bmod 32$, and 2^2 if $c \equiv 1 \bmod 8$ or $28 \bmod 32$. There is a missing factor in the infinite product (17) if and only if $c \equiv 1 \bmod 8$ or $28 \bmod 32$; and the factor is $(1-2^{1-s})^{-1}$ if $c \equiv 1 \bmod 16$ or $60 \bmod 64$, but $(1+2^{1-s})^{-1}$ if $c \equiv 9 \bmod 16$ or $28 \bmod 64$.

We can immediately derive from these the missing factors and functional equation for $L'(V,s)$ itself; but there seems to be no simplification in doing so.

6 THE FUNCTION L''

The $\alpha_{p\nu}$ associated with $L''(V,s)$ are the numbers (13) for $p \equiv 1 \bmod 4$ and p and $-p$ for $p \equiv 3 \bmod 4$; so $L''(V,s)$ depends on V only through the value of

$$c = a_0 a_1 a_2 a_3.$$

Moreover the product over the good primes is

$$\prod_{p\equiv 1} \{(1-\overline{\chi}(c)\pi^2 p^{-s})(1-\chi(c)\overline{\pi}^2 p^{-s})\}^{-1} \prod_{p\equiv 3} (1-p^{2-2s})^{-1} \tag{18}$$

and using the notation (14) this is equal to

$$\sum \alpha^2 \left(\frac{c}{\alpha}\right)_4^3 (\operatorname{Norm} \alpha)^{-s} \tag{19}$$

where the sum is over all α prime to c such that $\alpha \equiv 1 \bmod (2+2i)$. The product (18) is not changed by replacing c by $-4c$, and not significantly affected by multiplying c by a fourth power; so in what follows we shall always assume that c is fourth power free and on occasion we shall also assume that $c>0$.

The series (19) is not in a convenient form because of the congruence condition on α. For any odd α prime to c we therefore write

$$\alpha = i^n \alpha_0 \quad \text{with} \quad \alpha_0 \equiv 1 \bmod (2+2i);$$

then 4 times the expression (18) or (19) is equal to $\sum \alpha^2\phi(\alpha)(\mathrm{Norm}\,\alpha)^{-s}$ where

$$\phi(\alpha) = (-1)^n \left(\frac{c}{\alpha}\right)_4^3$$

and the sum is over all α in $\mathbf{Z}[i]$ prime to $2c$. Write

$$c = 2^m \epsilon c_0 \quad \text{with} \quad \epsilon = \pm 1,\ c_0 \equiv 1 \bmod 4. \tag{20}$$

Using the results at the end of section 3 we find that the odd part of the conductor of ϕ is the product of the odd primes dividing c; the even part is 8 if m is odd, 4 if $m \equiv 1 + \epsilon \bmod 4$, 2 if $m \equiv 1 - \epsilon \bmod 4$ and $c_0 \equiv 1 \bmod 8$, and 1 if $m \equiv 1 - \epsilon \bmod 4$ and $c_0 \equiv 5 \bmod 8$. Thus the conductor can always be written as (f) where $f > 0$ is in $\mathbf{Z}$. Moreover if f is odd, the definition of ϕ can be completed by

$$\phi(1+i) = i^{(c_0^2-1)/8}$$

together with multiplicativity. To mark the dependence of $L''(V,s)$ on c we therefore write

$$4L^*(c,s) = 4L''(V,s) = \sum \alpha^2\phi(\alpha)(\mathrm{Norm}\,\alpha)^{-s} \tag{21}$$

where now the sum is over all α in $\mathbf{Z}[i]$ prime to f. There is a missing factor in (18) if and only if f is odd, and the missing factor is then

$$\{1 - (-1)^{(c_0-5)/8} 2^{1-s}\}^{-1}.$$

The calculation of the functional equation follows Hecke [4]. The fundamental theta formula is

$$\sum_\mu \exp\{-\pi t(f\mu + u_1)(f\bar{\mu} + u_2)\}$$
$$= t^{-1} f^{-2} \sum_\nu \exp\{-\pi t^{-1} f^{-2} \nu\bar{\nu} + \pi i f^{-1}(\bar{\nu} u_1 + \nu u_2)\}$$

where $t > 0$, u_1 and u_2 are arbitrary complex numbers and μ, ν each run through the elements of $\mathbf{Z}[i]$. This can easily be derived from the equation for the classical theta function. Alternatively it holds for $u_2 = \bar{u}_1$, because in that case the right hand side is just the double Fourier series expansion of the left; and it therefore holds in general because both sides are analytic in u_1, u_2. Differentiating twice with respect to u_2 and setting $\bar{u}_2 = u_1 = u$, we obtain

$$\begin{aligned} &\sum (f\mu + u)^2 \exp\{-\pi t(f\mu + u)(f\bar{\mu} + \bar{u})\} \\ &\quad = -t^{-3} f^{-4} \sum \nu^2 \exp\{-\pi t^{-1} f^{-2} \nu\bar{\nu} + \pi i f^{-1}(\bar{\nu} u + \nu\bar{u})\}; \end{aligned} \tag{22}$$

this too can be obtained directly by Fourier series arguments.

Multiply this last equation by $\phi(u)$ and sum over a complete set of representatives of residue classes mod f prime to f. On the left we obtain

$$\sum \alpha^2\phi(\alpha)\exp(-\pi t\alpha\overline{\alpha})$$

where the sum is over all α in $\mathbf{Z}[i]$ prime to f. For fixed ν not prime to f, the summand on the right of (22) only depends on u mod $f/(f,\overline{\nu})$. But the sum of $\phi(u)$ over a set of residue classes mod f which are equal mod $f/(f,\overline{\nu})$ is 0 because f is the conductor of ϕ. Hence on the right we need only consider the terms with ν prime to f, and we obtain

$$\begin{aligned}-t^{-3}f^{-4}\sum\sum\phi(\overline{\nu}u)\exp(\pi i f^{-1}\mathrm{Tr}\overline{\nu}u)\nu^2\phi(\nu)\exp(-\pi t^{-1}f^{-2}\nu\overline{\nu})\\ = -t^{-3}f^{-4}W(\phi)\sum\nu^2\phi(\nu)\exp(-\pi t^{-1}f^{-2}\nu\overline{\nu})\end{aligned}$$

where

$$W(\phi)=\sum\phi(u)\exp(\pi i f^{-1}\mathrm{Tr}\,u),$$

the sum being taken over a set of representatives of the residue classes mod f prime to f in $\mathbf{Z}[i]$. If we define

$$\theta(t,\phi)=\sum\alpha^2\phi(\alpha)\exp(-\pi t f^{-1}\alpha\overline{\alpha})$$

then we obtain the functional equation

$$\theta(t,\phi)=-t^{-3}f^{-1}W(\phi)\theta(t^{-1},\phi).$$

It follows from this that $W(\phi)=\pm f$, but we can in fact do better:

Lemma 3 $W(\phi)=-f$ and $\theta(t,\phi)=t^{-3}\theta(t^{-1},\phi)$.

Proof We need only prove the first result. For this, write

$$f=\pm q_0q_1\ldots q_r$$

where q_0 is a power of 2 and the q_ν for $\nu>0$ are $\pm$ primes in $\mathbf{Z}$ with $q_\nu\equiv 1 \bmod 4$. Working with $8f$ instead of f, we can choose Q_ν in $\mathbf{Z}$ so that

$$1=Q_0+Q_1+\ldots+Q_r$$

where Q_0 is divisible by f/q_0 and Q_ν by $8f/q_\nu$ for $\nu>0$. In particular

$$Q_0\equiv 1 \bmod 8,\quad Q_\nu\equiv 1 \bmod q_\nu \quad \text{for} \quad \nu>0. \tag{23}$$

We obtain a complete set of values of u by writing

$$u = Q_0 u_0 + \ldots + Q_r u_r$$

where each u_ν runs through a set of representatives of the elements of $\mathbf{Z}[i]/(q_\nu)$ prime to q_ν; if $q_0 = 1$ we take $u_0 = \pm 1$ so that u is always odd. Moreover (23) implies $u \equiv u_0 \bmod 8$, so that the value of n in

$$u \equiv i^n \bmod (2+2i)$$

depends only on u_0. Note also that

$$\left(\frac{i}{q_\nu}\right)_4 = (-1)^{(q_\nu-1)/4} \tag{24}$$

for $\nu > 0$.

Now extend (20) by writing

$$c = 2^m \epsilon q_1^{m_1} \ldots q_r^{m_r}.$$

Using the results at the end of section 3 we obtain

$$\phi(u) = (-1)^{n(c_0+3)/4} \left(\frac{2}{u_0}\right)_4^{3m} \left(\frac{\epsilon}{u_0}\right) \prod \left(\frac{u_\nu}{q_\nu}\right)_4^{3m_\nu}.$$

Hence $W(\phi) = S_0 S_1 \ldots S_r$ where

$$S_0 = \sum (-1)^{n(c_0+3)/4} \left(\frac{2}{u_0}\right)_4^{3m} \left(\frac{\epsilon}{u_0}\right) \exp(\pi i q_0^{-1} \operatorname{Tr} u_0),$$

$$S_\nu = \sum \left(\frac{u_\nu}{q_\nu}\right)_4^{3m_\nu} \exp(\pi i q_\nu^{-1} \operatorname{Tr} u_\nu) \quad \text{for} \quad \nu > 0.$$

It remains to evaluate these sums.

Consider first S_ν for $\nu > 0$. We have $\left(\frac{u_\nu}{q_\nu}\right)_4 = 1$ for u_ν in $\mathbf{Z}$. In the terms with $\operatorname{Tr} u_\nu \equiv 0 \bmod 2q_\nu$ we can take u_ν pure imaginary; so by (24) these terms contribute

$$(|q_\nu| - 1)(-1)^{m_\nu(q_\nu-1)/4}$$

to S_ν. But the sum of the $\left(\frac{u_\nu}{q_\nu}\right)^{3m_\nu}$ over the u_ν with any other fixed value of $\operatorname{Tr} u_\nu \bmod 2q_\nu$ does not depend on that fixed value, because we can change the fixed value by multiplying the relevant u_ν by an element of $\mathbf{Z}$; hence that sum is $-(-1)^{m_\nu(q_\nu-1)/4}$ since the sum over all u_ν is 0. Thus finally we obtain

$$S_\nu = (-1)^{m_\nu(q_\nu-1)/4} \, |q_\nu|,$$

and therefore

$$S_1 \dots S_r = (-1)^{(c_0-1)/4} \mid q_1 \dots q_r \mid . \tag{25}$$

It remains to evaluate S_0. If $q_0 = 1$ then $S_0 = 1$, but then $c_0 \equiv 5 \bmod 8$. If $q_0 = 2$ then $c_0 \equiv 1 \bmod 8$ and $S_0 = -2$. If $q_0 = 4$ then $\epsilon i^m = -1$ and hence $S_0 = 4(-1)^{(c_0+3)/4}$. If $q_0 = 8$ then m is odd and hence $S_0 = 8(-1)^{(c_0+3)/4}$. Combining these results with (25) we obtain in all cases $W(\phi) = -f$, which proves the lemma.

The Mellin transform equation relating θ and L^* is

$$\begin{aligned} 4\Gamma(s)(f/\pi)^s L^*(c,s) &= \int_0^\infty t^{s-1}\theta(t,\phi)dt \\ &= \int_1^\infty (t^{s-1} + t^{2-s})\theta(t,\phi)dt \end{aligned}$$

and the right hand side is not changed if we replace s by $3 - s$; this gives analytic continuation and the functional equation for L^*. In particular

$$L^*(c,2) = \sum \frac{(2\pi\alpha\overline{\alpha} + f)\phi(\alpha)}{4f\overline{\alpha}^2} e^{-\pi\alpha\overline{\alpha}/f}$$

where the sum is taken over all α in $\mathbf{Z}[i]$ prime to f. This is a convenient formula for computation and, when combined with the theorem below, is the source of Table 3.

We now turn to the algebraicity of $L^*(c,2)$, which depends on the identity – easily derived from (21) –

$$4L^*(c,2) = f^{-2} \sum \phi(u)\wp(\overline{u}f^{-1}; 1, i)$$

where u runs through a set of representatives of the elements of $\mathbf{Z}[i]/(f)$ prime to f. It is convenient to scale the Weierstrass $\wp$-function so that it has periods ω, $i\omega$ where

$$\omega = 2^{1/2}\pi e^{-\pi/6} \prod (1 - e^{-2\pi n})^2 = 2.6220575...$$

is such that $\wp(z; \omega, i\omega)$ satisfies $\wp'^2 = 4\wp^3 - 4\wp$; thus

$$4f^2\omega^{-2}L^*(c,2) = \sum \phi(u)\wp(\omega\overline{u}f^{-1}; \omega, i\omega). \tag{26}$$

Lemma 4 With the notation above, $c^{1/4}\omega^{-2}L^*(c,2)$ is in $\mathbf{Q}(i)$.

This follows from standard results on class field theory over $\mathbf{Q}(i)$; for the detailed proof of an analogous result see [1], Corollary to Lemma 7.

Corollary If $c > 0$ then $c^{1/4}\omega^{-2}L^*(c,2)$ is in $\mathbf{Q}$.

For if $s > 2$ is real, the terms in (21) are either real or complex conjugate in pairs; hence $L^*(c,s)$ is real. Now let $s \to 2$.

Lemma 5 Let $\mu = f^{2/(f^2-1)}$ if $f \equiv 3 \bmod 4$ is prime, and $\mu = 1$ otherwise; then $\mu\wp(\omega\overline{u}f^{-1})$ is an algebraic unit.

Proof See [1], Lemmas 2 and 3.

Theorem Suppose $c > 0$ and $f > 3$; then $c^{1/4}fq_0\omega^{-2}L^*(c,2)$ is in $\mathbf{Z}$, where q_0 is the even part of f.

Proof By the Corollary to Lemma 4, $c^{1/4}fq_0\omega^{-2}L^*(c,2)$ is in $\mathbf{Q}$. Since $f > 2$, the terms on the right of (26) come in equal sets of four, and this compensates for the factor 4 on the left. It follows from Lemma 5 that the denominator of the expression we are interested in divides

$$c^{-1/4}(f/q_0)\mu.$$

The numerator of this is odd, and it can only be divisible by an odd prime p in $\mathbf{Z}$ if $p = f$ and $2/(f^2-1) \geq 1/4$ – that is, if $f = 3$. This proves the theorem.

It will be seen from Table 3 that $f = 2$, which corresponds to $c = 1$, is genuinely an exception and that $f = 3$, which corresponds to $c = 12$ or 108, is an exception in the former case but not in the latter.

The evidence of Table 3 suggests that, under the same restrictions,

$$2^{-r}c^{1/4}f\omega^{-2}L^*(c,2)$$

is in $\mathbf{Z}$, where r is the number of odd prime factors of f; but we have not attempted to prove this. However it is clearly consistent with the comments in the Introduction.

APPENDIX

Table 1. Cases for the set of lines (7).

Case	a_1/a_0	a_3/a_2	Not a fourth power	r	d
I	-1	-1		6	2
II	4	4		6	2
III	$-c^2$	$-c^2$	c^2	4	4
IV	$4c^2$	$4c^2$	c^2	4	4
V	-1	$-c^2$	c^2	3	4
VI	4	$4c^2$	c^2	3	4
VII	c	c	c^2	3	8
VIII	c_1c_2	$c_1c_2^{-1}$	$c_1^2, c_2^2, -c_1c_2, 4c_1c_2, -c_1c_2^{-1}, 4c_1c_2^{-1}$	2	8 or 16
IX	-1	4		5	2
X	$-c^2$	$4c^2$	c^2	3	4
XI	-1	$4c^2$	c^2	2	4
XII	$-c^2$	4	c^2	2	4
XIII	$-c$	$4c$	c^2	2	8
XIV	c_1c_2	$-4c_1c_2^{-1}$	$c_1^2, c_2^2, -c_1c_2, 4c_1c_2, -c_1c_2^{-1}, 4c_1c_2^{-1}$	1	8 or 16
XV	$-c^2$	c	c^2	2	8
XVI	$4c^2$	c	c^2	2	8
XVII	c_1	c_2	$-c_1c_2^2, 4c_1c_2^2, -c_1^2c_2, 4c_1^2c_2, c_1^2c_2^2$	1	8, 16 or 32

Table 2. The Néron-Severi group over $\mathbf{Q}$.

Case	a_1, a_2, a_3	Table 1 cases	n	Δ
1	$1, -1, -1$	IV,I,I	9	2^9
2	$4, -1, -4$	II,I,III	9	2^{11}
3	$1, 4, 4$	IV,II,II	9	2^{14}
4	$-1, c^2, -c^2$	I,IV,III	7	2^9
5	$4, -c^2, -4c^2$	II,III,III	7	2^{11}
6	$4, c^2, 4c^2$	II,IV,IV	7	2^{11}
7	$-1, c, -c$	I,VII,VII	6	-2^8
8	$4, c, 4c$	II,VII,VII	6	-2^{10}
9	$4, -1, -1$	VI,V,V	5	2^8
10	$-c_1^2, -c_2^2, c_1^2c_2^2$	III,III,IV	5	2^9
11	$c_1^2, c_2^2, c_1^2c_2^2$	IV,IV,IV	5	2^9
12	$1, 1, 4$	VI,VI,VI	5	2^{10}
13	$-1, -1, c^2$	V,V,VIII	4	-2^6
14	$-1, -4, c^2$	V,VIII,VI	4	-2^7
15	$c_1, c_2^2, c_1c_2^2$	VII,IV,VII	4	-2^8
16	$1, 4, c^2$	VIII,VI,VI	4	-2^8
17	$c_1, -c_2^2, -c_1c_2^2$	VII,III,VII	4	-2^8
18	$c, 4c, 4c^2$	VII,VII,VI	4	-2^9
19	$-1, c, -c^3$	V,VII,VII	4	-2^9
20	$-1, c_1, -c_1c_2^2$	V,VIII,VIII	3	2^5
21	$c_1, 4c_1, c_2^2$	VIII,VIII,VI	3	2^6
22	c_1, c_2, c_1c_2	VII,VII,VIII	3	2^7
23	c_1c_2, c_1c_3, c_2c_3	VIII,VIII,VIII	2	-2^4
24	$1, 4, -1$	X,IX,IX	8	-2^{12}
25	$-1, c^2, 4c^2$	IX,X,X	6	-2^{11}
26	$-1, c, 4c$	IX,XIII,XIII	5	2^9
27	$4, 4, -1$	XII,XII,XI	4	-2^9
28	$c_1^2, c_2^2, -4c_1^2c_2^2$	X,X,X	4	-2^{10}
29	$1, 1, -1$	XI,XI,XI	4	-2^{11}
30	$1, 4, -c^2$	XIV,XII,XII	3	2^6
31	$-1, 4, c^2$	XI,XII,XIV	3	2^7
32	$c_1, c_2^2, -4c_1c_2^2$	XIII,X,XIII	3	2^8
33	$1, -1, c^2$	XIV,XI,XI	3	2^8
34	$4, c, -c^3$	XII,XIII,XIII	3	2^8
35	$-1, c, 4c^3$	XI,XIII,XIII	3	2^9
36	$4, c_1, -c_1c_2^2$	XII,XIV,XIV	2	-2^4
37	$-1, c_1, 4c_1c_2^2$	XI,XIV,XIV	2	-2^5
38	$c_1, c_2, -4c_1c_2$	XIII,XIII,XIV	2	-2^6
39	$-c_1c_2, -c_1c_3, -c_2c_3$	XIV,XIV,XIV	1	2^2
40	$1, 1, 2c^2$	XVI,XVI,XVI	4	-2^8
41	$1, 1, -2c^2$	XVI,XVI,XVI	4	-2^8
42	$1, -4, 2c^2$	XVI,XV,XV	4	-2^{10}
43	$1, -4, -2c^2$	XVI,XV,XV	4	-2^{10}
44	$1, c, 4c^2$	XVII,XVI,XVI	3	2^6
45	$-4, c, -c^2$	XVII,XVI,XV	3	2^7
46	$1, c, -c^2$	XVII,XV,XV	3	2^8
47	$c_1, 4c_2^2, -c_1c_2$	XVII,XVI,XVII	2	-2^4
48	$c_1, -c_2^2, -c_1c_2$	XVII,XV,XVII	2	-2^5
49	c_1, c_2, c_3	XVII,XVII,XVII	1	2^2

Table 3. Values of $c^{1/4} f q_0 \omega^{-2} L^*(c, 2)$.

c	f	value	factors
1	2	$\frac{1}{2}$	2^{-1}
2	8	16	2^4
3	12	8	2^3
4	4	4	2^2
5	5	2	2
6	24	32	2^5
7	28	48	$2^4 3$
8	8	16	2^4
9	6	4	2^2
10	40	96	$2^5 3$
11	44	56	$2^3 7$
12	3	$\frac{2}{3}$	2.3^{-1}
13	13	2	2
14	56	192	$2^6 3$
15	60	64	2^6
16	2	1	
17	34	32	2^5
18	24	32	2^5
19	76	136	$2^3 17$
20	20	32	2^5
21	21	12	$2^2 3$
22	88	352	$2^5 11$
23	92	112	$2^4 7$
24	24	96	$2^5 3$
25	10	8	2^3
26	104	224	$2^5 7$
27	12	24	$2^3 3$
28	14	8	2^3
29	29	10	2.5
30	120	320	$2^6 5$
31	124	128	2^7
32	8	32	2^5
33	66	32	2^5
34	136	640	$2^7 5$
35	140	208	$2^4 13$
36	12	16	2^4
37	37	42	2.3.7
38	152	288	$2^5 3^2$
39	156	256	2^8
40	40	160	$2^5 5$
41	82	48	$2^4 3$
42	168	832	$2^6 13$
43	172	248	$2^3 31$
44	11	10	2.5
45	15	4	2^2
46	184	576	$2^6 3^2$
47	188	448	$2^6 7$
48	12	16	2^4
49	14	16	2^4

c	f	value	factors
50	40	160	$2^5 5$
51	204	208	$2^4 13$
52	52	128	2^7
53	53	42	2.3.7
54	24	96	$2^5 3$
55	220	320	$2^6 5$
56	56	128	2^7
57	114	112	$2^4 7$
58	232	608	$2^5 19$
59	236	760	$2^3 5.19$
60	30	32	2^5
61	61	18	2.3^2
62	248	1280	$2^8 5$
63	84	96	$2^5 3$
64	4	8	2^3
65	130	128	2^7
66	264	768	$2^8 3$
67	268	584	$2^3 73$
68	68	96	$2^5 3$
69	69	72	$2^3 3^2$
70	280	960	$2^6 3.5$
71	284	368	$2^4 23$
72	24	128	2^7
73	146	112	$2^4 7$
74	296		
75	60	112	$2^4 7$
76	19	6	2.3
77	77	40	$2^3 5$
78	312	960	$2^6 3.5$
79	316	960	$2^6 3.5$
80	5	4	2^2
81	6		
82	328		
83	332		
84	84	128	2^7
85	85	100	$2^2 5^2$
86	344		
87	348		
88	88	224	$2^5 7$
89	178	272	$2^4 17$
90	120	384	$2^7 3$
91	364		
92	46	72	$2^3 3^2$
93	93	20	$2^2 5$
94	376		
95	380		
96	24	64	2^6
97	194	224	$2^5 7$
98	56	192	$2^6 3$

Table 3 (continued). Values of $c^{1/4} f q_0 \omega^{-2} L^*(c,2)$.

c	f	value	factors	c	f	value	factors
99	132	352	$2^5 11$	105	210	160	$2^5 5$
100	20	48	$2^4 3$	106	424		
101	101	74	2.37	107	428		
102	408			108	3	2	2
103	412			109	109	42	2.3.7
104	104	672	$2^5 3.7$				

REFERENCES

[1] Birch, B. J. and Swinnerton-Dyer, H. P. F., Notes on elliptic curves II, *J. für reine angew Math* **218** (1965), 79–108.

[2] Cassels, J. W. S. and Fröhlich, A., Algebraic Number Theory (Academic Press, 1967).

[3] Coates, J. H. & Wiles, A., On the conjecture of Birch and Swinnerton-Dyer, *Invent. Math.* **39** (1977) 223–51.

[4] Hecke, E., Eine neue Art von Zetafunktionen und ihre Beziehungen zur Verteilung der Primzahlen II, *Math. Zeit.* **6** (1920), 11–51 = Werke, 249–89.

[5] Weil, A., Numbers of solutions of equation in finite fields, *Bull. Amer. Math. Soc.* **55** (1949), 497–508 = Works I, 399–410.

[6] Weil, A., Jacobi sums as Grossencharaktere, *Trans. Amer. Math. Soc.* **73** (1952), 487–95 = Works II, 63–71.

ON CERTAIN ARTIN L-SERIES

Dinakar Ramakrishnan[1]

In this note we strengthen (see *Theorem A* of section 1 below) the main result in appendix 7 of [Bu] (whose proof was based on a letter of J.-P. Serre). Our method is the same, except that we are able to appeal, in addition, to the existence (proved recently in [BHR]) of Galois conjugates of arbitrary *arithmetic* automorphic forms on $GL(2)$ over *totally real* number fields, along with results on the poles of the triple product L-functions ([Ik]) and on the symmetric square lifting for $GL(2)$ over number fields ([GeJ]). We begin by reviewing some basic facts (and conjectures) concerning Artin L-functions and automorphic forms of Galois type.

0. Background. Let $\overline{\mathbf{Q}}$ be the algebraic closure of $\mathbf{Q}$ in $\mathbf{C}$. For any number field $k \subset \overline{\mathbf{Q}}$, denote by G_k the Galois group of $\overline{\mathbf{Q}}$ over k, equipped with the usual profinite topology. Consider a continuous representation:

$$\sigma : G_k \to GL(V_{\mathbf{C}}), \tag{0.1}$$

where $V_{\mathbf{C}}$ is a vector space over $\mathbf{C}$ of dimension n. Then σ necessarily factors through a finite quotient of G_k, and it is completely reducible. Denote by $\mathcal{N}$ the Artin conductor of σ, and by $L(\sigma, s)$ the $\mathbf{C}$-valued L-series associated by E. Artin to σ ([A], [M]). Then we have an Euler product expansion over the *finite* places v of k (with norm Nv):

$$L(\sigma, s) = \prod_v L_v(\sigma, s) \tag{0.2}$$

with

$$L_v(\sigma, s)^{-1} = \det(1 - F_v T \mid V_{\mathbf{C}}{}^{I_v})_{|T = Nv^{-s}}$$

[1] Partially supported by grants from the NSF and the Sloan Foundation.

where F_v is the Frobenius at v, and I_v, the of inertia subgroup of the decomposition group at v. If v is unramified for σ, i.e., if it does not divide $\mathcal{N}$, then I_v acts trivially on $V_{\mathbb{C}}$, and so $L_v(\sigma, s)$ has degree n in Nv^{-s}.
It is known that $L(\sigma, s)$ converges absolutely in $\mathrm{Re}(s) > 1$ and admits a *meromorphic* continuation to the entire (complex) s-plane. Moreover, there is an "Euler factor at infinity": $L_\infty(\sigma, s)$, defined as a product of Gamma factors, such that $L^*(\sigma, s) \stackrel{\mathrm{def}}{=} L(\sigma, s) L_\infty(\sigma, s)$ satisfies a functional equation:

$$L^*(\sigma, 1-s) = \epsilon(\sigma, s) L^*(\check{\sigma}, s) , \tag{0.3}$$

where $\check{\sigma}$ denotes the contragredient representation of σ, and $\epsilon(\sigma, s)$ is an exponential function (depending on the norm of $\mathcal{N}$ and the discriminant) times a non-zero scalar involving the "Artin root number" $W(\sigma)$.
The Galois representation σ is said to satisy the *Artin conjecture* if the following holds:

(A) $\qquad L(\sigma, s)$ is holomorphic at every $s \neq 1$

This conjecture can be shown to be equivalent to the assertion that $L^*(\sigma, s)$ is holomorphic at every $s \neq 0, 1$.
The behavior of $L(\sigma, s)$ is completely understood at $s = 1$. One has:

$$-\mathrm{ord}_{s=1}\, L(\sigma, s) = \dim_{\mathbb{C}} \mathrm{Hom}_{G_k}(1, \sigma) \tag{0.4}$$

where $-\mathrm{ord}_{s=1}$ denotes the order of pole at $s = 1$. In particular, if σ is irreducible and non-trivial, $L(\sigma, s) \neq 0, \infty$ at $s = 1$.
Conjecture (A) is known to hold (at least) when every irreducible summand η of σ is one of the following two types:

(0.3) $\qquad (n, k)$: arbitrary *with* $\eta(G_k)$ nilpotent
(Artin, Brauer and Hecke),

and

$n = 2,\ k$: arbitrary *with* $\eta(G_k)$ solvable
(cf. [La1], [Tu]).

Actually, in all these positive examples, one also knows the truth of the *Langlands conjecture* (see [AC], chap.3, sec.7, for a proof in the *nilpotent* case), which says:

(L) $L(\sigma, s) = L(\pi_f, s)$, for an isobaric (see [La2], [JS]) automorphic representation $\pi = \pi_\infty \otimes \pi_f$ of $GL(n, \mathbb{A}_k)$ of conductor $\mathcal{N}$, which is cuspidal if σ is non-trivial and irreducible.

Such an identity of L-functions, when it holds, requires, in particular, the central character ω_π of π to be $\delta \circ \det$, where δ is the character of $\mathbb{A}_k^*/k^*$ associated to $\det(\sigma)$ by class field theory. One also conjectures that $L_\infty(\sigma, s)$ equals $L(\pi_\infty, s)$.
Given a cuspidal representation π(resp. π') of $GL(n, \mathbb{A}_k)$ (resp. $GL(m, \mathbb{A}_k)$), there exists a canonically associated (isobaric) automorphic representation $\pi \boxplus \pi'$ of $GL(n+m, \mathbb{A}_k)$ such that $L(\pi \boxplus \pi', s) = L(\pi, s)L(\pi', s)$. The conjectural map: $\sigma \mapsto \pi$ should be a functor A on the (semisimple) category $\mathcal{R}_k$ of continuous Complex representations of $G_{k'}$ sending $\sigma \oplus \sigma'$ to $\pi \boxplus \pi'$. Its image should be describable as follows. Say that an (isobaric) automorphic representation π of $GL(n, \mathbb{A}_k)$ is *of Galois type* iff, for every infinite place w of k, π_w is attached, by the archimedean correspondence ([La2]), to an n-dimensional representation σ_w of $\mathrm{Gal}(\overline{F}_w/F_w)$ (so that $L(\pi_w, s)$ equals the gamma factor attached to σ_w). Then $A(\mathcal{R}_k)$ should be the (full) subcategory $\mathcal{C}_k$ of the category of isobaric automorphic representations consisting of those of *Galois type*. (For further information on related matters, see the articles of L.Clozel and D.Blasius in "Automorphic Forms, Shimura Varieties, and L-functions", Ann Arbor Proceedings, edited by L.Clozel and J.S.Milne, Academic Press (1990).)
$L(\pi, s)$ is known to be entire for any (non-trivial) cuspidal representation π of $GL(n, \mathbb{A}_k)$. Furthermore, $L(\pi_\infty, s)$ has no zero anywhere, and it has no pole in the half plane $\{\mathrm{Re}(s) > 0\}$. Thus $L(\pi_f, s)$ is holomorphic when $L(\pi, s)$ is, and Langlands's conjecture implies Artin's conjecture. If $n \leq 3$ and $k = \mathbb{Q}$, this stronger conjecture (L) is seen, by using the converse theorem ([Li] and [JPSS]), to be equivalent to the following variant of conjecture (A):

(A′) Given any continuous irreducible $\sigma : G_k \to GL(n, \mathbf{C})$, there exists a positive integer M such that, for every $m \leq [n/2]$ and for any continuous, m-dimensional irreducible representation μ of G_k of conductor prime to M, the series $L(\sigma \otimes \mu, s)$ is holomorphic at any s in $\mathbf{C} - \{1\}$.

(This corresponds to condition (A) of [Se] for $n = 2$.)

When k is not $\mathbf{Q}$, it is not clear, even for $n \leq 3$, that Conjecture (A′) is equivalent to Conjecture (L), because, to use the converse theorem, one needs to (apriori) consider twists of σ by one-dimensional representations ν of the absolute Weil group W_k ([Ta]). When $k = \mathbf{Q}$, every such ν is a finite order character λ times $|\ \ |^t$, for some t in $\mathbf{C}$. (Here $|\ \ |$ denotes the normalized absolute value on W_k.) For any k, the finite order characters of W_k are in natural bijection with (continuous) one dimensional (complex) representations of G_k.
Now let $(\sigma, V_\mathbf{C})$ be as in (0.1). Since it factors through a finite group, we get a model V over a number field E, which is determined by the characteristic polynomials of $\sigma(\mathrm{Frob}_v)$. In the language of [De], V is a motive of rank n over k with coefficients in E. Let τ be any embedding of E in $\mathbf{C}$. Then, by conjugating by τ, one obtains a new continuous representation σ^τ of G_k on $V \otimes_{E,\tau} \mathbf{C}$. It is irreducible if σ is, and its contragredient is equivalent to $\check{\sigma}^\tau$. Suppose $L(\sigma, s) = L(\pi_f, s)$, for an automorphic representation $\pi = \pi_\infty \otimes \pi_f$ of $GL(n, \mathbb{A}_k)$. Then, for every τ in $\mathrm{Hom}(E, \mathbf{C})$, there should be an automorphic representation $\pi^{(\tau)}$ of $GL(n, \mathbb{A}_k)$ such that $L(\sigma^\tau, s) = L(\pi^{(\tau)}, s)$ (with the corresponding equality of the Gamma factors at infinity). In other words, there should be an *action of Aut* $\mathbf{C}$ on C_k which is *natural relative to A*.

1. Statement of the result. Let F be a *totally real* number field $\subset \overline{\mathbf{Q}}$. In this note we want to discuss the following

Theorem A. *Let K be a finite Galois extension of F defined by the kernel of a continuous homomorphism $\overline{\sigma}$ of G_F into $PGL(2, \mathbf{C})$ $(= GL(2, \mathbf{C})/\mathbf{C}^*)$ with $\overline{\sigma}(G_F)$ isomorphic to A_5. Assume that, for some lifting σ of $\overline{\sigma}$ to a (continuous) representation of G_F in $GL(2, \mathbf{C})$, there exists a positive integer*

M such that $L(\sigma \otimes \nu, s)$ is entire for every character ν of W_F of conductor prime to M. Then, every finite dimensional complex representation η of G_F factoring through Gal(K/F) satisfies Conjecture (A). *If dim$(\eta) \leq 4$, η satisfies* (L) *as well.*

For $F = \mathbf{Q}$ *and σ odd*, i.e., when complex conjugation does not act as a scalar via σ, the conclusion relating to Conjecture (A) was proved in [Bu], appendix 7, essentially following an unpublished letter of Serre.
It is a classical fact that every finite subgroup of $PGL(2, \mathbf{C})$ is either cyclic or dihedral or tetrahedral (A_4) or octahedral (S_4) or icosahedral (A_5). It is not solvable iff it is isomorphic to the alternating group A_5. One sees that, when $F = \mathbf{Q}$, $\overline{\sigma}(G_F)$ can be icosahedral for *both* odd and even types of σ. Recall that σ (or $\overline{\sigma}$) is said to be *even* when complex conjugation acts by a scalar. See, for example, [Bu], pp.136-141, where the conductors of *totally real* A_5 fields K (over $\mathbf{Q}$) corresponding to *even* $\overline{\sigma}$ are marked with an asterisk. It is a very interesting open problem to prove, in some example, the holomorphy of an *even icosahedral* $\overline{\sigma}$.

Crucial to our extension is the following result on Galois conjugation (proved by the author in collaboration with D. Blasius and M. Harris):

Theorem B. ([BHR]) *Let π be a cuspidal automorphic representation of $GL(2, \mathbb{A}_F)$ of Galois type. Then, for every τ in Aut $\mathbf{C}$, there exists a cuspidal automorphic representation $\pi^{(\tau)}$ of $GL(2, \mathbb{A}_F)$ of Galois type such that, for every finite place v not dividing the conductor of π, $L(\pi_v^{(\tau)}, s)$ equals $L(\pi_v, s)^\tau$.*

A word of explanation is needed here. The local factor $L(\pi_v, s)$ is the inverse of a polynomial $f(T)$ evaluated at $T = Nv^{-s}$, and τ in Aut $\mathbf{C}$ acts on $\mathbf{C}[T]$ in the obvious way. In the above theorem, we mean by $L(\pi_v, s)^\tau$ the function: $1/(f^\tau(Nv^{-s}))$. Needless to say, such a result is well known for *holomorphic* Hilbert modular forms.

In [BHR] it is proved in addition that, given any finite place v where π is unramified, $\epsilon(\pi_v^{(\tau)}, s)$ equals $\epsilon(\pi_v, s)^\tau$. It is known that an irreducible, admissible $\mathbf{C}$-representation of $GL(2)$ over a local field is determined by (the collection of one-dimensional twists of) its L- and ϵ- factors. Another way to approach this theorem is as follows. Let H be the C-vector space on which $GL(2, F_v)$ acts by π_v. For τ in Aut C, choose a τ-linear isomorphism t of H onto a space $H(\tau)$. Define a representation π_v^τ of $GL(2, F_v)$ on $H(\tau)$ by: $\pi_v^\tau(g)(w) = t(\pi_v(g)(t^{-1}(w)))$, for g in $GL(2, F_v)$ and w in $H(\tau)$. Then the conclusion of Theorem B is that there is a cuspidal automorphic representation $\pi^{(\tau)}$ such that, at every good v, $\pi_v^{(\tau)}$ is equivalent to π_v^τ. One knows (cf. [HC]) that the space of automorphic forms on $GL(2, \mathbb{A}_F)$, of fixed central character, infinity type and conductor, is *finite dimensional.* From this and Theorem B one gets the rationality of π_f over a *number field*, for any cuspidal π of Galois type on $GL(2, \mathbb{A}_F)$. When $F = \mathbb{Q}$, the fact that the unramified Hecke eigenvalues (of such a π) are algebraic numbers was first proved (for non-holomorphic forms) in [BCR 1,2], under a suitable ramification hypothesis. For some details of the basic method used, and for for a partially completed program for the inverse problem of associating Galois representations to π, see [BR]. For a more recent progress report, see G.Henniart's Seminaire Bourbaki article (to appear).

2. The argument. Let σ be as in the statement of *TheoremA*. It is a continuous, irreducible representation of G_F of conductor, say, $\mathcal{N}$. Put $G =$ Gal(K/F), where K is the finite Galois extension of F defined by the kernel of $\overline{\sigma}$. Let $\tilde{G} = \sigma(G_F)$, which can be seen to be a 2-fold central covering group of A_5, isomorphic to $SL(2, \mathbb{F}_5)$. We will view σ as an injective representation of $\tilde{G}$ in $GL(2, \mathbb{C})$. One knows (cf. [NS], II, §5, for example) that there are exactly two irreducibles of $\tilde{G}$ of dimension 2, both rational over $\mathbb{Q}(\sqrt{5})$ and Galois conjuagte to each other by the non-trivial automorphism τ of $\mathbb{Q}(\sqrt{5})$. Denote by $S^2(\sigma)$ the (three dimensional) symmetric square of σ. Then $S^2(\sigma)$ and $\sigma \otimes \sigma^\tau$ both descend to linear representations of G. The following is well known ([Bu], appendix 7).

Lemma 2.1. *Let η be an irreducible complex representation of G. Then η is equivalent to one of the following types of representations:*

(a) *trivial (one dimensional)*

(b) $S^2(\sigma^\beta)$, *with β in $\{id, \tau\}$*

(c) $\sigma \otimes \sigma^\tau$

(d) *monomial (five dimensional)*

We indicate a proof for completeness. From the character table of A_5, one sees that it has one irreducible, namely the trivial one, in dimension 1, two irreducibles in dimension 3, one irreducible in dimension 4, one in dimension 5, and none other. It is easy to verify that the five dimensional is monomial. The irreducibles of dimension 3 are rational over $\mathbf{Q}(\sqrt{5})$, and are τ-conjugate. It then suffices to show that $S^2(\sigma)$ and $\sigma \otimes \sigma^\tau$ are both irreducible as representations of $\tilde{G}$. First note that $\Lambda^2(\sigma)$ must be trivial, as $\tilde{G}$ has no other one dimensional. Since σ is irreducible, the trivial representation cannot occur in $S^2(\sigma)$ as well. Hence there is no one dimensional summand of $S^2(\sigma)$, forcing it to be irreducible. Finally, since A_5 has no two dimensional irreducible, if $\sigma \otimes \sigma^\tau$ is reducible, it must admit a one dimensional, necessarily trivial, summand. But this cannot be as σ and σ^τ are inequivalent. Done.

Now we begin the *proof of Theorem A*. Let σ, G, $\tilde{G}$ be as above. By the strengthened version ([Li]) of the Weil-Jacquet-Langlands converse theorem for $GL(2)$, the hypothesis (of Theorem A) implies the existence of a cuspidal automorphic reprsentation π of $GL(2, \mathbb{A}_F)$ such that, at every finite v, we have:

$$L_v(\sigma, s) = L(\pi_v, s) \tag{2.2}$$

In particular,

$$\omega_\pi = \det(\sigma) \tag{2.3}$$

where ω_π denotes (as usual) the central character of π.

Let $\Gamma_{\mathbf{R}}(s)$ denote $\pi^{-s/2}\,\Gamma(s/2)$. Then, at any infinite place w, the local factor $L_w(\sigma, s)$ equals *either* $\Gamma_{\mathbf{R}}(s)\Gamma_{\mathbf{R}}(s+1)$ *or* $\Gamma_{\mathbf{R}}(s)^2$ *or* $\Gamma_{\mathbf{R}}(s+1)^2$, and this depends (respectively) on whether the set of eigenvalues of complex conjugation (under σ) is $\{1, -1\}$ *or* $\{1, 1\}$ *or* $\{-1, -1\}$. (One says that σ_w is *odd* in the first case and *even* in the second and third cases.) Consequently,

(2.2) forces π to be of *Galois type*. Appealing to Theorem B (see [BHR], sec.7, for a proof), we then deduce the existence, for every τ in Aut C, of a cuspidal automorphic representation $\pi^{(\tau)}$ of *Galois type* such that:

$$\text{(2.4)} \qquad \pi_v^{(\tau)} = \pi_v^\tau, \quad \text{at every finite place } v \text{ where } \pi \text{ is } \textit{unramified},$$

and

$$\omega_{\pi^{(\tau)}} = \det(\sigma^\tau)$$

Lemma 2.5. *Let π, σ, τ be as above. Then we have:*

$$L(\pi_v^{(\tau)}, s) = L_v(\sigma, s), \quad \textit{at every place } v \textit{ (possibly infinite)}$$

Proof. Let S be the (finite) set of finite places of F where π is ramified. Let μ be a one dimensional representation of G_F, and let μ' be the corresponding idele class character of F. In the above construction, if we replace σ^τ by $\sigma^\tau \otimes \mu$, then $\pi^{(\tau)}$ will get replaced by $\pi^{(\tau)} \otimes \mu'$. (Since the contragredient of σ is also a one dimensional twist of σ by a character (unramified outside S), we also deduce that $\breve{\sigma}^\tau$ corresponds to $\breve{\pi}^{(\tau)}$.) We may (and we will) now choose μ to be trivial at infinity, and to be sufficiently ramified at S so that $L_v(\beta, s) = 1 = L(\gamma_v, s)$, at every v in S, for $\beta = \sigma^\tau \otimes \mu$ (resp. $\breve{\sigma}^\tau \otimes \overline{\mu}$), and $\gamma = \pi^{(\tau)} \otimes \mu'$ (resp. $\breve{\pi}^{(\tau)} \otimes \overline{\mu'}$). Note that, by (2.4), $L_v(\sigma^\tau \otimes \mu, s) = L(\pi_v^{(\tau)} \otimes \mu'_v, s)$ at every finite place v outside S. Similarly for the contragredients. Since the global L-functions of $\sigma^\tau \otimes \mu$ and $\pi^{(\tau)} \otimes \mu'$ both satisfy their own respective functional equations, the coincidence of the Euler factors at finite places yields the following:

$$\text{(2.6)} \qquad L_\infty(\sigma^\tau, s) \sim [L_\infty(\breve{\sigma}^\tau, 1-s)/L(\breve{\pi}_\infty^{(\tau)}, 1-s)]\, L(\pi_\infty^{(\tau)}, s)$$

where $\sim$ denotes equality up to multiplication by an invertible holomorphic function of the s-plane. Consider this identity in the region $\{\mathrm{Re}(s) \leq 0\}$, where the factors $L_\infty(\breve{\sigma}^\tau, 1-s)$ and $L(\breve{\pi}_\infty^{(\tau)}, 1-s)$ admit, by the standard properties of the Gamma function, neither a pole nor a zero. Thus we see that $L_\infty(\sigma^\tau, s) \sim L(\pi_\infty^{(\tau)}, s)$ in $\mathrm{Re}(s) \leq 0$. By definition, $L_\infty(\sigma^\tau, s)$ is of the form: $\Gamma_{\mathbf{R}}(s)^a\,\Gamma_{\mathbf{R}}(s+1)^b$, for some non-negative integers a, b such that $a+b = 2[F:\mathbb{Q}]$. Similarly, since $\pi^{(\tau)}$ is of Galois type, $L(\pi^{(\tau)}_\infty, s)$ is of the form: $\Gamma_{\mathbf{R}}(s)^c\,\Gamma_{\mathbf{R}}(s+1)^d$, for some non-negative integers c, d such that $c+d = 2[F:\mathbb{Q}]$. Recalling that $\Gamma(s)$ has no zero anywhere, and that it has simple poles at (exactly) the non-positive integers, we see that we must have: $a = c$ and $b = d$. Thus we get:

$$\prod_{v|\infty} L_v(\sigma^\tau, s) \;=\; \prod_{v|\infty} L(\pi_v^{(\tau)}, s) \tag{2.7}$$

Now fix any infinite place w, say, and modify μ to be the sign character at w, while remaining the same as before at other infinite places and at the finite places in S. Arguing as before, we get:

$$L_w(\sigma^\tau, s) \;\sim\; [L_w(\check{\sigma}^\tau, 1-s)/L(\check{\pi}_w^{(\tau)}, 1-s)]\; L(\pi_w^{(\tau)}, s) \tag{2.8}$$

Since the central character (resp. determinant) of $\pi^{(\tau)}$ (resp. σ^τ) is trivial at infinity, the L-factors at w of $\check{\pi}^{(\tau)}$ and $\pi^{(\tau)}$ (resp. $\check{\sigma}^\tau$ and σ^τ) coincide. Moreover, there are only three possibilities for $L_w(\sigma^\tau, s)$ and $L(\pi_w^{(\tau)}, s)$, namely: $\Gamma_{\mathbf{R}}(s)\,\Gamma_{\mathbf{R}}(s+1)$, $\Gamma_{\mathbf{R}}(s)^2$ and $\Gamma_{\mathbf{R}}(s+1)^2$.

Case (i): $L_w(\sigma^\tau, s) \;=\; \Gamma_{\mathbf{R}}(s)\,\Gamma_{\mathbf{R}}(s+1)$

Suppose $L(\pi_w^{(\tau)}, s) = \Gamma_{\mathbf{R}}(s+\delta)^2$, where $\delta = 0$ or 1. Then, by (2.8),

$$\Gamma_{\mathbf{R}}(s+1-\delta) \;\sim\; [\Gamma_{\mathbf{R}}(2-s-\delta)/\Gamma_{\mathbf{R}}(1-s+\delta]\,]\Gamma_{\mathbf{R}}(s+\delta)$$

This gives a contradiction because, in $\{Re(s) \le 0\}$, the left hand side has poles at $-1, -3, \ldots$ (resp $0, -2, -4, \ldots$), while the right hand side has poles at $0, -2, -4, \ldots$ (resp. $-1, -3, \ldots$) when δ is 0 (resp. 1). Thus $L(\pi_w^{(\tau)}, s) = \Gamma_{\mathbf{R}}(s)\,\Gamma_{\mathbf{R}}(s+1)$.

Case (ii): $L_w(\sigma^\tau, s) \;=\; \Gamma_{\mathbf{R}}(s+\delta)^2$, *with* $\delta = 0$ *or* 1.

Arguing as in case (i), but with σ^τ and $\pi^{(\tau)}$ interchanged, we see that $L(\pi_w^{(\tau)}, s)$ cannot be $\Gamma_{\mathbf{R}}(s)\,\Gamma_{\mathbf{R}}(s+1)$. Suppose it is $\Gamma_{\mathbf{R}}(s+\delta)^2$, instead. Then (2.8) becomes:

$$\Gamma_{\mathbf{R}}(s+1-\delta) \;\sim\; [\Gamma_{\mathbf{R}}(1-s-\delta)^2/\Gamma_{\mathbf{R}}(-s+\delta]^2\,]\;\Gamma_{\mathbf{R}}(s+1-\delta)^2$$

In $\{Re(s) \leq -1\}$, the left hand side has (double) poles at $-1, -3, \dots$ (resp $-2, -4, \dots$), while the right hand side has poles at $-2, -4, \dots$ (resp. -1, $-3, \dots$) when δ is 1 (resp. 0). We get a contradiction, and so $L(\pi_w^{(\tau)}, s)$ must equal $L_w(\sigma^\tau, s)$.

It remains to show that $L(\pi_v^{(\tau)}, s)$ equals $L_v(\sigma^\tau, s)$ at any *finite* place v in S. We can appeal to *Lemma* 4.9 of [DeS] to deduce this result. Instead we will give a somewhat different argument, using less about $\pi_v^{(\tau)}$.

Fix any place u in S, and choose a character μ of finite order, which is: (i) sufficiently ramified (in the sense it was used above) at any $v \neq u$ in S and (ii) trivial at u and at every infinite place w. Then we get:

$$(2.9) \qquad L_u(\sigma^\tau, s) \sim [L_u(\check{\sigma}^\tau, 1-s)/L(\check{\pi}_u^{(\tau)}, 1-s)]\, L(\pi_u^{(\tau)}, s)$$

We can write: (with $T = q^{-s}$, $q = Nu$)

$$L_u(\sigma^\tau, s)^{-1} = (1-\alpha T)(1-\alpha' T)$$

and

$$L(\pi_u^{(\tau)}, s)^{-1} = (1-\beta T)(1-\beta' T)$$

where α, α', β, β' are complex numbers with $|\alpha| = |\alpha'| = 1$. Then (2.9) yields:

$$\begin{aligned} &(1-\alpha T)(1-\alpha' T)(1-\overline{\beta} q^{-1}T^{-1})(1-\overline{\beta}' q^{-1}T^{-1}) \sim \\ (2.10) \qquad &(1-\beta T)(1-\beta' T)(1-\overline{\alpha} q^{-1}T^{-1})(1-\overline{\alpha}' q^{-1}T^{-1}) \end{aligned}$$

Suppose $\alpha = \alpha' = 0$. Then we are forced to have: $\beta = \beta' = 0$. The converse is clearly true as well. Suppose $\beta \neq 0$, but $\beta' = 0$. If α, α' are both non-zero, then, multiplying both sides of (2.10) by T^2 and comparing roots, we get: $\{\beta, q\alpha^{-1}, q\alpha'^{-1}\} = \{q\beta, \alpha, \alpha'\}$. Since $q \neq 1$, $\beta = \alpha$ or $\beta = \alpha'$. α' is then $q\alpha^{-1}$ or $q\alpha'^{-1}$, both contradicting the fact that α and α' are on the unit circle. We get a similar contradiction if $\beta = \alpha'$. Consequently, *either* $\alpha = 0$ and $\beta = \alpha'$ *or* $\alpha' = 0$ and $\beta = \alpha$. Either way, $L(\pi_u^{(\tau)}, s)$ equals $L_u(\sigma^\tau, s)$, when $\beta \neq 0$, $\beta' = 0$. The situation is the same when $\beta = 0$, $\beta' \neq 0$.

Conversely, consider the case when $\alpha \neq 0$, but $\alpha' = 0$. Suppose $\beta\beta' \neq 0$. Then, arguing as above, we get: $\{\alpha, q\beta^{-1}, q\beta'^{-1}\} = \{q\alpha, \beta, \beta'\}$. Then

$\alpha = \beta$ or $\alpha = \beta'$. Suppose $\alpha = \beta$. If $\beta^2 = 1$, then $\beta' = q{\beta'}^{-1}$, so that $|\beta'| = \sqrt{q}$. But, since the central character of $\pi_u^{(\tau)}$ is unitary, we must have $|\beta\beta'| = 1$, and so we have a contradiction. If $\beta^2 \neq 1$, then $\beta' = q\beta^{-1}$, which again leads to a contradiction. The situation is the same if $\alpha = \beta'$. Consequently, either either β or β' is zero, with the other being non- zero. We thus get $L(\pi_u^{(\tau)}, s) = L_u(\sigma^\tau, s)$, if $\alpha\alpha'\beta\beta' = 0$.
It remains to treat the case: $\alpha\alpha'\beta\beta' \neq 0$. Multiplying both sides of (2.10) by $\alpha\alpha'\beta\beta' T^2$, and equating roots, we get:

$$\{\alpha,\, \alpha',\, q\beta^{-1},\, q{\beta'}^{-1}\} = \{\beta,\, \beta',\, q\alpha^{-1},\, q{\alpha'}^{-1},\}$$

α cannot be $q\alpha^{-1}$ or $q{\alpha'}^{-1}$, because $|\alpha| = |\alpha'| = 1$. Similarly, α' cannot be $q\alpha^{-1}$ or $q{\alpha'}^{-1}$. So $\{\alpha, \alpha'\} = \{\beta, \beta'\}$ and $L(\pi_u^{(\tau)}, s) = L_u(\sigma^\tau, s)$. This concludes the proof of *Lemma* 2.5.

Now we continue with the *proof of Theorem A*. Let η be a representation of G_F, factoring through $G = \mathrm{Gal}(K/F)$. Because of the existence of the sum operation $\boxplus$ in the category of isobaric automorphic forms, we may assume η to be irreducible. If it is one dimensional, then the statement of Theorem A is a consequence of class field theory. If it is monomial, (A) is a theorem of Artin, Hecke and Brauer. We need to consider cases (b) and (c) of Lemma 2.1. For case (b), we appeal to the existence of the symmetric square lifting map: $\lambda \mapsto \mathrm{sym}^2(\lambda)$ from cusp forms on $GL(2)/F$ to automorphic forms on $GL(3)/F$ ([GeJ]). One associates this way an automorphic representation $\prod$ of $GL(3, \mathbb{A}_F)$ such that $L(\prod \otimes \mu', s) = L(\eta \otimes \mu, s)$, for every character μ of W_F with μ' being the idele class character of F associated to μ (by class field theory). As η is irreducible (by assumption), σ cannot be dihedral, since otherwise $S^2(\sigma^\tau)$ will be reducible for any τ in Aut C. This shows that, for any quadratic extension M of F, the standard L-series of the *base change* of $\pi^{(\tau)}$ does not factor into a product of two abelian L-series. Consequently, $L(\pi^{(\tau)}, s) \neq L(\chi, s)$, for any Grossencharacter χ of any quadratic extension of F; equivalently, $\pi^{(\tau)} \not\simeq \pi^{(\tau)} \otimes \delta$, for any quadratic idele class character δ of F ([LL]). In this case, $\mathrm{sym}^2(\pi^{(\tau)})$, and hence $\prod$, must be a *cuspidal* automorphic representation of $GL(3, \mathbb{A}_F)$ (cf. [GeJ]). In particular, $L(\eta, s)$ is entire.

Finally, suppose we are in case (c) (of Lemma 2.1). Let π (resp. $\pi^{(\tau)}$) be associated to σ (resp. σ^τ) as above. Since η is irreducible, σ^τ cannot be of the form: $\check{\sigma} \otimes \mu$, for some one dimensional μ. This implies (cf. [J]) that $L(\eta, s)$, which equals the (Rankin) $GL(2) \times GL(2)$ L-function: $L(\pi \times \pi^{(\tau)}, s)$, is holomorphic everywhere. This proves conjecture (A) for η. It remains to show *(L)*. First note that the above reasoning yields (by [J]) the holomorphy (in the entire s-plane) of $L(\eta \otimes \mu, s)$, for every one-dimensional representation μ of W_F. Next consider, for any cuspidal automorphic representation λ of $GL(2, \mathbb{A}_F)$, the (Garrett) $GL(2) \times GL(2) \times GL(2)$ L-function ([PSR]): $L(\pi \times \pi^{(\tau)} \times \lambda, s)$. (If λ would correspond to a two dimensional representation β of G_F, then this L-function would identify, at least outside the ramified and infinite places, with $L(\eta \otimes \beta, s)$.) Since $\pi^{(\tau)}$ is not equivalent to any one dimensional twist of the contragredient of π, the main theorem of [Ik] shows that $L(\pi \times \pi^{(\tau)} \times \lambda, s)$ has no pole anywhere in the s-plane. Then, by the converse theorem for $GL(4)$[2], due to Jacquet, Piatetski-Shapiro and Shalika, there exists a cuspidal automorphic representation $\pi \boxtimes \pi^{(\tau)}$ of $GL(4, \mathbb{A}_F)$ such that we have: $L(\pi \boxtimes \pi^{(\tau)}, s) = L(\pi \times \pi^{(\tau)}, s)$. This finishes the proof of Theorem A.

Concluding remarks. *(i)* For the lone five-dimensional monomial irreducible of A_5, we do not know how to associate a cuspidal automorphic representation of $GL(5, \mathbb{A}_F)$. Indeed, if one knows how to do base change for non-normal extensions of prime degree, one will get closer to Artin's conjecture. For non-normal cubic extensions, base change is known by [JPSS], and this is used in [Tu] for dealing with octahedral representations. *(ii)* To get the analog of *Theorem A* for the Artin L-series of $SL(2, \mathbb{F}_5)$ extensions, one will have to contend with the four dimensional irreducible δ which does not descend to a representation of A_5. (δ is a discrete series representation

[2] Unfortunately, there is no published proof of this theorem available so far. But this situation is expected to be remedied in the near future, possibly in a sequel to [Ik], especially because the results of [Ik] find some of their most striking applications to arithmetic when used in conjunction with the converse theorem for $GL(4)$. We request the indulgence of the reader on this point.

with rational character.) *(iii)* For any finite simple group of Lie type, one can in principle write down a minimal set of basic irreducibles from which the others are obtained by linear algebra operations, Galois conjugation, and by spin module constructions. Thus, (if and) when the appropriate lifting results for, and conjugation properties of, automorphic forms on $GL(n)$ of *Galois type* are established, one could reduce the collection of Galois representations for which one needs to verify Artin's conjecture. *(iv)* We can say nothing at all, at this point, when the base field F is not totally real.

Acknowledgement. We thank J.Buhler, V.K.Murty and D.Rohrlich for enlightening conversations.

BIBLIOGRAPHY

[AC] J.Arthur and L.Clozel, "Simple Algebras, Base Change, and the Advanced Theory of the Trace Formula", Annals of Math. Studies **120**, Princeton (1989).

[A] E.Artin, *Über eine nueu Art von L-Reihen*, Hamb. Abh. **1** (1923); *Zur Theorie der L-Reihen mit algemeinen Gruppencharakteren*, Hamb. Abh. **8** (1930), 292-306.

[BCR] D.Blasius, L.Clozel and D.Ramakrishnan, *Algébricité de l'action des opérateurs de Hecke sur certaines formes de Maass & Opérateurs de Hecke et formes de Maass: application de la formule des traces*, C.R.Acad. Sci. Paris, série I, **305** (1987), 705-708, & **306** (1988), 59-62.

[BHR] D.Blasius, M.Harris and D.Ramakrishnan, *Coherent cohomology, limits of discrete series, and Maass forms of Galois type*, preprint.

[BR] D.Blasius and D.Ramakrishnan, *Maass forms and Galois representations*, in "Galois Groups over $\mathbb{Q}$", edited by Y.Ihara, K.Ribet and J.-P.Serre, MSRI Publications **16**, Springer-Verlag (1989), 33-77.

[Bu] J.Buhler, "Icosahedral Galois Representations", Springer Lecture Notes **654** (1978).

[De] P.Deligne, *Périodes d'intégrales et valeurs de fonctions L*, Proc. Symp. Pure Math. XXXIII, part II, (1979).

[DeS] P.Deligne and J.-P.Serre, *Formes modulaires de poids 1*, Ann. Sci. ENS, 4e série, **7** (1984), 507-530.

[GeJ] S.Gelbart and H.Jacquet, *A relation between automorphic forms on GL(2) and GL(3)*, Ann. Sci. ENS, 4e série, 11 (1978), 471-542.

[HC] Harish-Chandra, "Automorphic Forms on Semisimple Lie Groups", Springer Lectur Notes **68**, Springer-Verlag, NY.

[Ik] T.Ikeda, *On the location of poles of triple L-functions*, preprint (1989).

[J] H.Jacquet, "Automorphic Forms on GL(2) II", Springer Lecture Notes **278** (1972).

[JPSS] H.Jacquet, I.I.Piatetski-Shapiro and J.Shalika, *Automorphic forms on GL(3)* I & II Annals of Math. **103** (1981), 169-212.

[JS] H.Jacquet and J.Shalika, *On Euler products and classification of automorphic representations* I & II, American Journal of Math. **103** (1981), 499-558 & 777-815.

[LL] J.-P.Labesse and R.P.Langlands, *L-Indistinguishability for SL(2)*, Can. J. Math. XXXI, 4 (1979), 726-785.

[La1] R.P.Langlands, "Base Change for GL(2)", Ann. Math. Studies **96** (1980).

[La2] R.P.Langlands, *Automorphic representations, Shimura varieties and Motives. Ein Märchen*, Proc. Symp. Pure Math. XXXIII, part II, AMS (1979), 205-246.

[Li] W-C.W.Li, *On converse theorems for GL(2) and GL(1)*, American Journal of Math. **103**, no. 5 (1981), 851-885.

[M] J.Martinet, *Character theory and Artin L-functions*, in "Algebraic Number Fields", edited by A.Fröhlich, Academic Press (1977), 1-87.

[NS] M.A.Naimark and A.I.Stern, "Theory of Group Representations," Grundlehren der math. Wiss. **246** (English translation), Springer-Verlag (1980)

[PSR] I.Piatetski-Shapiro and S.Rallis, "L-functions for the classical groups", Springer Lecture Notes **1254**, 1-52.

[Se] J.-P.Serre, *Modular forms of weight one and Galois representations*, in "Algebraic Number Fields", edited by A.Fröhlich, Academic Press (1977), 193-268.

[Ta] J.Tate, *Number theoretic background*, Proc. Symp. Pure Math. XXXIII, part II, AMS (1979).

[Tu] J.Tunnell, *Artin's conjecture for representations of octahedral type*, Bulletin of the AMS **5**, no. 2 (1981), 173-175.

Dinakar Ramakrishnan
Department of Mathematics
California Institute of Technology
Pasadena, CA 91125
U.S.A.

The one-variable main conjecture for elliptic curves with complex multiplication

KARL RUBIN*

Department of Mathematics, Ohio State University, Columbus, OH 43210, USA

INTRODUCTION

In a forthcoming paper [12] we will present a proof of the one- and two-variable "main conjectures" of Iwasawa theory for imaginary quadratic fields. This proof uses the marvelous recent methods of Kolyvagin [6], combined with ideas from [9] and [11] and a great deal of technical Iwasawa theory. Because it deals with the two-variable situation, with primes of degree two as well as those of degree one, and with all imaginary quadratic fields, the proof in [12] will necessarily be quite complicated and, at least at first glance, rather unintelligible.

The purpose of this paper is to present a proof of the one-variable main conjecture in the simplest setting (see §1 for the precise statement). That is, we consider only imaginary quadratic fields K of class number one, elliptic curves E defined over K with complex multiplication by K, and only primes of good reduction which split in K. This is the setting in which Coates and Wiles worked in [1] and [2]. These restrictions make it possible to simplify the proof considerably. However, the important ideas of the general proof do appear here, and even with these restrictions there are powerful applications (see Theorem 1.2).

Contents

---

*partially supported by NSF grants

1 STATEMENT OF THE MAIN CONJECTURE

Fix once and for all an imaginary quadratic field K of class number 1, an elliptic curve E defined over K with complex multiplication by the ring of integers $\mathcal{O}$ of K, and a rational prime $p > 3$ such that

(i) p splits into two distinct primes in K, say $p = \mathfrak{p}\mathfrak{p}^*$, and

(ii) E has good reduction at both $\mathfrak{p}$ and $\mathfrak{p}^*$.

For every n, $0 \le n \le \infty$, let $K_n = K(E_{\mathfrak{p}^{n+1}})$, the extension of K generated by the coordinates of the points annihilated by $\mathfrak{p}^{n+1}$ in $E(\bar{K})$. Put

$$\Delta = \mathrm{Gal}(K_0/K), \quad \Gamma = \mathrm{Gal}(K_\infty/K_0) \quad \text{and} \quad \mathcal{G} = \mathrm{Gal}(K_\infty/K).$$

Then (see §3 of [1]) K_n/K is totally ramified at $\mathfrak{p}$,

$$\Delta \cong (\mathcal{O}/\mathfrak{p})^\times, \quad \Gamma \cong 1+\mathfrak{p}\mathcal{O}_\mathfrak{p} \quad \text{and} \quad \mathcal{G} = \Delta \times \Gamma \cong \mathcal{O}_\mathfrak{p}^\times$$

where $\mathcal{O}_\mathfrak{p}$ is the completion of $\mathcal{O}$ at $\mathfrak{p}$. For every $n < \infty$ write A_n for the p-part of the ideal class group of K_n, $\mathcal{E}_n$ for the group of global units of K_n, and $\mathcal{C}_n$ for the group of elliptic units of K_n (see §2). Write U_n for the group of local units of the completion of K_n above $\mathfrak{p}$ which are congruent to 1 modulo the prime above $\mathfrak{p}$, and let $\bar{\mathcal{E}}_n$ and $\bar{\mathcal{C}}_n$ denote the closures of $\mathcal{E}_n \cap U_n$ and $\mathcal{C}_n \cap U_n$, respectively, in U_n. We also define

$$A_\infty = \varprojlim A_n, \quad \bar{\mathcal{E}}_\infty = \varprojlim \bar{\mathcal{E}}_n, \quad \bar{\mathcal{C}}_\infty = \varprojlim \bar{\mathcal{C}}_n, \quad \text{and} \quad U_\infty = \varprojlim U_n,$$

all inverse limits with respect to the norm maps. Let M_∞ be the maximal abelian p-extension of K_∞ which is unramified outside of the prime above $\mathfrak{p}$, and write $X_\infty = \mathrm{Gal}(M_\infty/K_\infty)$.

For any $\mathbf{Z}_p[\Delta]$-module Y and any character $\chi : \Delta \to \mathbf{Z}_p^\times$, define Y^χ to be the χ-component of Y, the maximal submodule on which Δ acts via χ. If we define

$$e_\chi = \frac{1}{p-1} \sum_{\tau \in \Delta} \chi^{-1}(\tau)\tau \in \mathbf{Z}_p[\Delta],$$

then $Y^\chi = e_\chi Y$ and for $y \in Y$ we will write $y^\chi = e_\chi y$ for the projection of y into Y^χ. Define the Iwasawa algebra

$$\mathbf{Z}_p[[\mathcal{G}]] = \varprojlim \mathbf{Z}_p[\mathrm{Gal}(K_n/K)],$$

and for every χ write $\Lambda = \Lambda_\chi = \mathbf{Z}_p[[\mathcal{G}]]^\chi$. Then A_∞^χ, U_∞^χ, $\overline{\mathcal{E}}_\infty^\chi$, $\overline{\mathcal{C}}_\infty^\chi$, and X_∞^χ are all finitely generated Λ-modules, and A_∞^χ, X_∞^χ, $U_\infty^\chi/\overline{\mathcal{C}}_\infty^\chi$ and $\overline{\mathcal{E}}_\infty^\chi/\overline{\mathcal{C}}_\infty^\chi$ are torsion Λ-modules ([5] §3.4 for A_∞^χ, [3] §III.1.3 for $U_\infty^\chi/\overline{\mathcal{C}}_\infty^\chi$, and the others follow).

Two Λ-modules are said to be pseudo-isomorphic if there is a map between them with finite kernel and cokernel. The well-known classification theorem states that any finitely generated torsion Λ-module Y is pseudo-isomorphic to a module of the form $\oplus\Lambda/f_i\Lambda$ for some $f_i \in \Lambda$, and the characteristic ideal $(\prod f_i)\Lambda$ is a well-defined invariant of Y which we will denote by char(Y).

The following theorem is one form of the "main conjecture" for this setting.

Theorem 1.1 *For all characters χ of Δ,*

$$\mathrm{char}(A_\infty^\chi) = \mathrm{char}(\overline{\mathcal{E}}_\infty^\chi/\overline{\mathcal{C}}_\infty^\chi) \textit{ and } \mathrm{char}(X_\infty^\chi) = \mathrm{char}(U_\infty^\chi/\overline{\mathcal{C}}_\infty^\chi).$$

This theorem will be proved in §6. The connection between Theorem 1.1 and the usual statement of the main conjecture is that $\mathrm{char}(U_\infty^\chi/\overline{\mathcal{C}}_\infty^\chi)$ is generated by a $\mathfrak{p}$-adic L-function attached to E and χ ([2] Theorem 1). Using this connection Theorem 1.1 has the following consequence (parts of which were already known; see [1] Theorem 1 and [10] Corollary C). Other consequences of the main conjecture follow from [7] Chapter IV.

Theorem 1.2 *Suppose* E *and* p *are as above.*
(i) *If* $L(E_{/K}, s) \neq 0$ *then* E(K) *is finite and the order of the* $\mathfrak{p}$*-part of the Tate-Shafarevich group* $\text{Ш}(E_{/K})$ *is as predicted by Gross' refinement* [4] *of the Birch and Swinnerton-Dyer conjecture.*
(ii) *If* E *is defined over* $\mathbf{Q}$ *and* $\mathrm{ord}_{s=1}L(E_{/\mathbf{Q}}, s) = 1$, *then* $\mathrm{rank}_{\mathbf{Z}}(E(\mathbf{Q})) = 1$ *and the order of the* p*-part of* $\text{Ш}(E_{/\mathbf{Q}})$ *is as predicted by the Birch and Swinnerton-Dyer conjecture.*

Proof. This uses results of Perrin-Riou and Gross and Zagier. See [7] Théorème 22 and [8] Corollaire 1.9. □

2 PRELIMINARIES

Let M denote a large power of p (to be specified later). Fix for §§2-4 a non-negative integer n and write $F = K_n = K(E_{\mathfrak{p}^{n+1}})$.

Lemma 2.1 F/K *is totally ramified at* $\mathfrak{p}$, *ramified at all primes where* E *has bad reduction, unramified at all other primes, and* $\mathrm{Gal}(F/K) \cong (\mathcal{O}/\mathfrak{p}^{n+1})^{\times}$.

Proof. See [1] Lemmas 4 and 5. □

Define $\mathcal{L} = \mathcal{L}_{F,M}$ to be the set of all primes ℓ of K satisfying

(i) ℓ splits completely in F/K, and

(ii) $N(\ell) \equiv 1 \pmod{M}$

where $N(\ell)$ is the norm of ℓ. Write $\mathcal{S} = \mathcal{S}_{F,M}$ for the set of squarefree integral ideals of K which are divisible only by primes $\ell \in \mathcal{L}$. For every $\mathfrak{a} \in \mathcal{S}$ write $G_{\mathfrak{a}} = \mathrm{Gal}(F(E_{\mathfrak{a}})/F)$.

Lemma 2.2 (i) *For every* $\ell \in \mathcal{L}$, $F(E_\ell)/F$ *is totally ramified at all primes above* ℓ, *unramified at all other primes, and* $G_\ell \cong (\mathcal{O}/\ell)^{\times}$.

(ii) If $\mathfrak{a} \in \mathcal{S}$ *then* $G_{\mathfrak{a}} = \prod_{\ell \mid \mathfrak{a}} G_\ell$ (*product over primes* ℓ *dividing* $\mathfrak{a}$).

(iii) *For every* $\mathfrak{a} \in \mathcal{S}$, $F(E_{\mathfrak{a}})$ *contains no nontrivial* p^{th}*-roots of unity.*

Proof. Since ℓ is unramified in F/K, E has good reduction at ℓ by Lemma 2.1. Therefore as in Lemma 2.1, $K(E_\ell)/K$ is totally ramified at ℓ and $\mathrm{Gal}(K(E_\ell)/K) \cong (\mathcal{O}/\ell)^{\times}$. By [1] Theorem 2, E has good reduction everywhere over F, so $F(E_\ell)/F$ is ramified only above ℓ. This proves (i) and (ii) follows easily. For the third assertion we need only observe that by Lemma 2.1 and (i), $\mathfrak{p}^*$ is unramified in $F(E_{\mathfrak{a}})/\mathbf{Q}$ (this is the only place where we need the hypothesis that E has good reduction at $\mathfrak{p}^*$). □

By Lemma 2.2, if $\mathfrak{a} \in \mathcal{S}$ and $\ell \mid \mathfrak{a}$ we can identify

$$G_\ell \cong \mathrm{Gal}(F(E_{\mathfrak{a}})/F(E_{\mathfrak{a}/\ell})) \subset G_{\mathfrak{a}}.$$

For every $\ell \in \mathcal{L}$ fix a generator σ_ℓ of G_ℓ and define

$$N_\ell = \sum_{\tau \in G_\ell} \tau \in \mathbf{Z}[G_\ell], \quad D_\ell = \sum_{i=1}^{N(\ell)-2} i\sigma_\ell^{\,i} \in \mathbf{Z}[G_\ell].$$

The operator D_ℓ is constructed to satisfy

$$(\sigma_\ell - 1)D_\ell = (N(\ell)-1) - N_\ell \text{ in } \mathbf{Z}[G_\ell]. \tag{1}$$

For $\mathfrak{a} \in \mathcal{S}$ define

$$D_\mathfrak{a} = \prod_{\ell \mid \mathfrak{a}} D_\ell \in \mathbf{Z}[G_\mathfrak{a}].$$

For every ideal $\mathfrak{g}$ of $\mathcal{O}$ let $\mathcal{S}(\mathfrak{g}) \subset \mathcal{S}$ be the subset

$$\mathcal{S}(\mathfrak{g}) = \{\mathfrak{a} \in \mathcal{S} : \mathfrak{a} \text{ is prime to } \mathfrak{g}\}.$$

We will sometimes write 1 for the trivial ideal $\mathcal{O} \in \mathcal{S}$. For every $\mathfrak{g}$ let $\mathcal{U}_F(\mathfrak{g})$ denote the set of functions $\alpha : \mathcal{S}(\mathfrak{g}) \to \overline{F}^\times$ such that for all $\mathfrak{a} \in \mathcal{S}(\mathfrak{g})$ and all primes $\ell \mid \mathfrak{a}$:

$$\alpha(\mathfrak{a}) \in F(E_\mathfrak{a})^\times, \tag{2a}$$

$$\alpha(\mathfrak{a}) \text{ is a global unit for } \mathfrak{a} \neq 1, \tag{2b}$$

$$\alpha(\mathfrak{a})^{N_\ell} = \alpha(\mathfrak{a}/\ell)^{\mathrm{Fr}_\ell - 1} \tag{2c}$$

where Fr_ℓ denotes the Frobenius of ℓ in $\mathrm{Gal}(F(E_{\mathfrak{a}/\ell})/K)$, and

$$\alpha(\mathfrak{a}) \equiv \alpha(\mathfrak{a}/\ell) \text{ modulo all primes above } \ell. \tag{2d}$$

Let $\mathcal{U}_F = \mathcal{U}_{F,M} = \amalg\, \mathcal{U}_F(\mathfrak{g})$, disjoint union over all ideals $\mathfrak{g}$ of $\mathcal{O}$. If $\alpha_1 \in \mathcal{U}_F(\mathfrak{g}_1)$ and $\alpha_2 \in \mathcal{U}_F(\mathfrak{g}_2)$ then $\alpha_1\alpha_2 \in \mathcal{U}_F(\mathfrak{g}_1\mathfrak{g}_2)$ so $\mathcal{U}_F$ is closed under multiplication. For $\alpha \in \mathcal{U}_F$ we will write $\mathcal{S}(\alpha)$ for the domain of α, i.e. $\mathcal{S}(\alpha) = \mathcal{S}(\mathfrak{g})$ if $\alpha \in \mathcal{U}_F(\mathfrak{g})$.

We now show how to obtain elements $\alpha \in \mathcal{U}_F$ from elliptic units. We follow the construction of elliptic units in Chapter II of [3] (see also §3 of [2]). Fix an embedding $\overline{K} \subset \mathbf{C}$ and an isomorphism $E(\mathbf{C}) \cong \mathbf{C}/L$ with some lattice $L \subset \mathbf{C}$. For every $\ell \in \mathcal{L}$ fix a point $x_\ell \in \mathbf{C}/L$ of order exactly ℓ. Let $\tau \in \mathbf{C}/L$ be an element of order exactly $\mathfrak{f}\mathfrak{p}^{n+1}$ for some ideal $\mathfrak{f}$ of $\mathcal{O}$, and let $\mathfrak{g}$ be an ideal of $\mathcal{O}$ prime to $6\mathfrak{f}\mathfrak{p}$. For every $\mathfrak{a} \in \mathcal{S}(\mathfrak{f}\mathfrak{g})$ define

$$\alpha_{\tau,\mathfrak{g}}(\mathfrak{a}) = \mathbf{N}_{F(E_{\mathfrak{f}\mathfrak{a}})/F(E_\mathfrak{a})}\Theta(\tau + \sum_{\ell \mid \mathfrak{a}} x_\ell;\, L, \mathfrak{g}),$$

where $\Theta(z; L, \mathfrak{g})$ is the function defined in §II.2.3 of [3]. Then $\alpha_{\tau,\mathfrak{g}} \in \mathcal{U}_F(\mathfrak{f}\mathfrak{g})$

(see [3] §II.2 and [10] §12). We define $\mathcal{E}_F = \mathcal{E}_n$, the elliptic units of $F = K_n$, to be the global units in the group generated by the $\alpha_{\tau,\mathfrak{g}}(1)$ for τ and $\mathfrak{g}$ as above. This proves the following.

Proposition 2.3 *If* $u \in \mathcal{E}_F$ *is an elliptic unit, then (for every* M*) there is an element* $\alpha \in \mathcal{U}_F$ *such that* $\alpha(1) = u$. □

3 PROPERTIES OF KOLYVAGIN'S SYSTEMS OF UNITS

Lemma 3.1 *If* $\alpha \in \mathcal{U}_F$ *and* $\mathfrak{a} \in \mathcal{S}(\alpha)$ *then*

$$\alpha(\mathfrak{a})^{D_\mathfrak{a}} \in [F(E_\mathfrak{a})^\times/(F(E_\mathfrak{a})^\times)^M]^{G_\mathfrak{a}}.$$

Proof. We prove this by induction on the number of primes dividing $\mathfrak{a}$. If $\mathfrak{a} = 1$ there is nothing to prove. For every $\ell \mid \mathfrak{a}$, by (1) and (2c)

$$\alpha(\mathfrak{a})^{D_\mathfrak{a}(\sigma_\ell - 1)} = \alpha(\mathfrak{a})^{D_{\mathfrak{a}/\ell}(N(\ell)-1-N_\ell)} \equiv \alpha(\mathfrak{a}/\ell)^{D_{\mathfrak{a}/\ell}(1-\mathrm{Fr}_\ell)} \pmod{(F(E_\mathfrak{a})^\times)^M}.$$

Since Fr_ℓ fixes F, our induction hypothesis shows

$$\alpha(\mathfrak{a}/\ell)^{D_{\mathfrak{a}/\ell}(1-\mathrm{Fr}_\ell)} \in (F(E_{\mathfrak{a}/\ell})^\times)^M.$$

These σ_ℓ generate $G_\mathfrak{a}$, so this proves the lemma. □

Proposition 3.2 *For every* $\alpha \in \mathcal{U}_F$ *there is a unique map*

$$\kappa = \kappa_\alpha : \mathcal{S}(\alpha) \to F^\times/(F^\times)^M$$

such that for every $\mathfrak{a} \in \mathcal{S}(\alpha)$, $\kappa(\mathfrak{a}) \equiv \alpha(\mathfrak{a})^{D_\mathfrak{a}} \pmod{(F(E_\mathfrak{a})^\times)^M}$.

Proof. There are canonical isomorphisms

$$F^\times/(F^\times)^M \cong H^1(\bar{F}/F, \mu_M) \cong H^1((\bar{F}/F(E_\mathfrak{a}), \mu_M)^{G_\mathfrak{a}} \cong [F(E_\mathfrak{a})^\times/(F(E_\mathfrak{a})^\times)^M]^{G_\mathfrak{a}},$$

the first and third from Kummer theory and the second from the inflation-restriction sequence of Galois cohomology, since $\mu_M \cap F(E_\mathfrak{a}) = 1$ by Lemma 2.2(iii). Thus the proposition follows from Lemma 3.1. □

Given an $\alpha \in \mathcal{U}_F$, each $\kappa(\mathfrak{a})$ gives a principal ideal of F (modulo M^{th}-powers of ideals) which can be viewed as a relation in the ideal class group of F. These relations will be used to bound the size of the ideal class group. To do this, we must understand the prime factorizations of these ideals and also how to choose $\mathfrak{a}$ so as to get useful relations.

Let $\mathcal{O}_F$ denote the ring of integers of F, and write $\mathcal{I} = \bigoplus_{\lambda} \mathbf{Z}\lambda$ for the group of fractional ideals of F, written additively. For every prime ℓ of K write $\mathcal{I}_\ell = \bigoplus_{\lambda \mid \ell} \mathbf{Z}\lambda$, so $\mathcal{I} = \bigoplus_{\ell} \mathcal{I}_\ell$, and if $y \in F^\times$ let $(y) \in \mathcal{I}$ denote the principal ideal generated by y, and $(y)_\ell \in \mathcal{I}_\ell$, $[y]_\ell \in \mathcal{I}_\ell/M\mathcal{I}_\ell$ the projections of (y). Note that $[y]_\ell$ is also well-defined for $y \in F^\times/(F^\times)^M$.

Proposition 3.3 *Suppose $\ell \in \mathcal{L}$. There is a unique* Gal(F/K)*-equivariant surjection*

$$\varphi_\ell : (\mathcal{O}_F/\ell\mathcal{O}_F)^\times \to \mathcal{I}_\ell/M\mathcal{I}_\ell$$

which makes the following diagram commute:

$$\begin{array}{ccc} & F(E_\ell)^\times & \\ x \mapsto x^{1-\sigma_\ell} \quad \swarrow & & \searrow \quad x \mapsto [x^{N_\ell}]_\ell \\ (\mathcal{O}_F/\ell\mathcal{O}_F)^\times & \xrightarrow[\varphi_\ell]{} & \mathcal{I}_\ell/M\mathcal{I}_\ell \end{array}$$

(For each λ of F *above ℓ and λ' of* $F(E_\ell)$ *above λ, we have identified $\mathcal{O}_{F(E_\ell)}/\lambda'$ with $\mathcal{O}_F/\lambda$.)*

Proof. Since $[F(E_\ell):F] = N(\ell)-1$ and all primes above ℓ are totally, tamely ramified in $F(E_\ell)/F$ (Lemma 2.2(i)), the vertical maps are both surjective and the kernel of the left-hand map, namely the subgroup

$$\{x \in F(E_\ell)^\times : \mathrm{ord}_{\lambda'}(x) \equiv 0 \pmod{N(\ell)-1} \text{ for all primes } \lambda' \mid \ell \text{ of } F(E_\ell)\},$$

is clearly contained in the kernel of the right-hand map. $\square$

One should regard the map φ_ℓ of Proposition 3.3 as a logarithm modulo ℓ. For $\ell \in \mathcal{L}$ we will also write φ_ℓ for the induced homomorphism

$$\varphi_\ell : \{y \in F^\times/(F^\times)^M : [y]_\ell = 0\} \to \mathcal{I}_\ell/M\mathcal{I}_\ell.$$

Proposition 3.4 (Kolyvagin [6]) *Suppose* $\alpha \in \mathcal{U}_F$, $\kappa = \kappa_\alpha$ *is the map defined in Proposition 3.2, and* $\mathfrak{a} \in \mathcal{S}(\alpha)$, $\mathfrak{a} \neq 1$.
(i) *If* $\ell \nmid \mathfrak{a}$, *then* $[\kappa(\mathfrak{a})]_\ell = 0$.
(ii) *If* $\ell \mid \mathfrak{a}$, *then* $[\kappa(\mathfrak{a})]_\ell = \varphi_\ell(\kappa(\mathfrak{a}/\ell))$.

Proof. By definition, $\kappa(\mathfrak{a}) \in \alpha(\mathfrak{a})^{D_\mathfrak{a}}(F(E_\mathfrak{a})^\times)^M$. By Lemma 2.2(i), $F(E_\mathfrak{a})/F$ is unramified outside of primes dividing $\mathfrak{a}$, so (i) follows from (2b). Suppose $\ell \mid \mathfrak{a}$, say $\mathfrak{a} = \mathfrak{b}\ell$. Then we can represent $\kappa(\mathfrak{a})$ and $\kappa(\mathfrak{b})$ by

$$\kappa(\mathfrak{a}) = \alpha(\mathfrak{a})^{D_\mathfrak{a}}\beta_\mathfrak{a}{}^M \quad \text{and} \quad \kappa(\mathfrak{b}) = \alpha(\mathfrak{b})^{D_\mathfrak{b}}\beta_\mathfrak{b}{}^M$$

where $\beta_\mathfrak{a} \in F(E_\mathfrak{a})^\times$ and $\beta_\mathfrak{b} \in F(E_\mathfrak{b})^\times$ satisfy

$$\beta_\mathfrak{a}{}^{1-\sigma} = (\alpha(\mathfrak{a})^{D_\mathfrak{a}(\sigma-1)})^{1/M} \quad \text{and} \quad \beta_\mathfrak{b}{}^{1-\sigma} = (\alpha(\mathfrak{b})^{D_\mathfrak{b}(\sigma-1)})^{1/M} \tag{3}$$

for all $\sigma \in \mathrm{Gal}(\bar{F}/F)$. (Note that the Mth-root is uniquely defined since $\mu_M \cap F(E_\mathfrak{a})^\times = 1$.) By (i) we may choose $\beta_\mathfrak{b}$ prime to ℓ.

Write $d = (N(\ell)-1)/M$ and let γ be any element of $F(E_\ell)^\times$ such that $[N_\ell\gamma]_\ell = [\kappa(\mathfrak{a})]_\ell$. It follows that $\mathrm{ord}_{\lambda'}(\beta_\mathfrak{a}) \equiv \mathrm{ord}_{\lambda'}(\gamma^d) \pmod{N(\ell)-1}$ for all primes λ' of $F(E_\mathfrak{a})$ above ℓ. Therefore, modulo any prime above ℓ, using (3), (1), (2c), and (2d),

$$\gamma^{(1-\sigma_\ell)d} \equiv \beta_\mathfrak{a}{}^{1-\sigma_\ell} = (\alpha(\mathfrak{a})^{D_\mathfrak{a}(\sigma_\ell-1)})^{1/M} = \alpha(\mathfrak{a})^{D_\mathfrak{b}d}(\alpha(\mathfrak{b})^{D_\mathfrak{b}(1-\mathrm{Fr}_\ell)})^{1/M}$$

$$\equiv \alpha(\mathfrak{b})^{D_\mathfrak{b}d}\beta_\mathfrak{b}{}^{\mathrm{Fr}_\ell-1} \equiv (\alpha(\mathfrak{b})^{D_\mathfrak{b}}\beta_\mathfrak{b}{}^M)^d \equiv \kappa(\mathfrak{b})^d.$$

Now applying the diagram of Proposition 3.3 with $\gamma \in F(E_\ell)^\times$ shows

$$[\kappa(\mathfrak{a})]_\ell = \varphi_\ell(\kappa(\mathfrak{b})). \qquad \square$$

4 AN APPLICATION OF THE CHEBOTAREV THEOREM

Theorem 4.2 below together with Proposition 3.4 will enable us to construct all the relations we need in the ideal class group of F. In the simpler case where $p\nmid\#(G)$ it already appears in the work of Thaine ([13] Proposition 4); the version below which we will need is essentially Theorem 5.5 of [9].

Lemma 4.1 *Write* $F' = F(\mu_M)$.

(i) $\mathrm{Gal}(F'/F) \cong \mathrm{Gal}(\mathbf{Q}(\mu_M)/\mathbf{Q})$ *and* F'/F *is totally ramified at all primes above* $\mathfrak{p}^*$.

(ii) *The map* $F^\times/(F^\times)^M \to F'^\times/(F'^\times)^M$ *is injective.*

Proof. The first assertion is immediate from the fact that $\mathfrak{p}^*$ is unramified in F/K by Lemma 2.1. We have

$$F^\times/(F^\times)^M \cong H^1(\bar{F}/F, \mu_M) \quad \text{and} \quad F'^\times/(F'^\times)^M \cong H^1(\bar{F}/F', \mu_M),$$

so the kernel of the map in (ii) is $H^1(F'/F, \mu_M) \cong H^1((\mathbf{Z}/M\mathbf{Z})^\times, \mathbf{Z}/M\mathbf{Z}) = 0$. □

Write A for the p-part of the ideal class group of F and $G = \mathrm{Gal}(F/K)$.

Theorem 4.2 *Suppose one is given* $\mathfrak{c} \in A$, *a finite* G-*submodule* W *of* $F^\times/(F^\times)^M$, *and a Galois-equivariant map* $\psi : W \to (\mathbf{Z}/M\mathbf{Z})[G]$. *Then there are infinitely many primes* λ *of* F *such that (writing* ℓ *for the prime of* K *below* λ)

(i) $\lambda \in \mathfrak{c}$,

(ii) $N(\ell) \equiv 1 \pmod M$ *and* ℓ *splits completely in* F/K,

(iii) $[w]_\ell = 0$ *for all* $w \in W$, *and there is a* $u \in (\mathbf{Z}/M\mathbf{Z})^\times$ *such that for all* $w \in W$, $\varphi_\ell(w) = u\psi(w)\lambda$.

Proof. Let H be the maximal unramified abelian p-extension of F, so that A is identified with Gal(H/F) by class field theory. Write $F' = F(\mu_M)$. We have the diagram below.

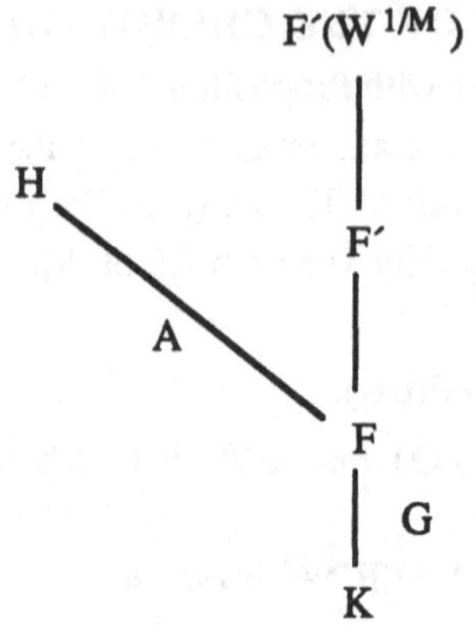

By Lemma 4.1(ii), Kummer theory gives a $\mathrm{Gal}(F'/K)$-equivariant isomorphism

$$\mathrm{Gal}(F'(W^{1/M})/F') \cong \mathrm{Hom}(W,\boldsymbol{\mu}_M). \tag{4}$$

Since $\boldsymbol{\mu}_M^{\mathrm{Gal}(F'/F)} = 1$, it follows that $\mathrm{Gal}(F'(W^{1/M})/F')$ has no nonzero quotients on which $\mathrm{Gal}(F'/F)$ acts trivially. But H is abelian over F, so $\mathrm{Gal}(F'/F)$ acts trivially on $\mathrm{Gal}(HF'/F')$, and thus $F'(W^{1/M}) \cap HF' = F'$. Further, by Lemma 4.1(i) there is no nontrivial unramified extension of F in F', so we conclude

$$F'(W^{1/M}) \cap H = F.$$

Fix a primitive M^{th}-root of unity ζ_M and define a map

$$\iota : (\mathbf{Z}/M\mathbf{Z})[G] \to \boldsymbol{\mu}_M \quad \text{by} \quad \iota(1) = \zeta_M \text{ and } \iota(\sigma) = 1 \text{ for } \sigma \in G,\ \sigma \neq 1.$$

Let $\gamma \in \mathrm{Gal}(F'(W^{1/M})/F')$ be the automorphism corresponding to $\iota \circ \psi \in \mathrm{Hom}(W,\boldsymbol{\mu}_M)$ under the isomorphism (4). Then by definition of the Kummer pairing, $\iota\circ\psi(w) = \gamma(w^{1/M})/w^{1/M}$ for all $w \in W$.

Since $F'(W^{1/M}) \cap H = F$ we can choose $\delta \in \mathrm{Gal}(HF'(W^{1/M})/F)$ such that δ restricts to γ on $F'(W^{1/M})$ and to $\mathfrak{c}$ on H. Let λ be a prime of F of degree 1 whose Frobenius in $\mathrm{Gal}(HF'(W^{1/M})/F)$ is the conjugacy class of δ, and such that all conjugates of λ are unramified in $HF'(W^{1/M})/K$. Since W is finite, the Chebotarev theorem guarantees the existence of infinitely many such λ. We must verify that λ satisfies (i), (ii) and (iii). Let ℓ be the prime of K below λ.

The identification of A with Gal(H/F) sends the class of λ to the Frobenius of λ, so (i) is immediate. Also, since δ is trivial on F', ℓ splits completely in $F(\mu_M)/K$ which proves (ii). The first assertion of (iii), that $[w]_\ell = 0$ for all $w \in W$, holds because ℓ is unramified in $F'(W^{1/M})/K$.

From the definition of φ_ℓ, $\mathrm{ord}_\lambda(\varphi_\ell(w)) = 0$ if and only if w is an Mth-power modulo λ. Also,

$$\mathrm{ord}_\lambda(\psi(w)\lambda) = 0 \Leftrightarrow \iota\circ\psi(w) = 1 \Leftrightarrow \gamma(w^{1/M})/w^{1/M} = 1$$

$$\Leftrightarrow w \text{ is an } M^{\text{th}}\text{-power modulo } \lambda.$$

Therefore there is a unit $u \in (\mathbf{Z}/M\mathbf{Z})^\times$ such that

$$\mathrm{ord}_\lambda(\varphi_\ell(w)) = u\,\mathrm{ord}_\lambda(\psi(w)\lambda) \text{ for all } w \in W.$$

It follows that the map

$$w \mapsto \varphi_\ell(w) - u\,\psi(w)\lambda$$

is a G-equivariant homomorphism into $\bigoplus_{\sigma\in G,\, \sigma\neq 1} (\mathbf{Z}/M\mathbf{Z})\lambda^\sigma$, which has no nonzero G-stable submodules. This proves (iii). □

5 TOOLS FROM IWASAWA THEORY

In this section we review the machinery from Iwasawa theory which will go into the proof of Theorem 1.1. Recall the notation of §1. We need to understand to what extent we can recover A_n, $\overline{\mathcal{E}}_n$, and $\overline{\mathcal{C}}_n$ from A_∞, $\overline{\mathcal{E}}_\infty$, and $\overline{\mathcal{C}}_\infty$, respectively. These questions were studied by Iwasawa in [5].

Fix a character χ of Δ and as in §1 write $\Lambda = \mathbf{Z}_p[[\mathcal{G}]]^\chi$. For every n let $\Gamma_n = \mathrm{Gal}(K_\infty/K_n)$, write $\mathcal{I}_n$ for the ideal of Λ generated by $\{\gamma-1 : \gamma \in \Gamma_n\}$ and define

$$\Lambda_n = \mathbf{Z}_p[\mathrm{Gal}(K_n/K)]^\chi = \Lambda/\mathcal{I}_n\Lambda.$$

If Y is a Λ-module we write $Y_{\Gamma_n} = Y/\mathcal{I}_nY = Y\otimes_\Lambda \Lambda_n$.

Theorem 5.1 (i) *For every* χ *and every* n *the projection map* $(A_\infty^\chi)_{\Gamma_n} \to A_n^\chi$ *is an isomorphism.*
(ii) *If* $\chi \neq 1$ *then* $\overline{\mathcal{C}}_\infty^\chi$ *is free of rank* 1 *over* Λ *and the projection map* $(\overline{\mathcal{C}}_\infty^\chi)_{\Gamma_n} \to \overline{\mathcal{C}}_n^\chi$ *is an isomorphism for every* n.
(iii) *If* $\chi \neq 1$ *then there is an ideal* $\mathfrak{a}$ *of finite index in* Λ *such that* $\mathfrak{a}$ *annihilates the kernel and cokernel of the projection map* $(\overline{\mathcal{E}}_\infty^\chi)_{\Gamma_n} \to \overline{\mathcal{E}}_n^\chi$ *for every* n.

Proof. The first statement is a standard result from Iwasawa theory , using the fact that K_∞/K_n is totally ramified at the unique prime above $\mathfrak{p}$ and unramified everywhere else. Assertion (ii) is in [3] §III.1.3, and (iii) follows from the Corollary in §5.4 of [5] (see for example [11] Lemma 1.2). □

For each character χ of Δ fix a generator $h_\chi \in \Lambda$ of $\mathrm{char}(\overline{\mathcal{E}}_\infty^\chi/\overline{\mathcal{C}}_\infty^\chi)$.

Corollary 5.2 *Suppose* $\chi \neq 1$. *There is an ideal* $\mathfrak{a}$ *of finite index in* Λ *such that for every* $\eta \in \mathfrak{a}$ *and every* n, *there is a map* $\theta_{n,\eta} : \overline{\mathcal{E}}_n^\chi \to \Lambda_n$ *such that*

$$\theta_{n,\eta}(\overline{\mathcal{C}}_n^\chi) = \eta h_\chi \Lambda_n.$$

Proof. By [3] §III.1.3, $\overline{\mathcal{E}}_\infty^\chi$ is a torsion-free, rank-one Λ-module. Therefore there is an injective homomorphism $\theta : \overline{\mathcal{E}}_\infty^\chi \to \Lambda$ with finite cokernel.

We first claim that $\theta(\overline{\mathcal{C}}_\infty^\chi) = h_\chi \Lambda$. Clearly θ induces a pseudo-isomorphism from $\overline{\mathcal{E}}_\infty^\chi/\overline{\mathcal{C}}_\infty^\chi$ to $\Lambda/\theta(\overline{\mathcal{C}}_\infty^\chi)$; since $\overline{\mathcal{C}}_\infty^\chi$ is free of rank one over Λ (Theorem 5.1(ii)),

$$\theta(\overline{\mathcal{C}}_\infty^\chi) = \mathrm{char}(\Lambda/\theta(\overline{\mathcal{C}}_\infty^\chi)) = \mathrm{char}(\overline{\mathcal{E}}_\infty^\chi/\overline{\mathcal{C}}_\infty^\chi) = h_\chi \Lambda.$$

Let $\mathfrak{a}$ be an ideal of Λ of finite index satisfying Theorem 5.1(iii). Fix an n and let θ_n be the homomorphism from $(\overline{\mathcal{E}}_\infty^\chi)_{\Gamma_n}$ to Λ_n induced by θ, and π_n the projection map from $(\overline{\mathcal{E}}_\infty^\chi)_{\Gamma_n}$ to $\overline{\mathcal{E}}_n^\chi$. For any $\eta \in \mathfrak{a}$ we can define a map $\theta_{n,\eta}$ from $\overline{\mathcal{E}}_n^\chi$ to Λ_n so that the following diagram commutes:

$$\begin{array}{ccc} (\overline{\mathcal{E}}_\infty^{\chi})_{\Gamma_n} & \xrightarrow{\theta_n} & \Lambda_n \\ \downarrow \pi_n & & \downarrow \eta \\ \overline{\mathcal{E}}_n^{\chi} & \xrightarrow[\theta_{n,\eta}]{} & \Lambda_n \end{array}$$

i.e., $\theta_{n,\eta}(u) = \theta_n(\pi_n^{-1}(\eta u))$. This is well-defined because η annihilates $\mathrm{coker}(\pi_n)$, and $\ker(\pi_n)$ is finite so $\ker(\pi_n) \subset \ker(\theta_n)$. Then by Theorem 5.1(ii),

$$\theta_{n,\eta}(\overline{\mathcal{E}}_n^{\chi}) = \eta\theta_n(\overline{\mathcal{E}}_\infty^{\chi}) = \eta h_\chi \Lambda_n. \qquad \square$$

Since A_∞^{χ} is a torsion Λ-module, A_∞^{χ} is pseudo-isomorphic to a module of the form $\bigoplus_{i=1}^{k} \Lambda/f_i\Lambda$ with nonzero $f_i \in \Lambda$. In particular writing $f_\chi = \prod_{i=1}^{k} f_i$, we have $\mathrm{char}(A_\infty^{\chi}) = f_\chi\Lambda$.

Corollary 5.3 *Let* $f_1,\ldots,f_k$ *be as above. There is an ideal* $\mathfrak{a}$ *of finite index in* Λ *and for every* n *there are classes* $\mathfrak{c}_1,\ldots,\mathfrak{c}_k \in A_n^{\chi}$ *such that the annihilator* $\mathrm{Ann}(\mathfrak{c}_i) \subset \Lambda_n$ *of* $\mathfrak{c}_i$ *in* $A_n^{\chi}/(\Lambda_n\mathfrak{c}_1 + \cdots + \Lambda_n\mathfrak{c}_{i-1})$ *satisfies*

$$\mathfrak{a}\,\mathrm{Ann}(\mathfrak{c}_i) \subset f_i\Lambda_n.$$

Proof. On torsion Λ-modules, the pseudo-isomorphism relation is reflexive, so there is an exact sequence

$$0 \to \bigoplus_{i=1}^{k} \Lambda/f_i\Lambda \to A_\infty^{\chi} \to Z \to 0$$

with a finite Λ-module Z. By Theorem 5.1(i) and a standard snake lemma argument, tensoring with $\Lambda_n = \Lambda/\mathcal{I}_n\Lambda$ yields

$$Z^{\Gamma_n} \to \bigoplus_{i=1}^{k} \Lambda_n/f_i\Lambda_n \to A_n^{\chi} \to Z_{\Gamma_n} \to 0.$$

Let $\mathfrak{a}$ be the annihilator of the finite module Z and choose $\mathfrak{c}_i$ to be the image in

A_n^χ of $1 \in \Lambda_n/f_i\Lambda_n$ under the map above. These $\mathfrak{c}_i$ have the desired property. □

Lemma 5.4 *Suppose* $\chi \neq 1$, *and let* f_χ *and* h_χ *be generators of* $\mathrm{char}(A_\infty^\chi)$ *and* $\mathrm{char}(\overline{\mathcal{E}}_\infty^\chi/\overline{\mathcal{C}}_\infty^\chi)$, *respectively. Then for every* n, $\Lambda_n/f_\chi\Lambda_n$ *and* $\Lambda_n/h_\chi\Lambda_n$ *are finite.*

Proof. From a pseudo-isomorphism $A_\infty^\chi \to \bigoplus_{i=1}^{k} \Lambda/f_i\Lambda$ we get, for every n, a map

$$A_n^\chi \cong (A_\infty^\chi)_{\Gamma_n} \to \bigoplus_{i=1}^{k} \Lambda_n/f_i\Lambda_n$$

with finite kernel and cokernel. In particular each $\Lambda_n/f_i\Lambda_n$ is finite, and it follows easily that $\Lambda_n/f_\chi\Lambda_n$ is finite for every n. Similarly for all n we have maps

$$(\overline{\mathcal{E}}_\infty^\chi/\overline{\mathcal{C}}_\infty^\chi)_{\Gamma_n} \to \Lambda_n/h_\chi\Lambda_n$$
$$(\overline{\mathcal{E}}_\infty^\chi/\overline{\mathcal{C}}_\infty^\chi)_{\Gamma_n} \to \overline{\mathcal{E}}_n^\chi/\overline{\mathcal{C}}_n^\chi$$

with finite kernel and cokernel (using Theorem 5.1 for the second map). Since $[\overline{\mathcal{E}}_n^\chi:\overline{\mathcal{C}}_n^\chi]$ is finite this completes the proof. □

6 PROOF OF THE MAIN CONJECTURE

For this section fix n and write $A = A_n$, $\mathcal{E} = \mathcal{E}_n$, and $\mathcal{C} = \mathcal{C}_n$. We will apply the results of the previous sections with $F = K_n$. If ℓ is a prime of K and $w \in F^\times$, recall that $(w)_\ell \in \mathcal{I}_\ell$ is the portion of the principal ideal (w) which is supported on the primes above ℓ and $[w]_\ell \in \mathcal{I}_\ell/M\mathcal{I}_\ell$ is its projection. For any character χ of Δ define $\mathcal{I}_\ell^\chi = (\mathcal{I}_\ell \otimes \mathbf{Z}_p)^\chi$. If λ is a prime of F above $\ell \in \mathcal{L}$ then $\mathcal{I}_\ell^\chi$ is free of rank one over Λ_n, generated by $\lambda^\chi = e_\chi\lambda$, and we define

$$v_\lambda = v_{\lambda,\chi} : F^\times \to \Lambda_n \quad \text{by} \quad v_\lambda(w)\lambda^\chi = e_\chi(w)_\ell.$$

We will write $\overline{v}_\lambda$ for the corresponding map from $F^\times/(F^\times)^M$ to $\Lambda_n/M\Lambda_n$ satisfying $\overline{v}_\lambda(w)\lambda^\chi = e_\chi[w]_\ell$.

Lemma 6.1 *Suppose we are given* $\beta \in (F^\times/(F^\times)^M)^\chi$, $\ell \in \mathcal{L}$, *a prime* λ *of* F *above* ℓ, *a set* S *of primes of* K, *and* $\eta, f \in \Lambda_n$. *Write* B *for the subgroup of the ideal class group* A *generated by the primes of* F *lying above primes in* S, $\mathfrak{c} \in A^\chi$ *for the class of* λ^χ *and* W *for the* Λ_n*-submodule of* $F^\times/(F^\times)^M$ *generated by* β. *If*

(a) $[\beta]_\mathfrak{q} = 0$ *for* $\mathfrak{q} \notin S \cup \{\ell\}$,

(b) $\#(A^\chi) \mid M$, *and* $\overline{v_\lambda}(\beta)$ *divides* $(M/\#(A^\chi))$ *in* $\Lambda_n/M\Lambda_n$,

(c) $\Lambda_n/f\Lambda_n$ *is finite, and*

(d) *the annihilator* $\mathrm{Ann}(\mathfrak{c}) \subset \Lambda_n$ *of* $\mathfrak{c}$ *in* A^χ/B^χ *satisfies* $\eta\,\mathrm{Ann}(\mathfrak{c}) \subset f\Lambda_n$,

then there is a Galois-equivariant map $\psi : W \to \Lambda_n/M\Lambda_n$ *such that*

$$f\psi(\beta) = \eta\overline{v_\lambda}(\beta).$$

Proof. Let γ be any lift of β to $F^\times$. Then

$$(\gamma) = (\gamma)_\ell + \sum_{\mathfrak{q} \neq \ell} (\gamma)_\mathfrak{q},$$

and by (a), $(\gamma)_\mathfrak{q} \in M\mathcal{I}_\mathfrak{q}$ if $\mathfrak{q} \notin S \cup \{\ell\}$. By (b) M annihilates A^χ and we conclude that $(\gamma)_\ell$ projects to 0 in A^χ/B^χ. But the projection of $(\gamma)_\ell$ is $v_\lambda(\gamma)\mathfrak{c}$ and therefore by (d), $\eta v_\lambda(\gamma) \in f\Lambda_n$. Write $\delta = \eta v_\lambda(\gamma)/f$; division by f is uniquely defined because of (c).

Define $\psi : W \to \Lambda_n/M\Lambda_n$ by $\psi(\beta^\rho) = \rho\delta$ for all $\rho \in \mathbf{Z}[\mathrm{Gal}(K_n/K)]$. This map has the desired property but we need to show that it is well-defined. Suppose $\beta^\rho = 1$, that is, $\gamma^\rho = x^M$ with $x \in F^\times$. Then in particular $\overline{v_\lambda}(\beta^\rho) = \rho\overline{v_\lambda}(\beta) = 0$. By (b) it follows that $\rho(M/\#(A^\chi))\Lambda_n \subset M\Lambda_n$, and therefore $\rho A^\chi = 0$. Then

$$\begin{aligned}(x) &= \sum_\mathfrak{q} (x)_\mathfrak{q} \\ &= \sum_{\mathfrak{q} \in S \cup \{\ell\}} M^{-1}(\gamma^\rho)_\mathfrak{q} + \sum_{\mathfrak{q} \notin S \cup \{\ell\}} \rho(M^{-1}(\gamma)_\mathfrak{q}) \\ &\equiv M^{-1}(\gamma^\rho)_\ell \pmod{\bigoplus_{\mathfrak{q} \in S} \mathcal{I}_\mathfrak{q}, \rho\mathcal{I}}.\end{aligned}$$

Since $\rho A^\chi = 0$ we conclude that $M^{-1}(\gamma^\rho)_\ell$ projects to 0 in A^χ/B^χ. Thus $M^{-1}v_\lambda(\gamma^\rho)\mathfrak{c} = 0$ in A^χ/B^χ, so by (d), $\rho\delta f = \eta v_\lambda(\gamma^\rho) \in Mf\Lambda_n$ and $\psi(\beta^\rho) = \rho\delta \in M\Lambda_n$. □

Theorem 6.2 *For every character* χ *of* Δ,

$$\mathrm{char}(A_\infty^\chi) \textit{ divides } \mathrm{char}(\overline{\mathcal{E}}_\infty^\chi/\overline{\mathcal{C}}_\infty^\chi).$$

Proof. If $\chi = 1$, this is true because $A_\infty^\chi = 0$ (by Theorem 5.1(i), combined with the fact that K has class number one). Thus we may assume $\chi \neq 1$.

Recall $\mathrm{char}(\overline{\mathcal{E}}_\infty^\chi/\overline{\mathcal{C}}_\infty^\chi) = h_\chi\Lambda$ and $\mathrm{char}(A_\infty^\chi) = f_\chi\Lambda$, where $f_\chi = \prod_{i=1}^{k} f_i$. Let $\mathfrak{c}_1,\ldots,\mathfrak{c}_k \in A^\chi$ be as in Corollary 5.3. Also choose one more class $\mathfrak{c}_{k+1}$ which can be any element of A^χ at all (for example $\mathfrak{c}_{k+1} = 0$). Fix an ideal $\mathfrak{a}$ of finite index in Λ satisfying both Corollaries 5.2 and 5.3, and let $\eta \in \mathfrak{a}$ be any element such that $\Lambda_m/\eta\Lambda_m$ is finite for all m (i.e., for all m, η is prime to $\gamma^{p^m}-1$, where γ is a topological generator of Γ). Let t be any power of p which annihilates both $\Lambda_n/\eta\Lambda_n$ and $\Lambda_n/h_\chi\Lambda_n$ (which is finite by Lemma 5.4), that is, $\eta \mid t$ and $h_\chi \mid t$ in Λ_n. Fix $M = \#(A^\chi)p^n t^{k+1}$.

By Theorem 5.1(ii), $(\mathcal{C}/(\mathcal{C}\cap\mathcal{E}^M))^\chi$ is cyclic over $\Lambda_n/M\Lambda_n$, and we fix a generator ξ. By Proposition 2.3 we can choose $\alpha \in \mathcal{U}_F$ such that $\alpha(1)$ projects to ξ in $\mathcal{E}/\mathcal{E}^M$. We have $\kappa = \kappa_\alpha$ as given by Proposition 3.2, and $\kappa(1) = \xi$. Let $\theta_{n,\eta} : \overline{\mathcal{E}}^\chi \to \Lambda_n$ be the map given by Corollary 5.2 with our choice of η, and $\bar\theta : \mathcal{E}/\mathcal{E}^M \to \Lambda_n/M\Lambda_n$ the reduced map. Without loss of generality we can normalize $\theta_{n,\eta}$ so that $\bar\theta(\xi) = \eta h_\chi$.

We will use Theorem 4.2 inductively to choose primes λ_i of F lying above ℓ_i of K for $1 \le i \le k+1$ satisfying:

$$\lambda_i \in \mathfrak{c}_i,\ \ \ell_i \in \mathcal{L}, \tag{5}$$

$$\bar v_{\lambda_1}(\kappa(\ell_1)) = u_1\eta h_\chi,\ \ f_{i-1}\bar v_{\lambda_i}(\kappa(\mathfrak{a}_i)) = u_{i-1}\eta\bar v_{\lambda_{i-1}}(\kappa(\mathfrak{a}_{i-1})) \text{ for } 2 \le i \le k+1 \tag{6}$$

where $\mathfrak{a}_i = \prod_{j\le i}\ell_j$ and $u_i \in (\mathbf{Z}/M\mathbf{Z})^\times$.

For the first step take $\mathfrak{c} = \mathfrak{c}_1$, $W = (\mathcal{E}/\mathcal{E}^M)^\chi$, and $\psi = \bar\theta : W \to \Lambda_n/M\Lambda_n$. Let λ_1 be a prime satisfying Theorem 4.2 with this data which lies above a prime $\ell_1 \in \mathcal{S}(\alpha)$. Then (i) and (ii) of Theorem 4.2 give (5). By Theorem 4.2(iii) and Proposition 3.4(ii), for some $u_1 \in (\mathbf{Z}/M\mathbf{Z})^\times$,

$$\overline{v}_{\lambda_1}(\kappa(\ell_1))\lambda_1{}^{\chi} = e_\chi[\kappa(\ell_1)]_{\ell_1} = e_\chi\varphi_{\ell_1}(\kappa(1)) = u_1\psi(\kappa(1))\lambda_1{}^{\chi}$$

$$= u_1\overline{\theta}(\xi)\lambda_1{}^{\chi} = u_1\eta h_\chi\lambda_1{}^{\chi}.$$

Since $(\mathcal{I}_{\ell_1}/M\mathcal{I}_{\ell_1})^{\chi}$ is free over $\Lambda_n/M\Lambda_n$, generated by $\lambda_1{}^{\chi}$, this proves the first equality of (6).

Now suppose $1 \le i \le k$ and we have chosen $\lambda_1,\ldots\lambda_i$ satisfying (5) and (6). We will define λ_{i+1}. Let $\mathfrak{a}_i = \prod_{j\le i}\ell_j$. By (6), $\overline{v}_{\lambda_i}(\kappa(\mathfrak{a}_i))$ divides $\eta^i h_\chi$ in $\Lambda_n/M\Lambda_n$, so by definition of M and $\overline{t}$, $\overline{v}_{\lambda_i}(\kappa(\mathfrak{a}_i))$ divides $(M/\#(A^{\chi}))$ in $\Lambda_n/M\Lambda_n$.

Let W_i be the Λ_n-submodule of $F^{\times}/(F^{\times})^M$ generated by $\kappa(\mathfrak{a}_i)^{\chi}$. Using Proposition 3.4(i), Lemma 5.4 and Corollary 5.3, we can apply Lemma 6.1 with $\beta = \kappa(\mathfrak{a}_i)^{\chi}$, $\ell = \ell_i$, $S = \{\ell_1,\ldots,\ell_{i-1}\}$ and $f = f_i$ to obtain a map $\psi_i : W_i \to \Lambda_n/M\Lambda_n$ such that

$$f_i\psi_i(\kappa(\mathfrak{a}_i)^{\chi}) = \eta\overline{v}_{\lambda_i}(\kappa(\mathfrak{a}_i)^{\chi}).$$

Now choose λ_{i+1} satisfying the conclusions of Theorem 4.2 with $\mathfrak{c} = \mathfrak{c}_i$, $W = W_i$, and $\psi = \psi_i$, lying above a prime $\ell_{i+1} \in \mathcal{S}(\alpha)$. Then (i) and (ii) of Theorem 4.2 give (5) for $i+1$. By Proposition 3.4(ii) and Theorem 4.2(iii) there is a $u_i \in (\mathbf{Z}/M\mathbf{Z})^{\times}$ so that

$$f_i\overline{v}_{\lambda_{i+1}}(\kappa(\mathfrak{a}_{i+1}))\lambda_{i+1}{}^{\chi} = f_i[\kappa(\mathfrak{a}_{i+1})^{\chi}]_{\ell_{i+1}} = f_i\varphi_{\ell_{i+1}}(\kappa(\mathfrak{a}_i)^{\chi})$$

$$= f_iu_i\psi_i(\kappa(\mathfrak{a}_i)^{\chi})\lambda_{i+1}{}^{\chi} = u_i\eta\overline{v}_{\lambda_i}(\kappa(\mathfrak{a}_i)^{\chi})\lambda_{i+1}{}^{\chi}.$$

This proves (6) for $i+1$.

Continue this induction process $k+1$ steps. Combining all of the relations (6) gives

$$\eta^{k+1}h_\chi = u\Big(\prod_{i=1}^{k} f_i\Big)\overline{v}_{\lambda_{k+1}}(\kappa(\mathfrak{a}_{k+1})) \quad \text{in } \Lambda_n/M\Lambda_n$$

for some $u \in (\mathbf{Z}/M\mathbf{Z})^{\times}$. Thus $f_\chi = \prod_{i=1}^{k} f_i$ divides $\eta^{k+1} h_\chi$ in $\Lambda_n/p^n\Lambda_n$. This holds for every n, so $f_\chi \mid \eta^{k+1} h_\chi$ in Λ.

To conclude the proof we need to remove the extra factor of η^{k+1}. Recall that $\mathfrak{a}$ is an ideal of finite index in Λ and η is any element of $\mathfrak{a}$ such that $\Lambda_n/\eta\Lambda_n$ is finite for every n. It is easy to verify that we can make two choices of η which are relatively prime. Since Λ is a unique factorization domain, it follows that f_χ divides h_χ. □

Proof of Theorem 1.1. Class field theory gives an exact sequence

$$0 \to U_\infty^\chi/\mathcal{E}_\infty^\chi \to X_\infty^\chi \to A_\infty^\chi \to 0.$$

Also we clearly have

$$0 \to \mathcal{E}_\infty^\chi/\mathcal{C}_\infty^\chi \to U_\infty^\chi/\mathcal{C}_\infty^\chi \to U_\infty^\chi/\mathcal{E}_\infty^\chi \to 0.$$

Since the characteristic ideal is multiplicative in exact sequences, Theorem 6.2 shows that $\mathrm{char}(X_\infty^\chi)$ divides $\mathrm{char}(U_\infty^\chi/\mathcal{C}_\infty^\chi)$ for every character χ of Δ. By a standard argument using the analytic class number formula to compare the Iwasawa invariants of X_∞^χ and $U_\infty^\chi/\mathcal{C}_\infty^\chi$ ([3] §III.2.1) it follows that $\mathrm{char}(X_\infty^\chi) = \mathrm{char}(U_\infty^\chi/\mathcal{C}_\infty^\chi)$ for every χ. Using again the two exact sequences above we see also that $\mathrm{char}(A_\infty^\chi) = \mathrm{char}(\mathcal{E}_\infty^\chi/\mathcal{C}_\infty^\chi)$ for every χ. □

REFERENCES

[1] Coates, J., Wiles, A.: On the conjecture of Birch and Swinnerton-Dyer. *Invent. Math.* **39** (1977) 223-251

[2] Coates, J., Wiles, A.: On p-adic L-functions and elliptic units. *J. Austral. Math. Soc.* **26** (1978) 1-25

[3] de Shalit, E.: The Iwasawa Theory of Elliptic Curves with Complex Multiplication. *Perspec. in Math.* **3**, Orlando: Academic Press (1987)

[4] Gross, B.: On the conjecture of Birch and Swinnerton-Dyer for elliptic curves with complex multiplication. In: Number Theory related to Fermat's Last Theorem, *Prog. in Math.* **26**, Boston: Birkhäuser (1982) 219-236

[5] Iwasawa, K.: On $\mathbf{Z}_l$-extensions of algebraic number fields. *Ann. of Math.* **98** (1973) 246-326

[6] Kolyvagin, V.A.: Euler systems. *To appear.*

[7] Perrin-Riou, B.: Arithmétique des courbes elliptiques et théorie d'Iwasawa. *Bull. Soc. Math. de France Suppl., Mémoire* **17** (1984)

[8] Perrin-Riou, B.: Points de Heegner et dérivées de fonctions L p-adiques. *Invent. Math.* **89** (1987) 455-510

[9] Rubin, K.: Global units and ideal class groups. *Invent. Math.* **89** (1987) 511-526

[10] Rubin, K.: Tate-Shafarevich groups and L-functions of elliptic curves with complex multiplication. *Invent. Math.* **89** (1987) 527-560

[11] Rubin, K.: On the main conjecture of Iwasawa theory for imaginary quadratic fields. *Invent. Math.* **93** (1988) 701-713

[12] Rubin, K.: The "main conjectures" of Iwasawa theory for imaginary quadratic fields. To appear in *Invent. Math.*

[13] Thaine, F.: On the ideal class groups of real abelian number fields. *Ann. of Math.* **128** (1988) 1-18

Remarks on special values of L-functions

ANTHONY J. SCHOLL*

INTRODUCTION

This article does not represent precisely a talk given at the symposium, but is complementary to [**DenS**]. Its purpose is to explain a setting in which the various conjectures on special values of L-functions admit a unified formulation. At critical points, Deligne's conjecture [**Del2**] relates the value of an L-function to a certain period, and at non-critical points, the conjectures of Beilinson [**Be1**] give an interpretation in terms of regulators. Finally, at the point of symmetry of the functional equation, there is the conjecture of Birch and Swinnerton-Dyer, generalised by Bloch [**Bl2**] and Beilinson [**Be2**], in which the determinant of the height pairing on cycles appears.

Both the periods and the regulators are constructed globally, and their definitions are in some sense archimedean. The height pairing, on the other hand, is defined as a sum of local terms. Our aim is to show how all of these objects—periods, regulators, and heights—may be interpreted as 'periods of mixed motives'.

That such a reformulation is possible in the case of regulators is clearly indicated in the letter of Deligne to Soulé [**Del3**]. Perhaps the only novel feature of our account is to regard the mixed motives as primary objects, rather than the Ext groups. It is appropriate to mention in this connection work of Anderson and of Harder [**H**], in which certain particular mixed motives arising in the study of the cohomology of Shimura varieties are investigated. These motives fit directly (and without assuming a grand conjectural framework) into our setting, although their connection with the K-theoretical formulation of Beilinson's conjectures remains obscure.

Section I recalls some of the properties of pure motives, and Deligne's period conjecture [**Del2**]. In section II we state a suitable generalisation of this

* Partially funded by NSF grant #DMS–8610730

conjecture to mixed motives. Although there does not as yet exist an entirely satisfactory definition of the category of mixed motives, Deligne [**Del4**] and Jannsen [**J**] have given an unconditional definition, based on absolute Hodge cycles.

In the third section we review the expected relation between extensions of motives and 'motivic cohomology' ([**Del3**], [**Be2**], [**J**]). We define a category of 'mixed motives over **Z**', in which the Ext-groups should correspond to the integral part of motivic cohomology (coming from the K-theory of regular schemes, projective over Spec **Z**). In the appendix to [**DenS**] some evidence for this relation is described.

In sections IV and V we describe the consequences of the period conjecture in the case of certain particular mixed motives ('universal extensions'). This shows how the conjecture includes the relevant parts of the conjectures of Beilinson and Birch–Swinnerton-Dyer as special cases. The comparison at the central point is somewhat more complicated than at the other points, and we only give a sketch; more details will appear later [**S**].

In the final section we show, in answer to a question raised by Deligne, that the period conjecture of section II contains no further information on L-values than the existing conjectures. We also repeat the calculations of §5 of [**Del2**] to show that it is compatible with the functional equation. Thus the period conjecture satisfies the most obvious consistency conditions.

An obvious gap in our account is the failure to allow motives with coefficients other than **Q**. However we hope that it is apparent that the same constructions can carried out be done word-for-word for motives with coefficients. The other restriction we have imposed—that the ground field be always **Q**—seems in contrast to be essential to our approach, mainly because of the 'peculiar' behaviour of the Riemann zeta function at $s=0$ and $s=1$.

As should be clear to the reader, this article is really only a naïve attempt to come to grips with Beilinson's conjectures in a motivic setting, and relies heavily on the ideas of [**Be1**], [**Be2**], [**Del3**] and [**J**]. I would like to thank particularly Peter Schneider and Uwe Jannsen for stimulating discussions and comments.

This article is an expanded version of talks given in the summer of 1988 in Luminy and Oberwolfach. The final version was completed during a stay at the Institute for Advanced Study. It is a pleasure to thank them for their hospitality.

I. PURE MOTIVES AND DELIGNE'S CONJECTURE

We first recall the notion of a (pure, or homogeneous) motive. For the moment it will not be too important which category of motives we consider–either Grothendieck motives (defined by algebraic correspondences modulo homological equivalence), or Deligne's category of motives defined by absolute Hodge cycles, would do. However in subsequent sections it will be important that the realisation functors should be faithful—this being the case in both categories mentioned, by construction. For simplicity we consider only motives defined over $\mathbf{Q}$, with coefficients in $\mathbf{Q}$.

Associated to a motive M over $\mathbf{Q}$ are its various realisations M_B, M_l, M_{DR} and the comparison isomorphisms

$$I_l : M_B \otimes \mathbf{Q}_l \xrightarrow{\sim} M_l, \qquad I_\infty : M_B \otimes \mathbf{C} \xrightarrow{\sim} M_{DR} \otimes \mathbf{C}.$$

M will be *pure of weight* w if the eigenvalues of an unramified Frobenius Frob_p acting on M_l have absolute value $p^{w/2}$, and if the Hodge filtration induces a Hodge structure on M_B which is pure of weight w. The example to bear in mind is of course $M = h^i(X)(m)$, where X is smooth and projective over $\mathbf{Q}$, and $w = i - 2m$.

The L-function $L(M,s)$ is the Euler product:

$$L(M,s) = \prod_p L_{(p)}(M,s)$$

where

$$L_{(p)}(M,s) = \det\left(1 - p^{-s}\,\mathrm{Frob}_p \,|M_l^{I_p}\right)^{-1}, \quad l \neq p$$

(conjecturally independent of l). We assume the existence of the meromorphic continuation and functional equation whenever necessary.

The period mapping I_∞^+ is defined as the composite

$$M_B^+ \otimes \mathbf{R} \hookrightarrow M_B \otimes \mathbf{C} \xrightarrow{\sim} M_{DR} \otimes \mathbf{C} \twoheadrightarrow M_{DR} \otimes \mathbf{R} \twoheadrightarrow \frac{M_{DR}}{F^0} \otimes \mathbf{R}$$

(this differs slightly from the notations of [**Del2**]). M is *critical* if I_∞^+ is an isomorphism. In this case its determinant is a well-defined element of $\mathbf{R}^*/\mathbf{Q}^*$, denoted $c^+(M)$.

Deligne's conjecture is that, for critical M,

$$L(M,0) \cdot c^+(M)^{-1} \in \mathbf{Q}.$$

Remarks Deligne also conjectures that $L(M,0) \neq 0$ if $w \neq -1$. We shall return to this point later. Of course, the period c^+ may be defined even when M is not critical, but only satisfies a rather mild restriction (see **[Del2]** §1.7), but for mixed motives this will not turn out to be the case. Finally we remind the reader that the notion of being critical depends only on the vanishing of certain Hodge numbers h^{pq}, $h^{p\pm}$.

II. MIXED MOTIVES

If X is an arbitrary scheme of finite type over $\mathbf{Q}$, then the cohomology groups $H^i(X)$, with respect to either de Rham, Betti or l-adic theory, have a natural filtration $W_\bullet$ (the weight filtration), compatible with the comparison isomorphisms and (in the case of l-adic cohomology) with the Galois action. The weight filtration induces a mixed Hodge structure on $H^i_B(X)$, and the Galois modules $\mathrm{Gr}^W_j H^i(X \otimes \overline{\mathbf{Q}}, \mathbf{Q}_l)$ are pure of weight j.

We now require the existence of a category $\mathcal{MM}_{\mathbf{Q}}$ of *mixed motives* over $\mathbf{Q}$. (As mentioned in the introduction, such a category has been constructed by Deligne and Jannsen.) This is to be an abelian category, containing $\mathcal{M}$ as a full subcategory. To each mixed motive E will be associated in a functorial way realisations E_B, E_{DR}, E_l. E_l is to be a finite-dimensional representation of $\mathrm{Gal}(\overline{\mathbf{Q}}/\mathbf{Q})$, and E_B, E_{DR} finite-dimensional vector spaces over $\mathbf{Q}$, equipped with an involution Φ_∞ and a decreasing filtration $F^\cdot$. There will be comparison isomorphisms

$$I_l : E_B \otimes \mathbf{Q}_l \xrightarrow{\sim} E_l, \qquad I_\infty : E_B \otimes \mathbf{C} \xrightarrow{\sim} E_{DR} \otimes \mathbf{C}.$$

Finally there is to be an increasing filtration $W^\cdot$ on E_B, stable under Φ_∞. The filtration induced (via I_l) on E_l is to be stable under the action of $\mathrm{Gal}(\overline{\mathbf{Q}}/\mathbf{Q})$, and the graded pieces $\mathrm{Gr}^W_j E_l$ are to be pure Galois representations of weight j. The filtration induced (by I_∞) on $E_{DR} \otimes \mathbf{C}$ is to be defined over $\mathbf{Q}$, and together with $F^\cdot$ and Φ_∞ is to define a mixed Hodge structure over $\mathbf{R}$. Lastly, the comparison isomorphisms I_l are to take the involution Φ_∞ to complex conjugation in $\mathrm{Gal}(\overline{\mathbf{Q}}/\mathbf{Q})$.

We may define the L-function of a mixed motive E in the same way as for a pure motive—notice that in general $L(E,s)$ and $\prod L(\mathrm{Gr}^W_j E, s)$ will differ by a finite number of Euler factors, as the passage to invariants under inertia is not an exact functor. There is one obvious case in which we have equality.

Definition E is a mixed motive over $\mathbf{Z}$ if the weight filtration on E_l splits over $\mathbf{Q}_p^{nr}$, for every l, p with $l \neq p$.

Remark Presumably one should expect that for a given p this condition need only be checked for one l.

The mixed motives over $\mathbf{Z}$ form a full subcategory $\mathcal{MM}_{\mathbf{Z}}$ of $\mathcal{MM}_{\mathbf{Q}}$, containing $\mathcal{M}$. They will play an important part in the next section.

If X is of finite type over $\mathbf{Q}$, then there should be mixed motives $h^i(X)$ in $\mathcal{MM}_{\mathbf{Q}}$. In general they will not be motives over $\mathbf{Z}$ (see the example below).

Returning to the case of an arbitrary mixed motive E, we define the period map

$$I_\infty^+ : E_B^+ \otimes \mathbf{R} \longrightarrow \frac{E_{DR}}{F^0} \otimes \mathbf{R}$$

in the same way as for pure motives.

Definition A mixed motive E is critical if I_∞^+ is an isomorphism. If this holds, define $c^+(E) = \det I_\infty^+$.

Remark It is obvious that if the pure motives $\mathrm{Gr}_j^W E$ are critical, then so is E, but the converse is far from true. For mixed motives, the notion of critical does not just depend on the Hodge numbers and the action of Φ_∞.

Conjecture A If E is critical, then $L(E,0) \cdot c^+(E)^{-1} \in \mathbf{Q}$.

Example (Trivial) Let X be the singular curve obtained from $\mathbf{G}_m/\mathbf{Q}$ by identifying the points 1, p for some prime p. Then $E = h^1(X)(1)$ is an extension:

$$0 \longrightarrow \mathbf{Q}(1) \longrightarrow E \longrightarrow \mathbf{Q}(0) \longrightarrow 0$$

Here E is the 1-motive $[\mathbf{Z} \xrightarrow{1 \mapsto p} \mathbf{G}_m]$, in the sense of Deligne [**Del1**]. It is easy to see (using the explicit realisations of 1-motives given in *loc. cit.*) that

$$L(E,s) = \zeta(s)\zeta(s+1)(1-p^{-s})$$

so that $L(E,0) = -\frac{1}{2}\log p$. Moreover E is critical, and

$$c^+(E) = \int_1^p \frac{dt}{t}.$$

So in this case Conjecture A holds. Notice that in this case we have obtained the leading term of an *incomplete* L-function as a period.

III. EXTENSIONS OF MOTIVES

To discuss conditions under which a motive is critical, and to predict orders of L-functions, we need Ext groups. Write $\mathrm{Ext}_{\mathbf{Q}}$ for the Ext groups in $\mathcal{MM}_{\mathbf{Q}}$, and $\mathrm{Ext}_{\mathbf{Z}}$ for the groups in $\mathcal{MM}_{\mathbf{Z}}$. We should have $\mathrm{Ext}^q_{\mathbf{Q}} = \mathrm{Ext}^q_{\mathbf{Z}} = 0$ unless $q=0$ or 1. If M, M' are motives over $\mathbf{Z}$, then $\mathrm{Ext}^0_{\mathbf{Q}}(M',M) = \mathrm{Ext}^0_{\mathbf{Z}}(M',M) = \mathrm{Hom}(M',M)$. Moreover $\mathrm{Ext}^1_{\mathbf{Z}}(M',M)$ is the subgroup of $\mathrm{Ext}^1_{\mathbf{Q}}(M',M)$ comprising the classes of those extensions

$$0 \longrightarrow M \longrightarrow E \longrightarrow M' \longrightarrow 0$$

such that for every p and every $l \neq p$, the extension E_l of Galois modules splits over $\mathbf{Q}_p^{nr}$. Conjecturally, the groups $\mathrm{Ext}_{\mathbf{Z}}$ will be finite-dimensional over $\mathbf{Q}$.

Suppose that X is smooth and proper over $\mathbf{Q}$, and that $M = h^i(X)(m)$, $N = M^\vee(1) \xrightarrow{\sim} h^i(X)(n)$ with $n = i+1-m$. Then we should have

$$\mathrm{Ext}^0_{\mathbf{Z}}(M,\mathbf{Q}(1)) = \mathrm{Ext}^0_{\mathbf{Z}}(\mathbf{Q}(0),N) = \mathrm{Hom}(\mathbf{Q}(-n),h^i(X)) = \begin{cases} 0 & \text{if } i \neq 2n \\ CH^n(X)/CH^n(X)^0 \otimes \mathbf{Q} & \text{if } i = 2n; \end{cases}$$

$$\mathrm{Ext}^1_{\mathbf{Z}}(M,\mathbf{Q}(1)) = \mathrm{Ext}^1_{\mathbf{Z}}(\mathbf{Q}(0),N) = \begin{cases} H^{i+1}_{\mathcal{M}}(X,\mathbf{Q}(n))_{\mathbf{Z}} & \text{if } i+1 \neq 2n \\ CH^n(X)^0 \otimes \mathbf{Q} & \text{if } i+1 = 2n. \end{cases}$$

Here $CH^n(X)$ is the Chow group of codimension n cycles on X modulo rational equivalence, and $CH^n(X)^0$ is the subgroup of classes of cycles homologically equivalent to zero. $H_{\mathcal{M}}$ denotes the motivic cohomology:

$$H^i_{\mathcal{M}}(X,\mathbf{Q}(j)) = (K_{2j-i}X \otimes \mathbf{Q})^{(j)}$$

and $H^*_{\mathcal{M}}(X,\cdot)_{\mathbf{Z}}$ is the image in $H^*_{\mathcal{M}}(X,\cdot)$ of the K-theory of a regular model for X, proper and flat over $\mathbf{Z}$. The $\mathrm{Ext}_{\mathbf{Q}}$ groups will be the given by the same rules, but with $H^*_{\mathcal{M}}(X,\cdot)_{\mathbf{Z}}$ replaced by $H^*_{\mathcal{M}}(X,\cdot)$. In the case $i = 2n-1$ the equality of the groups $\mathrm{Ext}^1_{\mathbf{Z}}$ and $\mathrm{Ext}^1_{\mathbf{Q}}$ would be a consequence of the monodromy-weight filtration conjecture, which implies that any extension of $H^{2n-1}(X \otimes \overline{\mathbf{Q}}, \mathbf{Q}_l)$ by $\mathbf{Q}_l(n)$ splits over $\mathbf{Q}_p^{nr}$ for every $p \neq l$.

In the case of the Tate motive, the above would imply that

$$\mathrm{Ext}^1_{\mathbf{Q}}(\mathbf{Q}(0),\mathbf{Q}(1)) = \mathbf{Q}^* \otimes_{\mathbf{Z}} \mathbf{Q};$$
$$\mathrm{Ext}^1_{\mathbf{Z}}(\mathbf{Q}(0),\mathbf{Q}(1)) = \mathbf{Z}^* \otimes_{\mathbf{Z}} \mathbf{Q} = 0.$$

(The extension in the example of the previous section corresponds to $p \otimes 1 \in \mathbf{Q}^* \otimes \mathbf{Q}$.)

Although we are far from having a good category of mixed motives in which the above isomorphisms hold, one can nevertheless unconditionally associate to elements of $H_{\mathcal{M}}$-groups explicit extensions of cohomology arising from the cohomology of non-compact or singular schemes. For example, let X be smooth and projective over $\mathbf{Q}$, and let ξ be a cycle on X of codimension n, homologically equivalent to zero. If $H(-)$ denotes (say) l-adic cohomology, then by pullback from the exact sequence

$$\begin{array}{ccccccc} 0 & \longrightarrow & H^{2n-1}(X)(n) & \longrightarrow & H^{2n-1}(X-|\xi|)(n) & \longrightarrow & H^{2n}_{|\xi|}(X)(n) \\ & & & & & & \uparrow \\ & & & & & & \mathbf{Q}\cdot\xi \end{array}$$

one obtains an extension of $\mathbf{Q}_l(0)$ by $H^{2n-1}(X)(n)$, whose class depends only on the rational equivalence class of ξ—see [**J**], §9. The general case can be treated in a similar way, using Bloch's description of motivic cohomology by means of higher Chow groups [**Bl1**]—see the appendix of [**DenS**] for a sketch.

In this context, the regulator maps arise from the realisation functors. For example, suppose that $n > \frac{i}{2}+1$. Then it is shown in [**Be2**] that there is a canonical isomorphism

$$H^{i+1}_{\mathcal{D}}(X/\mathbf{R},\mathbf{R}(n)) \xrightarrow{\sim} \mathrm{Ext}^1_{\mathbf{R}-\mathrm{Hdg}}(\mathbf{R},H^i(X)(n))$$

(where the second group is the Ext group in the category of real Hodge structures with an infinite Frobenius) and the regulator should then fit into a commutative square

$$\begin{array}{ccc} \mathrm{Ext}^1_{\mathbf{Z}}(\mathbf{Q},h^i(X)(n)) & \xrightarrow{\sim} & H^{i+1}_{\mathcal{M}}(X,\mathbf{Q}(n))_{\mathbf{Z}} \\ \downarrow{\scriptstyle \text{realisation}} & & \downarrow{\scriptstyle \text{regulator}} \\ \mathrm{Ext}^1_{\mathbf{R}-\mathrm{Hdg}}(\mathbf{R},H^i(X)(n)) & \xrightarrow{\sim} & H^{i+1}_{\mathcal{D}}(X/\mathbf{R},\mathbf{R}(n)) \end{array}$$

The conjectures of Birch–Swinnerton-Dyer, Tate and Beilinson on the orders of L-series at integer points can be simply stated in terms of Ext-groups:

Conjecture B Let E be a motive over $\mathbf{Z}$. Then

$$\mathrm{ord}_{s=0}L(E,s) = \dim\mathrm{Ext}^1_{\mathbf{Z}}(E,\mathbf{Q}(1)) - \dim\mathrm{Ext}^0_{\mathbf{Z}}(E,\mathbf{Q}(1)).$$

Remarks (i) In the case of motives over $\mathbf{Z}$ both sides of this conjectural identity should be additive in exact sequences, so the essential case of the conjecture is for a pure motive—in which case it it simply a restatement of the existing conjectures.

(ii) We can also write $\mathrm{Ext}^q_{\mathbf{Z}}(E,\mathbf{Q}(1)) = \mathrm{Ext}^q_{\mathbf{Z}}(\mathbf{Q}(0),E^\vee(1))$ to write the order of $L(E,s)$ in terms of the dual motive $E^\vee(1)$, in accordance with the principles of [**Del3**].

Now we can propose an algebraic criterion for a mixed motive to be critical.

Definition The mixed motive E over $\mathbf{Z}$ is *highly critical* if

$$\mathrm{Ext}^q_{\mathbf{Z}}(E, \mathbf{Q}(1)) = \mathrm{Ext}^q_{\mathbf{Z}}(\mathbf{Q}(0), E) = 0 \quad \text{for } q = 0, 1.$$

Conjecture C If E is highly critical, then it is critical.

IV. MOTIVES WITH TWO WEIGHTS

It should be clear that, starting from an arbitrary motive M, one should be able to construct some kind of 'universal extension' by sums of $\mathbf{Q}(0)$ and $\mathbf{Q}(1)$ to create a new motive E over $\mathbf{Z}$ which is highly critical, and whose L-function is of the form

$$L(E,s) = \zeta(s)^{?}\zeta(s+1)^{?}L(M,s).$$

Up to a nonzero rational, $L(E,0)$ will equal the leading coefficient of $L(M,s)$ at $s=0$, and conjecture A will then be applicable to E. We now describe the consequences of this, using the conjectural framework outlined above.

We first consider the case of a pure motive of weight $w \leq -2$, which we denote N. Since the weight filtration is increasing and $\mathcal{M}$ is supposed to be semisimple we have

$$\mathrm{Ext}^1_{\mathbf{Q}}(N, \mathbf{Q}(1)) = \mathrm{Ext}^0_{\mathbf{Q}}(\mathbf{Q}(0), N) = 0.$$

First assume that $\mathrm{Hom}(N, \mathbf{Q}(1)) = 0$, and let $\rho = \dim \mathrm{Ext}^1_{\mathbf{Z}}(\mathbf{Q}(0), N)$. Then the universal extension

$$0 \longrightarrow N \longrightarrow N^{\dagger} \longrightarrow \mathrm{Ext}^1_{\mathbf{Z}}(\mathbf{Q}(0), N) \otimes \mathbf{Q}(0) \longrightarrow 0$$

will have $\mathrm{Ext}^q_{\mathbf{Z}}(N^{\dagger}, \mathbf{Q}(1)) = \mathrm{Ext}^q_{\mathbf{Z}}(\mathbf{Q}(0), N^{\dagger}) = 0$ for $q = 0, 1$. (We are using the expected vanishing of $\mathrm{Ext}^1_{\mathbf{Z}}(\mathbf{Q}(0), \mathbf{Q}(1))$.) Conjecture C therefore implies that $N^{\dagger}$ is critical. Let us examine its periods.

The period map $I^+_{\infty}(N)$ is easily seen to be injective (from consideration of the Hodge numbers) whereas $I^+_{\infty}(\mathbf{Q}(0)) = 0$. Therefore $N^{\dagger}$ will be critical if and only if the connecting homomorphism:

$$\mathrm{Ext}^1_{\mathbf{Z}}(\mathbf{Q}(0), N) \xrightarrow{\partial_N} \frac{N_{DR} \otimes \mathbf{R}}{F^0 N_{DR} \otimes \mathbf{R} + N^+_B \otimes \mathbf{R}} = \mathrm{coker}\, I^+_{\infty}(N)$$

becomes an isomorphism when tensored with $\mathbf{R}$.

If $N = h^i(X)(n)$ and $n > 1+\frac{i}{2}$ then the first group is $H^{i+1}_{\mathcal{M}}(X, \mathbf{Q}(n))_{\mathbf{Z}}$, the second group is $H^{i+1}_{\mathcal{D}}(X_{\mathbf{R}}, \mathbf{R}(n))$ and the homomorphism ∂_N is the regulator map. The canonical $\mathbf{Q}$-structure on the target of ∂_N is the $\mathbf{Q}$-structure $\mathcal{D}_{i,n}$ (cf **[DenS]**, (2.3.1)). Since $L(N^{\dagger},0) = L(N,0)\zeta(0)^{\rho}$ is a non-zero rational multiple of $L(N,0)$, we see that conjectures A, B, C imply Deligne's reformulation (**[Del3]**, **[DenS]** 3.1) of Beilinson's conjectures for $L(h^i(X),n)$.

Now consider a pure motive M of weight ≥ 0 with $\mathrm{Hom}(\mathbf{Q}(0), M) = 0$. Dually to the above, there is a universal extension

$$0 \longrightarrow \mathbf{Q}(1)^{\rho} \longrightarrow \widetilde{M} \longrightarrow M \longrightarrow 0$$

where $\rho = \dim \mathrm{Ext}^1_{\mathbf{Z}}(M, \mathbf{Q}(1))$. The period map for $\widetilde{M}$ gives a connecting homomorphism

$$\partial_M : \ker I^+_{\infty}(M) \longrightarrow \mathbf{R}^{\rho}.$$

If we write $N = M^{\vee}(1)$ then there is a canonical isomorphism

$$\mathrm{coker}\, I^+_{\infty}(N) \xrightarrow{\sim} (\ker I^+_{\infty}(M))^{\vee}$$

(compare **[Del2]**, §5.1) in terms of which ∂_M and ∂_N are adjoint. Observe that the highest exterior powers of both sides have a natural $\mathbf{Q}$-structure. However the isomorphism does not respect these $\mathbf{Q}$-structures. In the case $N = h^i(X)(n)$, $M = h^i(X)(i+1-n)$ the natural $\mathbf{Q}$-structure on the right hand side can be seen to be Beilinson's $\mathbf{Q}$-structure ($\mathcal{B}_{i,n}$ in the notations of **[DenS]** 2.3.1). Therefore $c^+(\widetilde{M})$ is Beilinson's regulator. Since $L(\widetilde{M},s) = \zeta(s+1)^{\rho} L(M,s)$, the conjectures imply that the leading coefficient of $L(M,s)$ at $s=0$ is a nonzero rational multiple of $c^+(\widetilde{M})$, and we recover the original formulation of Beilinson's conjectures (**[DenS]**, 3.1.3).

If $\dim \mathrm{Hom}(\mathbf{Q}(0), M) = \sigma > 0$, then we should replace M by the quotient $M/\mathbf{Q}(0)^{\sigma}$. Since $\mathrm{Ext}^q_{\mathbf{Z}}(\mathbf{Q}(0), \mathbf{Q}(1)) = 0$ for $q = 1$, 2 we have

$$\mathrm{Ext}^1_{\mathbf{Z}}(M, \mathbf{Q}(1)) \xrightarrow{\sim} \mathrm{Ext}^1_{\mathbf{Z}}(M/\mathbf{Q}(0)^{\sigma}, \mathbf{Q}(1))$$

and we take $\widetilde{M}$ to be the universal extension of $M/\mathbf{Q}(0)^{\sigma}$ by $\mathbf{Q}(1)$. This corresponds precisely to the 'thickening' of the regulator which occurs in the Beilinson conjectures at the near-central point (i.e., $m = i/2$ and $n = 1+i/2$).

V. MOTIVES WITH THREE WEIGHTS

We now consider possible extensions of a motive M which is pure of weight -1. Let $G = \mathrm{Ext}^1_{\mathbf{Z}}(M, \mathbf{Q}(1))$ and $G' = \mathrm{Ext}^1_{\mathbf{Z}}(\mathbf{Q}(0), M)$. Then we obtain two universal extensions:

$$\begin{array}{ccccccccc} 0 & \longrightarrow & G^*\otimes\mathbf{Q}(1) & \longrightarrow & \widetilde{M} & \longrightarrow & M & \longrightarrow & 0 \\ 0 & \longrightarrow & M & \longrightarrow & M^\dagger & \longrightarrow & G'\otimes\mathbf{Q}(0) & \longrightarrow & 0 \end{array}$$

where $G^* = \mathrm{Hom}(G, \mathbf{Q})$.

Assuming $\mathrm{Ext}^1_{\mathbf{Z}}(\mathbf{Q}(0), \mathbf{Q}(1)) = \mathrm{Ext}^2_{\mathbf{Z}}(\mathbf{Q}(0), \mathbf{Q}(1)) = 0$, there is a unique mixed motive E over $\mathbf{Z}$ with

$$\mathrm{Gr}^W_0 E = G'\otimes\mathbf{Q}(0), \quad \mathrm{Gr}^W_{-1} E = M, \quad \mathrm{Gr}^W_{-2} E = G^*\otimes\mathbf{Q}(1)$$

and $\mathrm{Gr}^W_j E = 0$ for $j > 0$ or $j < -2$, such that

$$W_{-1}E = \widetilde{M}, \quad E/W_{-2}E = M^\dagger.$$

Setting $\rho = \dim_{\mathbf{Q}} G$, $\rho' = \dim_{\mathbf{Q}} G'$, we find that E is highly critical and that

$$L(E,s) = L(M,s)\cdot\zeta(s+1)^\rho\cdot\zeta(s)^{\rho'}.$$

Now M is itself critical. Examining the period mapping for E shows that conjecture C holds for E if and only if a certain connecting homomorphism

$$\Omega_M : G'\otimes\mathbf{R} \longrightarrow G^*$$

is an isomorphism, and $c^+(E)$ is then $c^+(M)\cdot\det\Omega_M$.

Now consider the particular case $M = h^{2n-1}(X)(n)$, $n \geq 1$. In this case we should have $G' = CH^n(X)^0\otimes\mathbf{Q}$, and $G = CH_{n-1}(X)^0\otimes\mathbf{Q}$. As a first attempt to construct E, choose $Y \subset X$ of codimension n and $Z \subset X$ of dimension $n-1$ such that $Y \cap Z = \emptyset$ and every cycle of codimension n (respectively dimension $n-1$) is rationally equivalent to a cycle supported in Y (resp. Z). Then in any of the various cohomology theories, $E' = H^{2n-1}(X - Y \mathrm{rel} Z)(n)$ has three nonzero steps in its weight filtration:

$$\begin{aligned} &\mathrm{Gr}^W_0 E' = H^{2n}_Y(X)^0(n) \overset{\mathrm{def}}{=} \ker\{H^{2n}_Y(X)(n) \longrightarrow H^{2n}(X)(n)\}; \\ &\mathrm{Gr}^W_{-1} E' = H^{2n-1}(X)(n); \\ &\mathrm{Gr}^W_{-2} E' = H^{2n-2}(Z)_0(n) \overset{\mathrm{def}}{=} \mathrm{coker}\{H^{2n-2}(X)(n) \longrightarrow H^{2n-2}(Z)(n)\}. \end{aligned}$$

By choosing suitable maps $G' \to H^{2n}_Y(X)^0(n)$ and $G\otimes H^{2n-2}(Z)_0(n) \to \mathbf{Q}(1)$, and taking the associated pullback and pushout, we obtain an object E'' with the correct graded pieces. The homorphism $\Omega_{E''}$ is essentially the infinite component of the height pairing between cycles supported in Y and in Z.

However E'' need not be a mixed motive over $\mathbf{Z}$. The obstruction is a certain element of $\mathrm{Ext}^1_{\mathbf{Q}}(\mathbf{Q}(0)^{\rho'}, \mathbf{Q}(1)^{\rho})$. If X satisfies some reasonable hypotheses (essentially those required to define tha global height pairing) this obstruction can be explicitly constructed in terms of local heights, and can be realised as a suitable sum of 1-motives $[\mathbf{Z} \xrightarrow{1 \mapsto p} \mathbf{G}_m]$, for various primes p (as described at the end of §II). Therefore in order to obtain E itself we must twist by the inverse of this obstruction. This will change the period mapping by appropriate multiples of $\log p$, and it turns out that Ω_E is the homomorphism attached to the complete height pairing (including the contributions from the finite primes). Consequently:

- Firstly, E is critical if and only if the global height pairing of Beilinson–Bloch–Gillet–Soulé ([**Be3**] [**Bl2**] [**GS**])

$$< \cdot, \cdot >: G \otimes G' \longrightarrow \mathbf{R}$$

  is non-singular;
- Secondly, if E is indeed critical, then

$$c^+(E) = c^+(h^{2n-1}(X)(n)) \cdot \det < \quad >.$$

In other words, conjectures A, B and C imply the Beilinson–Bloch generalisation of the Birch–Swinnerton-Dyer conjectures.

It is possible to rewrite the construction of the extensions of this and the previous section in a unified way in a (hypothetical) derived category of motives, but we shall not attempt to describe this here.

Finally we remark here that one can consider the p-adic periods of mixed motives in the same way as above. The essential point (which was shown to me by U. Jannsen) is that if the l-adic realisations of a motive E (pure or mixed) are unramified at p, then the p-adic representation E_p of $\mathrm{Gal}(\overline{\mathbf{Q}}_p/\mathbf{Q}_p)$ should be crystalline. In this case there is then a p-adic period map

$$I_p^+ : E_B^+ \otimes B_{\mathrm{cris}} \longrightarrow \frac{E_{DR}}{F^0} \otimes B_{\mathrm{cris}}$$

which should be an isomorphism if and only if E is critical. Now if we start with a pure motive M whose l-adic realisations are unramified at p, the same will be true of the various universal extensions constructed here.

By considering the associated period mappings, one then obtains B_{cris}-valued regulators and height pairings. (These have been directly constructed by P. Schneider.)

VI. SOME COMPATIBILITIES

In this section we show that the only consequences of conjecture (A) are those described above, and that the conjecture is compatible with the functional equation. In order to give the statements of the theorems below some actual meaning, we will take for $\mathcal{MM}_{\mathbf{Q}}$ the category of mixed motives defined by absolute Hodge cycles, as considered by Deligne and Jannsen [**J**]. We will restrict our attention to motives whose l-adic realisations are independent of l, so as to be able to discuss L-functions, and we shall assume the analytic continuation and functional equation for any L-functions that may arise. Of the remaining desirable properties this category might enjoy, we assume (only) the following:

Hypotheses

(a) $\mathrm{Ext}^1_{\mathbf{Q}}(\mathbf{Q}(0),\mathbf{Q}(1))$ is generated by the classes of 1-motives of type (iv) below.

(b) If M is pure of weight -1, then $\mathrm{ord}_{s=0} L(M,s) \geq \dim \mathrm{Ext}^1_{\mathbf{Z}}(M,\mathbf{Q}(1)) = \dim \mathrm{Ext}^1_{\mathbf{Q}}(M,\mathbf{Q}(1))$.

(c) If M is pure of weight ≤ -2 then the realisation map $\mathrm{Ext}^1_{\mathbf{Z}}(\mathbf{Q}(0),M)\otimes \mathbf{R} \to \mathrm{Ext}^1_{\mathbf{R}-\mathrm{Hdg}}(\mathbf{R}(0),M_{\mathbf{R}})$ is injective.

The first hypothesis implies that $\mathrm{Ext}^1_{\mathbf{Q}}(\mathbf{Q}(0),\mathbf{Q}(1)) = \mathbf{Q}^{(\mathcal{S})}$ where $\mathcal{S}$ denotes the set of rational primes. Hypothesis (b) includes a weak form of conjecture (B) for M. (See also III above.) Finally, (c) is equivalent to the statement that $I^+_\infty(M^\dagger)$ is injective, hence is a weak form of conjecture (C). (The reader who is sceptical of the general finiteness of groups such as Ш will be reassured that we do not require equality in (b).)

Theorem 1 Assume the truth of hypotheses (a)–(c) above. Let E be a critical mixed motive over $\mathbf{Q}$, with $L(E,0) \neq 0$. Then there is a filtration $K_\cdot$ of E with the following properties:

- The graded pieces $E_i = \mathrm{Gr}^K_i E$ are critical motives with $L(E_i,0) \in \mathbf{R}^*$;
- $\prod L(E_i,0)\cdot L(E,0)^{-1} \in \mathbf{Q}^*$;
- Each E_i is one of the following:
  - (i) An extension of a pure motive M of weight ≥ 0 by a sum of copies of $\mathbf{Q}(1)$, with $\mathrm{Hom}(\mathbf{Q}(0),M) = 0$;

(ii) An extension of a sum of copies of $\mathbf{Q}(0)$ by a pure motive M of weight ≤ -2 with $\mathrm{Hom}(\mathbf{Q}(1),M)=0$;

(iii) A motive whose nonzero graded pieces in the weight filtration are a sum of copies of $\mathbf{Q}(1)$, a pure motive of weight -1, and a sum of copies of $\mathbf{Q}(0)$;

(iv) The 1-motive $[\mathbf{Z}\xrightarrow{\phi}\mathbf{G}_m]$, $\phi(1)=p$.

- In cases (i)–(iii) E_i is a motive over $\mathbf{Z}$.

If x, $y\in\mathbf{R}$, we write $x\sim y$ if $x=ay$ for some $a\in\mathbf{Q}^*$.

Theorem 2 The hypotheses being as in Theorem 1, assume the truth of conjecture 6.6 of [**Del2**] (that every motive of rank 1 and weight 0 is an Artin motive). Suppose that $L(E^\vee(1),0)$ is also nonzero. Then $L(E,0)\cdot c^+(E)^{-1}\sim L(E^\vee(1),0)\cdot c^+(E^\vee(1))^{-1}$.

Before giving the proofs, we make some general observations. We say that a motive is supercritical (subcritical) if the period mapping is surjective (injective). Suppose $A\to B\to C$ is a short exact sequence of motives. Then by applying the snake lemma to the ladder

$$\begin{array}{ccccc} A_B^+\otimes\mathbf{R} & \longrightarrow & B_B^+\otimes\mathbf{R} & \longrightarrow & C_B^+\otimes\mathbf{R} \\ \downarrow & & \downarrow & & \downarrow \\ F^0\backslash A_{DR}\otimes\mathbf{R} & \longrightarrow & F^0\backslash B_{DR}\otimes\mathbf{R} & \longrightarrow & F^0\backslash B_{DR}\otimes\mathbf{R} \end{array}$$

we see that if B is supercritical, then C is supercritical and $\dim\ker I_\infty^+(C)\geq \dim\operatorname{coker} I_\infty^+(A)$; there is a dual statement if B is subcritical.

We recall the conjectural analytic continuation and functional equation of a (pure) motive. If A is pure of weight w and contains no direct factors of $\mathbf{Q}(0)$ or $\mathbf{Q}(1)$ then the conjectures imply:

- If $w\geq 0$ then A is supercritical and $\mathrm{ord}_{s=0}L(A,s)=\dim\ker I_\infty^+(A)$;
- If $w\leq -2$ then A is subcritical and $\mathrm{ord}_{s=0}L(A,s)=0$.

Now suppose that A is mixed of weight ≥ 0, and $\mathrm{Hom}(\mathbf{Q}(0),A)=0$. Then as $\prod L(\mathrm{Gr}_q^W A,s)$ is the product of $L(A,s)$ with a finite number of Euler factors, we have $\mathrm{ord}_{s=0}L(A,s)\geq\dim\ker I_\infty^+(A)$; likewise if A is mixed of weight ≤ -2 and $\mathrm{Hom}(A,\mathbf{Q}(-1))-0$ then $\mathrm{ord}_{s=0}L(A,s)\geq 0$. In each case equality holds if and only if the multiplicities of the eigenvalue 1 of Frob_p on $\mathrm{Gr}_\cdot^W(A_l)^{\mathcal{I}_p}$ and on $A_l^{\mathcal{I}_p}$ are equal.

Construction of the filtration
We first introduce a convenient notation. By the symbol

$$\left\{\begin{array}{c} A_k \\ \hline \vdots \\ \hline A_1 \end{array}\right.$$

we mean any mixed motive with an increasing filtration (not necessarily the weight filtration!) whose i^{th} graded constituent is isomorphic to A_i.

By virtue of the weight filtration, and the fact that pure motives are semisimple, we can write:

$$E = \left\{\begin{array}{c} A \\ \hline N = \left\{\begin{array}{c} \mathbf{Q}(0)^d \\ \hline M \\ \hline \mathbf{Q}(1)^e \end{array}\right. \\ \hline B \end{array}\right.$$

where M is pure of weight -1 (and therefore critical), A is mixed of weight ≥ 0 with $\mathrm{Hom}(\mathbf{Q}(0), A) = 0$, and B is mixed of weight ≤ -2 with $\mathrm{Hom}(B, \mathbf{Q}(1)) = 0$. Since E is critical, A and B are respectively supercritical and subcritical; let $f = \dim\ker I^+_\infty(A)$, $g = \dim\operatorname{coker} I^+_\infty(B)$. We have $f + d = e + g$.

By removing direct factors we can also write

$$N = \mathbf{Q}(0)^{d'} \oplus \mathbf{Q}(1)^{e'} \oplus M' \oplus C$$

with

$$M' = \left\{\begin{array}{c} X = \left\{\begin{array}{c} \mathbf{Q}(0)^{q'} \\ \hline M \end{array}\right. \\ \hline \mathbf{Q}(1)^{q''} \end{array}\right\} = Y \quad , \qquad C = \left\{\begin{array}{c} \mathbf{Q}(0)^m \\ \hline \mathbf{Q}(1)^n \end{array}\right.$$

and $\mathrm{Hom}(\mathbf{Q}(0), X) = \mathrm{Hom}(Y, \mathbf{Q}(1)) = \mathrm{Hom}(\mathbf{Q}(0), C) = \mathrm{Hom}(C, \mathbf{Q}(1)) = 0$. By hypothesis (a), the extension C is classified by some

$$\phi = (\phi_p) \in \mathrm{Hom}(\mathbf{Q}^m, \mathbf{Q}^n)^{(S)} = \mathrm{Ext}^1_{\mathbf{Q}}(\mathbf{Q}(0)^m, \mathbf{Q}(1)^n).$$

Kummer theory gives an isomorphism

$$\mathrm{Ext}^1_{\mathcal{I}_p}(\mathbf{Q}_l(0)^m, \mathbf{Q}_l(1)^n) \xrightarrow{\sim} \mathrm{Hom}(\mathbf{Q}_l^m, \mathbf{Q}_l^n);$$

under this isomorphism, the representation C_l of the inertia group at p is classified by the homomorphism $\phi_p : \mathbf{Q}_l^m \to \mathbf{Q}_l^n$.

Proposition Let R_p denote the rank of ϕ_p. Then:

(i) $\operatorname{ord}_{s=0} L(C,s) = \sum_{p\in\mathcal{S}} R_p - n$;
(ii) $\sum_{p\in\mathcal{S}} R_p \geq \max(m,n)$;
(iii) If C is critical and $L(C,0) \neq 0$ then $C = \bigoplus_p [\mathbf{Z} \xrightarrow{1\mapsto p} \mathbf{G}_m]^{R_p}$.

Proof (i) Consider the short exact sequence

$$0 \longrightarrow \mathbf{Q}_l(1)^n \longrightarrow C_l \longrightarrow \mathbf{Q}_l(0)^m \longrightarrow 0$$

of $\mathcal{I}_p$-modules. R_p is the dimension of the image of the boundary homomorphism

$$\mathbf{Q}_l(0) \longrightarrow \operatorname{Ext}^1_{\mathcal{I}_p}(\mathbf{Q}_l(0)^m, \mathbf{Q}_l(1)^m)$$

and therefore equals the codimension of the image of $(C_l)^{\mathcal{I}_p}$ in $\mathbf{Q}_l(0)^m$. Therefore the Euler factor of $L(C,s)$ at p is

$$(1-p^{-1-s})^{-n}(1-p^{-s})^{R_p-n}$$

whence

$$L(C,s) = \zeta(s)^m \zeta(s+1)^n \prod_p (1-p^{-s})^{R_p}$$

and (i) follows.

(ii) Write $U = \mathbf{Q}^m$, $V = \mathbf{Q}^n$. Since $\operatorname{Hom}(\mathbf{Q}(0), C) = \operatorname{Hom}(C, \mathbf{Q}(1)) = 0$, the homomorphisms

$$\phi' : U \xrightarrow{x\mapsto(\phi_p(x))} V^{(\mathcal{S})} \quad \text{and} \quad \phi'' : V^* \xrightarrow{x\mapsto({}^t\phi_p(x))} (U^*)^{(\mathcal{S})}$$

attached to the classifying map ϕ are injective. Since the images of these are contained in the subspaces

$$\bigoplus_p \phi_p(U) \subset V^{(\mathcal{S})}, \quad \bigoplus_p {}^t\phi_p(V^*) \subset (U^*)^{(\mathcal{S})}$$

we have the inequality (ii).

(iii) The period mapping for C is

$$\sum_{p\in\mathcal{S}} \log p \cdot \phi_p : U \otimes \mathbf{R} \longrightarrow V \otimes \mathbf{R}.$$

If C is critical and $L(C,0)$ is nonzero, $m=n=\sum R_p$. In this case the image of ϕ' is precisely $\bigoplus_p \phi_p(U)$, hence we can write $U=\bigoplus_p U_p$ in such a way that ϕ_p factors through the projection:

$$\phi_p : U = \bigoplus U_p \twoheadrightarrow U_p \hookrightarrow V.$$

For $\sum \log p \cdot \phi_p$ to be an isomorphism we must therefore have $V=\bigoplus_p \phi_p(U_p)$, which implies (iii). ■

Now let $q=\mathrm{ord}_{s=0} L(M,s)$. By hypothesis (b) for M and the properties of M' we have q', $q''\le q$. Also the conjectural analytic continuation and functional equation give

$$\mathrm{ord}_{s=0} L(A,s) \ge f \quad \text{and} \quad \mathrm{ord}_{s=0} L(B,s) \ge 0.$$

Because of the decomposition of N we can rearrange the filtration on E as:

$$E = \left\{ \begin{array}{c} A' = \left\{ \dfrac{A}{\mathbf{Q}(1)^{e'}} \right. \\ \hline M' \oplus C \\ \hline B' = \left\{ \dfrac{\mathbf{Q}(0)^{d'}}{B} \right. \end{array} \right. .$$

Then A' is supercritical and B' is subcritical. Therefore $f \ge e'$ and $g \ge d'$, with equality if and only if A' and B' are critical, respectively. This gives

$$\begin{aligned} 0 &\ge \mathrm{ord}_{s=0} L(E,s) \\ &\ge \mathrm{ord}_{s=0} L(A',s) + \mathrm{ord}_{s=0} L(M',s) + \mathrm{ord}_{s=0} L(C,s) + \mathrm{ord}_{s=0} L(B',s) \\ &\ge (f-e') + (q-q'') + (\textstyle\sum R_p - n) + 0 \\ &\ge 0. \end{aligned}$$

Therefore $L(E,0) \ne \infty$, and $L(E,0)$ is nonzero if and only if $f=e'$, $q=q''$ and $\sum R_p = n$. Then

$$\sum_{p\in S} R_p = n = e - f - g = d - g - q \le d - d' - q' = m.$$

By part (ii) of the proposition, this implies also $g=d'$, $q=q'$, $m=n$. Therefore A' and B' are critical. Hence so are M' and C. By part (iii) of the proposition, C is a sum of motives of type (iv). Moreover we have $\mathrm{ord}_{s=0} L(M',s)=0$. Now by hypothesis (b)

$$\mathrm{ord}_{s=0} L(M',s) \ge \mathrm{ord}_{s=0} L(Y,s) \ge \mathrm{ord}_{s=0} L(M,s) + q \ge 0$$

and the first inequality is an equality only if the extension of $\mathcal{I}_p$-modules $Y_l \to M_l \to \mathbf{Q}_l(0)^q$ splits. Also the hypothetical equality of $\mathrm{Ext}^1_{\mathbf{Z}}(M,\mathbf{Q}(1))$ and $\mathrm{Ext}^1_{\mathbf{Q}}(M,\mathbf{Q}(1))$ shows that the extension $\mathbf{Q}_l(1)^q \to Y_l \to M_l$ splits over $\mathcal{I}_p$, hence M' is a motive over $\mathbf{Z}$ of the type (iii).

It finally remains to analyze the motives A' and B'. Consider first B'. By the above we have $\mathrm{ord}_{s=0} L(B',s)=0$. Moreover as $\mathrm{Hom}(B',\mathbf{Q}(1))=0$ we have $\mathrm{ord}_{s=0} L(B,s)\geq 0$. Consider the extension $B_l \to B'_l \to \mathbf{Q}_l(0)^{d'_1}$ of modules under the inertia group at $p\neq l$. Since the local Euler factors for $\mathbf{Q}(0)$ each have a pole at $s=0$, this extension must split (otherwise $L(B',0)$ would vanish). Now let w be the highest weight of B, and write $B_1=\mathrm{Gr}^W_w(B)$, $Z=W_{w-1}(B)$. Then

$$B=\left\{\frac{B_1}{Z}\right.$$

and we can then rewrite B' as:

$$B'=\left\{\frac{\dfrac{\mathbf{Q}(0)^{d'}}{B_1}}{Z}\right.=\left\{\frac{\mathbf{Q}(0)^{d'-d'_1}\oplus\left\{\dfrac{\mathbf{Q}(0)^{d'_1}}{B_1}\right.}{Z}\right.=\left\{\frac{B'_1=\left\{\dfrac{\mathbf{Q}(0)^{d'_1}}{B_1}\right.}{Z'=\left\{\dfrac{\mathbf{Q}(0)^{d'-d'_1}}{Z}\right.}\right.$$

Continuing in this way we obtain a filtration:

$$B'=\left\{\begin{array}{c}B'_1\\ \hline \vdots\\ \hline B'_r\end{array}\right.,\quad B'_i=\left\{\frac{\mathbf{Q}(0)^{d'_i}}{B_i}\right.$$

where each B_i is pure, $\mathrm{Hom}(\mathbf{Q}(0),B'_i)=0$, and $\sum d'_i=d'$. By the above, each B'_i is a motive over $\mathbf{Z}$, so is a submotive of the universal extension $B^\dagger_i$ of section IV. The hypothesis (c) implies that $B^\dagger_i$ is subcritical. Therefore each B'_i is subcritical, hence critical (since B' itself is critical).

Now consider A'. In the same way we can write

$$A'=\left\{\begin{array}{c}A'_1\\ \hline \vdots\\ \hline A'_t\end{array}\right.,\quad A'_i=\left\{\frac{A_i}{\mathbf{Q}(0)^{e'_i}}\right.$$

with A_i pure, $\mathrm{Hom}(A'_i,\mathbf{Q}(1))=0$, and $\sum e'_i=e'$. Now

$$\mathrm{ord}_{s=0} L(A_i,s)=f_i=\dim\ker I^+_\infty(A'_i)$$

Therefore $\mathrm{ord}_{s=0} L(A'_i,s)\geq f_i-e'_i$, with equality if and only if the $(\mathrm{Frob}_p=1)$-eigenspace of $(A'_{i,l})^{\mathcal{I}_p}$ maps onto that of $(A_{i,l})^{\mathcal{I}_p}$. In view of the exact sequence

$$(A_{i,l})^{\mathcal{I}_p}\longrightarrow(A'_{i,l})^{\mathcal{I}_p}\longrightarrow H^1(\mathcal{I}_p,\mathbf{Q}_l(1)^{e'_i})=\mathbf{Q}_l(0)^{e'_i}$$

this holds if and only if A'_i is a motive over $\mathbf{Z}$.

Now

$$0 = \operatorname{ord}_{s=0} L(A', s) \geq \sum \operatorname{ord}_{s=0} L(A'_i, s) \geq \sum (f_i - e'_i) = f - e' = 0,$$

so we must have equality at each stage. Therefore each A'_i is a motive over $\mathbf{Z}$, and therefore is a quotient of the universal extension $\widetilde{A}_i$. So by hypothesis (c) each A'_i is supercritical, and therefore also critical. Also $e'_i = f_i$, whence $L(A'_i, 0) \in \mathbf{R}^*$. This concludes the proof of Theorem 1. ■

Proof of Theorem 2 Recall the definition of the constant $\delta(M)$ ([**Del2**], 1.7.3); we extend this definition to mixed motives as well. The proof of Proposition 5.1 of [**Del2**] then gives:

Proposition Let E be a critical (mixed) motive. Then $E^\vee(1)$ is also critical, and

$$c^+(E) \cdot c^+(E^\vee(1))^{-1} \sim (2\pi i)^{-d^-(E)} \cdot \delta(E)$$

(where $d^-(E) = \dim E_B^-$).

Now suppose that E satisfies the hypotheses of the theorem. Write $L^*(-)$ for the leading coefficient in the Laurent expansion of $L(-, s)$. By the proof of Theorem 1, we have

$$L(E, 0) = L^*(E) = \prod_q L^*(\operatorname{Gr}_q^W E) \cdot \prod_p (\log p)^{R_p}$$

and likewise

$$L(E^\vee(1), 0) = L^*(E^\vee(1)) = \prod_q L^*(\operatorname{Gr}_q^W E^\vee(1)) \cdot \prod_p (\log p)^{R_p}.$$

It therefore suffices to prove that for any pure motive M of weight w

$$L^*(M) \cdot L^*(M^\vee(1))^{-1} \sim (2\pi i)^{-d^-(M)} \cdot \delta(M)$$

which by under our assumptions (as in 5.6 of [**Del2**]) is equivalent to

$$L^*_\infty(M) \cdot L^*_\infty(M^\vee(1))^{-1} \sim (2\pi)^{d^-(M) + \frac{1}{2} w d(M)}.$$

This in turn is a trivial extension of Proposition 5.4 of [**Del2**]. ■

REFERENCES

[**Be1**] A. A. Beilinson; Higher regulators and values of L-functions. J. Soviet Math. **30** (1985), 2036–70

[**Be2**] A. A. Beilinson; Notes on absolute Hodge cohomology. Applications of algebraic K-theory to algebraic geometry and number theory (Contemporary Mathematics **55** (1986)), 35–68

[**Be3**] A. A. Beilinson; Height pairing on algebraic cycles. Current trends in arithmetical algebraic geometry, ed. K. Ribet (Contemporary Mathematics **67** (1987)), 1–24

[**Bl1**] S. Bloch; Algebraic cycles and higher K-theory . Advances in Math. **61** (1986) 267–304

[**Bl2**] S. Bloch; Height pairing for algebraic cycles. J. Pure Appl. Algebra **34** (1984), 119–45

[**Del1**] P. Deligne; Théorie de Hodge III. Publ. Math. IHES **44** (1974), 5–78

[**Del2**] P. Deligne; Valeurs de fonctions L et périodes d'intégrales. Proc. Symp. Pure Math. AMS **33** (1979), 313–46

[**Del3**] P. Deligne; Letter to C. Soulé. 20–1–1985

[**Del4**] P. Deligne; Le groupe fondamental de la droite projective moins trois points. Galois groups over **Q** (ed. Y. Ihara, K. Ribet, J.-P. Serre). MSRI publications **16**, 1989

[**DenS**] C. Deninger, A. J. Scholl; The Beilinson conjectures. This volume

[**GS**] H. Gillet, C. Soulé; Intersection sur les variétés d'Arakelov. C.R.A.S. t.299, sér.1, **12** (1984), 563–6

[**H**] G. Harder; Arithmetische Eigenschaften von Eisensteinklassen, die modulare Konstruktion von gemoschten Motiven und von Erweiterungen endlicher Galoismoduln. Preprint, 1989

[J] U. Jannsen; Mixed motives and algebraic K-theory. Lecture notes in math. **1400** (1990)

[S] A. J. Scholl; Height pairings and special values of L-functions. In preparation

Printed in the United States
By Bookmasters